Imperial College London
2407818375
AF843368

Advances in Ground-Source Heat Pump Systems

Related titles

Geothermal Energy: An Alternative Resource for the 21st Century
(ISBN: 978-0-44452-875-9)

Geothermal Reservoir Engineering, 2e
(ISBN: 978-0-12383-880-3)

Geothermal Power Plants, 3e
(ISBN: 978-0-08098-206-9)

Advances in Thermal Energy Storage Systems
(ISBN: 978-1-78242-088-0)

Renewable Heating and Cooling: Technologies and Applications
(ISBN: 978-1-78242-213-6)

Woodhead Publishing Series in Energy:
Number 100

Advances in Ground-Source Heat Pump Systems

Edited by

Simon J. Rees

ELSEVIER

AMSTERDAM • BOSTON • CAMBRIDGE • HEIDELBERG
LONDON • NEW YORK • OXFORD • PARIS • SAN DIEGO
SAN FRANCISCO • SINGAPORE • SYDNEY • TOKYO
Woodhead Publishing is an imprint of Elsevier

WP
WOODHEAD
PUBLISHING

Woodhead Publishing is an imprint of Elsevier
The Officers' Mess Business Centre, Royston Road, Duxford, CB22 4QH, UK
50 Hampshire Street, 5th Floor, Cambridge, MA 02139, USA
The Boulevard, Langford Lane, Kidlington, OX5 1GB, UK

Copyright © 2016 Elsevier Ltd. All rights reserved.

No part of this publication may be reproduced or transmitted in any form or by any means, electronic or mechanical, including photocopying, recording, or any information storage and retrieval system, without permission in writing from the publisher. Details on how to seek permission, further information about the Publisher's permissions policies and our arrangements with organizations such as the Copyright Clearance Center and the Copyright Licensing Agency, can be found at our website: www.elsevier.com/permissions.

This book and the individual contributions contained in it are protected under copyright by the Publisher (other than as may be noted herein).

Notices
Knowledge and best practice in this field are constantly changing. As new research and experience broaden our understanding, changes in research methods, professional practices, or medical treatment may become necessary.

Practitioners and researchers must always rely on their own experience and knowledge in evaluating and using any information, methods, compounds, or experiments described herein. In using such information or methods they should be mindful of their own safety and the safety of others, including parties for whom they have a professional responsibility.

To the fullest extent of the law, neither the Publisher nor the authors, contributors, or editors, assume any liability for any injury and/or damage to persons or property as a matter of products liability, negligence or otherwise, or from any use or operation of any methods, products, instructions, or ideas contained in the material herein.

British Library Cataloguing-in-Publication Data
A catalogue record for this book is available from the British Library

Library of Congress Cataloging-in-Publication Data
A catalog record for this book is available from the Library of Congress

ISBN: 978-0-08-100311-4 (print)
ISBN: 978-0-08-100322-0 (online)

For information on all Woodhead Publishing publications
visit our website at https://www.elsevier.com/

ELSEVIER Book Aid International Working together to grow libraries in developing countries
www.elsevier.com • www.bookaid.org

Publisher: Joe Hayton
Acquisition Editor: Sarah Hughes
Editorial Project Manager: Alex White
Production Project Manager: Omer Mukthar
Designer: Victoria Pearson Esser

Typeset by TNQ Books and Journals

Contents

List of contributors

P. Baudron Polytechnique Montréal, Montréal, QC, Canada

M. Bernier Polytechnique Montréal, Montréal, QC, Canada

J.M. Corberan Polytechnic University of Valencia, Valencia, Spain

B.C. Drijver IF Technology B.V., Arnhem, The Netherlands

F. Eppner Polytechnique Montréal, Montréal, QC, Canada

Z. Fang School of Thermal Engineering, Shandong Jianzhu University, Jinan, China

S. Gehlin Swedish Centre for Shallow Geothermal Energy, Lund, Sweden

L. Helsen University of Leuven (KU Leuven), Heverlee, Belgium; EnergyVille, Genk, Belgium

S. Javed Building Services Engineering, Chalmers University of Technology, Gothenburg, Sweden

S.W. Jeon Korea University, Seoul, Republic of Korea

Y. Kim Korea University, Seoul, Republic of Korea

J.S. Lee Korea University, Seoul, Republic of Korea

M. Li School of Energy Science and Engineering, Central South University, Changsha, China

D. Marcotte Polytechnique Montréal, Montréal, QC, Canada

M.S. Mitchell Oklahoma State University, Stillwater, OK, United States

A. Nguyen Polytechnique Montréal, Montréal, QC, Canada

P. Pasquier Polytechnique Montréal, Montréal, QC, Canada

S.J. Rees University of Leeds, Leeds, United Kingdom

Y. Rui University of Cambridge, Cambridge, United Kingdom

A.L. Snijders IFTech International B.V., Arnhem, The Netherlands

K. Soga University of Cambridge, Cambridge, United Kingdom

J.D. Spitler Oklahoma State University, Stillwater, OK, United States

C.P. Underwood University of Northumbria, Newcastle upon Tyne, United Kingdom

H.J.L. Witte Groenholland Geo-energysystems, Amsterdam, Netherlands

K. Zhu School of Thermal Engineering, Shandong Jianzhu University, Jinan, China

Woodhead Publishing Series in Energy

1 **Generating power at high efficiency: Combined cycle technology for sustainable energy production**
Eric Jeffs
2 **Advanced separation techniques for nuclear fuel reprocessing and radioactive waste treatment**
Edited by Kenneth L. Nash and Gregg J. Lumetta
3 **Bioalcohol production: Biochemical conversion of lignocellulosic biomass**
Edited by Keith W. Waldron
4 **Understanding and mitigating ageing in nuclear power plants: Materials and operational aspects of plant life management (PLiM)**
Edited by Philip G. Tipping
5 **Advanced power plant materials, design and technology**
Edited by Dermot Roddy
6 **Stand-alone and hybrid wind energy systems: Technology, energy storage and applications**
Edited by John K. Kaldellis
7 **Biodiesel science and technology: From soil to oil**
Jan C. J. Bart, Natale Palmeri and Stefano Cavallaro
8 **Developments and innovation in carbon dioxide (CO_2) capture and storage technology Volume 1: Carbon dioxide (CO_2) capture, transport and industrial applications**
Edited by M. Mercedes Maroto-Valer
9 **Geological repository systems for safe disposal of spent nuclear fuels and radioactive waste**
Edited by Joonhong Ahn and Michael J. Apted
10 **Wind energy systems: Optimising design and construction for safe and reliable operation**
Edited by John D. Sørensen and Jens N. Sørensen
11 **Solid oxide fuel cell technology: Principles, performance and operations**
Kevin Huang and John Bannister Goodenough
12 **Handbook of advanced radioactive waste conditioning technologies**
Edited by Michael I. Ojovan
13 **Membranes for clean and renewable power applications**
Edited by Annarosa Gugliuzza and Angelo Basile
14 **Materials for energy efficiency and thermal comfort in buildings**
Edited by Matthew R. Hall
15 **Handbook of biofuels production: Processes and technologies**
Edited by Rafael Luque, Juan Campelo and James Clark
16 **Developments and innovation in carbon dioxide (CO_2) capture and storage technology Volume 2: Carbon dioxide (CO_2) storage and utilisation**
Edited by M. Mercedes Maroto-Valer

17 **Oxy-fuel combustion for power generation and carbon dioxide (CO_2) capture**
Edited by Ligang Zheng
18 **Small and micro combined heat and power (CHP) systems: Advanced design, performance, materials and applications**
Edited by Robert Beith
19 **Advances in clean hydrocarbon fuel processing: Science and technology**
Edited by M. Rashid Khan
20 **Modern gas turbine systems: High efficiency, low emission, fuel flexible power generation**
Edited by Peter Jansohn
21 **Concentrating solar power technology: Principles, developments and applications**
Edited by Keith Lovegrove and Wes Stein
22 **Nuclear corrosion science and engineering**
Edited by Damien Féron
23 **Power plant life management and performance improvement**
Edited by John E. Oakey
24 **Electrical drives for direct drive renewable energy systems**
Edited by Markus Mueller and Henk Polinder
25 **Advanced membrane science and technology for sustainable energy and environmental applications**
Edited by Angelo Basile and Suzana Pereira Nunes
26 **Irradiation embrittlement of reactor pressure vessels (RPVs) in nuclear power plants**
Edited by Naoki Soneda
27 **High temperature superconductors (HTS) for energy applications**
Edited by Ziad Melhem
28 **Infrastructure and methodologies for the justification of nuclear power programmes**
Edited by Agustín Alonso
29 **Waste to energy conversion technology**
Edited by Naomi B. Klinghoffer and Marco J. Castaldi
30 **Polymer electrolyte membrane and direct methanol fuel cell technology Volume 1: Fundamentals and performance of low temperature fuel cells**
Edited by Christoph Hartnig and Christina Roth
31 **Polymer electrolyte membrane and direct methanol fuel cell technology Volume 2: *In situ* characterization techniques for low temperature fuel cells**
Edited by Christoph Hartnig and Christina Roth
32 **Combined cycle systems for near-zero emission power generation**
Edited by Ashok D. Rao
33 **Modern earth buildings: Materials, engineering, construction and applications**
Edited by Matthew R. Hall, Rick Lindsay and Meror Krayenhoff
34 **Metropolitan sustainability: Understanding and improving the urban environment**
Edited by Frank Zeman
35 **Functional materials for sustainable energy applications**
Edited by John A. Kilner, Stephen J. Skinner, Stuart J. C. Irvine and Peter P. Edwards
36 **Nuclear decommissioning: Planning, execution and international experience**
Edited by Michele Laraia
37 **Nuclear fuel cycle science and engineering**
Edited by Ian Crossland

38 **Electricity transmission, distribution and storage systems**
Edited by Ziad Melhem
39 **Advances in biodiesel production: Processes and technologies**
Edited by Rafael Luque and Juan A. Melero
40 **Biomass combustion science, technology and engineering**
Edited by Lasse Rosendahl
41 **Ultra-supercritical coal power plants: Materials, technologies and optimisation**
Edited by Dongke Zhang
42 **Radionuclide behaviour in the natural environment: Science, implications and lessons for the nuclear industry**
Edited by Christophe Poinssot and Horst Geckeis
43 **Calcium and chemical looping technology for power generation and carbon dioxide (CO_2) capture: Solid oxygen- and CO_2-carriers**
Paul Fennell and E. J. Anthony
44 **Materials' ageing and degradation in light water reactors: Mechanisms, and management**
Edited by K. L. Murty
45 **Structural alloys for power plants: Operational challenges and high-temperature materials**
Edited by Amir Shirzadi and Susan Jackson
46 **Biolubricants: Science and technology**
Jan C. J. Bart, Emanuele Gucciardi and Stefano Cavallaro
47 **Advances in wind turbine blade design and materials**
Edited by Povl Brøndsted and Rogier P. L. Nijssen
48 **Radioactive waste management and contaminated site clean-up: Processes, technologies and international experience**
Edited by William E. Lee, Michael I. Ojovan, Carol M. Jantzen
49 **Probabilistic safety assessment for optimum nuclear power plant life management (PLiM): Theory and application of reliability analysis methods for major power plant components**
Gennadij V. Arkadov, Alexander F. Getman and Andrei N. Rodionov
50 **The coal handbook: Towards cleaner production Volume 1: Coal production**
Edited by Dave Osborne
51 **The coal handbook: Towards cleaner production Volume 2: Coal utilisation**
Edited by Dave Osborne
52 **The biogas handbook: Science, production and applications**
Edited by Arthur Wellinger, Jerry Murphy and David Baxter
53 **Advances in biorefineries: Biomass and waste supply chain exploitation**
Edited by Keith Waldron
54 **Geological storage of carbon dioxide (CO_2): Geoscience, technologies, environmental aspects and legal frameworks**
Edited by Jon Gluyas and Simon Mathias
55 **Handbook of membrane reactors Volume 1: Fundamental materials science, design and optimisation**
Edited by Angelo Basile
56 **Handbook of membrane reactors Volume 2: Reactor types and industrial applications**
Edited by Angelo Basile

57 **Alternative fuels and advanced vehicle technologies for improved environmental performance: Towards zero carbon transportation**
Edited by Richard Folkson
58 **Handbook of microalgal bioprocess engineering**
Christopher Lan and Bei Wang
59 **Fluidized bed technologies for near-zero emission combustion and gasification**
Edited by Fabrizio Scala
60 **Managing nuclear projects: A comprehensive management resource**
Edited by Jas Devgun
61 **Handbook of Process Integration (PI): Minimisation of energy and water use, waste and emissions**
Edited by Jiří J. Klemeš
62 **Coal power plant materials and life assessment**
Edited by Ahmed Shibli
63 **Advances in hydrogen production, storage and distribution**
Edited by Ahmed Basile and Adolfo Iulianelli
64 **Handbook of small modular nuclear reactors**
Edited by Mario D. Carelli and Dan T. Ingersoll
65 **Superconductors in the power grid: Materials and applications**
Edited by Christopher Rey
66 **Advances in thermal energy storage systems: Methods and applications**
Edited by Luisa F. Cabeza
67 **Advances in batteries for medium and large-scale energy storage**
Edited by Chris Menictas, Maria Skyllas-Kazacos and Tuti Mariana Lim
68 **Palladium membrane technology for hydrogen production, carbon capture and other applications**
Edited by Aggelos Doukelis, Kyriakos Panopoulos, Antonios Koumanakos and Emmanouil Kakaras
69 **Gasification for synthetic fuel production: Fundamentals, processes and applications**
Edited by Rafael Luque and James G. Speight
70 **Renewable heating and cooling: Technologies and applications**
Edited by Gerhard Stryi-Hipp
71 **Environmental remediation and restoration of contaminated nuclear and NORM sites**
Edited by Leo van Velzen
72 **Eco-friendly innovation in electricity networks**
Edited by Jean-Luc Bessede
73 **The 2011 Fukushima nuclear power plant accident: How and why it happened**
Yotaro Hatamura, Seiji Abe, Masao Fuchigami and Naoto Kasahara. Translated by Kenji Iino
74 **Lignocellulose biorefinery engineering: Principles and applications**
Hongzhang Chen
75 **Advances in membrane technologies for water treatment: Materials, processes and applications**
Edited by Angelo Basile, Alfredo Cassano and Navin Rastogi
76 **Membrane reactors for energy applications and basic chemical production**
Edited by Angelo Basile, Luisa Di Paola, Faisal Hai and Vincenzo Piemonte

77 **Pervaporation, vapour permeation and membrane distillation: Principles and applications**
Edited by Angelo Basile, Alberto Figoli and Mohamed Khayet
78 **Safe and secure transport and storage of radioactive materials**
Edited by Ken Sorenson
79 **Reprocessing and recycling of spent nuclear fuel**
Edited by Robin Taylor
80 **Advances in battery technologies for electric vehicles**
Edited by Bruno Scrosati, Jürgen Garche and Werner Tillmetz
81 **Rechargeable lithium batteries: From fundamentals to applications**
Edited by Alejandro A. Franco
82 **Calcium and chemical looping technology for power generation and carbon dioxide (CO_2) capture**
Edited by Paul Fennell and Ben Anthony
83 **Compendium of hydrogen energy Volume 1: Hydrogen production and purificiation**
Edited by Velu Subramani, Angelo Basile and T. Nejat Veziroglu
84 **Compendium of hydrogen energy Volume 2: Hydrogen storage, transmission, transportation and infrastructure**
Edited by Ram Gupta, Angelo Basile and T. Nejat Veziroglu
85 **Compendium of hydrogen energy Volume 3: Hydrogen energy conversion**
Edited by Frano Barbir, Angelo Basile and T. Nejat Veziroglu
86 **Compendium of hydrogen energy Volume 4: Hydrogen use, safety and the hydrogen economy**
Edited by Michael Ball, Angelo Basile and T. Nejat Veziroglu
87 **Advanced district heating and cooling (DHC) systems**
Edited by Robin Wiltshire
88 **Microbial electrochemical and fuel cells: Fundamentals and applications**
Edited by Keith Scott and Eileen Hao Yu
89 **Renewable heating and cooling: Technologies and applications**
Edited by Gerhard Stryi-Hipp
90 **Small modular reactors: Nuclear power fad or future?**
Edited by Daniel T. Ingersoll
91 **Fuel flexible energy generation: Solid, liquid and gaseous fuels**
Edited by John Oakey
92 **Offshore wind farms: Technologies, design and operation**
Edited by Chong Ng & Li Ran
93 **Uranium for nuclear power: Resources, mining and transformation to fuel**
Edited by Ian Hore-Lacy
94 **Biomass supply chains for bioenergy and biorefining**
Edited by Jens Bo Holm-Nielsen and Ehiaze Augustine Ehimen
95 **Sustainable energy from salinity gradients**
Edited by Andrea Cipollina and Giorgio Micale
96 **Membrane technologies for biorefining**
Edited by Alberto Figoli, Alfredo Cassano and Angelo Basile
97 **Geothermal power generation: Developments and innovation**
Edited by Ronald DiPippo
98 **Handbook of biofuels' production: Processes and technologies (Second Edition)**
Edited by Rafael Luque, Carol Sze Ki Lin, Karen Wilson and James Clark

99 **Magnetic fusion energy: From experiments to power plants**
Edited by George H. Neilson
100 **Advances in ground-source heat pump systems**
Edited by Simon J. Rees
101 **Absorption-based post-combustion capture of carbon dioxide**
Edited by Paul Feron
102 **Advances in solar heating and cooling systems**
Edited by Ruzhu Wang and Tianshu Ge

Preface

It has been more than forty years since the first flurry of research and initial market deployment of ground-source heat pump (GSHP) technology. Although the technology is in some ways mature and design and analysis methods well established, there is, perhaps surprisingly, little available in the way of technical books on the subject. Although training programmes are readily accessible to installers in some countries, educational offerings at the university level remain few and far between. Although design guides and introductory texts have been produced for many years by organisations such as the International Ground Source Heat Pump Association, it is hard to argue that there is a recognised textbook on the technology suitable for university students and professional engineers. A textbook aimed at undergraduate students may come along in the near future but this book is not of that type. This book, like others in the series, is something of an expression of the state of the art in design practice and the current state of research. It is acknowledged that this is still a developing area of study – academic research publications in this area seem to have grown dramatically in recent years – but also that the area is mature enough that a book combining contributions from a group of established researchers and practitioners in the field provides an appropriate form of reference and educational resource.

In many countries both GSHP design and installation are still regarded as areas of specialist expertise and the technology is often not considered a conventional option amongst the various building heating and cooling system design alternatives. There is, I suggest, something of a knowledge gap amongst professionals such as heating, ventilation and air conditioning (HVAC) engineers who are responsible for designing heating and cooling systems with regard to GSHP technology. On the one hand, this may be understandable as such ground-coupled heating and cooling approaches cross the disciplinary boundaries between building HVAC and geology or geotechnics. On the other hand, speaking as someone who started his career working as a consulting engineer designing heating and cooling systems who later found his way into GSHP-related research, I do not see why this should remain the case – the principles of conduction heat transfer applied to ground heat exchangers are not a great mystery.

I think I can speak for my co-authors in saying that our hope is not simply to further disseminate research findings, but to provide something of a useful resource for an audience beyond the research community and that may go some way to promote better appreciation of what is sometimes a misunderstood technology. We also hope to look forward and point out forthcoming developments and continuing research questions.

In addition to these broad aims, the book seeks to firstly set out some of the underlying theory behind ground heat exchangers that are the heart of GSHP systems and review the principle approaches to their modelling, design and analysis. This is the focus of the first of three parts of the book. Analysis of closed-loop ground heat exchangers is the subject of a large body of research, as their successful design and operation is essential to many areas of application, and so the book deals with this topic from a number of perspectives in Part One. Design methods are first of all reviewed and supplemented by in-depth treatment of thermal resistance calculation. As there have been a number of advances in analytical modelling of closed-loop borehole heat exchangers in recent years, this is the focus of an additional chapter in Part One.

Sophisticated design methods do not necessarily add value to the design process or overall outcomes unless they are can be supported by good quality design parameter data. Uncertainty in estimates of ground thermal conductivity has long been recognized as a factor in overall design risk and a good deal of research effort has gone into developing thermal response testing methods and analysis techniques. Accordingly, it would be inappropriate to discuss design without also discussing in situ test methods and the Chapter "In situ estimation of ground thermal properties" is devoted to this topic. The Chapter "Horizontal and compact ground heat exchangers" is concerned with horizontal heat exchangers of the types found in residential systems and points out some of the complexities arising at shallow installation depths (despite the technology itself being relatively simple) and a number of technical challenges that are not fully explored. In recent years Energy Piles and related systems that seek to take advantage of earth-coupled structural elements have seen significant growth in application so that new opportunities for efficient heating and cooling have emerged. International research efforts focused on these types of heat exchanger continue to gather pace and engage the interests of geotechnical specialists and so it has been important to include a chapter on this topic to complete the first part of the book.

The natural environment offers a range of opportunities that can be taken advantage of as heat sources, sinks and storage mechanisms beyond what can be accessed by conduction of heat via closed-loop heat exchangers. Direct access to ground and surface water allows convection/advection flows to be taken advantage of and so both surface water and open-loop heat exchanger systems have always been appealing from the point of view of thermodynamic efficiency. It is hoped that the second part of the book, which includes chapters on surface water, open-loop and standing column well systems, will highlight the potential of these approaches and advances that have been made to deal with the related technical challenges and maximize their exploitation.

Of all the identifiable thermal energy storage technologies, it is only ground-coupled approaches that offer the real opportunity for storage over seasonal time scales. Open-loop wells can be taken advantage of in aquifer thermal energy storage (ATES) and closed-loop heat exchangers in borehole thermal energy storage (BTES) systems. ATES systems are discussed in the Chapter "Open-loop heat pump and thermal energy storage systems" and BTES systems in the Chapter "Standing column wells" to complete Part Two of the book.

The third part of the book is concerned, not with specific ground heat exchange approaches, but with integration of both ground heat exchangers and heat pumps to provide complete building heating and cooling systems. Hybrid systems are increasingly applied to cooling-dominated buildings and those with ground heat exchangers such as Energy Piles, and a further chapter on this topic is included in the third part of the book. No GSHP system is really complete without a heat pump. Although this book is not particularly aimed at refrigeration specialists – and many other books deal with the fundamentals of heat pump equipment and refrigeration in greater depth – ground-coupling introduces additional considerations of temperature ranges and heat exchanger design that are worthy of special consideration. Advances continue to be made in heat pump design for GSHP applications – not least hybrid or dual-source approaches – and so chapters on heat pump technology and heat pump modelling form a large part of Part Three. The final chapter deals with integration of geothermal heat exchangers and activated building structures – an approach that synthesizes the low energy benefits of both technologies.

Much of the early effort in GSHP research was spurred on by the 1973 oil crisis and the urgent need for improved energy efficiency. At the time of writing, those responsible for building design are facing the challenges of designing for a world where the need for energy efficient systems is compounded by the need for carbon emissions reductions and maximisation of renewable energy resources. GSHP systems, in not relying directly on fossil fuels and in being highly efficient, seem certain to be of enduring interest in coming decades and have potential to have significant impact in meeting these challenges. It is hoped that this book will make a contribution to understanding and further development of GSHPs. I finish by extending my thanks to all the authors who have contributed to the various chapters of the book and have helped to make it a resource available to others.

Simon J. Rees (Editor)
December 2015

An introduction to ground-source heat pump technology

1

S.J. Rees
University of Leeds, Leeds, United Kingdom

Abstract

Ground-source heat pumps (GSHPs, or geothermal heat pumps) have great appeal in offering levels of efficiency for building heat and cooling that are – both theoretically and practically – higher than other technologies. This chapter introduces the technology and reviews its historical development and current state of exploitation around the world. Current challenges and prospects are discussed.

1.1 Introduction to the technology

1.1.1 Heat pump principles

Heat pumps are a form of heat engine that uses mechanical work to transfer heat from a low temperature source to a higher temperature sink. There are a wide range of applications of heat pumps but in this context we are concerned with transferring heat between buildings and the external environment – either rejecting heat to the environment and cooling the building or extracting heat and heating the building. Although various forms of thermodynamic cycle can be used to move heat between source and sink, the predominant form is based on the vapour–compression cycle in which a refrigerant gas is evaporated, compressed and condensed in turn to transfer heat. The principle components in the cycle are shown in Fig. 1.1. The state of the refrigerant throughout the cycle is well illustrated in an enthalpy–pressure diagram such as that on the right of this figure. The enthalpy changes in the condenser and evaporator indicate the heat transfer rate per unit mass of refrigerant flowing. Chapter 'New trends and developments in ground-source heat pumps' provides an in-depth review of refrigeration technology applied in heat pumps adapted for ground-coupled applications.

The prime reason for the interest in using heat pumps to provide heating and cooling is that it takes less work to move heat from source to sink than it does to convert primary energy into heat. In other words, the power required is noticeably less than the heating or cooling delivered. This effect is quantified in classical thermodynamics by the coefficient of performance (COP). If Q_H is the quantity of heat delivered and W is the work required, then $COP = Q_H/W$. Classical thermodynamics also tells us that this has a theoretical maximum value (Carnot efficiency) limited by the absolute temperature of the source (T_C) and sink (T_H) so that, $COP_{max} = T_H/(T_H - T_C)$ in heating

Advances in Ground-Source Heat Pump Systems. http://dx.doi.org/10.1016/B978-0-08-100311-4.00001-7
Copyright © 2016 Elsevier Ltd. All rights reserved.

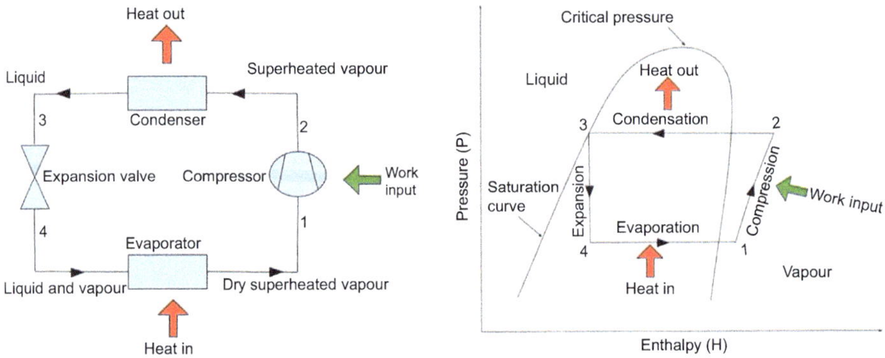

Figure 1.1 A conceptual model of a heat pump (left) and an idealized cycle represented on an enthalpy–pressure (right) (Naicker, 2016).

mode. This implies that higher efficiencies can be achieved if the source and sink temperature are close together – as is generally the case when considering buildings and their surrounding environment. It will be shown later (Section 1.1.3) that conditions are particularly favourable where the heat source takes the form of a geothermal heat exchanger.

The relatively high COP values that are achievable with GSHPs mean that the energy consumed is effectively leveraged so that much higher quantities of heating and cooling are delivered. This effect means that even with electricity derived from thermal power generation, where a lot of primary energy becomes waste heat at the power station, the heat exchanged with the building is greater than the primary energy consumed. This is illustrated in the upper Sankey diagram in Fig. 1.2. If one considers renewable energy sources (RES) of electricity (eg, wind power), where there is no wastage of primary energy at the source, the heat pump effect is even more significant and the carbon emissions associated with the heating and cooling demand can approach zero. RES are a limited resource and so when one considers provision of heating demands, a heat pump can be seen to be a valuable mechanism to leverage the available generation capacity (resistance heating being a poor use of renewable electricity in comparison). This is indicated in the lower Sankey diagram in Fig. 1.2.

1.1.2 Performance metrics

Strictly speaking, the heat-pump COP has to be quantified for a particular state or set of steady operating conditions. This metric is consequently useful when it comes to product testing and specification and is also sometimes termed the energy efficiency ratio (EER). When it comes to whole-system performance, we would like to consider a range of operating conditions and use a metric that represents performance when the heat pump is installed and forms part of a heating and cooling system. Choice of a suitable metric has long been a concern amongst researchers in the field. When Miriam Griffith published some of the first work on GSHPs (Griffith, 1957), she used the term performance energy ratio, and one of the audience members at the presentation of her paper commented, 'It is unfortunate that, having become accustomed to the term "coefficient

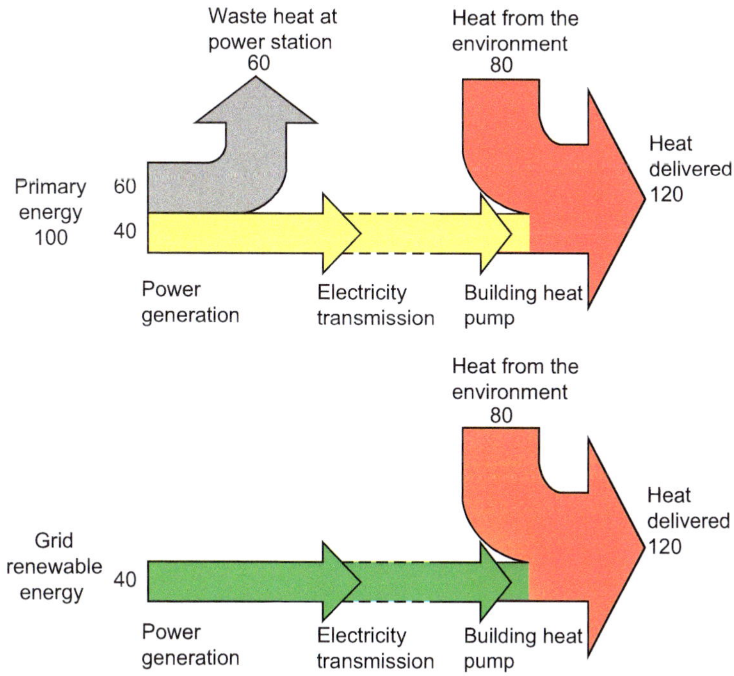

Figure 1.2 Sankey diagram representations of the energy flowing from source to delivery by a heat pump. A typical thermal power generation process is illustrated above, and power drawn from renewable sources below (an SPF of 3 is assumed).

of performance", we now have to think of "performance energy ratio"'. This term did not persist, but common agreement on a suitable metric was not found for some time.

One approach to deriving a seasonal metric is to use a weighted mean of COP values measured at a few different operating conditions. This gives rise to metrics such as seasonal energy efficiency ratio (SEER) for air conditioners and heating seasonal performance factor (HSPF) in the case of heat pumps. However, this can also be a source of confusion. The definition of SEER in the ARI standard 210/240 is not a simple ratio but has units BTU/Wh (ARI, 2008) and similarly for HSPF. This definition of SEER consequently differs from the European ESEER metric, which is a simple ratio of energies and is also weighted slightly differently. North American SEER values have values that are 3.41214 greater than the ratio as a consequence of the choice of units. Accordingly, some caution is required when interpreting manufacturers' data.

Another approach is to derive a seasonal metric from measurements of complete systems over one or more seasons. This is not useful for rating equipment but is more useful when making comparisons with other technologies or making realistic estimates of running costs or carbon emission savings. This type of metric is termed a seasonal performance factor (SPF) and can be applied to heating or cooling performance. Defining the exact meaning of this sort of metric is not as straightforward as it first seems, however (Gleeson and Lowe, 2013). In real systems there are a number of other sources of power demand that could be constituted to be part of the whole-system

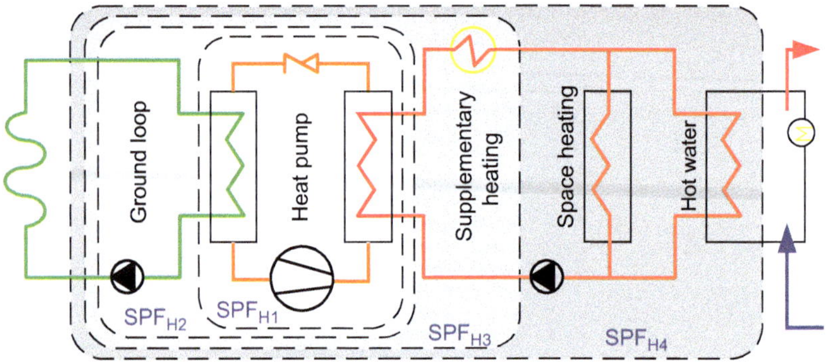

Figure 1.3 Seasonal performance factor scope according to SEPEMO definitions.

demand. For example, some heat pumps include a supplementary electric resistance heater to boost output at times of peak demand. Some packaged heat pump equipment incorporates the ground loop and possibly heating circulation pumps; in other systems these pumps are installed elsewhere. The complexities of agreeing on definitions of SPF was addressed by a European project SEPEMO that published definitions of SPF with defined scope (Zottl et al., 2012). The metrics are SPF_{H1} through SPF_{H4} depending on which supplementary electrical demands are included.

The definition of SPF_{H1} includes only the heat pump equipment (ie, compressor and controls) and no supplementary heaters or circulation pumps. SPF_{H2} additionally includes the ground loop circulating pump energy. SPF_{H3} furthermore includes any electric heater packaged with the heat pump and SPF_{H4} includes all circulating pumps and supplementary heaters. Consequently, values of SPF_{H1} are closest to the manufacturers' EER values and SPF_{H4} values are somewhat lower. The different system boundaries are indicated in Fig. 1.3. Full-load and part-load performance of heat pump equipment is discussed in detail in Section 13.2.

1.1.3 Heat sources and sinks

A number of different types of environmental heat source or sink have been used with heat pumps in building heating and cooling applications. An obvious approach is to extract or reject heat to the ambient air using a refrigerant-to-air heat exchanger with a suitable fan – forming an air-source heat pump (ASHP) system. Early workers in the heat pump field suggested, and indeed tried, a number of other forms of environmental heat exchange (Sumner, 1976). Ideas have included:

- the ambient air
- free flowing water courses – rivers and streams
- static open surface water sources – lakes and ponds
- the sea
- water wells (groundwater extraction)
- standing column wells (groundwater recirculation and extraction)
- the solid ground

- waste fluids – foul and rainwater drainage
- solar panels
- industrial or process waste heat

The ambient air is an attractive source/sink in that heat exchangers can be compact and with low thermal resistance so that systems can be relatively economic in terms of capital cost. However, the ambient air is not only more volatile than other forms of environmental source in terms of its temperature but also has the disadvantage of being moist with the consequence that frosting of the heat exchanger occasionally occurs at times of high demand, requiring reversal of operation or other energy-consuming measures to achieve defrosting. In cold climates there is also the concern that at subfreezing air temperatures the heat pump capacity can drop-off dramatically – exactly when maximum output is required.

Environmental water sources and the ground (ie, the soil and other geological formations below) have the advantage, compared to the ambient air, of having high heat capacity and consequently their natural temperature range through the year varies much less. The temperature of ground or water sources is consequently generally closer to room temperature than the ambient air. Consideration of Carnot efficiency, as noted earlier, tells us that the smaller difference between the source and sink temperatures means that higher system efficiencies are achievable. Free-flowing water sources are particularly attractive in that the source temperature is entirely driven by ambient conditions and is not influenced in the long term by heat exchange with the system. Similarly, open wells offer the opportunity of low thermal resistance to heat exchange and favourable ambient temperatures. It is consequently unsurprising that some of the earliest examples of heat pump applications for building heating used wells and rivers as heat sources. Water sources of various types are classed as GSHP systems and are, accordingly, treated in later chapters of this book.

Water sources are often not accessible as heat sources in urban environments and so other forms of ground coupling are more universally applicable and form the majority of GSHP system heat sources/sinks. Although it is feasible to circulate refrigerant through pipes buried in the ground, the environmental risks of leaking refrigerants means that this has fallen out of favour. Accordingly, the most common forms of ground heat exchanger (GHE) are closed-loops using a secondary heat exchange fluid such as an anti-freeze/water mixture or brine solution. Closed-loop GHEs come in many forms but the predominant types are vertical borehole devices or forms of shallow horizontal loop. The basic forms of GHE are illustrated in Fig. 1.4. Vertical borehole heat exchangers are discussed in detail in Chapters 'Vertical borehole ground heat exchanger design methods', 'Calculation of borehole thermal resistance' and 'In situ estimation of ground thermal properties' and horizontal heat exchangers in the Chapter 'Horizontal and compact ground heat exchangers'.

The advantageous nature of the ground as opposed to ambient air temperatures over the season, is well illustrated by the data in Fig. 1.5 that was recorded for a site in the moderate UK climate (Naicker, 2016). These daily mean temperatures show that the ground temperature in this case was more than 10K closer to room temperature than the corresponding air temperature during mid-winter. The thermodynamic advantages

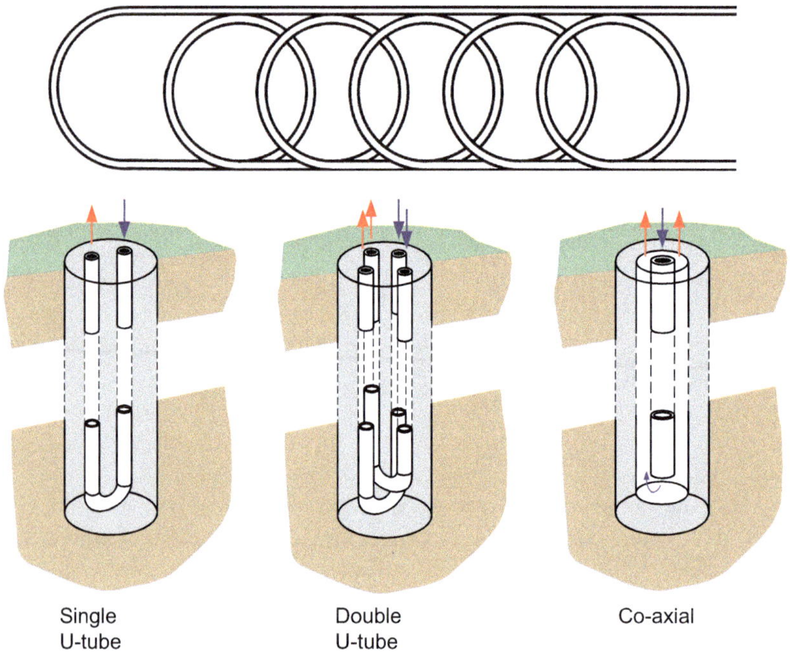

Figure 1.4 Outline diagrams of a horizontal 'Slinky' heat exchanger (above) and common vertical borehole heat exchanger types (below).

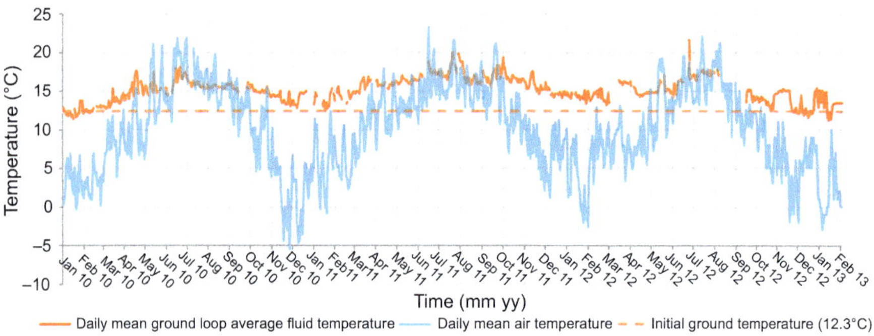

Figure 1.5 An example of measured ambient air and ground heat exchanger temperatures (Naicker, 2016).

of ground coupling, and favourable operating efficiencies that result, account for much of the growth of interest in GSHP technology. Although capital costs are not always favourable, low operating costs mean that life-cycle costs are often acceptable and carbon-emission savings welcomed. Essentially, GSHP systems are one of the most efficient forms of heating that can be conceived and it is easy to see why interest in the technology developed earlier in the 20th century and can be expected to persist.

1.2 Historical developments and industry growth

1.2.1 Early developments

The early and middle 19th century saw the development of the foundations of our understanding of heat engines and much discussion of the subject between scientists whose names are now synonymous with the subject of classical thermodynamics – the likes of Carnot, Joule, Rankine and Kelvin. Although the scientific literature reveals a number of discussions of applications of heat engines, William Thomson's (later Lord Kelvin) paper of 1852 gets credit for being the first to explicitly explore the possibilities of using heat engines to heat and cool buildings (Thomson, 1852). The 1852 paper discusses both air cycle devices and vapour-compression principles in general, without indicating what form a practical device may take.

By the first decade of the 20th century the Scottish engineer and University of Cambridge Professor, James Alfred Ewing, had published a number of editions of his book *The Steam Engine and other Heat Engines* that included sections on refrigeration (which had already been demonstrated in practical industrial applications) and also use of reverse heat engines to heat buildings. He stressed the thermodynamic advantages of such an approach – although practical implementation had yet to be demonstrated – and stated that 'Burning fuel to warm a room by a few degrees is a wasteful way to utilise heat', and he goes on to conclude, 'It is interesting, and may some day be useful, to recognize that even the most economical of the usual methods employed to heat buildings, with all their advantages in respect of simplicity and absence of mechanism, are in the thermodynamic sense spendthrift modes of treating fuel' (Ewing, 1910).

According to Wirth (1955), it was when a German translation of Ewing's book was published in 1910 that the Swiss engineer, Heinrich Zoelly, developed his ideas and was granted Swiss patent 59350 in February 1912 for a ground-coupled heat pump with an electrically driven compressor. Wirth goes on to explain that development of heat pump technology in Switzerland in the following few decades was limited to industrial heating and drying applications (Wirth, 1955) so that Zoelly's patent was not exploited for some time, however.

Haldane (1930) is sometimes credited with presenting the first results of application of heat pumps to buildings. His paper presents a number of ideas for practical heat pump applications and also presents results from a system he built in the late 1920s at his home in Scotland (Fig. 1.6). The ammonia vapour compression cycle machine was built using an electrically driven compressor (originally manufactured for food refrigeration systems) and used an external heat exchanger exposed to the air (without fan assistance) and radiators as the internal heat emitters. The air-cooled evaporator was supplemented by a second coil submerged in a tank of water through which fresh water was allowed to flow at times of peak demand. Haldane reported COPs in the range 2–3 depending on condensing temperature (Fig. 1.6) and used data from larger scale refrigeration systems to suggest that systems of the order of 100 kW capacity should be able to achieve COPs in the range 3–5. One of his most interesting proposals was for a water-source heat pump system (Fig. 1.7) that was intended to

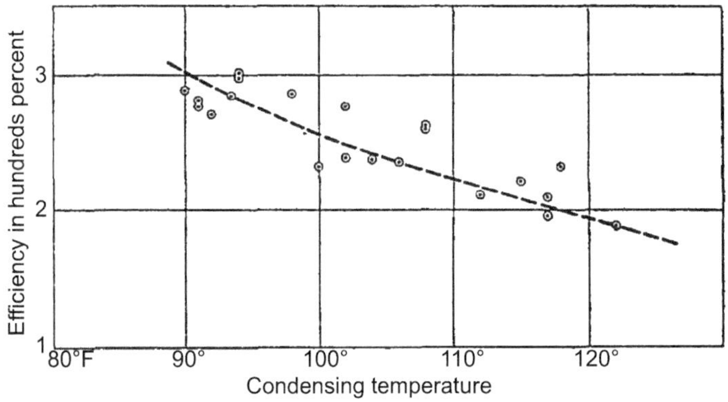

Figure 1.6 Results from Haldane's home system recorded between 1926 and 1928 (Haldane, 1930).

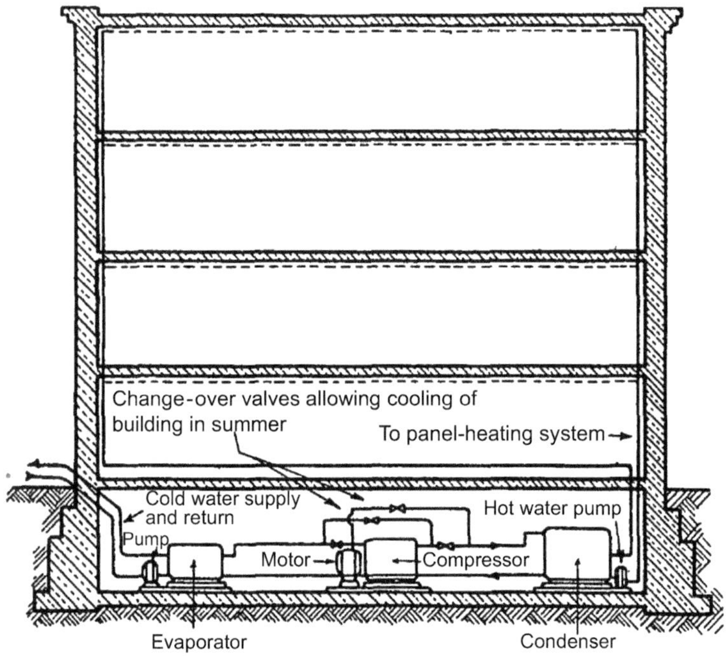

Figure 1.7 Haldane's proposed system for a water-source heat pump to deliver heat to low temperature panels in a commercial building (Haldane, 1930).

use a river or lake as a heat source and deliver its heat to a commercial building using low temperature panels.

When Penrod wrote his article in *American Scientist* in 1947 he was able to report data from several heat pump installations operating during the 1930s at a range of locations and different climates in the United States. These installations include both residential and commercial systems with reversible heat pumps using air and water sources. Some distinction was made between these and other air-conditioning installations in that they offered continuous or all year round air-conditioning. Penrod provides technical details of a number of systems, starting with an earth coil design, and provides calculations for a typical house showing the advantages of an earth coupled approach, in terms of COP, over the season. He concludes by suggesting an earth coupled system should give an overall (taking account of the fan) COP of 3.52 as opposed to 2.24 for an air-source system. Although he makes reference to a number of well-water heat pump systems (including one in Norwich, UK, by J.A. Sumner) the only example of an earth coupled system he gives was a domestic system reported by A.C. Crandall (1946). The system in question was installed in the home of one of the employees of Indianapolis Power and Light Company from which data was collected in the 1945–46 heating season and a COP value of 3.61 reported (Crandall, 1946). The house and system are illustrated in Figs. 1.8 and 1.9.

Figure 1.8 The house in Indiana and its ground-source heat pump system using earth coils tested during the 1945 heating season (Crandall, 1946).

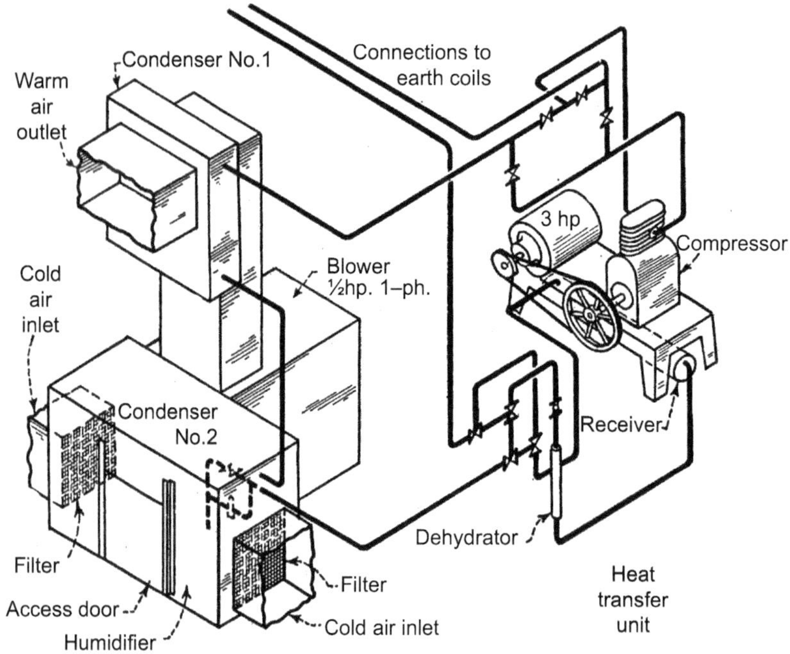

Figure 1.9 A diagram of the ground-source heat pump system at the house in Indiana (Crandall, 1946).

Penrod (1947) also comments on the initial commercial exploitation of reversible heat pump equipment in the United States. Packaged units were made under the Thermacooler and Airtopia brand names as reversible air-source heat pumps starting in the late 1930s. Some efforts were made by the Muncie Gear Works company of Indiana to take advantage of the well-source heat pump concept and developed their own Marvair unit and is reported to have shipped 50 units during 1947 (Penrod, 1947). The company also took out patents (Smith and Kemler, 1949) on a number of ideas to exploit water wells as heat sources – one such is illustrated in Fig. 1.10.

Many of the early heat pump installations in the United States in the 1930 and 1940s were sponsored by electric utility companies. In the immediate postwar period a number of research projects sponsored by these companies are reported that specifically studied earth-coupled designs, eg, Walton and Williams (1944), Hapley (1949). A particularly detailed study is reported in a small book by Coogan (1948) that is concerned with an experimental earth-coupled residential system in Connecticut. This system used an evaporator of copper pipe in parallel circuits that allowed testing with different pipe diameters (Figs. 1.11 and 1.12) and used an open compressor (Fig. 1.13). Coogan also carried out detailed lab tests of the soil thermal properties and collected data from a number of thermocouples in the soil surrounding the earth coil. It was also around this time that Ingersoll and his coauthors published some of the first theoretical analysis of ground heat transfer from buried pipes (Ingersoll and Plass, 1948).

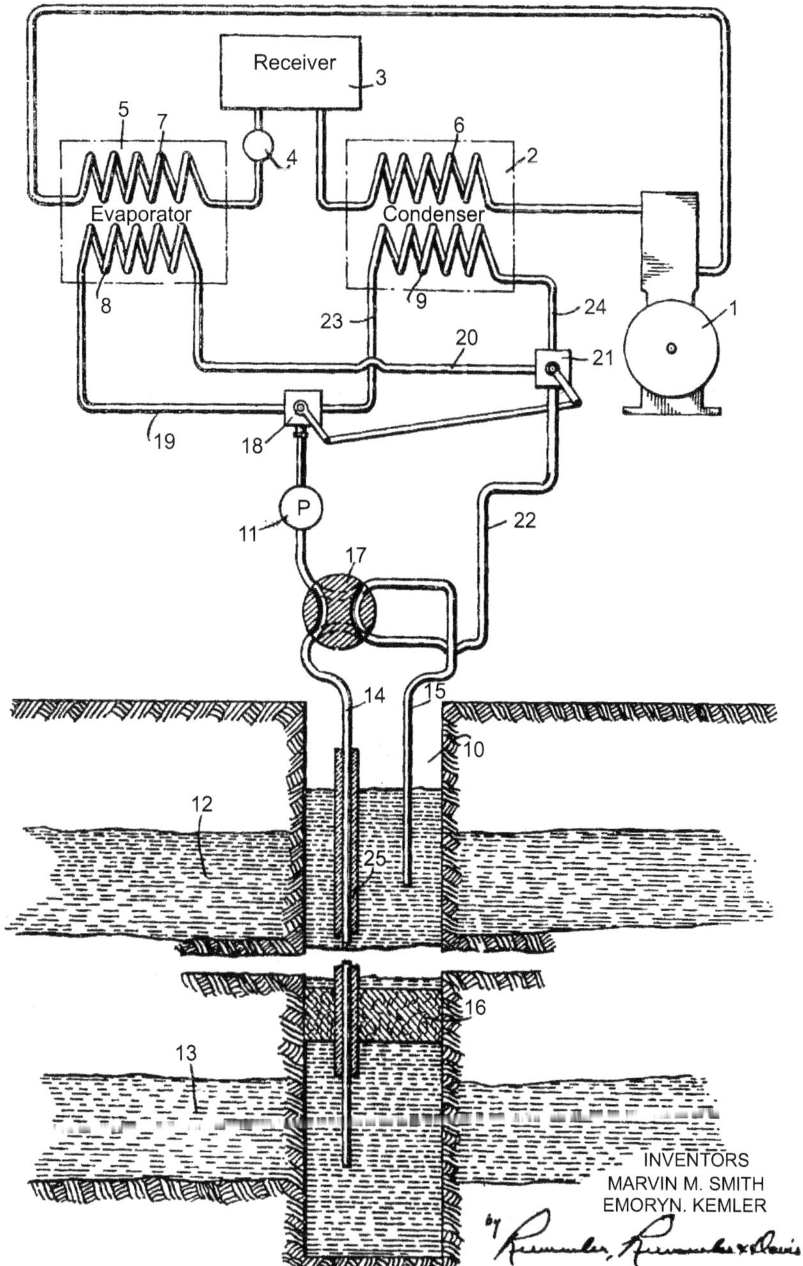

Figure 1.10 The diagram from a patented (filed 1946) water source heat pump design that circulates groundwater between levels in a well (Smith and Kemler, 1949).

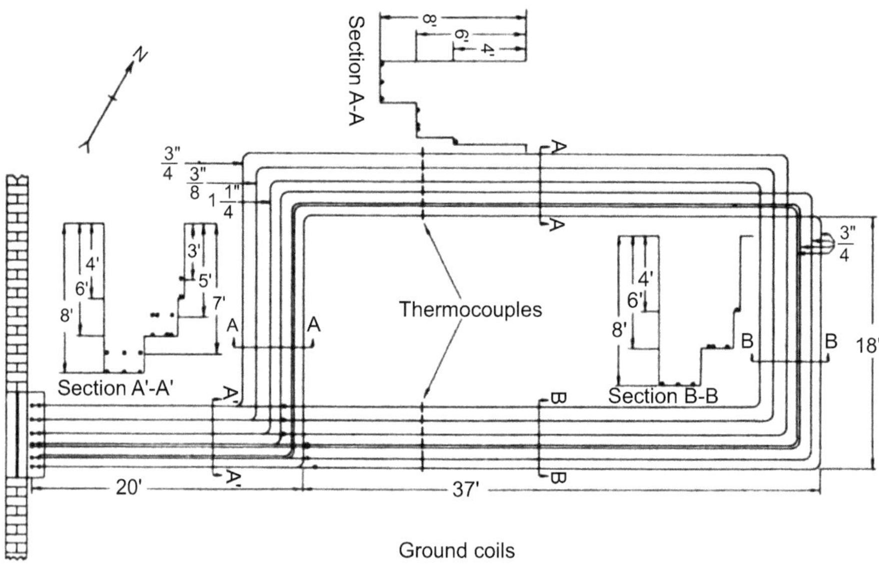

Figure 1.11 A diagram of the earth coil used in the experimental ground-source heat pump installation in Connecticut (Coogan, 1948).

Figure 1.12 The earth coil installation used in the experimental system in Connecticut pictured during installation (Coogan, 1948).

In the project reported by Coogan in the book *The Heat Pump in New England* and other early investigations (Vestal, 1949; Coogan, 1949; Penrod et al., 1950; Kidder and Neher, 1952) ground-coupled rather than air-coupled heat pumps were thought advantageous in view of the cold winter climate (and indeed this motivation for GSHP adoption persists). The work conducted in 1946–47 was at a time where the

Figure 1.13 The compressor used in the ground-source heat pump experiments in Connecticut (Coogan, 1948).

US electricity grid was much smaller in scale than now and demand for residential air conditioning was just starting to grow. The utility companies appeared to have been motivated by a desire to increase sales, compete with other forms of heating and also achieve better load factors than resistance heating. Comments on some of the papers (eg, as reported in the AIEE transactions) suggest there was also an issue of limited oil supply and so some motivation to avoid future growth in oil heating.

Andrews, writing in 1948 and speaking from a utility company point of view, states, 'the heat pump holds promise of permitting the industry to supply domestic-heating service in an economical manner. Up until now the electric utility industry has supplied very little of this market and we think that without the heat pump we are not likely ever to supply very much of it'. However, little is reported by way of further research or commercial exploitation of GSHP technology between this period and the 1970s. This may partly be a question of oil supply no longer being an issue but also the reliability of the compressor technology and durability of metal refrigerant pipework exposed to the ground. In his later book on heat pumps, Ambrose (1966) is rather sceptical about the use of GHEs due to reported problems with the ground surrounding the pipes drying out during cooling operation.

Outside of North America, one early piece of research worth noting is that carried out by the British Electricity Industry Research Association and reported by Miriam Griffith (1957). The experimental installation in this work differed from that reported by Coogan (1948) in one important aspect, and that is that the heat pump refrigeration circuit was separated from the GHE by a heat exchanger and brine was used as a secondary heat transfer fluid (as indicated in Fig. 1.14). The system was therefore closer to the modern GSHP concept than that reported by Coogan.

Miriam Griffith (1957) presented the paper at a number of meetings of the Institution of Electrical Engineers during 1957 and, although the research does not appear to have continued, the work received a good deal of interest and informal reports were

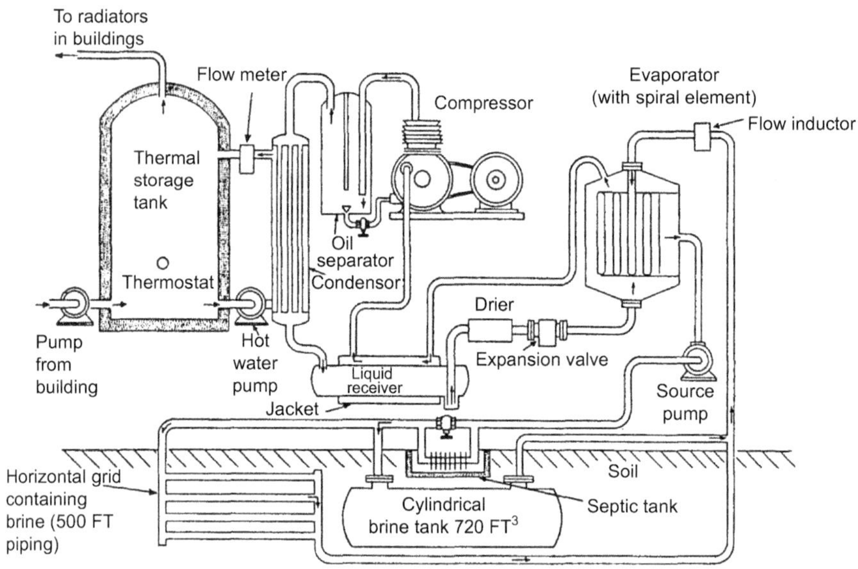

Figure 1.14 The experimental system at the Shinfield Field Research facility in the United Kingdom (Griffith, 1957).

recorded along with the paper (there were discussions of the implications of heat pump loads on the electricity grid and possible application in district heating, for example). A Mr I.A. Scholar makes the following report about the state of heat pump exploitation in Switzerland in 1956:

> *I read the annual report of the Municipality of Zurich on their heating published in the Neue Zürcher Zeitung. To my great surprise I found that the heating was done 1% by electricity, 9.6% by heat pumps and some 73% by coke. Owing to hydro-electric resources, electricity is much cheaper in Switzerland than anywhere else, and it is to the credit of the heat pump that almost ten times more heat was produced by that means than by electricity. The majority of municipal buildings which are heated in that fashion are either on or near the Limmatquay, the Limmat being the river flowing in this area. For that reason, the installation was first considered, and it was a good investment. While the Swiss are excellent engineers and always willing to experiment, before a municipality authorizes the expenditure of money it will need to be assured of some return; so this was not a venture undertaken lightly. They must have been satisfied with the first installation to have dropped some of the electrical ones in favour of the heat pump. Since the night load tariff in Zurich is very cheap and the cost of installation of the heat pumps is very high, the fact that they are installed means a great deal.*
>
> *Griffith (1957)*

Two speakers at these meetings who were also supporters of heat pump technology were J.W. MacFarlane and J.A. Sumner. Mr MacFarlane (also reported in MacFarlane, 1957) tells of a ground-coupled system he installed in his own Scottish home and operated from 1947. This was apparently a system using refrigerant Freon-12 and steel pipes buried in trenches below a basement as the GHE. The heat pump delivered warm air to the house and was later extended and adapted to also heat a hot water tank. Mr Sumner was the chief electrical engineer who had designed a heat pump system using river water as a heat source for his own Norwich Electricity Board office in the 1940s and who later installed a number of other systems in UK commercial buildings. He reported on a system installed in his own home in his later book *Domestic Heat Pumps* (Sumner, 1976).

Sumner, as an engineer working in the electricity industry, might have been expected to be in favour of further development of heat pumps. However, his accounts show he had a hard time persuading the electricity industry in the United Kingdom to take up the idea. He made the point on a number of occasions that, during a period of rapid growth of the national electrical energy system, a sensible route would be to invest in heat pumps and mitigate building new power generation capacity (Griffith, 1957; Sumner, 1976). He reports that this argument was countered by the suggestion that adopting more resistance electrical heating would bring beneficial sales of electricity. Given that the UK electricity was a centrally planned and publicly owned system at the time, this was an important debate. What transpired was that, although there was growth in electrical resistance heating in the United Kingdom, a large natural gas grid developed in the following decades and this became the predominant building heating fuel in the United Kingdom. It is only in the last few years that, in the face of climate change mitigation targets, it is recognized that there must be a move away from gas heating to electric heat pumps as a mater of national policy in the United Kingdom (DECC, 2012).

1.2.2 Technological and industrial development

International interest in GHSP applications did not gather pace until the energy crisis of the 1970s (Svec, 1987). In the later 1970s and the 1980s, research efforts grew both in North America and in Europe – mostly in the Scandinavian countries and Switzerland, eg, Rosenblad (1979), Hopkirk et al. (1988), Sanner (1996), Sanner et al. (2003), Zogg (2008) and Schär (1977). In Europe, research was focused on the design of vertical borehole heat exchangers whilst in the United States, although there was work on vertical borehole heat exchangers, there was also great interest in horizontal systems and significant work was undertaken at certain US national laboratories (Metz, 1979; Moi, 1980) and universities (Bose, 1982) in this period. It was also during this period that the International Ground Source Heat Pump Association (IGSHPA) was founded with headquarters at Oklahoma State University. This was an industrial and educational organization that facilitated research and dissemination of information but was also active in the training of installers and designers.

Some distinctions can be made between the studies in the 1980s and the early trials in the 1950s that have been noted earlier, in terms of the state of development of the technology. The growth of the residential air-conditioning market and the food refrigeration market meant that small compressor technology was well developed and competitively priced by this time. It is also true that suitable plastic pipe was available economically and so the risks associated with using refrigerant circulating in metal buried pipes could be avoided. It was consequently possible to develop reliable heat pumps and systems by the 1980s. There was steady growth in the uptake of GSHP technology in the United States, Sweden, and Switzerland from the 1980s followed by growth in Germany and Austria Sanner et al. (2003). By the end of the 1990s the industry in these countries was growing in a sustainable manner and the rate of installations was increasing. One of the policy measures that is thought to have benefitted market penetration in Sweden is the restriction of all new heating systems to fluid temperatures less than 55°C at the design outside temperature. This regulation was introduced in 1984 and has eased later adaptation of heating systems to heat pump operation (Karlsson et al., 2003).

The postwar growth the residential air-conditioning industry and development of a refrigeration skills-base and associated training and certification systems, has contributed to the uptake of GSHP technology in North America. There are also some important differences in residential heating and cooling between North American homes and those in other parts of the world such as Europe. A common arrangement in North American homes is to provide heating via a gas-fired furnace (air heater) and to provide cooling via an air conditioner with an evaporator in the same air duct system. Hot water is generally produced in a separate electrically heated tank or gas-fired water heater. If a home owner is considering a GSHP system (a water-to-air design being the common type) then a good economic case can be made as both the gas furnace and air conditioner are replaced by a single heat pump. If the property is away from the established gas grid, then adopting a GSHP also avoids the need for a second fuel source such as liquefied petroleum gas (LPG) or oil.

The situation with regard to home design in Europe and other parts of the world is rather different to that in North America in that residential air conditioning has been uncommon. The heat pump technology that developed in Northern Europe was intended for heating only and used water-to-water heat pumps supplying heat to hydronic heating systems (ie, radiators, convectors or underfloor heating; Spitler, 2015). Some of the packaged heat pump equipment that is designed for the European market also incorporates a hot water tank and heat exchanger (Karlsson et al., 2003). This arrangement is suited to the retrofit market but the need for compact heating and hot water equipment is also important in many European homes as they are often smaller and less likely to have basements than those in North America. The economic case that can be made for uptake in GSHP technology in Europe is different, in part, for these reasons and it is understandable that early growth in the technology in Europe was confined to countries with relatively cheap hydroelectricity and where air-source systems were disadvantaged by cold climatic conditions.

The early growth in GSHP deployment in United States, Scandinavia, and Switzerland (followed soon after by Germany and Austria) has meant that the technology was essentially validated and well proven for residential applications by the turn of the century. In 2000 it was reported that there were estimated to be 400,000 installations in the United States, 55,000 installations in Sweden, 30,000 in Canada, 24,000 in Switzerland, 19,000 in Austria and 18,000 in Germany (Lund and Freeston, 2001). It is notable that no GSHP systems were reported installed in China at this time and numbers of units in the other 19 countries providing reports, were relatively small. The US installation industry was growing rapidly so that more than 40,000 units were installed in 2000 and this annual rate of new installations was thought to be increasing by 10% each year.

1.2.3 Factors in market development

A number of technical and economic factors are thought to be important in the adoption of GSHP technology – particularly in the early stages of national installation industry growth. The technical factors that can favour GSHP systems stem from its fundamentally favourable thermodynamic efficiency and this is a significant factor in colder climates. Where properties do not have access to any natural gas grid, any sort of heat pump can be an attractive option. In these situations, an air-source heat pump can be one of the most economic options. However, the difficulties with frosting of outside heat exchangers and capacity falling at very low outside temperatures in the case of air-source heat pumps can mean a GSHP is the best technical solution. GSHP technology also has some advantage in that the final installation is unobtrusive and quiet when compared to air-source or conventional air-conditioning solutions.

Any detrimental technical factors associated with drilling, heat exchangers, materials and heat pump equipment are, from a national perspective, generally minimal – problems with limited single phase power supplies being a possible exception in some countries. At particular sites, there can be difficulties or uncertainties with drilling boreholes for heat exchangers but a range of drilling techniques are available to deal with most geological conditions. The available land (footprint) at a particular site can be an issue depending also on the density of the heating and cooling load. The question of available land is not usually an issue in residential systems using borehole heat exchangers as normally one is sufficient. However, where there is a high density of GSHP installations some account of adjacent systems has to be considered (Liuzzo-Scorpo et al., 2015).

Economic factors often have a significant impact on the deployment of GSHP technology, both at the national level and where particular projects are being considered. Drilling and heat exchanger costs have to be met in addition to the cost of packaged heat pump equipment. Consequently, where competition is from fossil-fuelled heating systems (and combinations of fossil fuel and air-cooled chillers in nonresidential applications) GSHP systems can be unattractive to developers due to the higher capital costs. Whether a lower life-cycle cost can be demonstrated is highly dependent on fuel and power costs and these can vary significantly between countries and over time. Life-cycle costs can be acceptable where financial incentives are available. Such incentives can take the form of tax rebates, energy company obligations, measured tariff or capacity-based schemes (Connor et al., 2013). In general, financial incentives have been shown to be influential in the early growth of national installation industries, and this lesson seems to be applicable in many countries and will be a relevant factor for some time. Incentives have been less significant in the growth in the USA, where drilling tends to be relatively economic, and in China where costs and the economic system are quite different to other nations.

The very earliest developments, trials and initial uptake of GSHP technology in many countries has been associated with the activities of research organizations, utility companies and early adopters of renewable energy – both small installation contractors and individual do-it-yourself installers. Progression beyond this stage is thought to be partly dependent on a number of contributing nontechnical factors (in addition to the economic support measures noted earlier). These include:

- public dissemination of information about the technology
- industrial training and skills development for designers and installers
- development of representative industry/trade bodies
- availability of drilling expertise and capacity
- clearly defined regulatory frameworks
- appropriate product and installation standards

The importance of these factors and the benefits of organized support for such developments is demonstrated in the countries that have established GSHP industries. The importance of such factors has also been noted in relation to other forms of renewable technology. The role of such developments in the case of the United Kingdom (where there is currently a state transition from early to sustainable stages of GSHP deployment) was commented on by Rees and Curtis (2014).

1.3 Current status and outlook

1.3.1 The current state of ground-source heat pump deployment

Consideration of the current status and outlook will show that the situation regarding national deployment of GSHP technology reported in 2000, has changed in a number of respects. It is suggested that it has taken something of a refocussing of energy policy in Europe and significant economic growth in China in the last decade to drive uptake of GSHP technology to the point where it is today.

Worldwide surveys of geothermal heat pump developments have been reported at five-year intervals since 1975 (eg, Freeston, 1996; Lund and Boyd, 2015) and, following the introduction of the RES Directive in Europe (European Commission, 2009), European Union (EU) Member State statistics are reported annually to the Eurostat Agency and also published by EurObserv'ER (2015) and so the current state of deployment of GSHP technology is relatively well understood. The current state of deployment in the United States, China and Europe is shown in Fig. 1.15 in terms of number of systems installed (more than 90% being residential). In Fig. 1.16 the data is shown on a per-capita basis.

The data from the World Geothermal Congress (Lund and Boyd, 2015) and EurObserv'ER (2015) in Fig. 1.15 show that, compared to the 2000 data noted earlier, growth has continued at some pace in the United States so that 1.4 million installations were reported in the 2015 worldwide survey. The second greatest number of GSHP installations can be found in China, and this indicates particularly dramatic growth as no installations were noted in 2000. France is another country

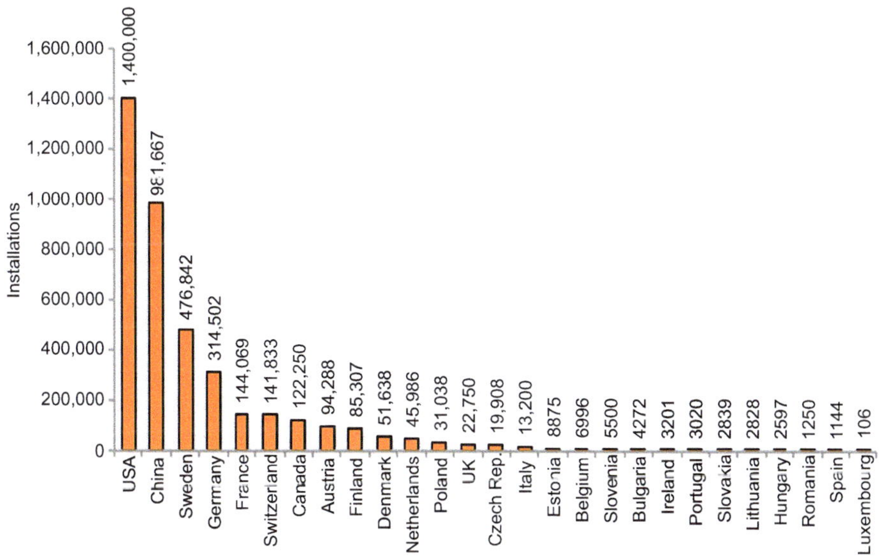

Figure 1.15 Ground-source heat pump installations in North America, China and Europe (EurObserv'ER, 2015; Lund and Boyd, 2015).

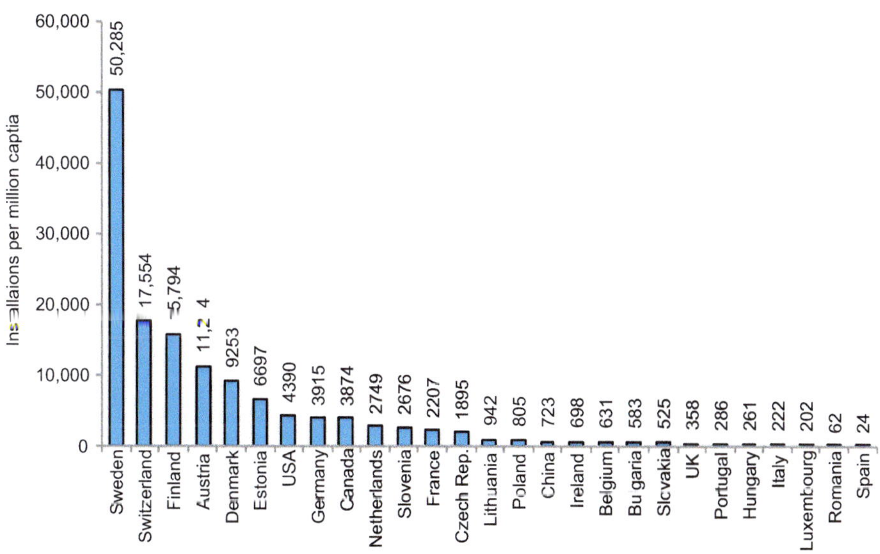

Figure 1.16 Ground-source heat pump installations in North America, China and Europe ranked on a per-capita basis (EurObserv'ER, 2015; Lund and Boyd, 2015).

amongst those with significant numbers of GSHP units that has shown dramatic growth in recent years – factors in this growth are discussed later in this chapter. When the data is compared on a per-capita basis, some normalization of the data is introduced and this gives a better indication of the penetration of the technology at the national level. It might be expected that Switzerland and the Scandinavian countries feature high in this ranking given early entry to the market in the last century and also their colder climates. Denmark has a more temperate climate than the other Scandinavian countries but still features high in this ranking. This suggests there may be potential for similar levels of uptake in other European countries with temperate climates that currently show low levels of uptake – Ireland, the United Kingdom and Belgium, for example – all other things being equal. Southern European nations like Spain and Italy are rather different to more northerly European countries by virtue of having greater needs for cooling. Although there are climatic similarities with parts of the United States that have high levels of GSHP uptake, home designs and consumer expectations of air conditioning are very different and uptake in these countries is modest in 2015.

1.3.2 Carbon emissions reduction

The turn of the century has seen significant changes in energy policy in a number of countries as some consensus on the need for climate change mitigation, and consequently better energy efficiency and renewable energy deployment, has emerged (Rosenow, 2012). This shift in energy policy in many countries has led to reform of building energy standards and also new incentive schemes and supporting initiatives to encourage uptake of building-integrated renewable energy technologies. Policy changes in European states has been driven, in part, by the Energy Performance of Buildings Directive (EPBD) (European Commission, 2002) and the RES Directive (European Commission, 2009). The EPBD required EU Member States to move to building energy efficiency standards based on allowable carbon emission rates and the RES Directive required commitments to plans for increased energy sources from a broad range of renewable technologies, including GSHP (the level of commitment to particular technologies varying between states significantly).

The degree to which a GSHP is a source of renewable thermal energy is, strictly speaking, dependent on the nature of the primary energy used to generate the electricity supplied to it. Given that a heat pump transfers more heat from the ground than is consumed by the compressor (Fig. 1.2), it can be said that the renewable element of the energy is at least that portion extracted from the ground. This is dependent on the system efficiency (eg, SPF_{H2}).

If, with a view to energy policy, one seeks to make a simplified comparison with systems that rely on local combustion of fuel such as gas, oil or biomass boilers in terms of overall energy efficiency, one approach is to consider the ratio of primary energy to heat delivered. The method of calculation applied in the RES Directive is to require an SPF_{H2} value greater than $1.15/\eta$, where η is the ratio of (at national grid level) electrical energy delivery to primary energy input and 15% distribution losses are assumed (Rees and Curtis, 2014). The value of η varies from country to country and over time (gradually increasing) but for the sake of uniformity and application over the 2010–20 timeframe, a conservative value of 0.455 has been agreed (European Commission, 2013). The means, after rounding, a minimum SPF_{H2} value of 2.5 is required for a heat pump system to be considered a renewable energy source for EU reporting purposes.

One way to quantify the impact of a fuel source is to examine its carbon intensity per unit of energy. This may be quantified as a Carbon Emission Factor that is an effective value that considers emissions from the power generation/conversion and transmission processes in terms of equivalent carbon dioxide emissions per unit of power (usually with units $kgCO_2/kWh$). Electricity Carbon Emission Factors are shown for a number of countries in Fig. 1.17. The carbon emission factors of heating fossil fuels varies little between countries and it becomes straightforward to compare system types in terms of carbon emissions and also calculate the minimum SPF value required for a heat pump to have a lower emission rate. In the United Kingdom, for example, a heat pump only has to have an SPF_{H4} value of greater than 2.21 to have lower emissions than a gas-fired boiler heating system (Rees and Curtis, 2014) and national trials in the United Kingdom have shown that this level of efficiency is readily attainable. The UK electricity carbon emissions factor (0.48 $kgCO_2/kWh$ in 2013) in this comparison is close to the EU median value.

The carbon emission factors associated with electricity can be expected to vary depending on both the efficiency of conversion and the mix of fuels that are used to generate power. This varies significantly from one country to another. It also varies over time – generally decreasing from year to year but also varying over a given day. Countries with lower values tend to have higher levels of renewable energy (eg, hydroelectric) and/or higher levels of nuclear power generation. Countries, such as Norway and Switzerland are in the former category and France is an example where the carbon emission factor of electricity is low due to 76% of power generation being from nuclear sources. The electricity carbon emissions factor in Sweden is relatively low because of a combination of nuclear and hydroelectric power generation. In such countries, the case for using heat pumps in preference to fossil fuel heating is very clear-cut. As the predominant heating fuel in France has been natural gas, the move towards higher uptake of heat pumps can result in significant reductions in building carbon emission rates. Development of the GSHP market in France has been rapid in the last decade.

1.3.3 Outlook

As in many countries, commitments have been made to long-term reductions in national carbon emissions and consequently decarbonization of the electricity systems, greater adoption of GSHP (and heat pump technology in general) can be seen to be a rational long-term policy measure. This aspiration is indeed reflected in the National Renewable Action Plans of EU Member States (which define 2020 targets) and other stated national strategies (DECC, 2012). In order for there to be overall reductions in emissions from heating it will be necessary to both decarbonize the electrical system (through changes in primary sources and distributed generation) and displace fossil fuel heating by electric heat pumps. This is a clear strategy for 2050 milestones but

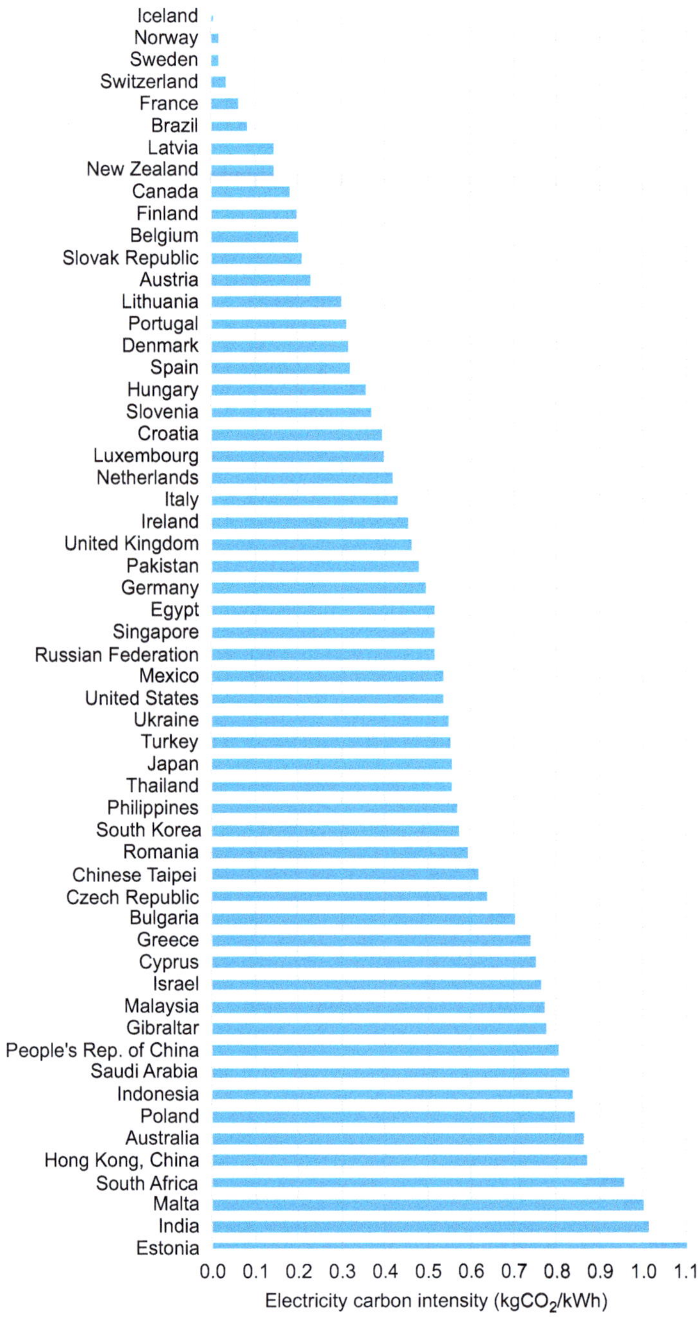

Figure 1.17 Electricity carbon factors.
2015 data for generation and distribution published by the UK Department of Climate Change.

in many countries the data shows (Fig. 1.15) that growth in the rate of GSHP installation, and so industrial capacity, would have to increase significantly.

There are other significant challenges in finding a development trajectory that will arrive at long-term carbon reduction targets. Not least of these challenges is achieving the changes to electricity grids that will be required to both deal with more variable renewable supplies and greater electrical demands from increasing numbers of heat pumps. GSHP technology can also partly address this challenge as it is amenable to control by smart grid systems and can facilitate elements of energy storage, however (Strbac et al., 2010).

A review of the development of GSHP technology and current levels of uptake shows that there is something of a mixed picture of sustained growth in some countries and the early stages of market development in others – with some countries in transition. In countries with markets at the early stages of development, one expects that it will be necessary for the sort of supporting nontechnical measures to come into action that are evident in successful markets, in order for sustained growth to be achieved. The efficiency levels that can be achieved through GSHP technology by virtue of thermodynamically advantageous conditions mean that there is certain to be continued deployment and development – indeed, it is hard to see other technologies reaching similar levels of efficiency.

Measures being introduced to decarbonize electricity and heat over the remainder of this century mean that energy policy is, broadly speaking, likely to be supportive of continued GSHP development. Although much remains uncertain, we can conclude that there are good reasons to take a positive view of GSHP technology and hope for further realization of its benefits. It is hoped that this book will play its part in the advancement of GSHP technology by encouraging the spread of knowledge as well as pointing the way to potential future developments.

References

Ambrose, E.R., 1966. Heat Pumps and Electric Heating. John Wiley and Sons, New York.

Andrews, S.W., 1948. The heat pump from the utility's point of view. Transactions of the American Institute of Electrical Engineers 67 (1), 562–564.

ARI, 2008. Standard 210/240 Performance Rating of Unitary Air-conditioning and Air-Source Heat Pump Equipment.

Bose, J.E., 1982. Earth coil/heat pump research at Oklahoma State University. In: Proceedings of the Sixth Heat Pump Technology Conference. Oklahoma State University, Tulsa.

Connor, P., Bürger, V., Beurskens, L., Ericsson, K., Egger, C., 2013. Devising renewable heat policy: overview of support options. Energy Policy 59, 3–16.

Coogan, C.H., 1948. The Residential Heat Pump in New England. Connecticut Light and Power Co, Waterbury, CT, USA.

Coogan, C.H., 1949. Summary of Heat Absorption Rates for an Experimental Ground Coil System. Technical report. University of Connecticut Engineering Experimental Station.

Crandall, A., November 9, 1946. House heating with earth heat pump. Electrical World 19, 94–95.

DECC, 2012. The Future of Heating: A Strategic Framework for Low Carbon Heat in the UK. Department of Energy and Climate Change, London, UK.
EurObserv'ER, 2015. Heat Pumps Barometer.
European Commision, 2002. Directive 2002/91/EC of the European Parliament and of the Council of 16 December 2002 on the energy performance of buildings. Official Journal of The European Union 65–71. L 1/65(4.1.2003).
European Commission, 2009. Directive 2009/28/EC of the European Parliament and of the Council of 23 April 2009 on the promotion of the use of energy from renewable sources and amending subsequently replealing Directives 2001/77/EC and 2003/30/EC. Official Journal of the European Union 16–62. L 140/16(5.6.2009).
European Commission, 2013. Decision (2013/114/EU) establishing the guidelines for Member States on calculating renewable energy from heat pumps from different heat pump technologies pursuant to Article 5 of Directive 2009/28/EC of the European Parliament and of the Council. Official Journal of the European Union. L 62/27.
Ewing, J.A., 1910. The Steam Engine and Other Heat-Engines, third ed. Cambridge University Press, Cambridge, UK.
Freeston, D., 1996. Direct uses of geothermal energy 1995. Geothermics 25 (2), 189–214.
Gleeson, C., Lowe, R., 2013. Meta-analysis of European heat pump field trial efficiencies. Energy and Buildings 66, 637–647.
Griffith, M.V., 1957. Some aspects of heat pump operation in Great Britain. Proceedings of the IEE Part A: Power Engineering 104 (15), 262–271.
Haldane, T., 1930. The heat pump—an economical method of producing low-grade heat from electricity. Journal of the Institution of Electrical Engineers 68 (402), 666–675.
Hapley, W.A., December 1949. Operating characteristics of heat pump ground coils. Edison Electric Institute Bulletin 14.
Hopkirk, R.J., Eugseter, W., Rybach, L., 1988. Vertical earth heat probes: measurements and prospects in Switzerland. In: Conférence internationale sur le stockage de l'énergie pour le chauffage et le refroidissement. 4. JIGASTOCK'88, pp. 367–371.
Ingersoll, L.R., Plass, H.J., 1948. Theory of the ground pipe heat source for the heat pump. ASHVE Transactions 47 (7), 339–348.
Karlsson, F., Axell, M., Fahlén, P., 2003. Heat Pump Systems in Sweden – Country Report for IEA HPP Annex 28. Technical report. SP Swedish Technical Research Institute, Borås, Sweden.
Kidder, A.H., Neher, J.H., 1952. Residential heat pump experiments in Philadelphia—earth as a heat source. Transactions of the American Institute of Electrical Engineers, Part II: Applications and Industry 71 (5), 343–350.
Liuzzo-Scorpo, A., Nordell, B., Gehlin, S., 2015. Influence of regional groundwater flow on ground temperature around heat extraction boreholes. Geothermics 56, 119–127.
Lund, J., Freeston, D., 2001. World-wide direct uses of geothermal energy 2000. Geothermics 30 (1), 29–68.
Lund, J.W., Boyd, T.L., 2015. Direct utilization of geothermal energy 2015 worldwide review. In: Proceedings of the World Geothermal Congress 2015. International Geothermal Association, Melbourne, Australia, pp. 1–31.
MacFarlane, J., 1957. South-West Scotland Sub-Centre: Chairman's address. Personal experiences with heat pumps. Proceedings of the IEE Part A: Power Engineering 104 (13), 37.
Mei, V.C., 1986. Horizontal Ground-Coil Heat Exchanger Theoretical and Experimental Analysis (ORNL/CON-193). Technical report. Oakridge National Laboratory.

Metz, P.D., 1979. Design, construction, and operation of the solar assisted heat pump ground coupled storage experiments at Brookhaven National Laboratory. In: 4th Annual Heat Pump Technology Conference, Stillwater, Oklahoma, 9 April, vol. 1.

Naicker, S., 2016. Performance Analysis of a Large-Scale Ground Source Heat Pump System (Ph.D). De Montfort University, Leicester, UK.

Penrod, E., 1947. Continuous air-conditioning with the heat pump. American Scientist 35 (4), 502–535.

Penrod, E., Knight, R., Baker, M., 1950. Earth heat pump research—Part II. Engineering Experimental Station Bulletin, University of Kentucky 5 (18).

Rees, S., Curtis, R., 2014. National deployment of domestic geothermal heat pump technology: observations on the UK experience 1995–2013. Energies 7 (8), 5460–5499.

Rosenblad, G., 1979. Earth heat pump system with vertical pipes for heat extraction and storage. In: Proceedings of Nordic Symposium of Earth Heat Pump Systems, Göteborg, pp. 102–110.

Rosenow, J., 2012. Energy savings obligations in the UK–a history of change. Energy Policy 49, 373–382.

Sanner, B., 1996. Some history of shallow geothermal energy use. In: International Summer School on Direct Application of Geothermal Energy. Institute or Applied Geosciences, Justus Liebig University, Giessen, Germany, pp. 1–10.

Sanner, B., Karytsas, C., Mendrinos, D., Rybach, L., 2003. Current status of ground source heat pumps and underground thermal energy storage in Europe. Geothermics 32 (4–6), 579–588.

Schär, O., 1977. Di Anwendung der Wämepumpe zur Heizung von Wohnbuaten mit Wämentnahme aus dem Erdreight (The use of the heat pump for heating dwelllings with heat drawn from the soil). Bulletin of the Association of Suisse Electrical Engineers 68 (4), 182–184.

Smith, M.M., Kemler, E.N., 1949. Heat Pump Using Deep Well for a Heat Source. Patent. US 2,461,449.

Spitler, J., 2015. Observations on ground-source heat pump systems in North America and Scandinavia. IEA Heat Pump Centre Newsletter 33 (3), 29–32.

Strbac, G., Gan, C.K., Aunedi, M., Stanojevic, V., Djapic, P., Dejvises, J., Mancarella, P., Hawkes, A., Pudjianto, D., Openshaw, D., Burns, S., West, P., Brogden, D., Creighton, A., Claxton, A., 2010. Benefits of Advanced Smart Metering for Demand Response Based Control of Distribution Networks. Technical Report April 2010. Imperial College, London.

Sumner, J.A., 1976. Domestic Heat Pumps. Prism Press, London.

Svec, O.J., 1987. Potential of ground heat source systems. International Journal of Energy Research 11 (4), 573–581.

Thomson, W., 1852. On the economy of the heating or cooling of buildings by means of currents of air. Proceedings of the Physical Society of Glasgow 3, [illegible].

Vestal, D.M., 1949. Heat pump buried coil design. Refrigeration Engineering 57 (6), 573–613.

Walton, E.H., Williams, C.A., 1944. Experience with a heat-pump system. Electrical Engineering 63 (10), 361.

Wirth, E., 1955. The evolution of the heat pump. Schweizerische Bauzeitung 73 (42), 647–651.

Zogg, M, 2008. History of Heat Pumps. Swiss Contributions and International Milestones. 9th International IEA Heat Pump Conference, pp. 20–22. May 2008, Zürich, Switzerland.

Zottl, A., Nordman, R., Miara, M., 2012. Benchmarking Method of Seasonal Performance. Technical report. SEPEMO-Build.

Part One

Closed-loop systems

Vertical borehole ground heat exchanger design methods

2

J.D. Spitler[1], *M. Bernier*[2]
[1]Oklahoma State University, Stillwater, OK, United States; [2]Polytechnique Montréal, Montréal, QC, Canada

Abstract

A critical component of any ground-source heat pump (GSHP) system is the ground heat exchanger (GHE). For many residential systems and nearly all commercial and institutional systems, the GHE will consist of one or more vertical boreholes with inserted tubing. Most commonly this is a single or double U-tube made of high-density polyethylene, but other configurations are possible. The complex nature of the ground heat transfer and the long time constant of the ground create some challenges in designing vertical borehole GHEs. This chapter covers methods for design of these GHEs.

2.1 Introduction

A fundamental question for designing ground-source heat pump (GSHP) systems is "How large a ground heat exchanger is required?" If the answer to this question is wrong, and too small a ground heat exchanger (GHE) is installed, the system may fail shortly after the beginning of operation or it may occur after some years of operation. If the answer to this question is wrong and too large a GHE is specified, the system will be unnecessarily expensive and a less energy efficient substitute may be chosen. Therefore, a "just right" solution—not too big and not too small—is sought. This is usually formulated as finding the GHE size that is just sufficient to not exceed minimum or maximum heat pump entering fluid temperatures (EFT).

Because the ground surrounding the GHE will have a time constant on the order of years, the required size of the GHE depends on the long-term heat rejection and extraction to and from the ground. Therefore, a reasonable estimate of the loads on the heat pump(s) over time must be made as part of the design.

Other important influences on the sizing process include the ground thermal properties (see chapter: "In situ estimation of ground thermal properties"), the thermal resistance of the borehole (see chapter: "Calculation of borehole thermal resistance"), and the characteristics of the heat pump (see chapters: "New trends and developments in ground-source heat pumps" and "Heat pump modelling").

Advances in Ground-Source Heat Pump Systems. http://dx.doi.org/10.1016/B978-0-08-100311-4.00002-9
Copyright © 2016 J.D. Spitler and M. Bernier. Published by Elsevier Ltd. All rights reserved.

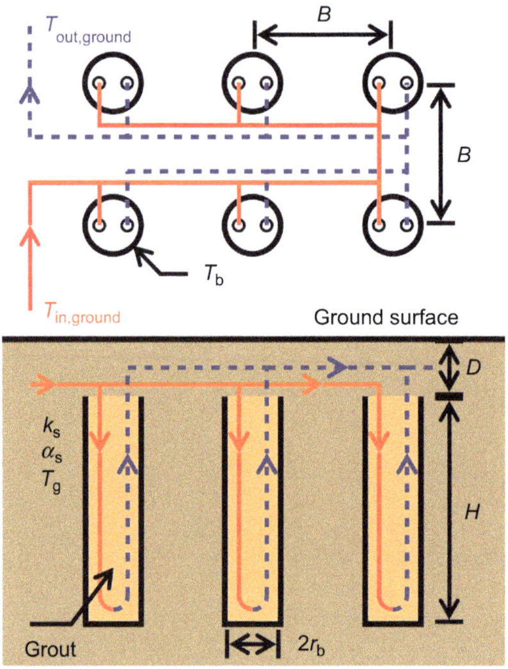

Figure 2.1 Six-borehole ground heat exchanger.

In this chapter, methods for sizing vertical borehole GHEs are discussed. Except for single borehole residential systems, most GHEs will consist of multiple boreholes. A six-borehole GHE is illustrated in Fig. 2.1. GHEs are commonly designed with uniform borehole-to-borehole spacing (B); uniform depth at the top of the GHEs (D); and uniform active heat transfer length (H). Of course, it is possible to design GHEs with irregular dimensions, but this adds complications to the analysis and discussion in this chapter will cover GHEs with regular dimensions.

2.2 Background: mathematical analysis of ground heat exchangers

This section briefly covers mathematical analyses of GHEs; applications of these methods are described in Section 2.3. Analytical methods treat the GHE as either a line source or cylinder source, giving the temperature response of a single borehole to a uniform heat rejection or extraction pulse over time (Section 2.2.1). Numerical methods (Section 2.2.2) have the possibility of treating time-varying heat inputs and/or multiple boreholes, but even with fast computers it is generally infeasible to utilize numerical methods in this way. Therefore, both types of analysis procedure commonly make use of the superposition principle (Section 2.2.3) to treat GHEs with time-varying loads and/or multiple boreholes. For design and simulation purposes, response functions (Section 2.2.4) are commonly used to reduce computational time requirements.

The behavior of GHEs is driven by variations in the inlet temperature and flow rate that occur during system operation. Inlet temperature and flow rate are consequently natural boundary conditions to apply in models used to simulate operation from hour to hour. However, for the purposes of system design, it is more useful to apply loads—imposed heat transfer rates (kW) or demands over a specific period (kWh)—as this is the type of design data that is usually available. This design data may be available in terms of loads imposed on the GHE or loads imposed on the heat pump. In the latter case additional information about the heat pump performance is used to derive the loads imposed on the GHE. Although design and simulation models can be very similar in approach, the formulation of the boundary conditions at the borehole or U-tube usually differ. Boundary conditions specified in terms of heat transfer rates or energy demands make it possible to apply superposition principles directly.

2.2.1 Analytical methods

Analytical methods treat the GHE as either a line source or cylinder source. The cylinder source solution may allow a closer representation of the borehole geometry and thermal properties when the grout thermal properties are different from the ground thermal properties. The finite line source (FLS) solution can better represent end effects and so gives a better representation of the GHE response for time periods that exceed, say, a year for a typical 100 m deep borehole.

2.2.1.1 Infinite line source[1]

The term “line source method” refers to the solution to a pure conduction heat transfer process involving an infinite line source (ILS) that begins generating heat continuously at time zero. This is often referred to as the Kelvin line source, crediting Lord Kelvin (Thomson, 1880, 1884) with the solution. However, anyone curious to review Kelvin’s works may be disappointed to not find the line source solution there. Rather, he gave a compendium of solutions based on Fourier’s point-source solution. The line source solution may be derived from that, but it was not until later (Whitehead, 1927) that a line source solution was developed (see also Banks, 2015). Whitehead’s solution was given in the form of an infinite series. Later, line source solutions were formulated by Ingersoll and Plass (1948) and Carslaw and Jaeger (1959).[2] The Carslaw and Jaeger line source solution gives the temperature T at a radius r of:

$$T = T_g + \frac{q}{4\pi k}\int_{\frac{r^2}{4\alpha t}}^{\infty}\frac{e^{-u}}{u}du = \frac{q}{4\pi k}\cdot E_1\left(\frac{r^2}{4\alpha t}\right) \qquad [2.1]$$

[1] Section 2.2.1.1 and the first part of Section 2.2.1.2 have been adapted from Spitler and Gehlin (2015).

[2] See pp. 261–262.

where: E_1 is the exponential integral,[3] T_g = Initial temperature of the medium, ground in this case (°C), q = Specific heat load (W/m), k = Thermal conductivity of the homogenous medium (W/(m K)), α = Thermal diffusivity of the homogenous medium (m^2/s), r = Radial distance from line source (m), and t = Time that the continuous line source has been applied (s).

Gautschi and Cahill (1964)[4] gave a series expansion of the exponential integral as

$$E_1(x) = -\gamma - \ln(x) - \sum_{n=1}^{\infty} \frac{(-1)^n x^n}{n \cdot n!} \tag{2.2}$$

Carslaw and Jaeger (1959, p. 262) also give a simplified form for large values of t:

$$T = T_g + \frac{q}{4\pi k} \cdot \left(\ln\left(\frac{4\alpha t}{r^2}\right) - \gamma \right) \tag{2.3}$$

with a maximum error in the temperature rise of 2% (Hellström, 1991, p. 151) when $t > 5r^2/\alpha$.

The above solutions cover a continuous and infinite line source. Application of these solutions to GHEs was suggested by a number of authors, eg, Ingersoll and Plass (1948). However, it is difficult to find a clear statement of how the line source should be applied to a borehole with a U-tube prior to Mogensen (1983). Mogensen's work incorporates several approximations that have since been commonly made:

- The line source solution is used to evaluate the temperature at the borehole wall.
- Heat transfer within the borehole is assumed to be in a quasi-steady state and can thus be described with an effective borehole thermal resistance that applies between the fluid in the U-tube and the borehole wall.
- The mean fluid temperature within the borehole is often taken to be the simple average of the inlet and outlet temperatures, though more detailed approaches have been proposed; eg, see Marcotte and Pasquier (2008), Beier (2011) and Zhang et al. (2014).

With these approximations, the temperature response to a constant heat pulse on a borehole can be represented using the detailed solution for the line source as:

$$T_f = T_g + \frac{q}{4\pi k^*} \cdot E_1\left(\frac{r^2}{4\alpha t}\right) + q \cdot R_b^* \tag{2.4}$$

where: T_f = Mean fluid temperature within the borehole (°C), T_g = Undisturbed ground temperature (°C), k^* = Effective ground thermal conductivity (W/m K), R_b^* = Effective borehole thermal resistance (K/(W/m)).

[3] Here, the naming convention of Gautschi and Cahill (1964) is used; Carslaw and Jaeger referred to this as E_i.
[4] See p. 229.

2.2.1.2 Infinite cylinder source

The so-called infinite cylinder source (ICS) is the solution to a pure conduction heat transfer process involving an infinite cylinder generating heat continuously starting at time zero. There are several forms of the cylinder source solution that vary with how the material inside the cylinder is treated. These include:

- A form (Carslaw and Jaeger, 1959, p. 338) where the inside of the cylinder is empty, with all heat flow going outward from the radius of the cylinder. A similar solution can be found in Ingersoll et al. (1951).
- A form (Carslaw and Jaeger, 1959, pp. 342–345) where the inside of the cylinder is a perfect conductor with a specified thermal capacitance. This form is discussed as being a good approximation for needle probes used to measure thermal conductivity.
- A form (Carslaw and Jaeger, 1959, p. 347) treating composite cylindrical regions where the inside of the cylinder has different thermal properties than the surrounding ground.

When the first form of the solution is used, the mean fluid temperature is given by:

$$T_f = T_g + \frac{q}{k^*} G(Fo) + q \cdot R_b^* \quad [2.5]$$

Where: Fo = Fourier number $\left(= \alpha t / r_b^2\right)$ and $G(Fo)$[5] is the ICS given by:

$$G(Fo) = \frac{1}{\pi^2} \int_0^\infty \left(e^{-\beta^2 Fo} - 1\right) \frac{J_0(\beta)Y_1(\beta) - Y_0(\beta)J_1(\beta)}{\beta^2\left(J_1^2(\beta) + Y_1^2(\beta)\right)} d\beta \quad [2.6]$$

Cooper (1976) and Bernier (2001) have suggested simplified forms of Eq. [2.6] which do not require an integration.

2.2.1.3 Finite line source

The ILS and ICS solutions are convenient and are relatively easy to calculate. However, they do not consider the two-dimensional nature of ground heat transfer associated with axial temperature gradients which starts to be significant after approximately a year for typical boreholes. When more accurate determination of ground heat transfer is required for design periods greater than a year, the so-called FLS solution should be used.

The FLS solution gives the temperature at a point (r, z) resulting from a constant heat transfer rate per unit length q originating from a finite line of length H located at a distance D from the ground surface. As shown in Fig. 2.2, a virtual mirror image

[5] Not to be confused with Eskilson's g-function.

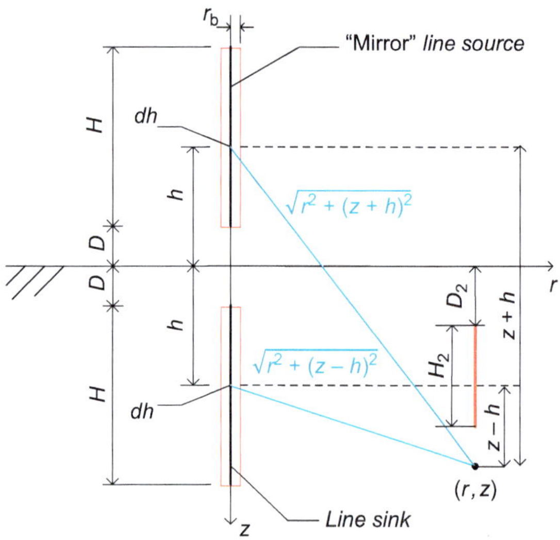

Figure 2.2 Geometrical parameters used to represent a finite line source and its mirror image.

of the real finite line is located above the ground surface to handle the boundary condition at the surface. For a constant surface temperature equal to the undisturbed ground temperature, T_g, the FLS solution is given by:

$$T(r,z,t) = T_g + \frac{q}{4\pi k^*} \int_D^{D+H} \left(\frac{\operatorname{erfc}\left(\frac{\sqrt{r^2+(z-h)^2}}{2\sqrt{\alpha t}}\right)}{\sqrt{r^2+(z-h)^2}} - \frac{\operatorname{erfc}\left(\frac{\sqrt{r^2+(z+h)^2}}{2\sqrt{\alpha t}}\right)}{\sqrt{r^2+(z+h)^2}} \right) dh \qquad [2.7]$$

where erfc is the complementary error function. For an adiabatic ground surface boundary condition, Eq. [2.7] has to be slightly modified and the minus sign between the two complementary error function terms is replaced by a plus sign.

Readers are referred to the works of Zeng et al. (2002), Diao et al. (2004), Katsura et al. (2008), Philippe et al. (2009), Marcotte et al. (2010), Abdelaziz et al. (2014) for applications of the FLS solution in the determination of two-dimensional temperature profiles around boreholes.

The integral mean temperature for a finite line of length H_2 positioned at a distance D_2 from the ground surface and located a certain distance r from the line sink at time t (see Fig. 2.2) is given by:

$$\overline{T}(r,t) = T_g + \frac{\frac{q}{4\pi k^*} \int_{D_2}^{D_2+H_2} \int_D^{D+H} \left(\frac{\operatorname{erfc}\left(\frac{\sqrt{r^2+(z-h)^2}}{2\sqrt{\alpha t}}\right)}{\sqrt{r^2+(z-h)^2}} - \frac{\operatorname{erfc}\left(\frac{\sqrt{r^2+(z+h)^2}}{2\sqrt{\alpha t}}\right)}{\sqrt{r^2+(z+h)^2}} \right) dh\, dz}{H_2}$$

[2.8]

With this equation it is possible to evaluate the thermal influence of one borehole segment on another. It forms the basis of the analytical determination of g-functions using the FLS (Cimmino and Bernier, 2014). The solution of Eq. [2.8] requires relatively long computational time because of the double integration. Lamarche and Beauchamp (2007) addressed this issue by reformulating the problem into a single integral for the specific case D and $D_2 = 0$ reducing calculation time significantly. Claesson and Javed (2011) also obtained a single integral solution for the case $D \geq 0$ (with $D_2 = D$ and $H_2 = H$). Cimmino and Bernier (2014) generalized Claesson and Javed's relationship for $D \geq 0$ and $D_2 \geq 0$ and $H_2 \neq H$.

2.2.2 Numerical methods

Numerical methods, eg, finite difference method, finite element method, finite volume method, are not usually feasible for design purposes. This is due to the widely varying length-scales and time-scales that are necessary to treat the heat transfer in the borehole and surrounding ground. Numerical methods have been used for development of response functions (Eskilson, 1987; Yavuzturk et al., 1999) and for research purposes. In addition, models for single boreholes that utilize custom resistance networks inside the borehole (Bauer et al., 2011; Zarrella et al., 2011; Pasquier and Marcotte, 2012; Godefroy and Bernier, 2014) have shown some promise, but are not yet used in design tools.

2.2.3 Basic principles of response superposition

Because it is generally not feasible to utilize detailed numerical methods, simpler analytical or response factor type methods are commonly used. These methods often rely on the superposition principle. Claesson and Dunand (1983) give an introduction to the application of the superposition principle, which relies on the heat conduction equation and boundary conditions being linear. Conditions that lead to nonlinear terms include freezing/thawing, time-varying surface convection coefficients, radiative heat exchange at the surface, and time-varying groundwater flows. So treatment of heat transfer to/from GHEs with the superposition principle necessarily neglects (or approximates) freezing/thawing, radiative exchange at the surface, and time-varying convection and groundwater flows. In practice, applications of the superposition principle to GHE design have generally neglected groundwater flow.

Superposition can be applied temporally, that is by deconvolution of the time-varying heat transfer rate to/from the ground (the "loads") into functions for which a solution can readily be determined. For example, in Fig. 2.3, hourly heat loads (the black line) are deconvolved into individual steps.

The responses to the individual step function heat loads can be determined in a number of different ways. In Fig. 2.4, the line source analytical solution has been utilized to determine the individual responses. DT 1 is the temperature response to the heat load labeled Step 1 in Fig. 2.3. The individual responses are convoluted or added together to give the response labeled DT Sum.

Likewise, the superposition principle can also be applied spatially. Fig. 2.5 shows the temperature response along a line that passes through two boreholes 5 m apart after

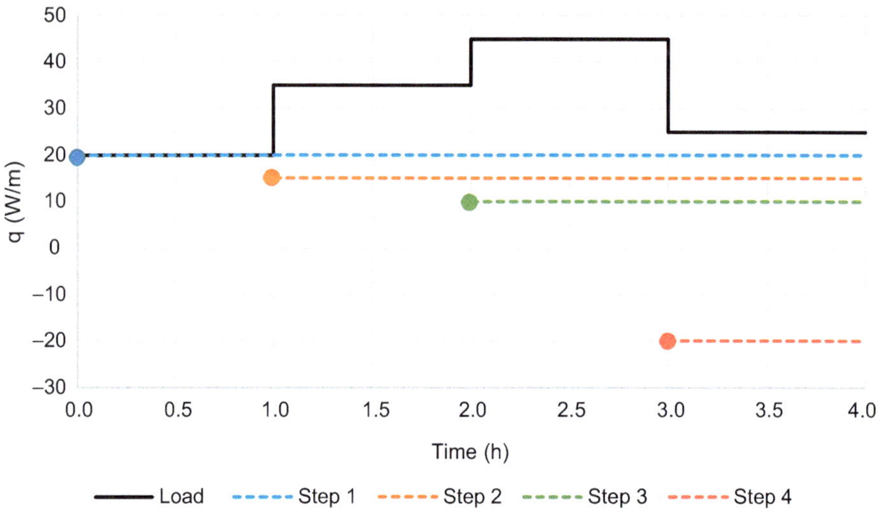

Figure 2.3 Hourly loads represented as step functions.
After Lazzarotto (2015).

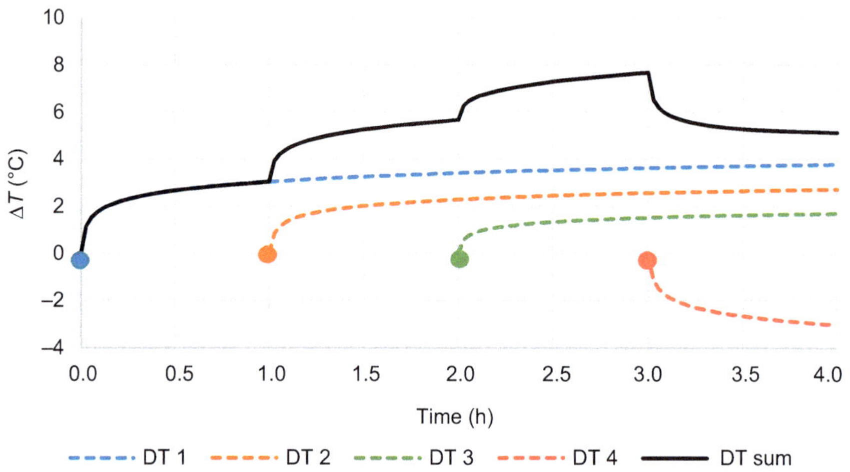

Figure 2.4 Individual responses to step pulses superimposed to give temperature response.
After Lazzarotto (2015).

being subjected to 50 W/m of heat rejection for a year. Again, the two individual responses can be added together.

2.2.4 *Response functions (g-functions)*

Superposition is a powerful tool for analysis of GHEs. Pioneering work by Prof. Johan Claesson and his PhD student, Per Eskilson, is summarized in Eskilson's (1987) PhD thesis. Their primary contributions were development of a nondimensionalized

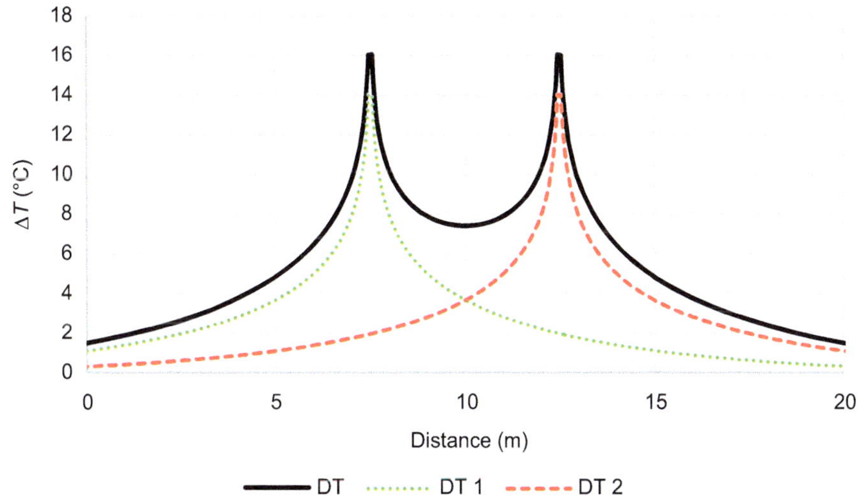

Figure 2.5 Example of spatial superposition for two boreholes 5 m apart.

response function, called the *g*-function, procedures for calculating the *g*-function, and development of a library of *g*-functions for different borehole configurations.[6]

The *g*-function gives the borehole wall temperature for a specific borehole configuration due to a constant heat input rate:

$$T_b = T_g + \frac{q}{2\pi k^*} \cdot g(t/t_s, r_b/H, B/H, D/H) \quad [2.9]$$

In this scheme, the time is nondimensionalized against the time scale:

$$t_s = \frac{H^2}{9\alpha} \quad [2.10]$$

Eskilson cataloged *g*-functions for specific configurations, specific values of the ratio *B*/*H*, and one value (0.0005) of the ratio r_b/H. Values for other ratios $\left(r_b^*/H\right)$ are approximated with:

$$g\left(\frac{t}{t_s}, \frac{r_b^*}{H}, \frac{B}{H}, \frac{D}{H}\right) = g\left(\frac{t}{t_s}, \frac{r_b}{H}, \frac{B}{H}, \frac{D}{H}\right) - \ln\left(\frac{r_b^*}{r_b}\right) \quad [2.11]$$

This correction is based on steady-state resistance[7] for the difference between the actual value of r_b^* and the value for which the *g*-functions are cataloged. Therefore, some caution may be warranted when significantly deviating from the ratio of 0.0005.

[6] By "borehole configuration" the geometric arrangement of the boreholes is meant; eg, 3 × 2 rectangular configuration or six boreholes in a line.

[7] See p. 10 of the first section of Eskilson's thesis.

The g-functions produced by Eskilson use an assumed value of D/H that is not clearly spelled out by Eskilson, though the numerical examples in the thesis all use a depth D of 4–5 m. It is worth noting that, for grouted boreholes, the depth D can represent the depth at which the horizontal connecting piping is buried and so the depth at which the GHE "starts." Or, for groundwater-filled boreholes, it can represent the depth of the water table. Above that depth, the borehole is filled with air and the borehole resistance is so high that the GHE is effectively inactive.

Methods for calculating g-functions are described in more detail in Section 2.4.2.

Each g-function is unique to one borehole field layout and one combination of the nondimensional parameters r_b/H, B/H, and D/H. The g-functions were documented in a series of nondimensional graphs similar to the one shown on Fig. 2.6 (for a 3×2 borehole field). The dashed line represents the response of a borehole field with infinite spacing and would be the same as that for a single borehole. Bernier (2014) provides a description of the various regions of a g-function graph.

Eskilson (1987) presented g-functions down to $\ln(t/t_s) = -4.5$. The corresponding time will vary with the borehole depth and ground thermal properties, but taking typical values as a 100-m depth and a thermal diffusivity of 0.1 m^2/day, $\ln(t/t_s) = -4.5$ corresponds to 123 days. This g-function data can be further extended to earlier times by using the line source approximation. However, this is limited by the accuracy of the line source at short times, say less than 12 h. Yavuzturk and Spitler (1999) extended the g-function data to much earlier times using a detailed numerical simulation of the borehole that represented the grout and pipes inside the borehole explicitly. The intended application was hourly and sub-hourly simulation of the GHE.

Once the g-function is calculated, it is used with a sequence of loads on the GHE to estimate the borehole wall temperature:

$$T_b = T_g + \sum_{i=1}^{n} \frac{(q_i - q_{i-1})}{2\pi k^*} \cdot g((t_n - t_{i-1})/t_s, r_b/H, B/H, D/H) \qquad [2.12]$$

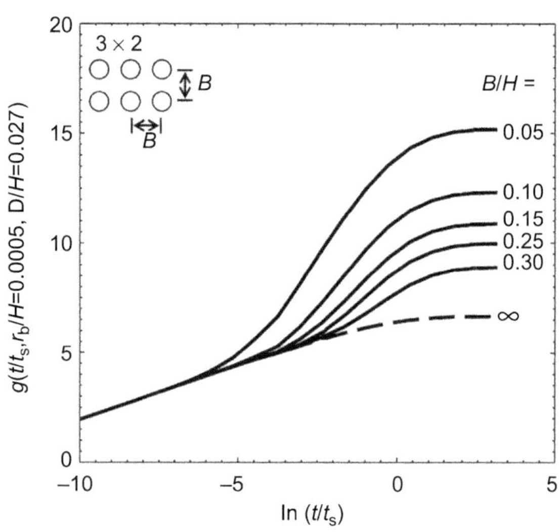

Figure 2.6 g-function curves for a 3×2 borehole field for $r_b/H = 0.0005$.

Where q = step heat rejection pulse (W/m), t = time (s), i = index denoting the end of a time step, n = index denoting current time step.

2.3 Design methodologies—overview

Engineers designing GHEs seek to find the minimum GHE size that will give satisfactory performance of the GSHP system over the life of the system. A common way to formulate the problem is for the designer to specify the ground thermal properties, number and arrangement of the boreholes, the heat pump loads and/or loads on the GHE, and the in-borehole design of the GHE. The in-borehole design would include such details as borehole diameter, single or double U-tube, tube position, and grout thermal properties. The design methodology would then give the required depth of the boreholes. If the depth is too shallow or too deep the designer would then adjust the number and arrangement of the boreholes or, perhaps, change the in-borehole design. Some programs, eg, Earth Energy Designer (BLOCON, 2015), can automatically decide the number and position of boreholes. In-borehole design and calculation of corresponding borehole resistances is the subject of chapter "Calculation of borehole thermal resistance."

2.3.1 *Classifying design methodologies*

In this section, different aspects of design methods are characterized.

2.3.1.1 *Direct and iterative solutions*

Design methodologies may be formulated such that the GHE size is given directly or such that the size is iteratively adjusted until it meets user-specified design criteria for minimum and maximum entering fluid temperature. Fig. 2.7 illustrates an iterative scheme as implemented in GLHEPRO (Spitler, 2000). The user enters information about the ground, borehole completion, the heat pump, and the loads. A computer

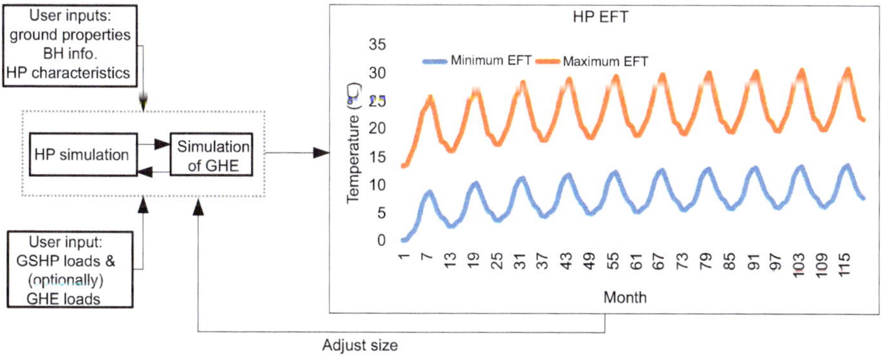

Figure 2.7 Sample iterative formulation.

simulation then computes the resulting monthly minimum and maximum heat pump EFT for a multiyear period. The size is then adjusted so that the user-specified design limits are not exceeded.

Time step resolution (levels 0–4)

The resolution of the time steps utilized by the design methodology has a significant effect (Cullin and Spitler, 2011; Cullin et al., 2015) on the accuracy of the predicted size requirements. Here, design methodologies are characterized on the time step resolution utilized by the design method. The time step resolution is important because early schemes that use fairly coarse time step resolution are less accurate. The levels are defined, as summarized below, with higher level corresponding to higher level of resolution.

Level 0: rules of thumb Rules of thumb relate GHE length to building peak heating or cooling loads or to installed capacity of equipment, typically expressed as W/m or ft/ton. Although such rules of thumb are highly desired by practitioners and there have been a number of such rules of thumb promulgated, there is little possibility for rules of thumb to adequately design GHEs for a wide range of commercial buildings (Underwood and Spitler, 2007; Spitler and Cullin, 2008). This is primarily related to the long time constant of the ground surrounding the GHE as well as the highly variable relationship between peak heating and cooling loads and annual heating and cooling loads. The time constant of the ground surrounding a commercial GHE is typically on the order of years. Practically speaking, this may mean that the heat pump EFT rise (or fall) over time so that each year's peak EFT is greater than the last, with the greatest change coming in the first three or four years as shown in Fig. 2.7. This rise or fall over time has nothing to do with the peak load and everything to do with the annual loads. Therefore, rules of thumb that try to relate the size of GHE to the peak loads are destined to fail, unless certain special conditions are present. These special conditions include:

- A strong relationship between peak loads and annual loads that is typically only the case for envelope-dominated buildings in a given location.
- Reasonably constant ground thermal properties between locations for which the rule of thumb would be applied.

Nevertheless, general guidelines have been established by certain countries. Pahud (2002) reports that the German VDI 4640 guideline provides specific heat extraction rates as a function of the ground thermal conductivity for two levels of equivalent full load hours (EFLH) of operation of the heat pump. A similar approach is used in Switzerland (Schweizerischer Ingenieur- und Architektenverein, 2010). In addition, corrections factors are provided for several borehole configurations (one to four boreholes) with various borehole spacing. In the United Kingdom, the Department of Energy and Climate Change (2011) issued look-up tables which give the maximum power to be extracted per unit length as a function of the average ground thermal conductivity, the full load equivalent hours, and the mean ground temperature. As noted by Sailer et al. (2015), these guidelines or standards are typically applicable for small systems (<30 kW) operating in heating-only conditions.

Level 1: two pulses Some early design methods, such as that described by Bose et al. (1985), OSU (1988), and Kavanaugh (1991) utilized a single heating season and single cooling season design period, the length of which was to be determined by the designer.

Level 2: six pulses The Kavanaugh (1995) design method, which has also appeared in the ASHRAE Handbook for a number of years, uses three pulses for heating and three for cooling—an annual pulse, a monthly pulse, and a peak pulse of user-selected duration.

Level 3: monthly and monthly peak pulses Several design tools (Hellström and Sanner, 1997, Spitler, 2000) rely on a representation of the loads as monthly total cooling and heating loads, upon which monthly peak cooling and heating loads are superimposed. The monthly peak loads are specified as a rectangular pulse with the magnitude and duration. Cullin and Spitler (2011) describe this as a hybrid time step approach, using a monthly time step for the average loads and a time step of one or more hours for the peak loads.

Level 4: hourly simulation In theory, it is possible to size the GHE using an hourly simulation in an iterative fashion. For example, this has been done for research purposes. Cullin and Spitler (2011) and Nagano et al. (2006) describe a design program which utilizes hourly time steps.

2.3.1.2 Ground heat transfer methodology

Methods for analyzing the ground heat transfer vary widely among design methods. These include ILS, ICS, FLS, and *g*-functions. Methods using *g*-functions may also vary with the approach used to calculate the *g*-functions. The effect of using different methods on the resulting size varies depending on the loads, borehole configuration, and other factors. For example, sizing for an imbalanced system where long-term heat build-up is an issue may be more sensitive to the ground heat transfer methodology than sizing for a relatively balanced system. Using the uniform heat flux assumption when computing *g*-functions (Malayappan and Spitler, 2013) gives larger GHE sizes than using the uniform borehole wall temperature assumption.

2.3.1.3 Treatment of borehole thermal resistance

Borehole thermal resistance is treated in chapter "Calculation of borehole thermal resistance" of this book. Methods for computing borehole thermal resistance include shape factors and multipole methods.

2.3.1.4 Treatment of the heat pump

Some design tools account for the heat pump with a simple polynomial model that relates the ratio of heat rejection to cooling provided and heat extraction to heating provided as a function of heat pump entering fluid temperature. This allows the temperature dependence of efficiency (ie, Coefficient of Performance (COP)) and consequently imposed loads, to be taken account of explicitly in the design calculations. In other cases, the user simply enters the expected heat rejection and extraction to/from the ground.

2.3.1.5 Summary table

Based on the classification scheme described above, a number of historical and currently available design tools are categorized in Table 2.1. This table is not meant to present all available tools but is limited to design tools for which information about the calculation algorithms is available.

2.3.2 Important inputs

2.3.2.1 Undisturbed ground temperature

The average ground temperature to which the GHE will be exposed has an important effect on its required size—the required size is often inversely proportional to the difference between the ground temperature and the maximum or minimum heat pump entering fluid temperature. This is usually called the undisturbed ground temperature: that is, the ground temperature before it's disturbed by the GHE. Maps have been drawn (Collins, 1925; Chang, 1958) showing undisturbed ground temperatures for a range of locations. However, these maps do not take into account disturbances from past urban development or local geological conditions, so measurement (Gehlin and Nordell, 2003) of the ground temperature is recommended.

2.3.2.2 Ground thermal properties

Ground thermal properties are location-dependent and in situ measurement of the thermal properties is recommended for GHEs that exceed a few boreholes. Such testing also results in local measurement of the undisturbed ground temperature. In situ thermal response testing is the subject of chapter "In situ estimation of ground thermal properties."

2.3.2.3 Sensitivity of length to design temperature limits

The designer chooses the design limits on the heat pump EFT. However, this choice is not arbitrary; choosing maximum values that are too high or minimum values that are too low can lead to poorly performing systems. Conversely, choosing design limits that are close to the undisturbed ground temperature can result in unfeasibly large and expensive GHEs. The optimal choice of design temperature limits can be explored with simulation-based design tools.

Several factors may influence the designer's choice of temperature limits. Kavanaugh (2008) recommends maximum heat pump EFT 11–19°C above the undisturbed ground temperature and minimum heat pump EFT 6–8°C below the undisturbed ground temperature. In other cases, local regulations may limit the allowable

Table 2.1 **Summary of design methodologies**

	Solution approach		Time resolution level					Ground heat transfer methodology					Heat pump accounted for		Borehole thermal resistance		
Method	**Direct**	**Iterative**	**0**	**1**	**2**	**3**	**4**	**ILS**	**FLS**	**ICS**	***g*-functions**	**Numerical**	**Yes**	**No**	**Shape factor**	**Multipole method**	**Other**
Bose et al. (1985)	●			●				●					●				●[1]
OSU (1988)	●			●				●					●				●[1]
Kavanaugh (1991)	●			●						●			●				●[1]
Kavanaugh (1995)	●				●					●			●				●[1]
GLHEPRO (Spitler, 2000)		●				●					●		●			●	
Nagano et al. (2006)		●					●			●			●				●[2]
Philippe et al. (2010)		●		●							●[5]		●				●
GLD (Gaia geothermal, 2014)		●[3]			●[4]			●		●	●		●		●		
ASHRAE Handbook (2015)	●			●						●			●		●		
ASHRAE Handbook (2015) alt.[6]		●		●							●		●		●		
EED (BLOCON, 2015)		●				●					●			●		●	

[1]Only accounts for pipe thermal resistance.
[2]Uses a numerical method.
[3]It's unclear if any of the formulations use a direct solution.
[4]Instead of a single peak load of user-specified duration, GLD (2014) uses a day made up of one 12-h pulse and three 4-h pulses.
[5]Uses *g*-functions to calculate the temperature penalty.
[6]Two methods are proposed in the 2015 edition of the ASHRAE Handbook.

deviations in temperature. A further consideration are the limits that are stated by the heat pump equipment manufacturer (cut-out limits). More generally, the choice of temperature limits will have some impact on seasonal efficiency. Consequently, if the designer has a particular efficiency target in mind, temperature limits may need to be adjusted to more conservative levels to reach the system efficiency target.

2.3.2.4 Loads: imbalanced or balanced

If the annual heat rejection to the ground is significantly imbalanced, eg, much more heat is rejected than extracted, long-term heat build-up in the ground can be an issue for systems with multiple boreholes. The extent to which this is a concern is proportional to the number and density[8] of boreholes. In general, long-term heat build-up can be mitigated by either making the GHE large enough or adding supplemental heat rejection. Systems with supplemental heat rejection or injection are referred to as hybrid ground-source heat pump (HGSHP) systems; see Chapter 13.

2.4 *g*-function-based methods

2.4.1 Using g-functions

g-function-based design methods utilize the iterative solution scheme described in Section 2.3.1.1. The size of the GHE is adjusted to be as small as possible while not exceeding the designer-specified minimum and maximum heat pump EFT. Eq. [2.12] gave the temperature at the borehole wall. In order to determine the heat pump entering fluid temperature, the most common approximation involves first using the effective borehole thermal resistance to determine the mean fluid temperature:

$$T_{\mathrm{f}} = T_{\mathrm{b}} + q_{\mathrm{i}} R_{\mathrm{b}}^{*} \quad [2.13]$$

Then, the entering fluid temperature to the heat pump (ie, exiting fluid temperature from the GHE) is:

$$T_{\mathrm{entering}} = \frac{q_{\mathrm{i}}}{2\dot{m}c_{\mathrm{p}}} + T_{\mathrm{f}} \quad [2.14]$$

Computing the entering fluid temperature in this way is an approximation that is effectively the same as assuming the fluid temperature through the borehole changes linearly. This is certainly an approximation, but one which gives reasonable accuracy for typical North American GHEs that are less than 100 m deep. As boreholes get significantly deeper and/or short-circuiting effects become significant the accuracy of this approximation will suffer. Several authors (Marcotte and Pasquier, 2008;

[8] "Density" here refers to how closely the boreholes are packed.

Beier, 2011; Acuña, 2013) have called attention to this, though none have proposed a solution that is readily implementable in a simulation.

The simulations used in the iterative approach rely on a library of *g*-functions that cover different configurations (eg, 3 × 2 rectangular, 3 × 3 L-shape). For each configuration, *g*-functions for several different B/H spacings are stored. The B spacing is fixed during the design process and H is adjusted iteratively, resulting in B/H spacings that are not in the library. A logarithmic interpolation is used to develop *g*-functions for these intermediate B/H spacings. Malayappan and Spitler (2013) showed that this interpolation scheme may lead to a few percent error.

2.4.2 Methods for calculating g-functions

g-functions form the basis of many design software tools. Most of these tools rely on a database of precalculated *g*-functions which are obtained for specific values of r_b/H and B/H. Originally, numerical tools such as the superposition borehole model were used to obtain *g*-functions (Eskilson, 1986). There have also been efforts (Monzó et al., 2015) to determine *g*-functions numerically using established computational fluid dynamics software tools.

For reasons of convenience and computational speed, determination of *g*-functions using analytical solutions may be preferred over generation of *g*-functions with numerical methods. The FLS solution is recommended for these situations.

The analytical determination of *g*-functions has been performed using various boundary conditions at the borehole wall. Cimmino and Bernier (2014) identified three boundary conditions: BC-I, BC-II, and BC-III. Eskilson's *g*-functions were derived using BC-III. These conditions are illustrated in Fig. 2.8 for a 3 × 1 borehole field. The complexity of the solution and the computational time increases significantly from BC-I to BC-III.

For BC-I, the heat extraction rate is uniform over the length of the boreholes and is the same for each borehole. This results in unequal mean borehole wall temperatures as shown on the temperature profile graph on the right. The average of the mean borehole wall temperatures is then used in the evaluation of the *g*-function. For BC-II, a uniform heat extraction rate along the length of the boreholes is assumed and the average temperature along the length of the boreholes is equal for all boreholes. As shown in Fig. 2.8, this leads to unequal heat extraction rates among boreholes. The average of these heat extraction rates is then used in the evaluation of the *g*-function. Finally, for BC-III, the borehole wall temperature is uniform along the length of the boreholes and is the same for each borehole. Thus, the heat extraction rates are not uniform over the length of the boreholes and vary from segment to segment in a borehole and from borehole to borehole.

The calculation of the *g*-function using the FLS can be achieved in three steps. First, the FLS (Eq. [2.8]) is used to calculate the temperature distribution around individual boreholes (or borehole segments for BC-III) due to heat extraction from any borehole (or borehole segments for BC-III). Then, spatial superposition is used to calculate the total temperature variation at the borehole walls by superposing the contribution of all boreholes (or segments). Finally, temporal superposition is used

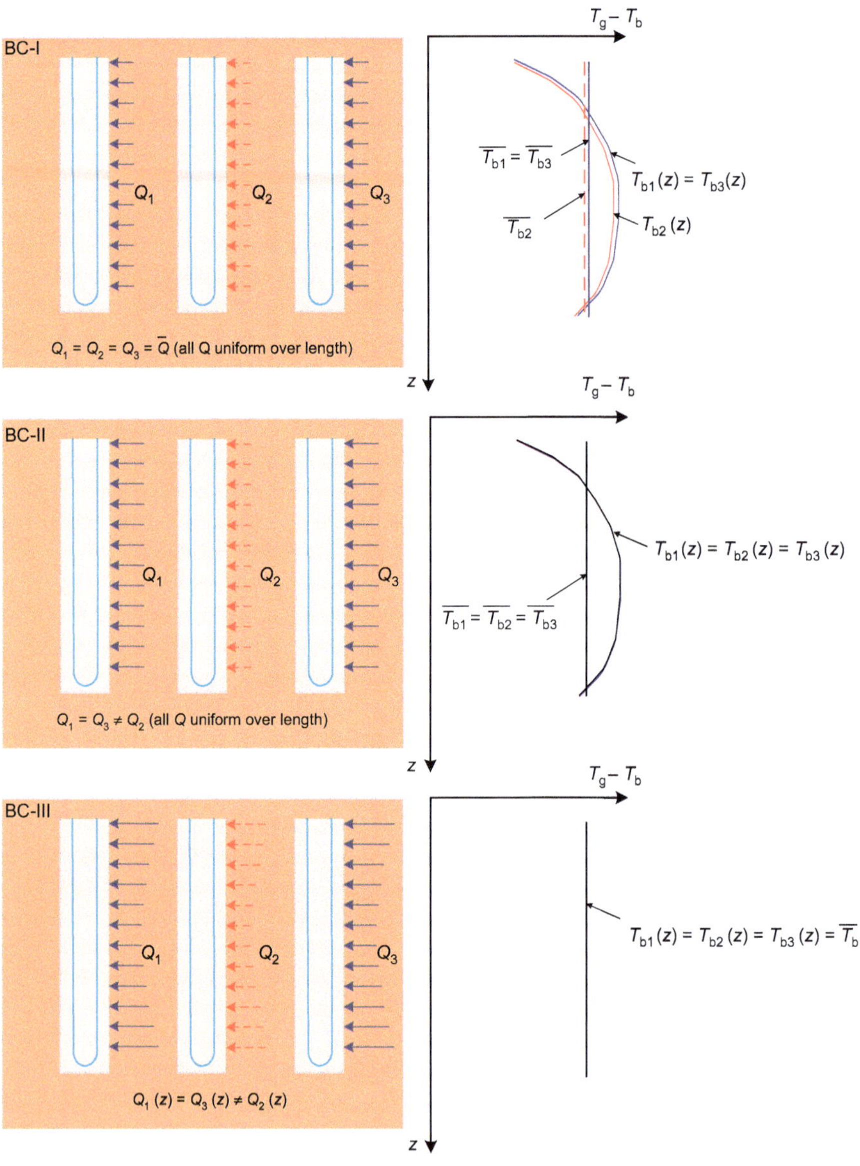

Figure 2.8 Boundary conditions for *g*-function computation.

to account for the time variation of the heat extraction rates of individual boreholes (or segments). A system of equations is built at each time step to calculate the heat extraction rates of the boreholes and the borehole wall temperature common to all boreholes. Cimmino and Bernier (2013, 2014) have shown how to execute these steps and how to perform temporal superposition in the spectral domain using Laplace transforms.

2.5 ASHRAE handbook method

Borehole field sizing methods rely on the principles of temporal and spatial superposition. Spatial superposition is used to determine g-functions or to evaluate so-called temperature penalties caused by borehole thermal interaction (see Section 2.5.3). Temporal superposition is required to account for changing ground loads over the expected duration of the system. In the ASHRAE sizing method, the required borehole field length is based on three successive heat transfer rates with different time periods. This will be illustrated here with consecutive heat transfer rates: Q_1 from $t = 0$ to t_1; Q_2 from t_1 to t_2; and Q_3 from t_2 to t_3.

Using the temporal superposition concepts presented earlier, the borehole temperature, T_b, is given by:

$$T_b - T_g = \frac{Q_1 \Gamma(t_3 - 0) + (Q_2 - Q_1)\Gamma(t_3 - t_1) + (Q_3 - Q_2)\Gamma(t_3 - t_2)}{H} \quad [2.15]$$

where Γ is the thermal response factor. Analytical solutions such as those identified in this chapter (ILS, ICS, and FLS) as well as g-functions can be used as response factors.

Eq. [2.15] can also be written as:

$$T_b - T_g = \frac{Q_1[\Gamma(t_3) - \Gamma(t_3 - t_1)] + Q_2[\Gamma(t_3 - t_1) - \Gamma(t_3 - t_2)] + Q_3\Gamma(t_3 - t_2)}{H} \quad [2.16]$$

Fig. 2.9 illustrates an application of Eq. [2.15]. In this case three consecutive heat transfer rates per unit length ($Q_1/H = 15$, $Q_2/H = 45$, $Q_3/H = 85$ W/m) are used over three different time periods with $t_1 = 3650$, $t_2 = 3680$, and $t_3 = 3680.167$ days. These

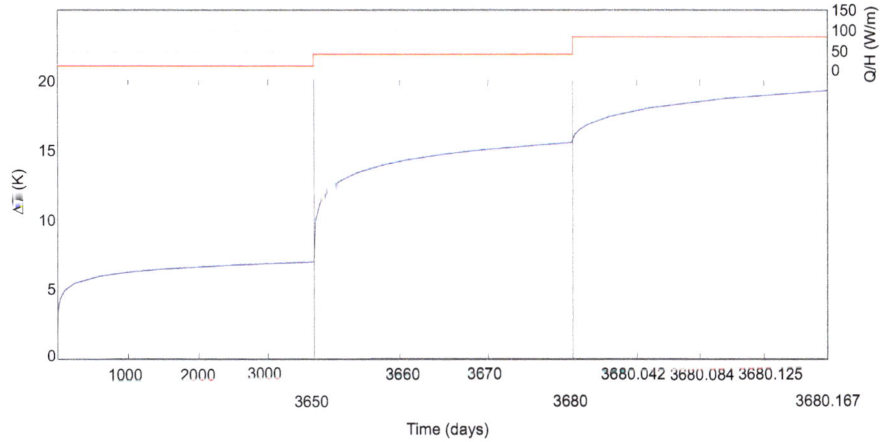

Figure 2.9 Evaluation of $(T_b - T_o)$ for three consecutive constant heat transfer rates.

time periods (10 years, 1 month, 4 hours) are typically used in the ASHRAE sizing procedure. Note that the x-axis has been divided into three sub-axes each with its own scale. For this example, the borehole wall temperature is about 7, 15.5, and 19.3°C higher than the undisturbed ground temperature after 10 years, 10 years + 1 month, and 10 years + 1 month + 4 hours, respectively.

If a steady-state condition is assumed between the fluid and the borehole wall at $t = t_3$, then:

$$T_f - T_b = \frac{Q_3 R_b^*}{H} \quad [2.17]$$

where R_b^*, the effective borehole resistance, can be calculated using the techniques outlined in chapter "Calculation of borehole thermal resistance." Using Eq. [2.17], Eq. [2.16] now becomes:

$$H = \frac{Q_1[\Gamma(t_3) - \Gamma(t_3 - t_1)] + Q_2[\Gamma(t_3 - t_1) - \Gamma(t_3 - t_2)] + Q_3\left[\Gamma(t_3 - t_2) + R_b^*\right]}{T_f - T_g} \quad [2.18]$$

Eq. [2.18] applies to single boreholes. However, it can also be used for borehole fields provided that borehole thermal interference is taken into account either by using g-functions or a temperature penalty.

Most borehole field sizing methods are based on Eq. [2.18]. Sizing methods will differ in the number of Q values they superimpose and in the way they calculate the thermal response factor Γ. The ASHRAE Handbook method uses three values of Q and the ICS is used for Γ.

Eqs. [2.19] and [2.20] below are often referred to as the ASHRAE borehole field sizing method. They first appeared in the 1995 ASHRAE Handbook—Applications and in a paper by Kavanaugh (1995), and they are still used in the 2015 ASHRAE Handbook—Applications. Earlier versions of these equations were presented by Kavanaugh (1988, 1991). In his thesis, Eskilson (1987) also uses a three pulse scenario similar to the one used in the ASHRAE sizing equation. Examples of the use of these equations are provided in the book by Kavanaugh and Rafferty (2015) and in the 2015 ASHRAE Handbook—Applications. Recently, the 2015 ASHRAE Handbook—Applications has also incorporated an alternative design method based on g-functions. This method is covered in Section 2.6.3.

$$L_c = \frac{q_a R_{ga} + (q_{lc} - W_c)(R_p + PLF_m R_{gm} + R_{gd} F_{sc})}{T_g - \frac{(T_{wi} + T_{wo})}{2} - T_p} \quad [2.19]$$

$$L_h = \frac{q_a R_{ga} + (q_{lh} - W_h)(R_p + PLF_m R_{gm} + R_{gd} F_{sc})}{T_g - \frac{(T_{wi} + T_{wo})}{2} - T_p} \quad [2.20]$$

where F_{sc} is the short circuit heat loss factor (-), L_c, L_h are the required borehole field lengths for cooling and heating (m), PLF_m is the part load factor during the design month (-), q_a is the net annual average heat transfer to the ground (W), q_{lc}, q_{lh} are the building peak block cooling and heating loads (W), R_p is the thermal resistance of the pipe (m·K/W), R_{ga}, R_{gm}, R_{gd} are the effective thermal resistances of the ground for the annual, monthly, and daily pulses (m·K/W), T_g is the undisturbed ground temperature (°C), T_{wi}, T_{wo} are the design heat pump inlet and outlet fluid temperatures (°C), T_p is the temperature penalty for interference of adjacent bores (°C), W_c,W_h are the power inputs to the heat pump at peak cooling and heating conditions (W).

The required borehole field length is the larger of the two lengths, L_c and L_h. The values of L_c and L_h can be regarded as the total borehole field lengths required to avoid exceeding the minimum and maximum values of T_{wi}, the inlet temperature to the heat pump (ie, the outlet temperature from the GHE) at the worst ground load conditions. It is assumed that these conditions occur after the borehole has been subjected to three successive heat transfer rates, q_a, $(q_{lc} - W_c)PLF_m$ (or $(q_{lh} - W_c)PLF_m$), and $(q_{lc} - W_c)\cdot(\text{or}(q_{lh} - W_c))$ each with its own time period. ASHRAE (2015) recommends the following time periods for these heat transfer rates: 10 years, 1 month, and 4 hours, respectively.

Thus, with respect to the generic equation developed earlier (Eq. [2.18]), q_a, $(q_{lc} - W_c)PLF_m$, and $(q_{lc} - W_c)$ correspond to Q_1, Q_2, and Q_3, respectively.

The sign convention in Eqs. [2.19] and [2.20] is as follows. The value of q_a is negative when more heat is extracted from the ground than is injected on an annual basis; the peak building load (q_{lc} or q_{lh}) is negative (positive) when the building is in cooling (heating) mode, the temperature penalty, T_p, is negative (positive) when q_a is negative (positive), and W_c and W_h are always positive. L_c and L_h are given by:

$$L_c = n_c \times H_c$$
$$L_h = n_h \times H_h \qquad [2.21]$$

where, n_c, n_h are the number of boreholes required in cooling and heating (-), H_c, H_h are the lengths of individual boreholes required in cooling and heating (m).

2.5.1 Ground loads

Typically, calculations of the required borehole field length would proceed as follows. First, the peak building block heating and cooling loads, q_{lc} and q_{lh} are determined. Since the required borehole field lengths are critically dependent on these peak loads and on their duration it is important to evaluate them as accurately as possible. These loads are block loads which are determined by adding the loads of all thermal zones in a building during a given time step (usually 1 h). One common mistake is to determine q_{lc} and q_{lh} by adding the installed heat pump capacity. This most often lead to borehole field oversizing as building diversity is not accounted for.

The required compressor powers at peak loads, W_c and W_h, are usually given by heat pump manufacturers. The values of W_c and W_h are typically evaluated for inlet temperatures corresponding to the design heat pump inlet temperatures (ie, at T_{wi}). The differences ($q_{lc} - W_c$) and ($q_{lh} - W_h$) represent the peak ground loads in cooling and heating, respectively. In accordance with the sign convention, the first difference is negative and the second one is positive.

Then, the values of PLF_m have to be evaluated. They correspond to the heat pump run time fraction $(0 - 1)$ during the design month when the peak ground load occurs. Thus, the values of $(q_{lc} - W_c)PLF_m$ and $(q_{lh} - W_h)PLF_m$ represent the average monthly ground loads during the design month.

Finally, the net annual average heat transfer to the ground, q_a, needs to be evaluated. One way to estimate this quantity is to use the concept of Equivalent Full Load Hours (EFLH) in heating (EFLH_h) and in cooling (EFLH_c). With this approach, q_a can be obtained as follows:

$$q_a = \frac{(q_{lc} - W_c)\text{EFLH}_c + (q_{lh} - W_h)\text{EFLH}_h}{8760} \qquad [2.22]$$

Typical values of EFLH_h and EFLH_c were determined by Carlson and Thornton (2002) for different building types and climates. The evaluation of q_a serves two purposes. First, it is used in the denominator in Eqs. [2.19] and [2.20]. Second, as will be shown below, it is used in the evaluation of the temperature penalty, T_p.

2.5.2 Effective ground thermal resistance

The ASHRAE sizing equation contains three effective ground thermal resistances, R_{ga}, R_{gm}, R_{gd}. With reference to the generic formulation (Eq. [2.18]), these values are given by:

$$R_{ga} = [\Gamma(t_3) - \Gamma(t_3 - t_1)]\ ;\ R_{gm} = [\Gamma(t_3 - t_1) - \Gamma(t_3 - t_2)]\ ;$$
$$R_{gd} = \Gamma(t_3 - t_2) \qquad [2.23a,b,c]$$

In the ASHRAE sizing method, the infinite cylinder solution (Eq. [2.6]) is used as the thermal response factor. Thus, R_{ga}, R_{gm}, R_{gd} are given by:

$$R_{ga} = \frac{G(Fo_{t_3}) - G(Fo_{t_3} - Fo_{t_1})}{k^*} \qquad [2.24a]$$

$$R_{gm} = \frac{G(Fo_{t_3} - Fo_{t_1}) - G(Fo_{t_3} - Fo_{t_2})}{k^*} \qquad [2.24b]$$

$$R_{gd} = \frac{G(Fo_{t_3} - Fo_{t_2})}{k^*} \qquad [2.24c]$$

where t_1, t_2, and t_3 are equal to 3650, 3680, and 3680.167 days, respectively.

The values R_{ga}, R_{gm}, R_{gd} depend on the time periods; the effective ground thermal conductivity, k^*; the thermal diffusivity, α; and the borehole radius, r_b. Philippe et al. (2010)

have calculated R_{ga}, R_{gm}, R_{gd} for a wide range of values of r_b and α, and they proposed a simple correlation to calculate them. Furthermore, they showed that the ranges of variation for $R_{ga}\cdot k^*$, $R_{gm}\cdot k^*$, and $R_{gd}\cdot k^*$ are 0.379–0.382, 0.30–0.38, 0.10–0.30, respectively, when $t_1 = 10$ years, $t_2 - t_1 = 1$ month, and $t_3 - t_2 = 6$ hours.

The evaluation of the three effective ground thermal resistances with the ICS solution implies that heat transfer from the boreholes is one-dimensional in the radial direction. This is convenient as the solution is independent of the borehole length. However, axial effects are important after a certain number of years. This results in two-dimensional heat transfer and the validity of the ICS becomes questionable (Philippe et al., 2009). Only R_{ga} suffers from this limitation and since its weight on the determination of L is typically small, an error on the determination of R_{ga} has limited consequences on the accuracy of the length determination.

2.5.3 Temperature penalty

When there is more than one borehole, thermal interaction among closely spaced boreholes exits after a certain period of time. There are several ways to account for this phenomenon. Bose et al. (1985) suggested modifying the effective ground thermal resistances of single boreholes to include thermal interference. Eskilson (1987) accounted for thermal interference by using spatial superposition of the thermal response of boreholes in a field to obtain g-functions.

The ASHRAE sizing equation uses a different approach to account for thermal interference by correcting the undisturbed ground temperature. The resulting temperature penalty, T_p, is tabulated in the ASHRAE Handbook (2015) for a limited number of borehole field configurations and annual ground thermal imbalance (q_a). These values are based on a calculation procedure developed by Kavanaugh and Rafferty (1997) assuming 10 years of operation and a worst-case scenario where moisture evaporation and underground water movement are considered negligible. It is essentially a one-dimensional method that calculates the amount of heat "trapped" in a borehole field using the ILS solution at various radii from the boreholes. This original method was slightly modified by Kavanaugh and Rafferty (2015) to account for heat transfer from the bottom of the borehole field. This method of calculating T_p has been criticized by a number of authors (Bernier et al., 2008; Fossa, 2011; Capozza et al., 2012) and has been shown to underestimate the temperature penalty.

Fossa (2011) and Bernier et al. (2008) recommend instead evaluating T_p using g-functions. Fossa recommends the following equation:

$$T_p = \frac{q_a}{k^* \cdot L}\left[\frac{g_N(t)}{2\pi} - G(t)\right] \qquad [2.25]$$

while Bernier et al. (2008) recommend a slightly different formulation:

$$T_p = \frac{q_a}{2\pi k^* \cdot L}\left[g_N(t) - g_1(t)\right] \qquad [2.26]$$

where g_N and g_1 are the g-functions for the entire field (composed of N boreholes) and for a single borehole, respectively; G is the ICS solution for one borehole. The time t at which T_p is evaluated is typically on the order of 10 years. Bernier et al. (2008) and Fossa (2011) provide correlations to approximate T_p for a wide range of conditions without the need to calculate g-functions. Finally, the value of T_p is zero for single boreholes according to Eq. [2.26]. On the other hand, Fossa's formulation (Eq. [2.25]) leads to a nonzero value of T_p for single boreholes. In effect, this corrects the inaccuracy of R_{ga} for long time periods as reported earlier.

2.5.4 Pipe resistance

Eqs. [2.19] and [2.20] include a pipe resistance term, R_p. This term accounts for the thermal resistance between the fluid and the borehole wall. It first appeared in the 1995 version of the ASHRAE Handbook. Though the concept of borehole thermal resistance had been formulated earlier (Mogensen, 1983), it had not yet found its way into the ASHRAE sizing method.

The original value of R_p was evaluated using the concept of equivalent diameter which had been proposed by a number of authors (Claesson and Dunand, 1983; Bose et al., 1985). As illustrated in Fig. 2.10, the borehole geometry characterized by a bore diameter $2r_b$ and two pipes of diameter D (single U-tube configuration) is replaced by one single pipe with an equivalent diameter D_b and an equivalent pipe resistance, R_p. Typical values of D_b and R_p can be found in the 1995 edition of the ASHRAE Handbook and in the book by Kavanaugh and Rafferty (1997).

When the equivalent borehole diameter method is used, the various effective ground thermal resistances, R_{ga}, R_{gm}, R_{gd} are evaluated with the ICS solution (Eq. [2.6]) using the equivalent diameter D_b and not the real borehole diameter.

In 2003 (ASHRAE, 2003), the equivalent pipe resistance was replaced with the borehole thermal resistances calculated using the formulation proposed by Remund (1999). However, this thermal resistance is still referred to as a thermal resistance of pipe and the values of R_{ga}, R_{gm}, R_{gd} are still based on the outside pipe diameter. This issue was resolved in the 2011 version of the Handbook (ASHRAE, 2011) with R_b replacing R_p in Eqs. [2.19] and [2.20].

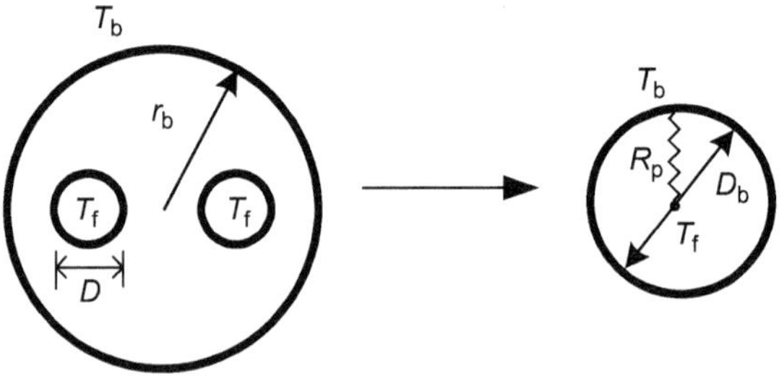

Figure 2.10 Schematic representation of the equivalent diameter concept.

2.5.5 Short circuit heat loss factor

The short circuit heat loss factor, F_{sc}, included in Eqs. [2.19] and [2.20], has been introduced in the original version of these equations in 1995 (ASHRAE, 1995) and is still present in the 2015 edition of the handbook (ASHRAE, 2015). The original intent of this term was to account for thermal short-circuiting in the U-bend (Kavanaugh, 1991). Initially, thermal short-circuiting took the form of a temperature difference correction factor applied to the temperature difference across a borehole (Kavanaugh, 1985; Deerman and Kavanaugh, 1991). The change from a temperature difference correction factor to a multiplication factor (F_{sc}) has apparently not been documented. Values of F_{sc} reported by Kavanaugh (1991) and Kavanaugh and Rafferty (2015) range from 1.05 for one U-tube to 1.01 for three U-tubes in a series. It is interesting to note that F_{sc} is applied as a multiplication factor to the effective ground thermal resistance for the daily pulse, R_{gd}, and not to the pipe thermal resistance, R_p, as one could have expected.

Since it is possible to account for thermal short-circuiting directly in the value of R_b^*, the use of F_{sc} is no longer needed.

2.6 Improvements to the ASHRAE sizing method

The main advantage of the ASHRAE sizing equations is that calculations are relatively simple as values of R_{ga}, R_{gm}, R_{gd} are independent of the borehole length. For single boreholes, $T_p = 0$, and the determination of L is obtained directly by solving Eq. [2.19] or [2.20]. For borehole fields, one can estimate the value of T_p from the tabulated values in the ASHRAE handbook and determine L without the need to iterate.

2.6.1 Different source for T_p

If more precise values of T_p are required, then the correlations proposed by Fossa (2011) or Bernier et al. (2008) should be used. However, for these cases, the length has to be determined iteratively as T_p depends on H which is unknown a priori. As shown by Philippe et al. (2010) these iterations can be easily performed with a spreadsheet.

2.6.2 Simpler form of the sizing equation

Eqs. [2.19] and [2.20] are somewhat cumbersome as they mix building and ground loads, which may confuse some users. It is possible to write these equations in terms of ground loads in a more compact form. This leads to the following equations:

$$L_c = \frac{q_a R_{ga} + q_{m,c} R_{gm} + q_{h,c} R_{gd} + q_{h,c} R_b^*}{T_g - \frac{(T_{wi} + T_{wo})}{2} - T_p} \quad [2.27]$$

$$L_h = \frac{q_a R_{ga} + q_{m,h} R_{gm} + q_{h,h} R_{gd} + q_{h,h} R_b^*}{T_g - \frac{(T_{wi} + T_{wo})}{2} - T_p} \qquad [2.28]$$

where $q_{m,c}$ and $q_{m,h}$ are the average monthly ground loads during the design month in cooling and in heating, respectively, and $q_{h,c}$ and $q_{h,h}$ are the peak ground loads in cooling and heating, respectively. The F_{sc} term has been dropped since thermal short circuit between pipes can be incorporated in R_b^*.

2.6.3 Effective ground thermal resistances with g-functions

It is possible to avoid the use of T_p by using g-functions as thermal response factors in Eqs. [2.23a–c]. With this approach, the three effective ground thermal resistances take the following form:

$$R_{ga,g} = \frac{g_{(t_3-t_0)} - g_{(t_3-t_1)}}{2\pi k^*} \qquad [2.29a]$$

$$R_{gm,g} = \frac{g_{(t_3-t_1)} - g_{(t_3-t_2)}}{2\pi k^*} \qquad [2.29b]$$

$$R_{gd,g} = \frac{g_{(t_3-t_2)}}{2\pi k^*} \qquad [2.29c]$$

where $g_{(t_i-t_j)}$ is the g-function evaluated for a time period corresponding to $(t_i - t_j)$ where $t_0 = 0$. The index "g" has been added to R_{ga}, R_{gm}, R_{gd} to indicate that they are based on g-functions and to distinguish them from the values obtained earlier (Eqs. [2.23a–c]) with the ICS solution. The ASHRAE sizing equation based on g-functions takes the following form:

$$L_c = \frac{q_a R_{ga,g} + q_{m,c} R_{gm,g} + q_{h,c} R_{gd,g} + q_{h,c} R_b^*}{T_g - \frac{(T_{wi} + T_{wo})}{2}} \qquad [2.30]$$

$$L_h = \frac{q_a R_{ga,g} + q_{m,h} R_{gm,g} + q_{h,h} R_{gd,g} + q_{h,h} R_b^*}{T_g - \frac{(T_{wi} + T_{wo})}{2}} \qquad [2.31]$$

Eqs. [2.30] and [2.31] form the basis of the alternative sizing method presented in the ASHRAE Handbook (2015). They have to be solved iteratively as $R_{ga,g}$, $R_{gm,g}$, $R_{gd,g}$, depending on borehole length which is unknown a priori. The other difficulty in using Eqs. [2.30] and [2.31] is the relative unavailability of g-functions in the open literature. Cimmino and Bernier (2013) have tried to alleviate this shortage by proposing a tool to generate g-functions for any borehole field configuration. Using this last technique, Ahmadfard and Bernier (2014) have solved Eqs. [2.30] and [2.31] with g-functions calculated on the fly in each iteration.

2.7 Design of hybrid ground-source heat pump systems

HGSHP systems (see Chapter 12) are suitable for situations with moderately to severely imbalanced annual heating and cooling loads for which conventional GSHP systems would require excessively large GHEs. In addition to the GHE, supplemental heat rejection may be provided by cooling towers, fluid coolers, lake loops, or other devices. For cases with significant heating loads, supplemental heat may be provided by solar collectors or boilers.

Kavanaugh (1998) described a method for designing HGSHP systems that is an extension of the ASHRAE Handbook method for sizing GHEs used by conventional GSHP systems. Since this method provides required GHE lengths for heating and for cooling, when designing an HGSHP system, the shorter length is chosen. Then, the supplemental heat source or sink can be sized based on the difference between the two design lengths.

Xu (2007) developed an optimization-based procedure using a detailed simulation that minimized the life cycle cost of the system by adjusting GHE length and cooling tower size. Although the method is ideal from an accuracy standpoint, it requires far too much computation time to be feasible for use in a consulting engineer's office.

Chiasson and Yavuzturk (2009a,b) developed dimensionless parameters that describe the HGSHP system. The total GHE length and annual heat rejection or extraction capacities of the supplemental equipment can be determined from the dimensionless parameters. They used detailed simulations to develop correlations for the dimensionless parameters.

Hackel et al. (2009) developed a design procedure for designing cooling-dominated HGSHP systems. From a simulation-based parametric study of four building types and six locations, a set of first-order equation fits were generated to provide the ideal GHE size and cooling tower capacity in order to minimize the 20-year life cycle cost. The GHE size is directly proportional to the peak heating load and inversely proportional to the difference between the heat pump minimum EFT constraint and the undisturbed ground temperature. The cooling tower capacity is proportional (130%) to the peak unmet cooling load.

Cullin and Spitler (2010) described an optimization approach which finds the combination of GHE length and supplemental heat rejection or heat injection capacities that meets both the maximum and minimum heat pump EFT. This is based on two general observations of GSHP system design:

1. The GHE first cost tends to dominate the system life cycle cost. Therefore, designs that minimize the GHE size tend to have the lowest life cycle cost.
2. Systems that minimize the GHE size and make maximum usage of the supplemental heat source/sink tend to, as a result, reach both the maximum and minimum heat pump entering fluid temperature limits.

Cullin and Spitler (2010) compared four systems designed with all four methods; the Xu method gives the lowest life cycle costs but can take many hours of computing time to optimize a single design. The Cullin and Spitler method gives the next best results and took on the order of a minute of computing time with a circa-2010 desktop computer. The other methods are all essentially instantaneous from a computing time standpoint but give results with slightly to moderately higher life cycle costs. All methods, though, were successful in that the minimum and maximum design heat pump EFT were met.

2.8 Conclusions and recommended future work

As shown in this chapter, several methods and tools are available to design vertical GHEs. Differentiating characteristics have been identified, and a number of current and historical design methods and tools have been cataloged. Analytical solutions such as the ILS, infinite cylinder source, and FLS can provide insight into GHE behavior. However, these methods have limitations and should be used carefully. The FLS solution can be used to evaluate the thermal interaction from one borehole segment to another which can then be used to determine *g*-functions analytically.

Spatial and temporal superposition are key concepts and are used in most design methods. Spatial superposition is used to evaluate borehole-to-borehole thermal interference while temporal superposition allows the deconvolution of the time-varying heat transfer rates into functions for which a solution can readily be determined.

Two families of design methods have been presented: *g*-function-based methods and the ASHRAE design method. Originally, the *g*-functions were computed numerically; newer ways to calculate them using the FLS are presented in this chapter. The various boundary conditions at the borehole wall that can be used to generate *g*-functions are discussed. The evolution of the ASHRAE design method from its first appearance in 1995 to its current state is reviewed, and possible improvements to the method are presented.

The various design tools have reached a certain maturity. However, validations with experimental data or extensive intermodel comparisons are scarce in the literature. Several authors have performed intermodel comparison studies comparing the results from different tools (Shonder et al., 1999, 2000) or the underlying simulations (Spitler et al., 2009). In one study, Cullin et al. (2015) performed a validation study based on four existing GHE installations in the United States, the United Kingdom, and Spain. Much like the Shonder et al. comparisons all systems had relatively balanced heat rejection and extraction, so none of the systems represented a case with significant heat build-up. Two design methods were compared: a simulation-based design tool (Spitler, 2000) which utilized numerically generated *g*-functions and the ASHRAE Handbook method (ASHRAE, 2011). The simulation-based design tool gave markedly better results than the handbook method and a sensitivity study attributed much of the difference to the time resolution of the two methods.

Validation of GHE design tools is inherently complicated by the long timescale of the ground surrounding the GHE. However, despite these difficulties, it is clear that more validation work and intermodel comparisons are needed to fully understand the differences among the current tools (Spitler and Bernier, 2011). Long-term monitoring of borehole fields with significant annual imbalances would be particularly useful.

Current design tools assume that the ground heat transfer may be modeled as pure heat conduction in a homogeneous ground. This may lead to conservative designs for systems that experience significant ground water movement or freezing/thawing. However, there are no known design tools that account for these phenomena. Likewise, unsaturated moisture transport, inhomogeneities in the ground, and near-surface effects are all untreated in current design tools. These effects may also be important in some situations. Incorporation of these phenomena will improve the capabilities of engineers to design economically optimal GHEs. Further research in methods for practically treating these phenomena is recommended.

References

Abdelaziz, S.L., Ozudogru, T.Y., Olgun, C.G., Martin Ii, J.R., 2014. Multilayer finite line source model for vertical heat exchangers. Geothermics 51, 406–416.

Acuña, J., 2013. Distributed Thermal Response Tests – New Insights on U-pipe and Coaxial Heat Exchangers in Groundwater-Filled Boreholes (Ph.D. doctoral thesis). KTH.

Ahmadfard, M., Bernier, M., 2014. An alternative to ASHRAE's design length equation for sizing borehole heat exchangers. In: ASHRAE Annual Meeting. Seattle, Washington. June 28–July 2.

ASHRAE, 1995. Chapter 32-Geothermal Energy. ASHRAE Handbook – Applications. ASHRAE, Atlanta, Georgia.

ASHRAE, 2003. Chapter 32-Geothermal Energy. ASHRAE Handbook – Applications. ASHRAE, Atlanta, Georgia.

ASHRAE, 2011. Chapter 34-Geothermal Energy. ASHRAE Handbook – Applications. ASHRAE, Atlanta, Georgia.

ASHRAE, 2015. Chapter 34-Geothermal Energy. ASHRAE Handbook – Applications. ASHRAE, Atlanta, Georgia.

Banks, D., 2015. Horatio Scott Carslaw and the origins of the well function and line source heat function. Scottish Journal of Geology 51, 100–104.

Bauer, D., Heidemann, W., Müller-Steinhagen, H., Diersch, H.J.G., 2011. Thermal resistance and capacity models for borehole heat exchangers. International Journal of Energy Research 35 (4), 312–320.

Beier, R.A., 2011. Vertical temperature profile in ground heat exchanger during in-situ test. Renewable Energy 36 (5), 1578–1587.

Bernier, M., 2014. Sizing and simulating bore fields using thermal response factors. In: 11th IEA Heat Pump Conference. Montreal, Canada. May 12–16.

Bernier, M.A., 2001. Ground-coupled heat pump system simulation. ASHRAE Transactions 107 (1), 605–616.

Bernier, M.A., Chahla, A., Pinel, P., 2008. Long term ground temperature changes in geo-exchange systems. ASHRAE Transactions 114 (2), 342–350.

BLOCON, 2015. EED v3.2.

Bose, J.E., Parker, J.D., McQuiston, F.C., 1985. Design/Data Manual for Closed-Loop Ground-Coupled Heat Pump Systems. ASHRAE, Atlanta, Georgia.

Capozza, A., De Carli, M., Zarrella, A., 2012. Design of borehole heat exchangers for ground-source heat pumps: a literature review, methodology comparison and analysis on the penalty temperature. Energy and Buildings 55, 369–379.

Carlson, S.W., Thornton, J.W., 2002. Development of equivalent full load heating and cooling hours for GCHPs. ASHRAE Transactions 108 (2), 88–97.

Carslaw, H.S., Jaeger, J.C., 1959. Conduction of Heat in Solids, second ed. Clarendon Press, Oxford.

Chang, J.-H., 1958. Ground Temperature: Bluehill Meteorological Observatory. Harvard University.

Chiasson, A.D., Yavuzturk, C., 2009a. A design tool for hybrid geothermal heat pump systems in cooling-dominated buildings. ASHRAE Transactions 115 (2), 74–87.

Chiasson, A.D., Yavuzturk, C., 2009b. A design tool for hybrid geothermal heat pump systems in heating-dominated buildings. ASHRAE Transactions 115 (2), 60–73.

Cimmino, M., Bernier, M., 2013. Preprocessor for the generation of g-functions used in the simulation of geothermal systems. In: 13th International IBPSA Conference, Chambery, France.

Cimmino, M., Bernier, M., 2014. A semi-analytical method to generate g-functions for geothermal bore fields. International Journal of Heat and Mass Transfer 70, 641–650.

Claesson, J., Dunand, A., 1983. Heat Extraction from the Ground by Horizontal Pipes – A Mathematical Analysis. Swedish Council for Building Research, Stockholm, 215 p.

Claesson, J., Javed, S., 2011. An analytical method to calculate borehole fluid temperatures for time-scales from minutes to decades. ASHRAE Transactions 117 (2), 279–288.

Collins, W.D., 1925. Temperature of Water Available for Industrial Use in the United States. Water Supply Paper. USGS.

Cooper, L.Y., 1976. Heating of a cylindrical cavity. International Journal of Heat and Mass Transfer 19, 575–577.

Cullin, J.R., Spitler, J.D., 2010. Comparison of simulation-based design procedures for hybrid ground source heat pump systems. In: System Simulation in Buildings 2010. Liège, Belgium. December 13–15.

Cullin, J.R., Spitler, J.D., 2011. A computationally efficient hybrid time step methodology for simulation of ground heat exchangers. Geothermics 40 (2), 144–156.

Cullin, J.R., Spitler, J.D., Montagud, C., Ruiz-Calvo, F., Rees, S.J., Naicker, S.S., Konečný, P., Southard, L.E., 2015. Validation of vertical ground heat exchanger design methodologies. Science and Technology for the Built Environment 21 (2), 137–149.

Deerman, J.D., Kavanaugh, S.P., 1991. Simulation of vertical U-tube ground-coupled heat pump systems using the cylindrical heat source solution. ASHRAE Transactions 97, 287–295.

Department of Energy and Climate Change, 2011. Microgeneration Installation Standard: MCS 022: Ground Heat Exchanger Look-up Tables – Supplementary Material to MIS 3005 Issue 1.0 (London, UK).

Diao, N.R., Zeng, H.Y., Fang, Z.H., 2004. Improvement in modeling of heat transfer in vertical ground heat exchangers. HVAC&R Research 10 (4), 459–470.

Eskilson, P., 1986. Superposition Borehole Model: Manual for Computer Code. University of Lund, Lund, Sweden.

Eskilson, P., 1987. Thermal Analysis of Heat Extraction Boreholes (Ph.D. doctoral thesis). University of Lund.

Fossa, M., 2011. The temperature penalty approach to the design of borehole heat exchangers for heat pump applications. Energy and Buildings 43 (6), 1473–1479.

Gaia Geothermal, 2014. Grout Loop Design™ User's Manual.

Gautschi, W., Cahill, W.F., 1964. Exponential integral and related functions. In: Abramowitz, M., Stegun, I. (Eds.), Handbook of Mathematical Functions with Formulas, Graphs, and Mathematical Tables. National Bureau of Standards, Washington, D.C.

Gehlin, S.E.A., Nordell, B., 2003. Determining undisturbed ground temperature for thermal response test. ASHRAE Transactions 109 (1), 151–156.
Godefroy, V., Bernier, M., 2014. A simple model to account for thermal capacity in boreholes. In: 11th IEA Heat Pump Conference. Montréal, Canada.
Hackel, S., Nellis, G., Klein, S., 2009. Optimization of cooling-dominated hybrid ground-coupled heat pump systems. ASHRAE Transactions 115 (1), 565–580.
Hellström, G., 1991. Ground heat storage. Thermal analyses of duct storage systems (Ph.D. doctoral thesis). University of Lund, Sweden.
Hellström, G., Sanner, B., 1997. EED – Earth Energy Designer. Version 1. User's Manual.
Ingersoll, L.R., Plass, H.J., 1948. Theory of ground pipe heat source for the heat pump. Heating Piping and Air Conditioning 20, 119–122.
Ingersoll, L., Adler, F., Plass, H., Ingersoll, A., 1951. Theory of Earth Heat Exchangers for the Heat Pump. ASHVE Transactions 57, 167–188.
Katsura, T., Nagano, K., Takeda, S., 2008. Method of calculation of the ground temperature for multiple ground heat exchangers. Applied Thermal Engineering 28 (14–15), 1995–2004.
Kavanaugh, S., 1995. A design method for commercial ground-coupled heat pumps. ASHRAE Transactions 101 (Pt. 2), 1088–1094.
Kavanaugh, S.P., 1985. Simulation and Experimental Verification of Vertical Ground-coupled Heat Pump Systems (Ph.D. doctoral thesis). Oklahoma State University.
Kavanaugh, S.P., 1988. Ground and water source heat pump performance and design for southern climates. In: Fifth Symposium on Improving Building Systems in Hot and Humid Climates, Houston, Texas.
Kavanaugh, S.P., 1991. Ground and Water Source Heat Pumps – A Manual for the Design and Installation of Ground-Coupled, Ground Water and Lake Water Heating and Cooling Systems in Southern Climates. University of Alabama, Tuscaloosa, AL.
Kavanaugh, S.P., 1998. A design method for hybrid ground-souce heat pumps. ASHRAE Transactions 104 (2), 691–698.
Kavanaugh, S., 2008. A 12-step method for closed-loop ground-source heat-pump design. ASHRAE Transactions 114 (2), 328–337.
Kavanaugh, S.P., Rafferty, K., 1997. Ground-source Heat Pumps – Design of Geothermal Systems for Commercial and Institutional Buildings. ASHRAE, Atlanta, Georgia.
Kavanaugh, S.P., Rafferty, K., 2015. Geothermal Heating and Cooling – Design of Ground-source Heat Pump Systems. ASHRAE, Atlanta, Georgia.
Lamarche, L., Beauchamp, B., 2007. A new contribution to the finite line-source model for geothermal boreholes. Energy and Buildings 39 (2), 188–198.
Lazzarotto, A., 2015. Developments in Ground Heat Storage Modeling (doctoral thesis). KTH.
Malayappan, V., Spitler, J.D., 2013. Limitations of using uniform heat flux assumptions in sizing vertical borehole heat exchanger fields. In: Clima 2013. Prague. June 16–19.
Marcotte, D., Pasquier, P., 2008. On the estimation of thermal resistance in borehole thermal conductivity test. Renewable Energy 33 (11), 2407–2415.
Marcotte, D., Pasquier, P., Sheriff, F., Bernier, M., 2010. The importance of axial effects for borehole design of geothermal heat-pump systems. Renewable Energy 35 (4), 763–770.
Mogensen, P., 1983. Fluid to duct wall heat transfer in duct system heat storages. In: International Conference on Subsurface Heat Storage in Theory and Practice, Stockholm, Sweden, Swedish Council for Building Research, June 6–8.
Monzó, P., Mogensen, P., Acuña, J., Ruiz-Calvo, F., Montagud, C., 2015. A novel numerical approach for imposing a temperature boundary condition at the borehole wall in borehole fields. Geothermics 56, 35–44.

Nagano, K., Katsura, T., Takeda, S., 2006. Development of a design and performance prediction for the ground source heat pump system. Applied Thermal Engineering 26, 1578–1592.
OSU, 1988. Closed-Loop/Ground-Source Heat Pump Systems Installation Guide.
Pahud, D., 2002. Geothermal Energy and Heat Storage. Scuola Universitaria Professionale della Svizzera Italiana, Canobbio, Switzerland.
Pasquier, P., Marcotte, D., 2012. Short-term simulation of ground heat exchanger with an improved TRCM. Renewable Energy 46, 92–99.
Philippe, M., Bernier, M., Marchio, D., 2009. Validity ranges of three analytical solutions to heat transfer in the vicinity of single boreholes. Geothermics 38 (4), 407–413.
Philippe, M., Bernier, M.A., Marchio, D., 2010. Sizing calculation spreadsheet: vertical geothermal borefields. ASHRAE Journal 52 (7), 20–28.
Remund, C.P., 1999. Borehole thermal resistance: laboratory and field studies. ASHRAE Transactions 105 (1), 439–445.
Sailer, E., Taborda, D.M.G., Keirstead, J., 2015. Assessment of design procedures for vertical borehole heat exchangers. In: Fortieth Workshop on Geothermal Reservoir Engineering. January 26–28.
Schweizerischer Ingenieur- und Architektenverein, 2010. Erdwärmesonden. SIA 384/6:2010. Zurich.
Shonder, J.A., Baxter, V., Thornton, J., Hughes, P., 1999. A new comparison of vertical ground heat exchanger design methods for residential applications. ASHRAE Transactions 102 (2), 1179–1188.
Shonder, J.A., Baxter, V.D., Hughes, P.J., Thornton, J.W., 2000. A comparison of vertical ground heat exchanger design software for commercial applications. ASHRAE Transactions 106 (Pt. 1), 831–842.
Spitler, J., Bernier, M., 2011. Ground-source heat pump systems: the first century and beyond. HVAC&R Research 17 (6), 891–894.
Spitler, J., Cullin, J., 2008. In: Misconceptions Regarding Design of Ground-Source Heat Pump Systems the World Renewable Energy Congress, Glasgow, Scotland, July 20–25.
Spitler, J., Cullin, J., Bernier, M., Kummert, M., Cui, P., Liu, X., Lee, E., Fisher, D., 2009. Preliminary intermodel comparison of ground heat exchanger simulation models. In: Effstock 2009. Stockholm, Sweden. June 14–17.
Spitler, J.D., Gehlin, S.E.A., 2015. Thermal response testing for ground source heat pump systems – an historical review. Renewable and Sustainable Energy Reviews 50, 1125–1137.
Spitler, J.D., 2000. GLHEPRO – a design tool for commercial building ground loop heat exchangers. In: Fourth International Heat Pumps in Cold Climates Conference, Aylmer, Québec, August 17–18, 2000.
Thomson, W., 1880. Heat. Being the Article Contributed to the Encyclopaedia Britannica. Adam and Charles Black, Edinburgh.
Thomson, W., 1884. Compendium of the Fourier Mathematics for the Conduction of Heat in Solids, and the Mathematically Allied Physical Subjects of Diffusion of Fluids, and Transmission of Electric Signals Through Submarine Cables. In: Article 72. Mathematical and Physical Papers, vol. II. University Press, Cambridge, pp. 41–60.
Underwood, C.P., Spitler, J.D., 2007. Analysis of vertical ground loop heat exchangers applied to buildings in the UK. Building Services Engineering Research and Technology 28, 133–159.
Whitehead, S., August 19, 1927. Determining temperature distribution – a contribution to the evaluation of the flow of heat in isotropic media. The Electrician 225–226.

Xu, X., 2007. Simulation and Optimal Control of Hybrid Ground Source Heat Pump Systems (Ph.D. doctoral thesis). Oklahoma State University.

Yavuzturk, C., Spitler, J.D., 1999. A short time step response factor model for vertical ground loop heat exchangers. ASHRAE Transactions 105 (2), 475–485.

Yavuzturk, C., Spitler, J.D., Rees, S.J., 1999. A transient two-dimensional finite volume model for the simulation of vertical U-tube ground heat exchangers. ASHRAE Transactions 105 (2), 465–474.

Zarrella, A., Scarpa, M., De Carli, M., 2011. Short time step analysis of vertical ground-coupled heat exchangers: the approach of CaRM. Renewable Energy 36 (9), 2357–2367.

Zeng, H.Y., Diao, N.R., Fang, Z.H., 2002. A finite line-source model for boreholes in geothermal heat exchangers. Heat Transfer – Asian Research 31 (7), 558–567.

Zhang, L., Zhang, Q., Huang, G., Du, Y., 2014. A p(t)-linear average method to estimate the thermal parameters of the borehole heat exchangers for in situ thermal response test. Applied Energy 131 (0), 211–221.

Calculation of borehole thermal resistance

3

S. Javed[1], J.D. Spitler[2]
[1]Building Services Engineering, Chalmers University of Technology, Gothenburg, Sweden;
[2]Oklahoma State University, Stillwater, OK, United States

Abstract

The borehole thermal resistance—that is, the thermal resistance between the fluid in the U-tube and the borehole wall—is a key performance characteristic of a closed-loop borehole ground heat exchanger (GHE). Lower borehole thermal resistance leads to better system performance. Since the original identification of borehole thermal resistance by Mogensen (1983), there have been numerous methods proposed for calculating this. In this chapter, we present methods for calculating local borehole thermal resistance and effective borehole thermal resistance which account for short-circuiting between the upward- and downward-flowing legs of the GHE. We also compare a wide range of simple, relatively easy-to-calculate methods to a detailed reference method and recommend suitable simplified methods.

3.1 Introduction

Ground heat exchangers (GHEs) reject heat to the ground and extract heat from the ground. Because the ground is an infinite medium with infinite thermal capacity, heat transfer to and from the ground is necessarily transient. This is particularly the case when one is considering long timescales and consequently large distances from and between vertical borehole heat exchangers (as in the design methods discussed in chapter "Vertical borehole ground heat exchanger design methods"). Nevertheless, when considering shorter length and time scales, it is often convenient to analyze the heat transfer process using thermal resistances or quasi-resistance values. This is particularly the case for heat transfer between the heat carrier fluid in the pipes of the U-tube and the borehole wall, which often approaches a quasi-steady condition. For this heat transfer, a local borehole thermal resistance can be defined:

$$R_b = \frac{T_{f,l} - T_b}{q_b} \quad [3.1]$$

where $T_{f,l}$ is the local mean[1] fluid temperature (°C), T_b is the borehole wall temperature (°C), q_b is the heat transfer rate (from the borehole to the ground) per unit length (W/m).

[1] When it comes to analyzing a whole borehole, there is more than one possible formulation of the mean fluid temperature. But here it may just be taken as the average fluid temperature of the two legs of the U-tube at a specific depth in the borehole.

Advances in Ground-Source Heat Pump Systems. http://dx.doi.org/10.1016/B978-0-08-100311-4.00003-0
Copyright © 2016 S. Javed and J.D. Spitler. Published by Elsevier Ltd. All rights reserved.

The borehole thermal resistance is important not only for analysis purposes but also as a key performance characteristic of the GHE: the lower the thermal resistance, the better the performance. In the past 30 years, many innovations have been proposed to reduce borehole thermal resistance (eg, thermally enhanced grout, thermally enhanced HDPE pipe, and numerous configurations besides the single U-tube). To date, only improvements that take little additional time to install, such as thermally enhanced grout, have received widespread acceptance. Though more sophisticated configurations can reduce the borehole thermal resistance, if they take too long to install, it is often more economically feasible to simply drill deeper or drill more boreholes.

The concept of borehole thermal resistance was introduced by Mogensen (1983) along with a method for analyzing a thermal response test (Spitler and Gehlin, 2015) to estimate the borehole thermal resistance. This brings us to several important distinctions, all of which are related to each other, but which are prone to misunderstanding. Eq. [3.1] gives the definition of local borehole resistance. Mogensen (1983) suggested determining the borehole thermal resistance for an already-installed borehole using a thermal response test. When using the simplified line source solution, the effective thermal conductivity and effective borehole thermal resistance can be determined[2] using

$$T_f = T_0 + \frac{q_b}{4\pi\lambda}\left[\ln\left(\frac{4\alpha t}{r_b^2}\right) - \gamma\right] + q_b R_b^* \qquad [3.2]$$

where T_f = simple mean fluid temperature, the average of the inlet and outlet temperatures (°C), T_0 = undisturbed ground temperature (°C), λ = effective ground thermal conductivity (W/m K), and R_b^* = effective borehole thermal resistance (m K/W).

The key concept is that the effective borehole thermal resistance is defined based on the simple mean fluid temperature rather than an actual averaged-over-the-depth fluid temperature. Hellström (1991)[3] derived relationships between the effective borehole thermal resistance $\left(R_b^*\right)$ and the local borehole thermal resistance (R_b) for two idealized boundary conditions—uniform borehole wall temperature (UBW) and uniform heat flux (UHF) over the borehole—neither of which applies perfectly. A mean value between the two expressions can be taken as a best approximation. Both expressions are formulated to give the ratio of the effective borehole thermal resistance $\left(R_b^*\right)$ to the local or 2-D borehole thermal resistance (R_b).

Then, the effective borehole thermal resistance can be defined for the whole borehole as

$$R_b^* = \frac{T_f - T_b}{q_b} \qquad [3.3]$$

[2] See Spitler and Gehlin (2015) for a complete explanation.

[3] See Section 8.6 of Hellström (1991). See also Claesson and Hellström (2011).

What causes the difference between the local and effective borehole thermal resistances? All vertical GHEs are affected by short-circuiting—that is, heat transfer between the upward-flowing and downward-flowing legs of the heat exchanger. Though this is often negligible for typical boreholes, the short-circuiting effect increases as either the depth or conductance between the legs increases and/or as the mass flow rate in the heat exchanger decreases. In other words, there is a direct relationship between the short-circuiting heat transfer and the number of transfer units (NTU) as there would be for any heat exchanger. (With regards to short-circuiting heat transfer, we prefer a very low effectiveness.) Another way of describing this is to say that the short-circuiting heat transfer increases with increasing residence time. The short-circuiting heat transfer causes the averaged-over-the-depth mean fluid temperature to deviate from the simple mean fluid temperature.

The fact that the averaged-over-the-depth mean fluid temperature deviates from the simple mean fluid temperature led Marcotte and Pasquier (2008) to develop an approximate expression for the mean fluid temperature (the p-linear average) which they proposed to use with thermal response tests in order to estimate the borehole thermal resistance. Indeed, to the degree that the p-linear average temperature is an accurate measure of the actual mean fluid temperature, using it should allow estimation of R_b rather than R_b^*. Marcotte and Pasquier also claimed that using the simple mean fluid temperature resulted in an overprediction of R_b, which when used in a design calculation[4] would result in overestimation of the required GHE size. However, it is important that the temperature profile model be consistent between the method used to analyze the thermal response test and the method used to perform[5] the design, as noted by Beier (2011). Furthermore, as current design methods use the simple mean fluid temperature, the effective borehole thermal resistance determined with the simple mean fluid temperature is consistent with the existing design methods. Beier (2011) also developed an improved temperature profile model, then characterized the error associated with using the simple mean fluid temperature or the p-linear average fluid temperature as a function of several nondimensional parameters. Other authors have followed with proposals[6] for

[4] Another consideration is the degree to which borehole thermal resistance in a single borehole, where the U-tube placement is often uncontrolled or poorly controlled, can be taken to be representative of other boreholes to be installed in the borehole field.

[5] As of this writing, most commercially available design tools use the simple mean fluid temperature, so inputting the effective borehole thermal resistance would be appropriate. However, this may not always be clear from the program documentation, and in fact the difference between the effective borehole thermal resistance and the local borehole thermal resistance is no more than a few percent for typical North American boreholes, of the order of 80-m deep, used with ground-source heat pump (GSHP) systems.

[6] The problem of whether to use an effective borehole thermal resistance coupled with a simple mean fluid temperature or a local borehole thermal resistance coupled with an improved estimate of the actual mean fluid temperature is roughly analogous to whether, for conventional heat exchangers, to use the log mean temperature difference (LMTD) method or the effectiveness-NTU method. Both methods will give the same result; in some situations one method or the other is more convenient. The quantity $1 - R_b/R_b^*$ could be considered a type of effectiveness for short-circuiting that would be analogous to effectiveness for a standard heat exchanger.

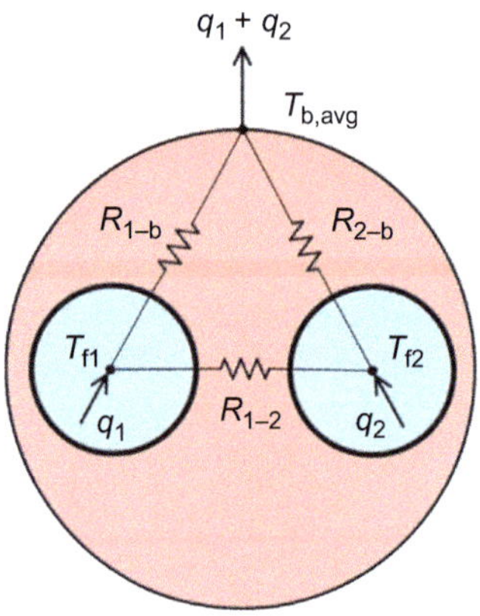

Figure 3.1 Δ resistance network for a borehole with a single U-tube.

calculating mean fluid temperature or measuring it (Acuña, 2013) with distributed thermal response tests. It is regrettable that there has been significant confusion over the different approaches for accounting for short-circuiting and we hope this chapter can help reduce the confusion.

If an averaged-over-the-depth fluid temperature is calculated, the local borehole resistance applied for the entire borehole can be defined as

$$R_b = \frac{\overline{T_f} - T_b}{q_b} \quad [3.4]$$

where: $\overline{T_f}$ is the actual averaged-over-the-depth mean fluid temperature (°C).

To consider short-circuiting heat transfer, a resistance network like that shown in Fig. 3.1 can be formulated. Other networks can be formulated (eg, see Fig. 3.2) and any such network is an approximation to reality. The network shown in Fig. 3.1 neglects any heat transfer between the tubes through the surrounding ground.[7] Terminology for internal thermal resistances in boreholes has not been consistent and is complicated by the fact that different resistance networks have different internal resistance elements. Terms such as "shunt resistance" have been used inconsistently in the past, and we have therefore chosen to refer to the total thermal resistance between the two tubes as the "total internal thermal resistance" denoted by the symbol R_a.

For the network shown in Fig. 3.1, the total internal thermal resistance is defined as

$$R_a = \frac{R_{1-2}(R_{1-b} + R_{2-b})}{R_{1-2} + R_{1-b} + R_{2-b}} \quad [3.5]$$

[7] Consider an example where this might be important: eg, when the surrounding ground has high thermal conductivity; the grout thermal conductivity is low; and the tubes are positioned near the borehole wall. In that case, a significant part of the short-circuiting heat transfer may travel through the surrounding rock.

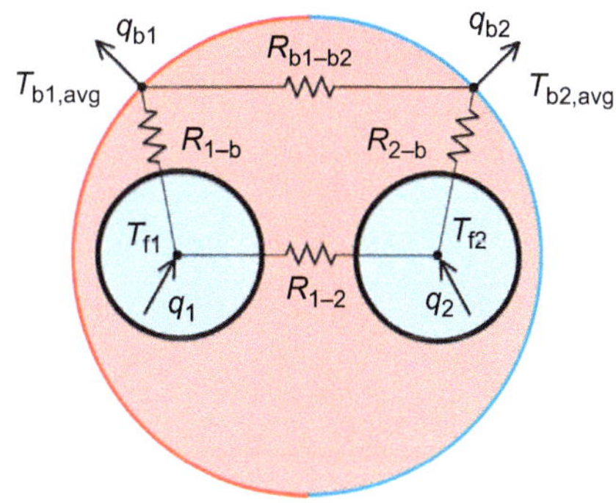

Figure 3.2 Four-resistance network for a borehole with a single U-tube.

However, the reader should be aware that there are other possible network formulations for which different expressions would be derived.

To summarize, we have the following definitions:

- Local borehole thermal resistance, R_b, is the resistance from the fluid in the U-tube to the borehole wall at a specific depth in the borehole as defined in Eq. [3.1]. If the averaged-over-the-depth mean fluid temperature is determined, it can also be used as shown in Eq. [3.3]. In this case, the short-circuiting is accounted for in the mean fluid temperature.
- Effective borehole thermal resistance, R_b^*, is the effective resistance between the fluid in the U-tube (characterized by the simple mean of the inlet and outlet temperatures) and the borehole wall, as shown in Eq. [3.2]. In this case, the short-circuiting is accounted for in the value of R_b^*.
- The total internal thermal resistance, R_a, is the resistance between the two legs of the U-tube, or the upward and downward-flowing legs. This is a local value, but is used to determine R_b^* from R_b. Other internal thermal resistances such as those shown in Fig. 3.1 are discussed in Section 3.2.
- The direct coupling resistance, R_{1-2}, shown in Fig. 3.1, is a consequence of the network representation and is not a directly measurable physical resistance. It is, however, useful in determining the relationship between local borehole thermal resistance and effective borehole thermal resistance.

This chapter will cover determination of values for these resistances.

3.2 Borehole resistance

The concept of borehole thermal resistance is often visualized as a series sum of resistances. The total borehole or fluid-to-ground resistance (R_b) is assumed to be comprised of two major parts: pipe resistance (R_p) and grout resistance (R_g). The pipe resistance includes convective resistance of the fluid in the pipe (R_{pic}) and the conductive resistance of the pipe (R_{pc}). The pipe resistance is typically calculated for a single pipe and then divided by total number of pipes:

$$R_b = R_g + \frac{R_p}{N} = R_g + \frac{R_{pc} + R_{pic}}{N} \qquad [3.6]$$

where N is 2, 4, or 6 for single, double, or triple U-tube, respectively. The variable N accounts for the fact that pipe resistances are in parallel.

Alternatively, the borehole thermal resistance may be presented in form of a thermal network. For a single U-tube, the steady-state thermal resistance of a borehole heat exchanger is often represented by a three-resistance or Δ thermal network of Fig. 3.1. This network is based on the following relations between heat flows q_1 and q_2 and fluid temperatures T_{f1} and T_{f2}:

$$q_1 = \frac{T_{f1} - T_{b,avg}}{R_{1-b}} + \frac{T_{f1} - T_{f2}}{R_{1-2}} \qquad [3.7]$$

$$q_2 = \frac{T_{f2} - T_{b,avg}}{R_{2-b}} + \frac{T_{f2} - T_{f1}}{R_{1-2}} \qquad [3.8]$$

where R_{1-b} and R_{2-b} are thermal resistances between pipe 1 or 2 and borehole wall, respectively, and R_{1-2} is the fluid-to-fluid thermal resistance between pipes 1 and 2. These resistances include the thermal resistances of the fluid and the pipe.

The total borehole resistance (R_b) can be determined by setting the fluid temperatures $T_{f1} = T_{f2}$ and solving for two parallel resistances R_{1-b} and R_{2-b}.

$$R_b = \frac{R_{1-b}\, R_{2-b}}{R_{1-b} + R_{2-b}} \qquad [3.9]$$

It is also possible to determine the total internal resistance (R_a) between the two pipes by solving for parallel resistances R_{1-2} and the sum of series resistances R_{1-b} and R_{2-b}.

$$R_a = \frac{R_{1-2}(R_{1-b} + R_{2-b})}{R_{1-2} + R_{1-b} + R_{2-b}} \qquad [3.10]$$

Eq. [3.10] can be rearranged to obtain the direct coupling resistance R_{1-2} between two pipes:

$$R_{1-2} = \frac{R_a(R_{1-b} + R_{2-b})}{R_{1-b} + R_{2-b} - R_a} \qquad [3.11]$$

If the two pipes are assumed to be in geometrical and thermal symmetry with two resistances $R_{1-b} = R_{2-b}$ then Eqs. [3.9]–[3.11] are reduced to the following.

$$R_b = \frac{R_{1-b}}{2} = \frac{R_{2-b}}{2} \qquad [3.12]$$

$$R_a = \frac{2R_{1-2}\, R_{1-b}}{2R_{1-b} + R_{1-2}} \qquad [3.13]$$

$$R_{1-2} = \frac{2R_a\, R_{1-b}}{2R_{1-b} - R_a} \quad [3.14]$$

As can be inferred from Fig. 3.1, the Δ resistance model calculates the borehole thermal resistances using a single value of average temperature at the borehole wall. This scenario neglects the fact that heat travels from one pipe to the other through the surrounding rock. This is most significant when the surrounding rock has higher thermal conductivity than the grout and the pipes are near the borehole wall. For this reason, some researchers have used a thermal network with four resistances and an average temperature for each half of borehole perimeter as shown in Fig. 3.2. This resistance network is based on the following relations between heat flows q_1 and q_2 and fluid temperatures T_{f1} and T_{f2}.

$$q_1 = \frac{T_{f1} - T_{b1}}{R_{1-b}} + \frac{T_{f1} - T_{f2}}{R_{1-2}} \quad [3.15]$$

$$q_2 = \frac{T_{f2} - T_{b2}}{R_{2-b}} + \frac{T_{f2} - T_{f1}}{R_{1-2}} \quad [3.16]$$

$$q_{b1} = \frac{T_{f1} - T_{b1}}{R_{1-b}} + \frac{T_{b2} - T_{b1}}{R_{b1-b2}} \quad [3.17]$$

$$q_{b2} = \frac{T_{f2} - T_{b2}}{R_{2-b}} + \frac{T_{b1} - T_{b2}}{R_{b1-b2}} \quad [3.18]$$

3.3 Fluid-to-pipe wall resistance

The fluid-to-pipe resistance, sometimes referred simply as pipe resistance, is the sum of the conductive heat transfer resistance of the pipe wall and the convective heat transfer resistance of the fluid flowing inside the pipe.

The conductive thermal resistance of the U-tube pipe dimensioned as shown in Fig. 3.3 can be found using the well-known expression for steady-state heat conduction through an annular region:

$$R_{pc} = \frac{1}{2\pi\lambda_p} \ln\left(\frac{r_{po}}{r_{pi}}\right) \quad [3.19]$$

The convective resistance of the fluid can be calculated using Eq. [3.20]:

$$R_{pic} = \frac{1}{2\pi\, r_{pi}\, h_{pi}} \quad [3.20]$$

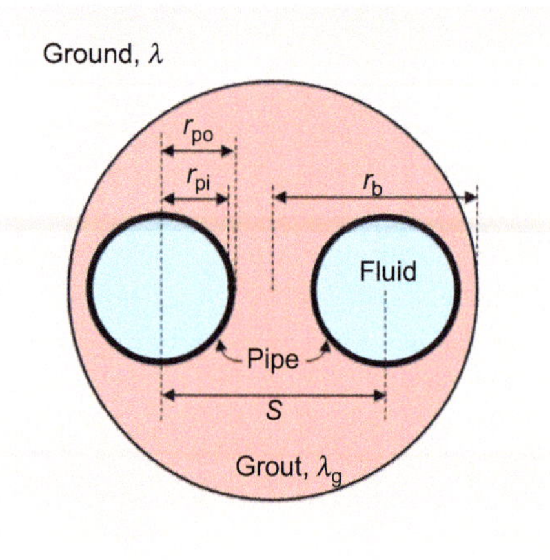

Figure 3.3 A borehole with a single U-tube showing the dimensions defining pipe size and spacing.

where h_{pi} (W/m^2 K) is the convective heat transfer coefficient usually determined from a convection correlation, which provides the Nusselt number:

$$h_{pi} = \frac{Nu_{pi}\,\lambda_f}{2r_{pi}} \qquad [3.21]$$

A range of correlations to calculate the Nusselt number are available depending on flow regime, aspect ratio, and thermal boundary conditions. Fuller descriptions are available in any introductory heat transfer textbook. We will only give a single correlation for fully developed turbulent flow, generally considered to be one of the most accurate correlations. The Gnielinski (1976) correlation covers turbulent flows down to transition-to-laminar conditions:

$$Nu_{pi} = \frac{(f/8)(\text{Re} - 1000)Pr}{1 + 12.7(f/8)^{1/2}\left(Pr^{2/3} - 1\right)} \qquad [3.22]$$

The friction factor for smooth pipe walls, as found in U-tubes, can be determined by the Petukhov (1970) equation for smooth pipe surface:

$$f = [0.79\ \ln(\text{Re}) - 1.64]^{-2} \qquad [3.23]$$

For fully developed laminar flow, asymptotic values of Nusselt numbers can be determined analytically for boundary conditions of UHF (4.364) and UBW temperature (3.657). Since neither condition is perfectly met in a GHE, it seems appropriate to use an approximate mean value of 4. For purposes of simulation where smooth numerical representations are desirable, it is also desirable to provide correlations that make a smooth transition between turbulent and laminar regimes.

3.4 Grout resistance

There exist a number of different methods and approaches for calculating pipe-to-borehole wall resistance, usually referred to as grout[8] resistance. These can be broadly classified into three categories: (1) equivalent radius methods, (2) empirical formulas, and (3) multipole method–based approaches. These are discussed in the following in detail.

3.4.1 *Equivalent radius methods*

The equivalent radius methods calculate the grout thermal resistance by assuming U-tube as an equivalent radius pipe as shown in Fig. 3.4. The grout resistance can then be calculated considering one-dimensional radial heat conduction from the pipe outer wall to the borehole wall:

$$R_g = \frac{1}{2\pi\lambda_g} \cdot \ln\left(\frac{r_b}{r_{eq}}\right) \tag{3.24}$$

As can be seen from Eq. [3.24], the grout resistance calculated from this equation depends on the particular value of the equivalent radius used for the calculation. However, there is no single definitive formulation of equivalent radius that can provide grout resistance values of accuracy comparable to those from more advanced methods. Nevertheless, several formulations have been presented that attempt to determine the

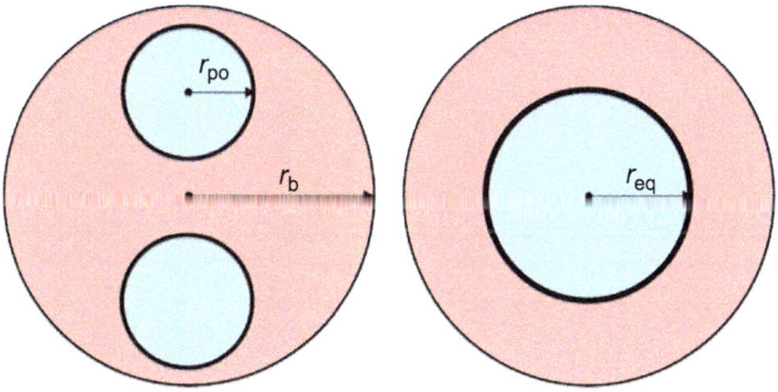

Figure 3.4 Actual and simplified borehole geometries.

[8] To be clear, the term "grout resistance" may be a slight misnomer; this resistance can be affected by the thermal conductivity of the ground surrounding the borehole.

equivalent radius of a U-tube. Shonder and Beck (2000) used the equivalent radius expression of Claesson and Dunand (1983) given in Eq. [3.25] to calculate the grout resistance.[9] The expression is quite simple and does not account for shank spacing between the U-tube pipes. Bose et al. (1985) addressed this issue and suggested an alternate equivalent radius expression that accounts for the effects of shank spacing. Their expression is given in Eq. [3.26]. Another expression considering shank spacing has been presented by Gu and O'Neal (1998). Their expression is given in Eq. [3.27].

$$r_{eq} = \sqrt{2}\, r_{po} \qquad [3.25]$$

$$r_{eq} = \sqrt{r_{po}\, s} \qquad [3.26]$$

$$r_{eq} = 0.414\, r_{po} + 0.5\, s \qquad [3.27]$$

3.4.2 Empirical formulas

The second broader approach to calculate the borehole thermal resistance is by using empirical relationships derived from curve fitting of experimental and/or numerical data. Due to obvious difficulties in obtaining sufficient experimental data to cover all possible contexts, most of the empirical formulas have been developed using numerical data generated from simulations of steady-state two-dimensional heat conduction in boreholes. The best-fit formulas typically have the following general form:

$$R_g = \frac{1}{2\pi\lambda_g} f(\theta_1, \theta_2, \theta_3, \sigma) \qquad [3.28]$$

where θ_1, θ_2, and σ are dimensionless parameters defined as

$$\theta_1 = \frac{s}{2r_b} \qquad [3.29]$$

$$\theta_2 = \frac{r_b}{r_{po}} \qquad [3.30]$$

$$\theta_3 = \frac{r_{po}}{s} = \frac{1}{2\theta_1\theta_2} \qquad [3.31]$$

$$\sigma = \frac{\lambda_g - \lambda_s}{\lambda_g + \lambda_s} \qquad [3.32]$$

[9] This expression is also sometimes erroneously attributed to Bose et al. (1985).

One of the most popular empirical methods was developed by Paul (1996); see also Remund (1999). The method calculates grout thermal resistance using the well-established concept of two-dimensional conduction shape factor as

$$R_g = \frac{1}{S_b \lambda_g} \tag{3.33}$$

where S_b is the borehole shape factor, which is determined from Eq. [3.34] using empirical coefficients B_0 and B_1.

$$S_b = B_0 \left(\frac{r_b}{r_{po}} \right)^{B_1} \tag{3.34}$$

Borehole shape factors for three configurations of Fig. 3.5 have been investigated by Paul (1996). Configuration A considers two legs of the U-tube in direct contact with each other in the center of the borehole. Configuration B represents U-tube legs spaced equally from one another and from the borehole wall. Configuration C has U-tube legs placed against opposite sides of the borehole. For these configurations, values of coefficients B_0 and B_1 have been back calculated from experimental data. These values are given in Table 3.1.

Bauer et al. (2011) also used the concept of shape factors to calculate borehole thermal resistance. However, unlike Paul (1996), who developed specific shape factors for borehole geometry, Bauer et al. (2011) modified shape factor of eccentric cylinders of equal length (Bergman et al., 2011; Holman, 1986) and introduced corrections based on numerical data. Their expression for grout thermal resistance modified to account for both legs of the U-tube is

$$R_g = \frac{1}{4\pi\lambda_g} \left[(1.601 - 0.888\, \theta_1) \cosh^{-1}\left(\frac{\theta_2}{2} + \frac{1}{2\theta_2} - \frac{\theta_1}{4\theta_3} \right) \right] \tag{3.35}$$

Sharqawy et al. (2009) developed the borehole thermal resistance expression presented in Eq. [3.36]. The authors suggest that the expression is highly accurate within $2.5 \leq \theta_2 \leq 7$ and $0.2 \leq \theta_1 \leq 0.8$ ranges of dimensionless geometrical parameters.

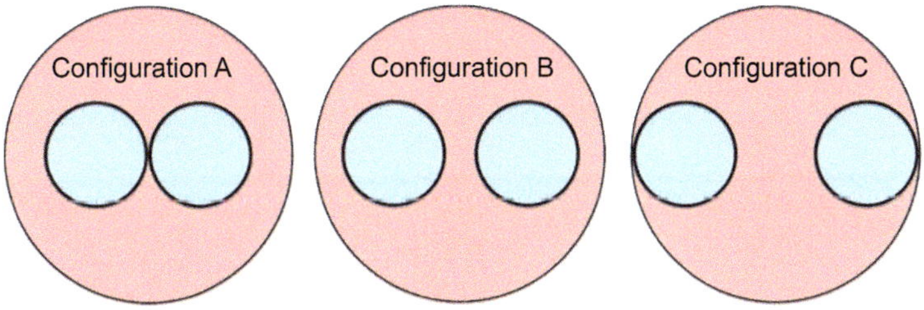

Figure 3.5 Borehole configurations studied by Paul (1996).

Table 3.1 Values of coefficients of Eq. [3.34]

Configuration	B_0	B_1
A	20.10	−0.9447
B	17.44	−0.6052
C	21.91	−0.3796

$$R_g = \frac{1}{2\pi\lambda_g}[-1.49\,\theta_1 + 0.656 \ln \theta_2 + 0.436] \quad [3.36]$$

Another best-fit correlation to determine grout resistance has been suggested by Liao et al. (2012). The expression is given by Eq. [3.37].

$$R_g = \frac{1}{2\pi\lambda_g}\left[-0.50125 \ln \theta_1 + 0.51248 \ln \theta_2 + 0.51057\sigma \ln\left(\frac{1}{1-\theta_1^4}\right) - 0.36925\right] \quad [3.37]$$

The major difference between the two above approaches is in the formulation of mathematical model to obtain the numerical data. Sharqawy et al. (2009) imposed a uniform temperature boundary condition at the borehole wall whereas Liao et al. (2012) had the same condition far out in the ground. In other words, the expression suggested by Sharqawy et al. (2009) does not consider the thermal conductivity of the surrounding ground when calculating the grout resistance. It has been shown that the accuracy of Eq. [3.36] decreases with increasing shank spacing (Lamarche et al., 2010; Al-Chalabi, 2013). The model is, however, reasonably accurate for calculating grout resistance of boreholes with pipes in close contact.

Another empirical expression for the grout resistance has been suggested by Abd Elatty et al. (2012). The expression is given as Eq. [3.38]. The authors suggest a validity range of $0.4 \leq \theta_1 \leq 0.7$ for their expression. However, the accuracy hasn't been verified independently.

$$R_g = \frac{1}{2\pi\lambda_g}\left[(-0.185\,\theta_1 + 0.0663)\ln\left(\frac{\lambda}{\lambda_g}\right) + (0.545 \ln \theta_2 - 0.335 \ln \theta_1 - 0.353)\right] \quad [3.38]$$

3.4.3 Multipole method–based approaches

The third main method to calculate borehole thermal resistance is through multipole method–based approaches. A common confusion regarding the multipole method is that it is generally thought of as the first-order multipole equation for two symmetrical pipes that is actually presented in Hellström (1991) but is typically attributed to Bennet et al. (1987). In reality, the multipole method is a complex algorithm which can be used to calculate thermal resistance for any number of arbitrarily placed pipes in a borehole. The algorithm has been implemented in Fortran and Mathcad codes and is used in borehole sizing programs including Earth Energy Designer (EED) (BLOCON, 2015) and GLHEPRO (Spitler, 2000). The multipole method has been compared with several numerical methods with remarkable accuracy (Al-Chalabi, 2013; Lamarche et al., 2010; Liao et al., 2012; Go et al., 2014; He, 2012). It is regarded as one of the most accurate methods to calculate borehole thermal resistance.

Here, we will first present a brief methodological framework of the multipole method, and subsequently present the multipole-based analytical expressions to calculate borehole resistance. The multipole method uses a combination of line heat sources and multipoles to solve the steady-state heat transfer problem in a composite region for any prescribed set of heat flows q_n from N number of pipes. The corresponding temperature field, given by Eq. [3.39], is a linear combination of solutions of N line sources (W_0) and J multipoles (W_j).

$$T\,(x,y) = T_{\text{b,avg}} + \Re\left[\sum_{n=1}^{N}\frac{q_n}{2\pi\lambda_{\text{g}}}\,W_0(z,z_n) + \sum_{n=1}^{N}\sum_{j=1}^{J}P_{n,j}\,W_j(z,z_n)\right] \qquad [3.39]$$

Here, $T_{\text{b,avg}}$ is the average temperature at the borehole wall, $\Re$ denotes the real part, and $P_{n,j}$ are the strength factors of the multipoles. Line source and multipole solutions used in Eq. [3.39] are given by Eqs. [3.40] and [3.41], respectively.

$$W_0(z,z_n) = \begin{cases} \ln\left(\dfrac{r_{\text{b}}}{z-z_n}\right) + \sigma\ln\left(\dfrac{r_{\text{b}}^2}{r_{\text{b}}^2 - \bar{z}\cdot z_n}\right), & |z| \le r_{\text{b}} \\[2ex] (1+\sigma)\ln\left(\dfrac{r_{\text{b}}}{z-z_n}\right) + \sigma\dfrac{1+\sigma}{1-\sigma}\ln\left(\dfrac{r_{\text{b}}}{z}\right), & |z| \ge r_{\text{b}} \end{cases} \qquad [3.40]$$

$$W_j(z,z_n) = \begin{cases} \left(\dfrac{r_{\text{pn}}}{z-z_n}\right)^j + \sigma\left(\dfrac{r_{\text{pn}}\,\bar{z}}{r_{\text{b}}^2 - \bar{z}\cdot z_n}\right)^j, & |z| \le r_{\text{b}} \\[2ex] (1+\sigma)\ln\left(\dfrac{r_{\text{pn}}}{z-z_n}\right)^j, & |z| \ge r_{\text{b}} \end{cases} \qquad [3.41]$$

$$\text{where:} \quad z = x + i\cdot y, \quad z_n = x_n + i\cdot y_n, \quad \bar{z} = x - i\cdot y. \qquad [3.42]$$

These solutions satisfy the two-dimensional heat equation in composite regions inside and outside the borehole as well as the continuous temperature and radial heat flow boundary conditions at the borehole wall. The first line of Eq. [3.40] shows the line source solution for the borehole region. Two heat sources are considered for pipe n: an original line source of strength q_n at $z = z_n$, and a second line source of strength $\sigma \cdot q_n$ at a mirror point $z = r_b^2/z_n$ outside the borehole. The second line of Eq. [3.40] shows the line source solution for the region surrounding the borehole. One line source of strength $(1-\sigma)\cdot q_n$ is considered at $z = z_n$ and a second line source of strength $\sigma \cdot q_n$ is considered at $z = 0$. The multipole solutions of Eq. [3.41], which are based on the jth derivative of the Eq. [3.40] with respect to complex variable z_n, follow a similar structure. A detailed description of these solutions can be found in Claesson and Hellström (2011), and Bennet et al. (1987).

The values of the strength of the complex-valued multipoles, $P_{n,j}$, are determined by the boundary conditions around the periphery of the pipes. The difference between the fluid temperature in the pipe and the grout temperature just the outside of the pipe is to be equal to the radial heat flux in the grout at the pipe times the thermal resistance over the pipe from fluid to grout. Let ψ_m denote the polar angle from the center of pipe m, then the boundary condition is to be satisfied for all polar angles: $-\pi < \psi_m \leq \pi$.

The temperature field $T(x,y)$, Eq. [3.39], may be expressed in the local polar coordinates from the center z_m of any pipe m. This involves quite a lot of mathematics using complex variables. The line sources and multipoles from all other pipes are expanded in suitable power series of the local polar coordinates of pipe m. The boundary condition of pipe m gives an exact Fourier expansion for the required fluid temperature T_{fm}. There is a constant term and Fourier coefficients for $\cos(j\cdot\psi_m)$ and $\sin(j\cdot\psi_m)$. The constant term gives an expression for the fluid temperature T_{fm} as a function of the prescribed heat fluxes q_n and multipoles $P_{n,j}$. For each multipole order $j = 1, 2, \ldots,$ the boundary condition gives a relation between heat fluxes q_n and multipoles $P_{n,j}$.

The Fourier expansion must be truncated at a chosen order $j = J$. This means the Fourier coefficients in the boundary condition at the pipes up $\cos(J\cdot\psi_m)$ and $\sin(J\cdot\psi_m)$ are zero. This gives an equation system with $N\cdot J$ unknown complex numbers that determines the multipoles $P_{n,j}$ as functions of the prescribed heat fluxes q_n. The equation system is readily solved by a very rapid and robust iterative method.

We will now present the explicit multipole expressions for calculating borehole thermal resistance. For a single U-tube (ie, two pipes), Eq. [3.39] can be written in the following form:

$$T_{f1} = T_{b,avg} + q_1 \widehat{R}^J_{1-1} + q_2 \widehat{R}^J_{1-2} \quad [3.43]$$

$$T_{f2} = T_{b,avg} + q_2 \widehat{R}^J_{2-2} + q_1 \widehat{R}^J_{2-1} \quad [3.44]$$

For zeroth-order multipoles, the corresponding linear coefficients $\widehat{R}^0_{m-n}$ of heat fluxes are

$$\widehat{R}^0_{1-1} = \widehat{R}^0_{2-2} = \frac{1}{2\pi\lambda_g}\left[\beta + \ln\left(\frac{\theta_2}{\left(1-\theta_1^2\right)^{\sigma}}\right)\right] \quad [3.45]$$

$$\widehat{R}^{0}_{1-2} = \widehat{R}^{0}_{2-1} = \frac{1}{2\pi\lambda_g}\left[\ln\left(\frac{1}{2\theta_1\left(1+\theta_1^2\right)^{\sigma}}\right)\right] \quad [3.46]$$

$$\text{where:} \quad \beta = 2\pi\lambda_g R_p \quad [3.47]$$

For each higher order multipole, an additional correction term is added to the zeroth-order coefficients. For first-order multipoles (ie, $J = 1$), the corresponding linear coefficients $\widehat{R}^{1}_{m-n}$ become

$$\widehat{R}^{1}_{1-1} = \widehat{R}^{1}_{2-2} = \widehat{R}^{0}_{1-1} - \frac{1}{4\pi\lambda_g}\left[B_1^+ + B_1^-\right] \quad [3.48]$$

$$\widehat{R}^{1}_{1-2} = \widehat{R}^{1}_{2-1} = \widehat{R}^{0}_{2-1} - \frac{1}{4\pi\lambda_g}\left[B_1^+ - B_1^-\right] \quad [3.49]$$

$$\text{where:} \quad B_1^+ = \left(\frac{1-\beta}{1+\beta}\right)\left[\frac{\theta_3^2\left(1-4\sigma\theta_1^4-\theta_1^4\right)^2}{\left(1-\theta_1^4\right)^2+\theta_3^2\left(\frac{1-\beta}{1+\beta}\right)\left(\left(1-\theta_1^4\right)^2+16\sigma\theta_1^4\right)}\right] \quad [3.50]$$

$$B_1^- = \left(\frac{1-\beta}{1+\beta}\right)\left[\frac{\theta_3^2\left(1+4\sigma\theta_1^2-\theta_1^4\right)^2}{\left(1-\theta_1^4\right)^2+\theta_3^2\left(\frac{1-\beta}{1+\beta}\right)\left(\left(1-\theta_1^4\right)^2+8\sigma\theta_1^2\left(1+\theta_1^4\right)\right)}\right] \quad [3.51]$$

Eqs. [3.43] and [3.44] can now be rearranged to the following form to represent the Δ network for two pipes in a borehole as shown in Fig. 3.1.

$$q_1 = \frac{\left(\widehat{R}^{J}_{2-2} - \widehat{R}^{J}_{1-2}\right)\left(T_{f1} - T_{b,avg}\right)}{\widehat{R}^{J}_{1-1}\widehat{R}^{J}_{2-2} - \widehat{R}^{J}_{1-2}\widehat{R}^{J}_{2-1}} + \frac{\widehat{R}^{J}_{1-2}\left(T_{f1} - T_{f2}\right)}{\widehat{R}^{J}_{1-1}\widehat{R}^{J}_{2-2} - \widehat{R}^{J}_{1-2}\widehat{R}^{J}_{2-1}} \quad [3.52]$$

$$q_2 = \frac{\left(\widehat{R}^{J}_{1-1} - \widehat{R}^{J}_{1-2}\right)\left(T_{f2} - T_{b,avg}\right)}{\widehat{R}^{J}_{1-1}\widehat{R}^{J}_{2-2} - \widehat{R}^{J}_{1-2}\widehat{R}^{J}_{2-1}} + \frac{\widehat{R}^{J}_{1-2}\left(T_{f2} - T_{f1}\right)}{\widehat{R}^{J}_{1-1}\widehat{R}^{J}_{2-2} - \widehat{R}^{J}_{1-2}\widehat{R}^{J}_{2-1}} \quad [3.53]$$

The thermal resistances of the Δ network can now be expressed using linear coefficients from Eqs. [3.52]–[3.53] and Eqs. [3.7]–[3.8].

$$R_{1-b} = \frac{\widehat{R}^{J}_{1-1}\widehat{R}^{J}_{2-2} - \widehat{R}^{J}_{1-2}\widehat{R}^{J}_{2-1}}{\widehat{R}^{J}_{2-2} - \widehat{R}^{J}_{1-2}} \quad [3.54]$$

$$R_{2-\mathrm{b}} = \frac{\widehat{R}^J_{1-1}\widehat{R}^J_{2-2} - \widehat{R}^J_{1-2}\widehat{R}^J_{2-1}}{\widehat{R}^J_{1-1} - \widehat{R}^J_{1-2}} \qquad [3.55]$$

$$R_{1-2} = \frac{\widehat{R}^J_{1-1}\widehat{R}^J_{2-2} - \widehat{R}^J_{1-2}\widehat{R}^J_{2-1}}{\widehat{R}^J_{1-2}} \qquad [3.56]$$

When the two pipes are symmetric and $R_{1-\mathrm{b}} = R_{2-\mathrm{b}}$, then the borehole resistance from Eq. [3.12] takes the following form:

$$R_{1-\mathrm{b}} = R_{2-\mathrm{b}} = \widehat{R}^J_{1-1} + \widehat{R}^J_{1-2} \qquad [3.57]$$

$$R_{\mathrm{b}} = \frac{R_{1-\mathrm{b}}}{2} = \frac{R_{2-\mathrm{b}}}{2} = \frac{\left(\widehat{R}^J_{1-1} + \widehat{R}^J_{1-2}\right)}{2} \qquad [3.58]$$

The borehole resistance for zeroth-order multipoles can now be calculated from Eqs. [3.45]−[3.46] and Eqs. [3.57]−[3.58]. The zeroth-order multipole solution doesn't have any multipole correction. In the literature, the zeroth-order multipole solution is often referred to as the line source solution or Hellström's solution after the author outlining the solution. Using dimensionless parameters of Eqs. [3.29]−[3.32], the zeroth-order multipole solution can be written as

$$R_{\mathrm{b}} = \frac{1}{4\pi\lambda_{\mathrm{g}}}\left[\beta + \ln\left(\frac{\theta_2}{2\theta_1\left(1-\theta_1^4\right)^{\sigma}}\right)\right] \qquad [3.59]$$

Similarly, borehole resistance for first-order multipoles can be calculated from Eqs. [3.48]−[3.49] and Eqs. [3.57]−[3.58]. This solution is often referred to as "the" multipole method. However, it must be reemphasized that it is only a first-order multipole solution for a specific case: the single U-tube with the legs placed symmetrically in the borehole. It can be seen that the first part of Eq. [3.60] is the zeroth-order or line source solution. The second part is the correction originating from the first-order multipoles. Each higher order multipole will add a new and more complicated correction term.

$$R_{\mathrm{b}} = \frac{1}{4\pi\lambda_{\mathrm{g}}}\left[\beta + \ln\left(\frac{\theta_2}{2\theta_1\left(1-\theta_1^4\right)^{\sigma}}\right) - \frac{\theta_3^2\left(1-\frac{4\sigma\,\theta_1^4}{1-\theta_1^4}\right)^2}{\frac{1+\beta}{1-\beta} + \theta_3^2\left(1+\frac{16\sigma\,\theta_1^4}{\left(1-\theta_1^4\right)^2}\right)}\right] \qquad [3.60]$$

Eqs. [3.59] and [3.60] are often used in literature to calculate borehole resistance values. These equations are explicit expressions for zeroth- and first-order multipole expansions for two pipes in geometrical and thermal symmetry. Borehole

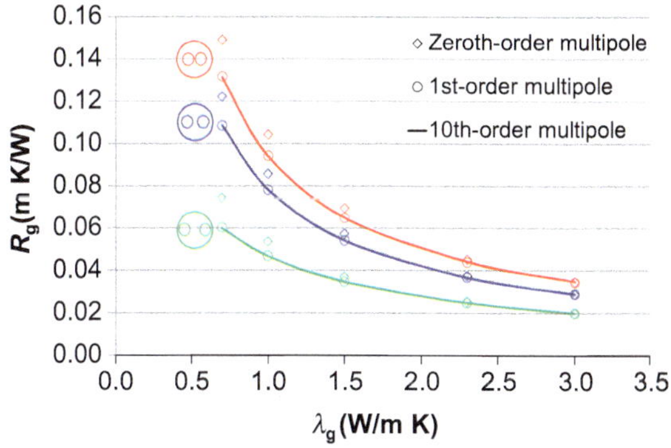

Figure 3.6 Grout thermal resistance for $r_b = 0.055$ m, $r_{po} = 0.020$ m, $\lambda = 3.0$ W/m K, and $R_p = 0.050$ m K/W. Different pipe shank spacing (A, B, or C) is indicated in different colors.

resistance for unsymmetrical cases can be obtained from Eqs. [3.45]–[3.46] and Eqs. [3.54]–[3.56] for zeroth-order multipoles and Eqs. [3.48]–[3.49] and Eqs. [3.54]–[3.56] for first-order multipoles. For all other cases, including those with more than two pipes, borehole resistance has to be calculated from the multipole program as explicit formulas for such cases have not been derived.

An important fact to be highlighted is that the multipole equations give the borehole resistance and not the grout resistance. That is, the resistance includes the resistance of the pipes. Some authors have simplified the first-order multipole expression by setting β to zero. However, for subtle physical reasons[10] this introduces some error. A related misunderstanding is the effect of the soil or ground surrounding the borehole. It, too, can have an effect on the overall grout resistance, particularly for cases with the tubes near the borehole wall and high ground thermal conductivity.

Fig. 3.6 presents a comparison of grout resistances obtained from zeroth-order, first-order, and tenth-order multipole expansions for a single U-tube with pipes in geometrical and thermal symmetry. Three alternate pipe positions of Fig. 3.5 have been investigated. It can be seen that the zeroth-order multipole expansions overpredict the thermal resistances for lower grout conductivity cases. The difference between resistance values calculated using zeroth- and tenth-order multipole expansion is between 10% and 25% for grout conductivities ≤ 1 but decreases rapidly for higher values of grout conductivity. On the other hand, the difference in resistance values determined from first- and tenth-order multipole expansions is very small and rarely exceeds 1%. Another interesting observation is that the error in thermal resistances calculated from zeroth- and first-order multipole expansions is lowest for Configuration B, with average pipe spacing. Between Configuration A and Configuration C, where the pipes are, respectively, close to each other and close to the borehole wall, the thermal resistances obtained from first-order multipole expansion have errors of

[10] The surface of the fluid at the inside of the inner pipe wall convective resistance is assumed to be isothermal. If R_p and β are set to zero, then the outer pipe wall is isothermal. Particularly for cases where the pipes are close together, this introduces some error.

similar magnitude for any specific grout conductivity value. Finally, we conclude from this comparison that the first-order multipole method has sufficient accuracy (~1%) for all practical applications involving computation of thermal resistance for typical grouted boreholes with geometrical symmetry.

3.4.4 Comparison of different methods to calculate grout resistance

In this section, we compare previously published methods to the tenth-order multipole method. Figs. (3.7–3.9) show comparisons for configurations A, B, and C, respectively, of Fig. 3.5. The methods vary in their accuracy with some methods being more accurate in one range of grout conductivities than others and being more accurate with one configuration than another. While it might be interesting to reflect on which assumptions in each method lead to better or worse accuracy, it may suffice to say that none of the methods provide outstanding accuracy over the entire range. Configuration C, where the pipes are touching the borehole wall, is the most challenging case. The Bauer et al. (2011) method cannot calculate this case; the Bose et al. (1985) and Gu and O'Neal (1998) methods give reasonable accuracy (maximum error 15%) for the A and B configurations, but have large errors (as high as 35–45%) for the C configuration. Taking all combinations of grout conductivity and configuration into account, only the Liao et al. (2012) and Abd Elatty et al. (2012) methods never exceed errors of 20%.

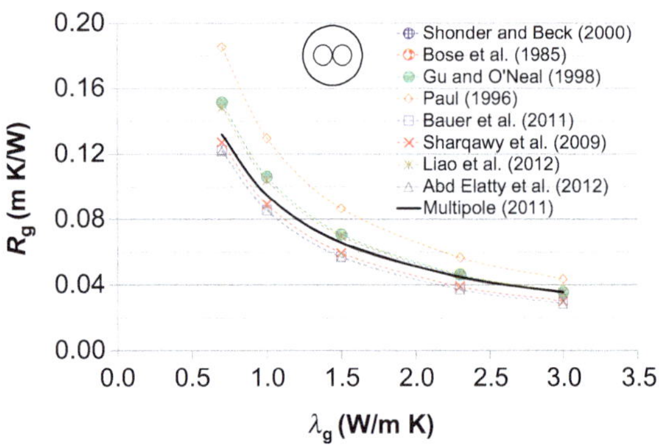

Figure 3.7 Grout thermal resistance for $r_b = 0.055$ m, $r_{po} = 0.020$ m, $\lambda = 3.0$ W/m K and $R_p = 0.050$ m K/W and pipe spacing A (closely spaced).

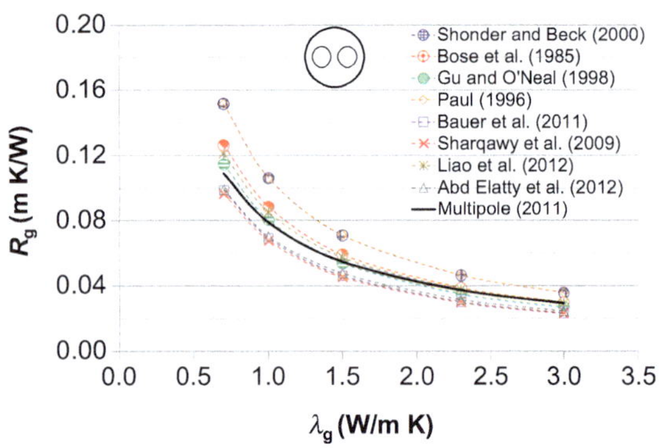

Figure 3.8 Grout thermal resistance for $r_b = 0.055$ m, $r_{po} = 0.020$ m, $\lambda = 3.0$ W/m K and $R_p = 0.050$ m K/W and pipe spacing B (moderately spaced).

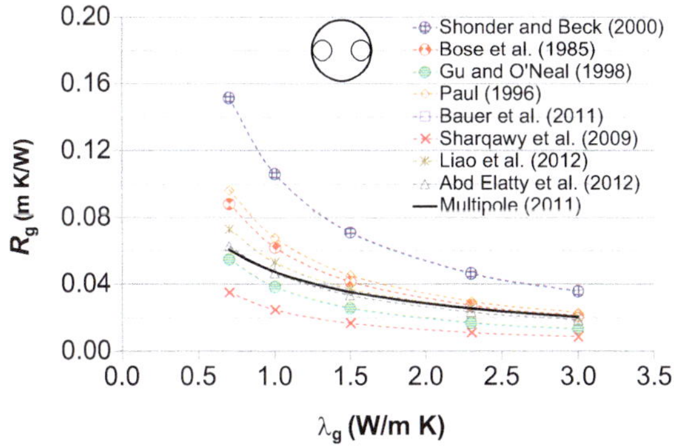

Figure 3.9 Grout thermal resistance for $r_b = 0.055$ m, $r_{po} = 0.020$ m, $\lambda = 3.0$ W/m K and $R_p = 0.050$ m K/W and pipe spacing C (widely spaced).

The first-order multipole method expression given in Eq. [3.60] is significantly more accurate than any of the other previously published methods. Although more complex than the other methods to implement, it is still relatively easy to implement and fast to compute. On balance, given its superior accuracy, it seems advisable to use it instead of other published methods.

3.5 Internal fluid-to-fluid resistance

As will be discussed in Section 3.6, heat transfer between the legs of a U-tube ("short-circuiting") can be important for deeper boreholes. This heat transfer is controlled by the fluid-to-fluid or total internal thermal resistance defined in Eq. [3.10]. The multipole algorithm can directly compute R_a. However, explicit zeroth- and first-order expressions for these resistances have also been developed for cases with single U-tubes with the two legs in geometrical and thermal symmetry with $R_{1-b} = R_{2-b}$. For these cases, Eq. [3.10] takes the form of Eq. [3.13].

Eq. [3.61] gives the zeroth-order multipole expression derived by Hellström (1991) for total internal fluid-to-fluid borehole resistance:

$$R_a = \frac{1}{\pi\lambda_g}\left[\beta + \ln\left(\frac{\left(1+\theta_1^2\right)^{\sigma}}{\theta_3\left(1-\theta_1^2\right)^{\sigma}}\right)\right] \qquad [3.61]$$

Eq. [3.62] gives the first-order multipole expression (Hellström, 1991) for total internal borehole resistance:

$$R_a = \frac{1}{\pi\lambda_g}\left[\beta + \ln\left(\frac{\left(1+\theta_1^2\right)^{\sigma}}{\theta_3\left(1-\theta_1^2\right)^{\sigma}}\right) - \frac{\theta_3^2\left(1-\theta_1^4+4\sigma\,\theta_1^2\right)^2}{\left(\frac{1+\beta}{1-\beta}\right)\left(1-\theta_1^4\right)^2 - \theta_3^2\left(1-\theta_1^4\right)^2 + 8\sigma\,\theta_1^2\,\theta_3^2\left(1+\theta_1^4\right)}\right] \qquad [3.62]$$

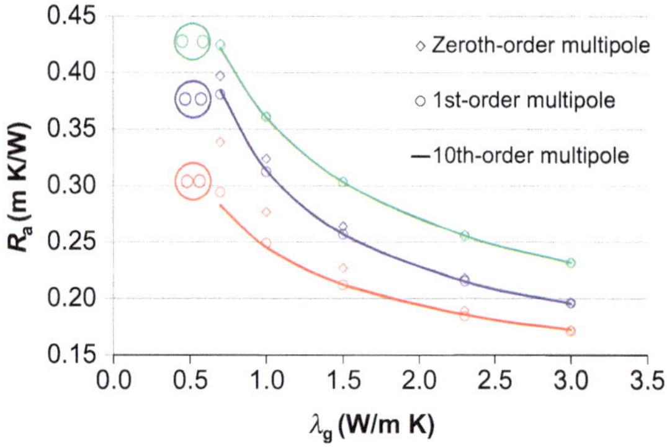

Figure 3.10 Total internal borehole thermal resistance for $r_b = 0.055$ m, $r_{po} = 0.020$ m, $\lambda = 3.0$ W/m K, and $R_p = 0.050$ m K/W. Different pipe shank spacing (A, B, or C) is indicated in different colors.

Fig. 3.10 presents a comparison of total internal resistances obtained from zeroth-order, first-order, and tenth-order multipole expansions for a single U-tube in geometrical and thermal symmetry. The comparisons are made for the three alternate U-tube positions of Fig. 3.5. As expected, the results show that the total internal resistance is lowest for Configuration A, ie, when the U-tube pipes are in direct contact in the center of the borehole, and highest for Configuration C, when the pipes are placed at opposite ends of the borehole outer boundary. Another interesting observation from Fig. 3.10 is that the differences between multipole solutions of different orders decrease as the pipes move from Configuration A to Configuration C. For Configuration A, where the U-tubes pipes are in direct contact with each other in center of the borehole, the difference between zeroth-order and tenth-order multipole resistances is over 13% for grout conductivities less than 1 W/m K. The difference between first-order and tenth-order multipoles for the same configuration and grout conductivity range is less than 5%, however. Outside that range, the difference between first-order and tenth-order multipole resistances is no more than 1%. Therefore, the first-order multipole expression is recommended as a pretty good approximation to the tenth-order multipole calculation for single U-tubes.

An alternate way of finding the total internal resistance R_a is from the four-resistance network of Fig. 3.2. For the U-tube in geometrical symmetry, the total internal resistance of this network can be written in the following form:

$$R_a = \frac{R_{1-2}(2R_{1-b} + R_{b1-b2})}{2R_{1-b} + R_{1-2} + R_{b1-b2}} \quad [3.63]$$

Eq. [3.63] for the R_a of four-resistance network is reduced to Eq. [3.13] for the R_a of Δ network when the resistance R_{b1-b2} is zero and $T_{b1} = T_{b2}$. Unlike for Δ networks, no exact explicit formulas to calculate R_a of the four-resistance network are currently available. However, the best-fit correlations of Liao et al. (2012) can be used to calculate the values of resistances R_{1-2} and R_{b1-b2}. These values can then be used in Eq. [3.63] to determine R_a.

$$R_{1-2} = \frac{1}{\lambda_g}\left(0.0305\,\theta_2 + 0.002\,\sigma^2 + 0.0065\,\sigma - 0.0265\right)e^{(0.107\,\sigma + 4.742)\theta_1} \quad [3.64]$$

$$R_{b1-b2} = \frac{1}{\lambda_g}\left(0.2592\,\sigma + 0.9414 - \frac{0.252\,\sigma + 0.7942}{\ln\theta_2}\right)e^{\left(0.185\,\sigma - 0.934 + \frac{1.755 - 0.1653\,\sigma}{\ln\theta_2}\right)\frac{1}{\theta_1}} \quad [3.65]$$

According to Liao et al. (2012) the above correlations are relatively accurate for $-0.214 \le \theta_1 \le 0.85$, $2.5 \le \theta_2 \le 7.0$ and $-0.2 \le \sigma \le 0.6$. When calculating R_a form Eqs. [3.63]–[3.65] for the cases shown in Fig. 3.10, errors as high as over 20% are observed even within the suggested validity range. The errors outside the validity range are even higher.

An interesting observation when comparing the Δ and four-resistance network of Figs. 3.1 and 3.2, respectively, is the behavior of the direct coupling resistance R_{1-2}. This resistance is connected to the total internal resistance R_a through Eq. [3.13] for the Δ network and Eq. [3.63] for the four-resistance network. Due to the mathematical formulation of the Δ network, the resistance R_{1-2} can sometimes have a negative value. The total internal resistance R_a is always positive, though. However, the value of resistance R_{1-2} in a four-resistance network is always positive. There has been some confusion regarding the usage of different R_{1-2} expressions. The R_{1-2} expression determined for a particular network has sometimes been mistakenly used to calculate the direct coupling resistance of some other network. It should be highlighted that the R_{1-2} expressions are network-specific and cannot be used interchangeably. The total internal resistance, though, is a physical quantity that can be readily compared between different thermal networks.

Some researchers (Bauer et al., 2011; Lamarche et al., 2010; Beier, 2011; Kavanaugh, 1985) have calculated the total internal borehole resistance using an expression based on the classical shape factor for two-dimensional heat conduction between two long cylinders in an infinite medium:

$$R_a = \frac{1}{2\pi\lambda_g}\left[2\beta + \cosh^{-1}\left(\frac{0.5 - \theta_3^2}{\theta_3^2}\right)\right] \quad [3.66]$$

This approximation collapses when the pipes are touching to give $R_a = 2R_p$, resulting in errors between 40% and 65%. For the range of cases shown in Fig. 3.10, this expression gives errors as high as 12% for Configuration B and 48% for Configuration C. Therefore, its use cannot be recommended.

3.6 Effective borehole thermal resistance

The borehole thermal resistance calculation methods discussed in Sections 3.3 and 3.4 are for one- or two-dimensional heat transfer in a cross-section perpendicular to the borehole axis. Hence, the thermal resistance values obtained from Eqs. [3.6] and [3.9] actually give local borehole thermal resistance at any depth in the borehole. Fluid temperature in a heat exchanger pipe varies along the borehole depth. Consequently, there is heat exchange between downward-flowing and upward-flowing pipes. As discussed in Section 3.1, this phenomenon, often referred to as thermal short-circuiting, can be accounted for either by using an effective mean fluid temperature or by using an effective borehole thermal resistance. In this section, we cover methods for computing the effective borehole thermal resistance. The difference between the local and effective borehole thermal resistances can vary from being negligible (often the case in typical North American boreholes) to being quite significant.

It is quite possible to calculate the effective borehole thermal resistance from theoretical models. However, some approximations have to be made to keep the calculations tractable—either borehole wall temperature or heat flux has to be treated as uniform along the borehole. As the actual case lies somewhere in between, a practical approach is to calculate the effective borehole thermal resistance based on both approaches and then use the arithmetic mean of the two values as the effective borehole thermal resistance.

For the UHF case, the effective borehole resistance can be calculated using the following expression from Claesson and Hellström (2011):

$$R_b^* = R_b + \frac{1}{3R_a}\left(\frac{H}{\rho_f c_f V_f}\right)^2 \qquad [3.67]$$

The above equation to calculate the effective borehole thermal resistance for UHF condition depends on the depth of the borehole, flow rate of the circulating fluid, and the total internal resistance between U-tube legs given by Eq. [3.13]. We can see that the borehole depth appears in the numerator of the equation whereas the flow rate appears in the denominator of the equation. This reaffirms that effective borehole resistance increases with borehole depth and deceasing flowrates.

For the UBW temperature case, the effective borehole resistance can be calculated using the methods suggested by Claesson and Hellström (2011) and Zeng et al. (2003). The method proposed by Claesson and Hellström (2011) directly applies a correction factor on the local borehole thermal resistance to account for fluid temperature variations along the U-tube pipes and is given as

$$R_b^* = R_b\,\eta \coth \eta \qquad [3.68]$$

$$\text{where:}\quad \eta = \frac{H}{\rho_f c_f V_f}\,\frac{1}{2R_b}\sqrt{1+\frac{4R_b}{R_{1-2}}} \qquad [3.69]$$

For values of η smaller than 1, Eq. [3.68] can be approximated as

$$R_b^* \approx R_b + \frac{1}{3R_{1-2}}\left(\frac{H}{\rho_f c_f V_f}\right)^2 + \frac{1}{12R_b}\left(\frac{H}{\rho_f c_f V_f}\right)^2 \quad \eta \leq 1 \tag{3.70}$$

R_{1-2} is determined from Eq. [3.14] and R_{1-b}, required by Eq. [3.14] can be determined from Eq. [3.12] after the borehole resistance R_b has been determined.

The method suggested by Zeng et al. (2003) to calculate the effective borehole resistance for uniform borehole condition is given as

$$R_b^* = \frac{H}{\rho_f c_f V_f}\left(\frac{1}{\varepsilon} - \frac{1}{2}\right) \tag{3.71}$$

where ε is the efficiency of the borehole that can be calculated using the following expression proposed by Diao et al. (2004). The resistance coefficients $\widehat{R}^0_{1-1}$ and $\widehat{R}^0_{1-2}$ in the following equation can be calculated from Eqs. [3.45] and [3.46], respectively.

$$\varepsilon = \frac{2\tanh\left(\dfrac{H}{\rho_f c_f V_f\sqrt{\left(\widehat{R}^0_{1-1}\right)^2 - \left(\widehat{R}^0_{1-2}\right)^2}}\right)}{\sqrt{\dfrac{\widehat{R}^0_{1-1} + \widehat{R}^0_{1-2}}{\widehat{R}^0_{1-1} - \widehat{R}^0_{1-2}}} + \tanh\left(\dfrac{H}{\rho_f c_f V_f\sqrt{\left(\widehat{R}^0_{1-1}\right)^2 - \left(\widehat{R}^0_{1-2}\right)^2}}\right)} \tag{3.72}$$

Either Eq. [3.68] or Eq. [3.71] can be used to calculate the effective borehole thermal resistance for the UBW temperature condition. These equations appear to have different mathematical forms but are actually identical.

The effect of thermal short-circuiting between the U-tube legs has been illustrated in Figs. 3.11 and 3.12. In Fig. 3.11, the borehole depth has been varied for a borehole with a flow rate of 1.8 m^3/h of water and a total internal thermal resistance of 0.257 m K/W. The local borehole thermal resistance, R_b, is compared to R_b^* calculated with the UHF and UBW temperature approximations. The solid line represents the mean value of the two approximations. The local borehole thermal resistance does not change with depth, but the effective thermal resistance increases exponentially for deeper boreholes. This is due to the increased short-circuiting caused by the increased total conductance between the two legs. For this case, the effective borehole resistance is approximately 4% higher than the local borehole resistance for a 100-m deep borehole. The difference between local and effective

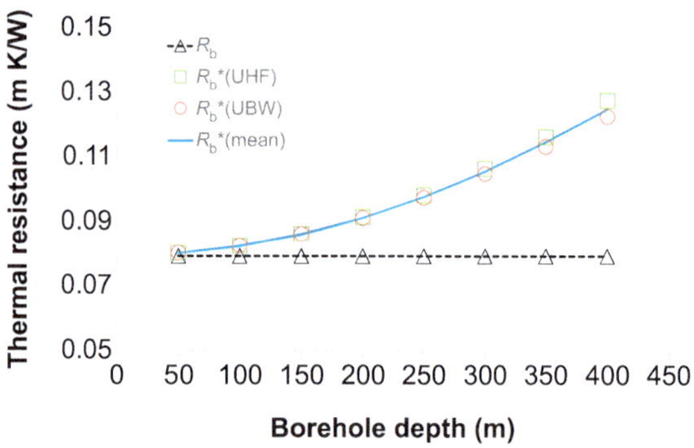

Figure 3.11 R_b^* for different borehole depths with $R_a = 0.257$ m K/W and $V_f = 1.8$ m^3/h.

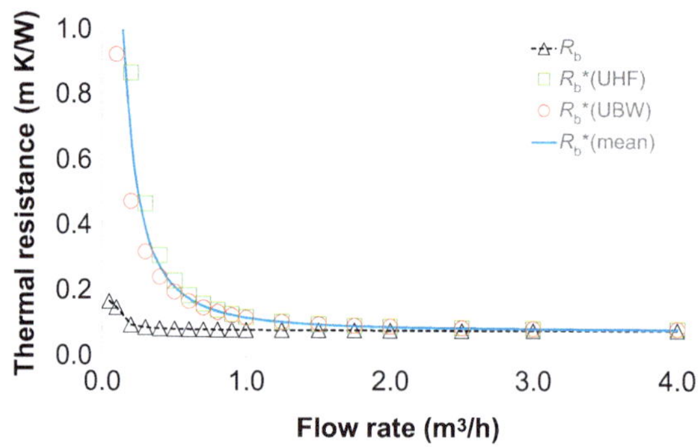

Figure 3.12 R_b^* for varying flow rate of heat carrier fluid and $H = 200$ m.

thermal resistances increases to 15 and 32–34% for 200-m and 300-m deep boreholes, respectively.

In Fig. 3.12, borehole thermal resistances are shown for a 200-m deep borehole with varying flow rates. Here, the local borehole thermal resistance values do change, particularly as the flow rate drops into the laminar flow regime. The increased convective resistance leads to increased borehole resistance and increased internal resistance. However, the effect of decreased thermal capacitance of the flow or increased residence time is very strong and the effective borehole thermal resistance increases exponentially with decreasing flow rate. For the cases shown in Fig. 3.12, the effective borehole resistance is about 12% higher than the local borehole resistance at 2 m^3/h. The difference increases to over 43% at 1 m^3/h and over 100% at 0.6 m^3/h. At laminar flows, ie, flowrates smaller than 0.2 m^3/h, the effective borehole resistance values are over 500 times higher than the local resistance values.

Another important observation from Figs. 3.11 and 3.12 is that the effective borehole resistance for UHF and UBW temperature are very similar for flows larger than 1 m^3/h and borehole depths up to 300 m. However, the differences between the two approaches become pronounced at low flow rates and with deeper boreholes. However, as neither of UHF and UBW temperature boundary conditions are perfectly met in a borehole, a mean value of the two cases is often used in practice.

3.7 Other heat exchanger types

Beyond single U-tubes, double U-tubes and coaxial tubes are also used as borehole heat exchangers. It is relatively simple to calculate the grout resistance for a coaxial heat exchanger. The steady-state heat conduction formula of Eq. [3.24] for the annular region can be used by setting the equivalent radius to the actual outer radius of the coaxial heat exchanger. On the other hand, the calculation of grout resistance for a double U-tube is quite complicated. There is also a general dearth of fully analytical models for the calculation of borehole thermal resistance with two or more U-tubes.

Grout resistance for boreholes with any prescribed number of U-tube pipes can be directly obtained from the multipole method. The method, as explained earlier, requires solving $N \cdot J$ complex equations iteratively, where N is the number of U-tube legs and J is the number of multipoles. Hence, for a double U-tube the number of equations to be solved is four for first-order multipoles, 20 for fifth-order multipoles, and 40 for tenth-order multipoles. For a triple U-tube configuration, the number of equations to be solved increases to six for first-order multipoles, 30 for fifth-order multipoles, and 60 for tenth-order multipoles. Manual calculations tend to become both complicated and time consuming especially when higher order multipole solutions are sought. As a result, unlike for the single U-tube configuration, there are currently no explicit first-order multipole expressions to calculate grout resistance for double or triple U-tube configurations. However, the multipole method implemented as Fortran and Mathcad codes can be used to calculate the grout resistance in a matter of a few seconds for any order of multipoles and any number of U-tubes.

It is, however, still possible to use line-sources, ie, zeroth-order multipoles, to obtain grout resistance for double or triple U-tube configurations. For a borehole with double U-tube placed in geometrical symmetry, as shown in Fig. 3.13, the linear

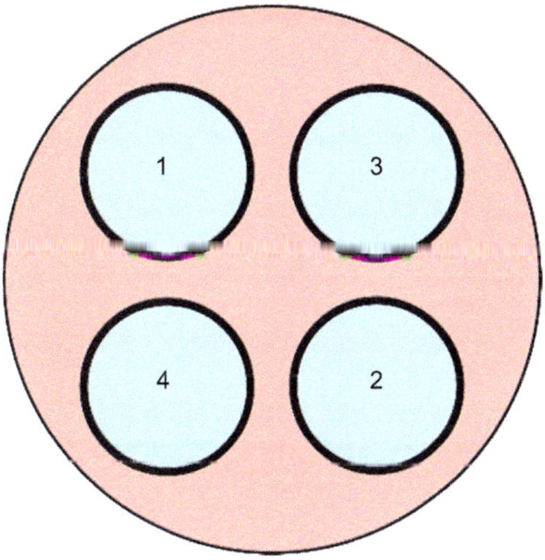

Figure 3.13 Double U-tube configuration.

coefficients $\widehat{R}^0_{m-n}$ obtained from the multipole method take the following form when presented using the dimensionless parameters defined in Eqs. [3.29]−[3.32].

$$\widehat{R}^0_{1-1} = \frac{1}{2\pi\lambda_g}\left[\beta + \ln\left(\frac{\theta_2}{\left(1-\theta_1^2\right)^\sigma}\right)\right] \quad [3.73]$$

$$\widehat{R}^0_{1-2} = \frac{1}{2\pi\lambda_g}\left[\ln\left(\frac{1}{2\theta_1\left(1+\theta_1^2\right)^\sigma}\right)\right] \quad [3.74]$$

$$\widehat{R}^0_{1-3} = \frac{1}{2\pi\lambda_g}\left[\ln\left(\frac{1}{\sqrt{2}\theta_1\left(1+\theta_1^4\right)^{\frac{\sigma}{2}}}\right)\right] \quad [3.75]$$

The fluid-to-borehole wall resistance and the borehole resistance for a double U-tube configuration in geometrical symmetry can then be calculated as

$$R_{1-b} = R_{2-b} = R_{3-b} = R_{4-b} = \widehat{R}^0_{1-1} + \widehat{R}^0_{1-2} + 2\,\widehat{R}^0_{1-3} \quad [3.76]$$

$$R_b = \frac{R_{1-b}}{4} = \frac{R_{2-b}}{4} = \frac{R_{3-b}}{4} = \frac{R_{4-b}}{4} = \frac{\left(\widehat{R}^0_{1-1} + \widehat{R}^0_{1-2} + 2\,\widehat{R}^0_{1-3}\right)}{4} \quad [3.77]$$

The following zeroth-order multipole expression of borehole resistance[11] can now be obtained from Eqs. [3.73]−[3.75] and [3.76]−[3.77] for double U-tubes in geometrical symmetry.

$$R_b = \frac{1}{8\pi\lambda_g}\left[\beta + \ln\left(\frac{\theta_2}{4\theta_1^3\left(1-\theta_1^8\right)^\sigma}\right)\right] \quad [3.78]$$

Bauer et al. (2011) have presented an expression for calculating grout resistance of boreholes with double U-tubes. Similar to their single U-tube expression of Eq. [3.35], the double U-tubes expression of Bauer et al. is also based on shape factor of eccentric cylinders of equal length (Bergman et al., 2011; Holman, 1986) corrected with numerical data. Their expression for grout resistance modified to account for four legs of a double U-tube is

$$R_g = \frac{1}{8\pi\lambda_g}\left[\left(3.098 - 4.432\,\theta_1 + 2.364\,\theta_1^2\right)\cosh^{-1}\left(\frac{\theta_2}{2} + \frac{1}{2\theta_2} - \frac{\theta_1}{4\theta_3}\right)\right] \quad [3.79]$$

[11] Not previously published, as far as we know.

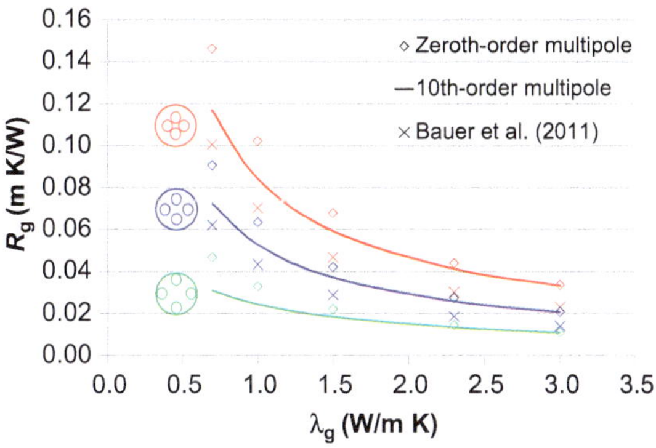

Figure 3.14 Grout thermal resistance for $r_b = 0.070$ m, $r_{po} = 0.020$ m, $\lambda = 3.0$ W/m K, and $R_p = 0.050$ m K/W calculated for double U-tubes of differing spacing.

The zeroth-order multipole expression and the Bauer et al. (2011) expressions are compared to the tenth-order multipole algorithm results in Fig. 3.14. The three configurations are similar but not identical to the A, B, and C configurations for single U-tubes. The pipe resistances have been subtracted from the borehole resistances to give the grout resistances. The Bauer et al. (2011) method gives values of zero for the grout resistance when the pipes are touching the walls, so these results have been left off the chart. Other than that case, the two simpler expressions compare to the tenth-order multipole algorithm reasonably well when the grout thermal conductivities are high. But for lower values, which are more common in practice, neither method performs very well. First-order multipole expressions for both R_b and R_a should be derived for double U-tubes.

Zeng et al. (2003) have studied the effective borehole resistance of double U-tubes in detail. They have presented the following general solution to determine the effective thermal resistance for double U-tube configuration in geometric symmetry:

$$R_b^* = \frac{H}{4\,\rho_f c_f V_f}\left(\frac{1+\Theta_f''}{1-\Theta_f''}\right) \quad [3.80]$$

where Θ_f'' is the dimensionless fluid temperature. It can be determined from the explicit expressions developed by Zeng et al. (2003) for various possible series and parallel configurations of double U-tube. The expressions are based on the zeroth-order linear coefficients of Eqs. [3.73–3.75].

3.8 Groundwater-filled boreholes

In Scandinavia, it is common practice to construct borehole heat exchangers by lowering a U-tube in a groundwater-filled borehole. To protect the environment, the borehole is cased from the surface to the bedrock and capped with a well top. Where the bedrock and groundwater levels are close to the surface, this can be a very effective design with relatively low borehole thermal resistance. Gehlin (2002) measured borehole thermal resistances of 0.05–0.06 m K/W for typical conditions in groundwater-filled boreholes in Luleå. Javed (2010) tested nine water-filled boreholes in close proximity and measured borehole thermal resistance values between 0.05 and 0.07 m K/W. Under a much wider range of conditions, Spitler et al. (2016) measured borehole thermal resistances between 0.05 and 0.1 m K/W for a single borehole in Gothenburg.

This relatively wide range of borehole thermal resistances brings us to one of the key characteristics of groundwater-filled boreholes: due to natural convection in the borehole annulus, the resistance varies with both heat transfer rate and annulus temperature. Higher heat transfer rates lead to higher temperature differences and hence higher differences in density leading to increased velocity of the recirculating flow within the annulus and decreased thermal resistances. Annulus temperature also has an influence on the natural convection. As water temperature approaches the maximum density point near 4°C, the derivative of density with respect to temperature approaches zero, resulting in reduced effect of buoyancy and increased thermal resistances. These effects have been demonstrated (Kjellsson and Hellström, 1997; Gustafsson and Gehlin, 2008) in both field and laboratory experiments.

In order to calculate borehole thermal resistance for a groundwater-filled borehole, it is necessary to find the convective resistances in the annulus shown in Fig. 3.15. Spitler et al. (2016) have developed correlations to find the convective coefficients at the outside-pipe wall and borehole wall for a single U-tube placed in a groundwater-filled borehole. From these convection coefficients, the convective resistances R_{poc} and R_{bc} shown in Fig. 3.15 may be determined. The inside pipe convective resistances and pipe conductive resistances are calculated in the same way as for grouted borehole. When all the individual resistances have been calculated, the total borehole thermal resistance can be readily determined.

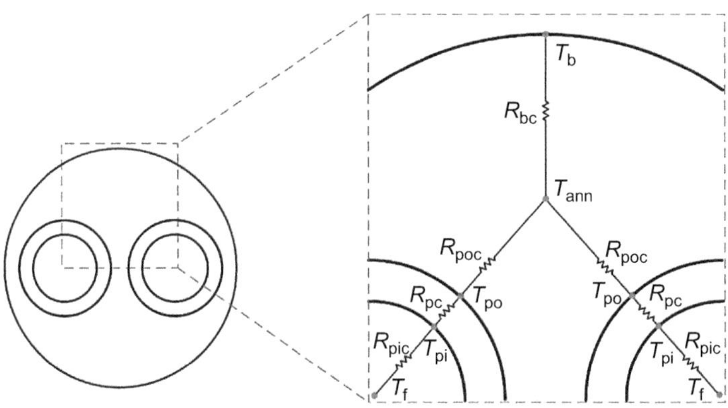

Figure 3.15 Y-Resistance network for groundwater-filled borehole. Additional resistances (R_{poc} and R_{bc}) and a node representing the surrounding fluid (temperature T_{ann}) are used to represent heat transfer in the annulus.

3.9 Conclusions

In this chapter, we have presented methods for calculating local borehole thermal resistance and effective borehole thermal resistance, which accounts for short-circuiting between the upward- and downward-flowing legs of the GHE. Some authors have developed alternative formulations that attempt to estimate and use the actual mean fluid temperature combined with the local borehole thermal resistance. However, use of the effective borehole thermal resistance with the simple mean fluid temperature is consistent with existing design tools and many simulation tools and is therefore our preferred approach.

We have presented an overview of the multipole algorithm, which when used with higher-order multipoles is very accurate and may be considered as a reference method by which other methods may be evaluated. It can be used for single U-tubes, double U-tubes, triple U-tubes, and other unusual configurations, eg, one central downward-flowing tube surrounded by multiple smaller upward-flowing tubes. It is not trivial to implement, though it has been implemented in Fortran and Mathcad.

For single U-tubes, many expressions have been proposed for calculating the borehole thermal resistance and the internal thermal resistance. These methods are compared to the multipole algorithm and most of the algorithms have accuracies that are not satisfactory over the entire range of tube placement and grout thermal conductivity. However, the explicit first-order multipole expressions originally presented by Hellström (1991) are shown to have accuracies within 1% for calculation of borehole thermal resistance and within 5% for calculation of the internal thermal resistance and are therefore recommended for practical use.

For double U-tubes and other more complex configurations, the multipole algorithm is recommended. Two simple analytical expressions are presented here for grout resistance, but their accuracy is rather low. It may be possible to derive explicit first-order multipole equations for double U-tubes, and development of such expressions for both R_b and R_a would be highly desirable.

Nomenclature

c_f	Specific heat of the circulating fluid in the U-tube, J/kg K
f	Friction factor; see Eq. [3.23], dimensionless
h_{pi}	Convection coefficient at the inside pipe wall; see Eq. [3.21], W/m^2 K
H	Depth of the borehole, m
j	Order of multipole
J	Number of multipoles
N	Number of pipes in the borehole; $N = 2$ for single U-tube
Nu_{pi}	Nusselt number for the inner pipe wall; see Eq. [3.22]

Continued

$P_{n,j}$	Strength factor of multipole, °C
Pr	Prandtl number, dimensionless
q_b	Heat rejection rate per unit length of borehole, W/m
q_n	Heat rejection rate per unit length of pipe n, W/m
r_b	Radius of the borehole, m
r_{eq}	Equivalent radius of the U-tube legs, m
r_{pi}	Inner radius of the pipe making up the U-tube, m
r_{po}	Outer radius of the pipe making up the U-tube, m
R_a	Total internal borehole thermal resistance; see Eq. [3.10], m K/W
R_b	Local or average borehole thermal resistance between fluid in U-tube(s) to borehole wall, m K/W
R_b^*	Effective borehole thermal resistance, m K/W
R_{b1-b2}	Borehole wall to borehole wall thermal resistance; see Fig. 3.2, m K/W
R_g	Grout thermal resistance; resistance between outer pipe wall of U-tube to borehole wall, m K/W
R_{m-n}	Thermal resistance between U-tube legs m and n, m K/W
$\widehat{R}_{m-n}^{J}$	Thermal resistance coefficient for J multipoles, m K/W
R_{n-b}	Thermal resistance between U-tube leg n and borehole wall, m K/W
R_p	Total fluid-to-pipe resistance for a single pipe—one leg of the U-tube, m K/W
R_{pc}	Conductive thermal resistance for a single pipe—one leg of the U-tube; see Eq. [3.19], m K/W
R_{pic}	Inner convective thermal resistance for a single pipe—one leg of the U-tube; see Eq. [3.20], m K/W
R_{poc}	Outer convective thermal resistance for single pipe in a groundwater-filled borehole—one leg of the U-tube; see Fig. 3.15, m K/W
R_{bc}	Convective thermal resistance at the borehole wall in a groundwater-filled borehole; see Fig. 3.15, m K/W
Re	Reynolds number, dimensionless
s	Shank spacing, ie, center-to-center distance between two legs of the U-tube; see Fig. 3.3, m
S_b	Borehole shape factor; see Eq. [3.34], dimensionless
t	Time, seconds
T_b	Mean temperature at the borehole wall, °C
T_f	Mean fluid temperature inside the U-tube, °C
T_{fm}	Fluid temperature in U-tube leg m; $m = 1,\ldots, N$, °C

T_{pi}	Temperature at the inner pipe wall, °C
T_{po}	Temperature at the outer pipe wall, °C
V_f	Volume flow rate of the circulating fluid in the U-tube, m^3/s
z	Complex coordinate; see Eq. [3.42], m
α	Thermal diffusivity of the ground, m^2/s
β	Dimensionless thermal resistance of one U-tube leg; see Eq. [3.47]
γ	Euler's constant; equal to 0.5772
ε	Borehole efficiency; see Eq. [3.72]
η	Factor; see Eq. [3.69]
θ_1	Dimensionless parameter; see Eq. [3.29]
θ_2	Dimensionless parameter; see Eq. [3.30]
θ_3	Dimensionless parameter; see Eq. [3.31]
λ	Thermal conductivity of the ground, W/m K
λ_f	Thermal conductivity of the fluid in the U-tube, W/m K
λ_g	Thermal conductivity of the grout, W/m K
λ_p	Thermal conductivity of the pipe, W/m K
ρ_f	Density of the circulating fluid in the U-tube, kg/m^3
σ	Thermal conductivity ratio; see Eq. [3.32], dimensionless

Acknowledgment

Professor Johan Claesson of Lund University and Chalmers University of Technology has been a pioneer in this field, having developed much of the mathematical analysis for analyzing the complex conduction heat transfer problems related to borehole thermal resistance and GHEs. We greatly appreciate his assistance in reviewing the treatment of multipole solutions in this chapter.

References

Abd Elatty, A.S., El-Nahhas, F.M., El-Mossallamy, Y.M., El-Baz, A.M., 2012. Effect of surrounding soil thermal conductivity on the effective pipe-to-borehole thermal resistance in vertical ground heat exchangers. In: International Engineering Conference on Design and Innovation in Infrastructure, June 18–20, Jordan.

Acuña, J., 2013. Distributed Thermal Response Tests – New Insights on U-pipe and Coaxial Heat Exchangers in Groundwater-filled Boreholes (Ph.D. doctoral thesis). KTH.

Al-Chalabi, R., 2013. Thermal Resistance of U-tube Borehole Heat Exchanger System: Numerical Study (Master's thesis). University of Manchester.

Bauer, D., Heidemann, W., Müller-Steinhagen, H., Diersch, H.J., 2011. Thermal resistance and capacity models for borehole heat exchangers. International Journal of Energy Research 35 (4), 312–320.

Bergman, T.L., Lavine, A.S., Incropera, F.P., Dewitt, D.P., 2011. Introduction to Heat Transfer, sixth ed. John Wiley & Sons, Inc.

Beier, R.A., 2011. Vertical temperature profile in ground heat exchanger during in-situ test. Renewable Energy 36 (5), 1578–1587.

Bennet, J., Claesson, J., Hellström, G., 1987. Multipole Method to Compute the Conductive Heat Flows to and between Pipes in a Composite Cylinder. Notes on Heat Transfer 3. University of Lund.

BLOCON, 2015. Earth Energy Designer (EED)v3.2.

Bose, J.E., Parker, J.D., McQuiston, F.C., 1985. Design/data Manual for Closed-loop Ground-coupled Heat Pump Systems. American Society of Heating, Refrigerating, and Air-Conditioning Engineers.

Claesson, J., Dunand, A., 1983. Heat Extraction from the Ground by Horizontal Pipes: A Mathematical Analysis. Swedish Council for Building Research.

Claesson, J., Hellström, G., 2011. Multipole method to calculate borehole thermal resistances in a borehole heat exchanger. HVAC&R Research 17 (6), 895–911.

Diao, N.R., Zeng, H.Y., Fang, Z.H., 2004. Improvement in modeling of heat transfer in vertical ground heat exchangers. HVAC&R Research 10 (4), 459–470.

Gehlin, S., 2002. Thermal Response Test: Method Development and Evaluation (Ph.D. doctoral thesis). Luleå University of Technology.

Gnielinski, V., 1976. New equations for heat and mass-transfer in turbulent pipe and channel flow. International Chemical Engineering 16 (2), 359–368.

Go, G.H., Lee, S.R., Yoon, S., Park, H., Park, S., 2014. Estimation and experimental validation of borehole thermal resistance. KSCE Journal of Civil Engineering 18 (4), 992–1000.

Gu, Y., O'Neal, D.L., 1998. Development of an equivalent diameter expression for vertical U-tubes used in ground-coupled heat pumps. ASHRAE Transactions 104 (2), 347–355.

Gustafsson, A., Gehlin, S., 2008. Influence of natural convection in water-filled boreholes for GCHP. ASHRAE Transactions 114 (1), 416–423.

He, M., 2012. Numerical Modelling of Geothermal Borehole Heat Exchanger Systems (Ph.D. doctoral thesis). De Montfort University.

Hellström, G., 1991. Ground Heat Storage – Thermal Analyses of Duct Storage Systems – Theory (Ph.D. doctoral thesis). University of Lund.

Holman, J.P., 1986. Heat Transfer, sixth ed. McGraw-Hill, Singapore.

Javed, S., 2010. Design of Ground Source Heat Pump Systems – Thermal Modelling and Evaluation of Boreholes (Licentiate thesis). Chalmers University of Technology.

Kavanaugh, S.P., 1985. Simulation and Experimental Verification of Vertical Ground-coupled Heat Pump Systems (Ph.D. doctoral thesis). Oklahoma State University.

Kjellsson, E., Hellström, G., 1997. Laboratory study of the heat transfer in a water-filled borehole with a single U-pipe. In: Megastock 1997. The 7th International Conference on Thermal Energy Storage, Sapporo, Japan.

Lamarche, L., Kajl, S., Beauchamp, B., 2010. A review of methods to evaluate borehole thermal resistances in geothermal heat-pump systems. Geothermics 39 (2), 187–200.

Liao, Q., Zhou, C., Cui, W., Jen, T.C., 2012. New correlations for thermal resistances of vertical single U-tube ground heat exchanger. Journal of Thermal Science and Engineering Applications 4 (3), 031010.

Marcotte, D., Pasquier, P., 2008. On the estimation of thermal resistance in borehole thermal conductivity test. Renewable Energy 33 (11), 2407–2415.

Mogensen, P., 1983. Fluid to duct wall heat transfer in duct system heat storages. In: International Conference on Subsurface Heat Storage in Theory and Practice. Swedish Council for Building Research, Stockholm, Sweden.

Paul, N.D., 1996. The Effect of Grout Thermal Conductivity on Vertical Geothermal Heat Exchanger Design and Performance (Master's thesis). South Dakota University.

Petukhov, B.S., 1970. Heat transfer and friction in turbulent pipe flow with variable physical properties. Advances in Heat Transfer 6 (503), 503–564.

Remund, C.P., 1999. Borehole thermal resistance: laboratory and field studies. ASHRAE Transactions 105 (1), 439–445.

Sharqawy, M.H., Mokheimer, E.M., Badr, H.M., 2009. Effective pipe-to-borehole thermal resistance for vertical ground heat exchangers. Geothermics 38 (2), 271–277.

Shonder, J.A., Beck, J.V., 2000. Field test of a new method for determining soil formation thermal conductivity and borehole resistance. ASHRAE Transactions 106 (1), 843–850.

Spitler, J.D., 2000. GLHEPRO – a design tool for commercial building ground loop heat exchangers. In: Fourth International Heat Pumps in Cold Climates Conference, Aylmer, Québec.

Spitler, J.D., Gehlin, S.E.A., 2015. Thermal response testing for ground source heat pump systems – an historical review. Renewable and Sustainable Energy Reviews 50 (0), 1125–1137.

Spitler, J.D., Javed, S., Kalskin Ramstad, R., 2016. Natural Convection in Groundwater-filled Boreholes Used as Ground Heat Exchangers. Applied Energy 164 (2016), 352–365.

Zeng, H., Diao, N., Fang, Z., 2003. Heat transfer analysis of boreholes in vertical ground heat exchangers. International Journal of Heat and Mass Transfer 46 (23), 4467–4481.

In situ estimation of ground thermal properties

4

H.J.L. Witte
Groenholland Geo-energysystems, Amsterdam, Netherlands

Abstract

The ground thermal conductivity and borehole thermal resistance (chapter: Calculation of borehole thermal resistance) are essential parameters for the design of a ground-source heat pump (GSHP) system (chapter: Vertical borehole ground heat exchanger design methods). As it is hardly possible to obtain sufficiently accurate values of these parameters from geological or soil-profile information, thermal response tests have been developed that provide measurements in situ. These tests have become part of the standard toolkit of the shallow geothermal energy systems engineer.

In this chapter, we introduce the main theory and basic principles of such a test with guidelines for test parameter selection and some examples. This chapter primarily deals with estimation of properties for vertical borehole heat exchanger applications. In situ test methods appropriate to shallow ground heat exchanger applications are discussed in chapter "Horizontal and compact ground heat exchangers."

4.1 Introduction

For the design and construction of a borehole heat exchanger system, several parameters are of importance. Information about the intended energy exchange (total heating and cooling load as well as maximum heat rates), available surface area, and layout of the borehole heat exchanger field (eg, single borehole, open or closed rectangular layout, and line layout) need to be known or estimated in advance (see chapter: Vertical borehole ground heat exchanger design methods). Obviously, the characteristics of the ground at the location are of major importance. Before one can even start to design a borehole heat exchanger system one needs to know at least:

1. The general geology and sequence (stratigraphy). This will give the designer an idea about drilling conditions and economic drilling depth. Also, this will allow the designer to define the correct backfilling of the borehole, for instance, taking into consideration the sealing of impermeable layers that separate aquifers.
2. The thermal ground parameters such as thermal conductivity, heat capacity, and vertical temperature distribution.
3. Thermal characteristics of the borehole (thermal resistance)
4. Assessment of occurrence and strength of ground water flow. If there is a considerable ground water flow present, the usual design methods do not apply as they only consider heat conduction.

Advances in Ground-Source Heat Pump Systems. http://dx.doi.org/10.1016/B978-0-08-100311-4.00004-2
Copyright © 2016 Elsevier Ltd. All rights reserved.

A first assessment can be based on a desktop study, but for larger systems or for systems being installed in a new area, it is advisable to drill a test borehole and carry out a thermal response test (TRT) with which a number of the above parameters can be measured accurately.

Although this chapter will deal mainly with the technique of TRTs, it should be emphasized that the information obtained from the test borehole is very valuable by itself. Often a usual production drilling technique is used, but it is advisable to put some additional effort into collecting information about the actual stratigraphy of the location itself.

In this chapter we will discuss the theory and basic analysis of TRT data, estimating the uncertainty associated with the measurement of the key thermal parameters as well as the design of the TRT.

The purpose of a TRT (Austin, 1998; Gehlin, 1998; van Gelder, 1999) is to measure the equivalent thermal conductivity of the ground volume tested and thermal resistance of the borehole heat exchanger. This sounds fairly straightforward, but in fact it is not. With regard to the thermal conductivity one has to realize that first of all the different soil types (such as clay, sand, gravel, chalk, or other rock types) that occur in sequence all have a specific thermal conductivity and the test on the complete borehole yields a value that is an aggregate thermal conductivity value for the complete profile. Moreover, the thermal conductivity of the ground will vary as a function of space and time because the ground is not a homogeneous medium but exhibits variations in composition at different spatial scales. Even in one single test this may affect the result: as the temperature gradient propagates through the ground with time the actual ground volume that is tested increases and the equivalent thermal parameters vary according to its evolution. In an extreme case, for instance a test on a steeply inclined geology such as glacial push ridges, this will lead to inconclusive tests as no final estimate of the ground thermal conductivity is possible simply because the approximation of a quasi-constant value does not apply.

Nevertheless, in many settings even a single TRT provides valuable information for the designer, as long as the above and other fundamental assumptions made by the test procedure are considered and evaluated.

4.2 Background and development

During the 1980s and 1990s, ground-source heat pump (GSHP) systems became more and more widespread. A GSHP system is used to heat or cool building with very high energy efficiency for every unit of energy expended (by compressor and pumps) four or five units of thermal energy are transferred. These systems gained popularity because of economic reasons (significant savings on energy bills) but also because they contribute significantly to reduction of primary energy use and greenhouse gas emissions. In the 1980s, advances had been made in the software that form the basis for the calculation of the thermal response of the ground source system to energy loadings—especially in Sweden where the superposition borehole model (Eskilson,

1986) was developed and the thermal behavior of the borehole heat exchanger itself was detailed (Claesson, 1985; Eskilson, 1987; Hellström, 1991).

Around the same time a field test for the measurement of thermal resistance of borehole heat exchangers was proposed by Mogenson (1983). The theory of measuring conductivity of different materials was already established, as the conductivity of different gases, fluids, and solids has been a topic of research ever since Newton formulated his law of cooling. The first experiments into the conductivity of different gases, using the transient hot wire method, was probably reported by Priestley in 1781 (Assael et al., 2010), finally resulting in the transient hot wire needle proposed by Stålhane and Pyk in 1931. Subsequent authors (see Assael et al., 2010 for an overview) made many improvements and the transient hot wire technique is now able to measure the conductivity of different materials with very high accuracy, the uncertainty being better than 1% (Assael et al., 2010). Nevertheless measurements of ground thermal conductivity over greater depths had not been attempted before.

4.2.1 Field tests

In 1995 the first practical field tests were constructed in the United States (Austin, 1998) and Sweden (Eklöf and Gehlin, 1996; Gehlin, 1998). These test machines used direct electrical heating to apply a more or less constant heat rate to a borehole heat exchanger while making measurements of the temperature evolution of the circulating fluid. In 1997 a test machine was developed in the Netherlands that used a reversible heat pump, allowing tests to be conducted with either heat extraction or heat injection (van Gelder et al., 1999; Witte et al., 2002). The description of the methodology presented here is mainly derived from the excellent discussions presented by Gehlin (1998) and Austin (1998).

The main concept of the test is to measure the temperature evolution of a borehole heat exchanger under a thermal load. The rate of temperature change with time is then used to estimate the thermal conductivity of the ground volume surrounding the borehole heat exchanger. Depending on different assumptions and simplifications, different analytical or numerical procedures can be used to analyze the test results and obtain values for different thermal parameters (thermal conductivity, thermal heat capacity, borehole thermal resistance).

A simplified schematic of a TRT test using electrical resistance heating or using an air-to-water heat pump with controlled power rate is shown in Fig. 4.1.

4.2.2 The line source model

The first method is based on Kelvin's line source (1882) that was applied by Ingersoll and Plass (1948) to ground-source heat exchangers, according to Carslaw and Jaeger (1959):

$$\Delta T(r,t) = \frac{Q}{4\pi\lambda}\int_{\frac{r^2}{4\alpha t}}^{\infty}\frac{e^{-u}}{u}du = \frac{Q}{4\pi\lambda}E_1\left(r^2/4\alpha t\right) \qquad [4.1]$$

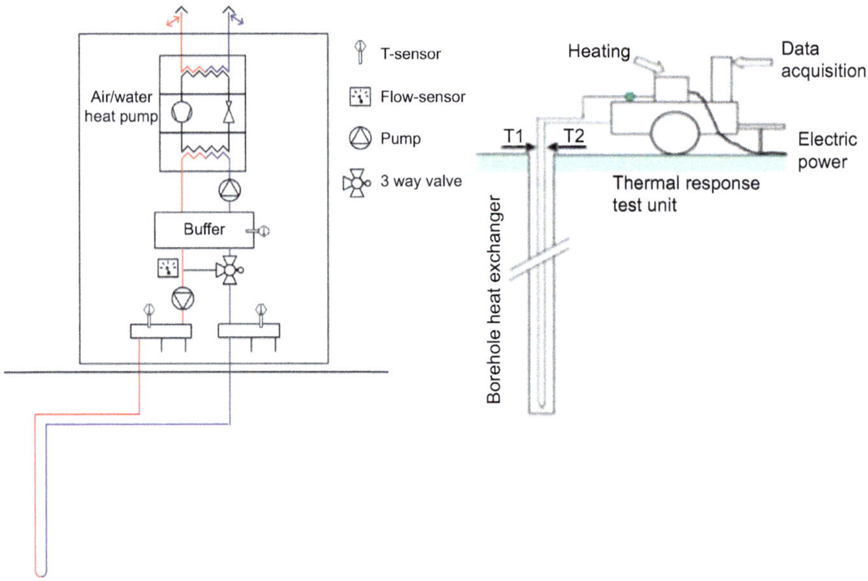

Figure 4.1 Schematic of typical TRT, system with electrical heating *(right: after Gehlin, S., 1998. Thermal Response Test, In Situ Measurements of Thermal Properties in Hard Rock (Licentiate thesis). LULEÅ University of Technology; permission for reproduction by author gratefully acknowledged)* and based on an air-to-water heat pump *(left: after Witte, H.J.L., van Gelder, A.J., Spitler, J.D., 2002. In-situ measurement of ground thermal conductivity: the Dutch perspective. ASHRAE Transactions, 108 (1), ©ASHRAE www.ashrae.org).*

The thermal diffusivity is the ratio between the thermal conductivity and the volumetric heat capacity:

$$\alpha = \frac{\lambda}{\rho C_p} \qquad [4.2]$$

The exponential integral in the line source equation can be approximated by a series expansion. However, for large values of $\alpha t/r^2$, the exponential integral can be further approximated by (Mogenson, 1983)

$$E_1 = \ln\left(\frac{4\alpha t}{r^2}\right) - \gamma \qquad [4.3]$$

This solution yields an error (when compared to the full line source equation) of 2.5% if $\alpha t/r^2 > 20\%$ and 10% if $\alpha t/r^2 > 5$. The line source model and the approximations to its solution are also discussed in chapter "Vertical borehole ground heat exchanger design methods," Section 2.1.2.

As we calculate the temperature evolution from the borehole radius but measure the temperature of the fluid circulating in the heat exchanger pipes inside the borehole, the borehole and heat exchanger need to be accounted for. To include the temperature difference between the borehole heat exchanger and the ground a thermal resistance term, the borehole thermal resistance R_b, is introduced. Finally, the initial temperature of the subsurface needs to be accounted for as well. The final simplified line source equation then becomes:

$$T_f(t) = \frac{Q}{4\pi\lambda}\left[\ln\left(\frac{4\alpha t}{r^2}\right) - \gamma\right] + Q \cdot R_b + T_g \quad [4.4]$$

In Eq. [4.4], if the heat rate Q is constant, only $T_f(t)$ and $\ln(t)$ are variables, the other terms are constant. By writing:

$$k = \frac{Q}{4\pi\lambda} \quad [4.5]$$

we can obtain a general linear equation:

$$y = k \cdot x + c \quad [4.6]$$

By plotting the temperature change with the natural logarithm of time, the slope of the line can readily be obtained by linear regression and plugged into the equation to yield the estimate of thermal conductivity:

$$\lambda = \frac{Q}{4\pi k} \quad [4.7]$$

with k the slope coefficient of the regression in Eq. [4.6].

If the undisturbed ground temperature has been measured and the thermal conductivity has been estimated the borehole resistance, considered constant, can be obtained:

$$R_b = Q(m - T_g) - \left[\frac{1}{4\pi\lambda}\left\{\ln\left(\frac{4\alpha}{r^2}\right) - \gamma\right\}\right] \quad [4.8]$$

The line source solution is easy to implement in a spreadsheet program and is widely used for the evaluation of TRT data. However, it is sensitive to disturbances, such as temperature fluctuations or nonconstant heat rates.

One solution to nonconstant heat rates with the line source is to decompose the varying heat rate in constant steps and superpose the individual solutions. Other solutions to varying heat rates use a numerical model with parameter estimation (Austin, 1998; Yavuzturk, 1999) or deconvolution of the temperature response (Beier and Smith, 2003).

4.2.3 The cylinder source model

The line source solution presented above is a simplification of the cylinder source solution. The cylinder source model approximates the borehole heat exchanger as an infinite cylinder with a constant heat flux. The heat exchanger pipes are considered to be adequately represented by an equal diameter cylinder. The cylinder source solution for a constant heat flux is given by

$$T(r,t) = \frac{Q}{\lambda}\cdot G(z,p)\begin{cases} z = \dfrac{\alpha t}{r^2} \\ p = \dfrac{r}{r_0} \end{cases} \quad [4.9]$$

$G(z, p)$ is the cylinder source function (Ingersoll et al., 1954):

$$G(z,p) = \frac{1}{\pi^2}\int_0^\infty f(\beta)\delta\beta \quad [4.10]$$

where:

$$f(\beta) = \left(e^{-\beta^2 z} - 1\right)\frac{[J_0(p\beta)Y_1\beta - Y_0 p\beta J_1\beta]}{\beta^2\left[J_1^2\beta + Y_1^2\beta\right]} \quad [4.11]$$

Here J_0, J_1, Y_0, and Y_1 are Bessel functions of the first and second kind.

Deerman and Kavanaugh (1991) as well as Kavanaugh and Rafferty (1997) apply the cylinder source to design borehole heat exchangers. By reversing the calculation process the equivalent thermal conductivity of the ground is obtained by an iterative procedure.

The cylinder source method is also widely used with the needle probe method. The cylinder needle probe method was first proposed by Schleiermacher (1888) and later also by Stålhane and Pyk (1931). De Vries and Peck (1958) further developed the probe method, which has been applied to find the thermal conductivity of soils among others. The probe cylinder solution is given by

$$T(t) = T_g + \frac{Q}{4\pi\lambda_g H}\left(2h + \ln\left(\frac{4\tau}{C}\right) - \frac{4h - \alpha_1}{2\alpha_1\tau} + \frac{\alpha_1 - 2}{2\alpha_1\tau}\cdot\ln\left(\frac{4\tau}{C}\right) + \ldots\right) \quad [4.12]$$

where:

$$h = 2\pi\lambda_g R_b; \quad \alpha_1 = \frac{2\pi r_b^2 c_g}{c_{cyl}}\tau = \frac{\alpha_g}{r_b^2}; \quad C = e^\gamma \quad [4.13]$$

The heat capacity of the cylinder that represents the borehole, c_{cyl}, is calculated by

$$c_{cyl} = c_{fluid}\pi r_{fluid}^2 + c_{fill}\pi\left(r_b^2 - r_{fluid}^2\right) \qquad [4.14]$$

The main advantage of the cylinder solution is that, in comparison with the line source method, it gives a much smaller error at short times.

4.2.4 Numerical models

In addition to analytical solutions several authors have used numerical models to evaluate test results. With a numerical model it is, for instance, possible to account for varying heat rates (Austin, 1998; Yavuzturk, 1999), to explicitly account for the thermal conductivity and heat capacity (Wagner and Clauser, 2005), to estimate concurrently more parameters (Shonder, 1999), or to investigate effects of ground water flow (Witte, 2002, 2006a,b; Signorelli et al., 2006).

A review of the application of numerical models is presented by Monzó (2011) and in the final report of the International Energy Agency (IEA) Annex 21 on TRTs.

4.3 Test assumptions

With the tests, several assumptions concerning the test conditions are made. For the line source solution, these assumptions are:

- the heat transport in the ground is by conduction only;
- the thermal conductivity in the tested ground volume is isotropic and constant in time and space;
- there is no axial heat transport;
- there is no effect of heat capacity in the borehole;
- the borehole heat exchanger is accurately approximated by a line source;
- there is, after an initial transient state, a steady state borehole resistance;
- the heat rate is constant.

The fundamental assumptions depend on the solution method chosen; the cylinder source solution, for instance, does include the borehole heat capacity in the equation while for a numerical model the heat flux is not necessarily constant and the model may be able to include ground water flow as well.

4.4 Equivalence with pumping tests

As noted by several authors, there is a clear equivalence between TRTs and hydraulic pump tests. With a hydraulic pump test the hydraulic transmissivity of an aquifer is being tested by a pumping test with measurements of drawdown in time (Theis, 1935).

In fact, the Darcy equation is of the same mathematical form as the Fourier equation of heat conduction, and both equations lead to the same differential equation that describes the time dependent response.

Hydraulic test methods have been applied to TRTs by Poppei et al. (2008) and Raymond et al. (2011a). In the analysis the recovery phase, where the temperature returns to the natural background temperature, can be used, which allows the analysis of tests with varying heat rates. A disadvantage is that the time required to run the test increases when recovery data is included. As Poppei et al. (2008) show, hydraulic tests employ diagnostic plots that can be very useful in test evaluation. By using a semilogarithmic derivative plot of thermal conductivity over time, minimum test time as well as maximum test time can be established while running the test.

4.5 Thermal response test in practice

The first question that arises when planning for a TRT is what borehole heat exchanger will be used for the test and what test parameters will be selected. If the main purpose of the test is to get a precise estimate of thermal conductivity of the vertical soil profile the borehole design and test conditions will be different than when the test should measure the response of a borehole as it will be used in the final system. In the first case the effects of the borehole itself should be as small as possible, while in the second case the installation should be similar to the one that will be used on the system itself. Based on the purpose of the test the different test conditions can be selected.

4.6 Desktop study

A general idea concerning the geology, geohydrology, and ground thermal parameters needs to be obtained. As a starting point for this, information about the expected geology can be collected from nearby bore-logs, for instance, obtained from the geological survey. These may be used to find an economic drilling depth but also to provide some a priori estimates of thermal conductivity and heat capacity values. If there are several strata with different thermal conductivities and heat capacities, a depth-weighted average value can be calculated.

An assessment of the expected ground water table and possible ground water movement needs to be made. If there is a possibility of significant ground water movement, this needs to be carefully considered as it invalidates one of the main assumptions of the TRT and the test strategy and evaluation method may need to be adjusted (Chiasson et al., 2000).

Another important parameter is the undisturbed ground temperature. Although this may be estimated using the average air temperature and geothermal heat flux, the result may differ significantly from the actual site conditions. In the Netherlands, for instance, the average (yearly) ambient temperature is between 8 and 10°C, but actual average ground temperatures range between 8 and 14°C depending on location, with sites located in urban areas showing considerably higher ground temperatures due to the "heat island" effect such as that evident in London (UK). Another aspect is the near surface temperature gradient. When the test is conducted at the end of summer, temperatures in the top 10 m will be much higher than average while at the end of winter the reverse is true (see chapter: Horizontal and compact ground heat exchangers, Fig. 5.7). Depending on the type of test (heat injection or heat extraction) and total length of the heat exchanger this affects the results. In winter a heat injection test will overestimate the thermal conductivity while in summer it will tend to underestimate the thermal conductivity. Therefore, the vertical temperature gradient in the borehole always needs to be measured on site just before the test commences.

4.7 Test borehole design and installation

After the desktop study a first assessment of the site thermal parameters has been made. This will aid in selecting the test parameters and also guide the borehole design. For a location that is going to be used for a ground source application, the drilling conditions can be inferred, backfilling chosen, and approximate target depth set.

The borehole diameter depends on the drilling technique and actual drill rig used. It also depends on the type of intended heat exchanger; a double U-tube heat exchanger will need a larger borehole diameter than a concentric-type heat exchanger. In addition, using a mud-rotary direct circulation drilling technique, augering technique, or cased pulsed/sonic drilling method will affect borehole diameter. It is important that during the drilling the borehole diameter is maintained as constant as possible. Having the drill string being stationary while circulating water, for instance, may result in caving at that depth and irregular borehole diameter.

During the drilling, important additional information about the local geological profile can be obtained and, even with drilling techniques less suited for accurate sampling, collecting information about the composition of the different layers is important. Information about the level of the natural ground water table should be noted.

The backfilling, together with the borehole diameter, affects the borehole thermal resistance. For a low borehole resistance and good heat exchange with the surrounding ground a high conductivity backfill material is preferred. In saturated conditions coarse sand or pea gravel can be used. When the borehole traverses several water-bearing formations, it may be necessary to seal at least the impermeable layers with a low conductivity sealing clay (bentonite). It may even be necessary (eg, if the actual geology is not known or due to regulations) to completely seal the borehole with a sealing clay or clay/cement mix. In any case, it is important to fully backfill the borehole and prevent bridging of backfill material where, at greater depths, the borehole remains filled with water or is even dry. Subsidence should also be avoided. In practice this means that the borehole should never be backfilled with the drill cuttings and that the backfilling should be from bottom to top, preferably using a tremie pipe for slurries or a backfill pipe for grainy materials.

The depth of the borehole and heat exchanger installation is usually established beforehand. However, when selecting the depth of the borehole the local geology should be considered and during the test drilling it may be necessary to adjust the final borehole depth.

4.8 Selecting test parameters

The main test parameters that need to be selected are fluid flow rate and temperature difference, together with the fluid heat capacity and density, as these also determine the required heat transfer rate. The main consideration is to achieve as small an uncertainty of the parameters of interest as possible. This usually means selecting a high temperature difference. Together with the fluid flow rate the required temperature difference also determines the heat rate. When considering the combination of temperature difference, flow rate, and resulting heat rate, selection of the proper parameter settings becomes more difficult and several things need to be considered:

- With a high heat rate a larger volume of ground is included in the same time period which will make the result more representative. However, in a low conductivity soil the temperature will also rise more quickly and will rise higher so that the limits of the test machine or installation will be reached more quickly (for instance, freezing in an extraction test or too high temperatures in an injection test). This may limit the total test duration.
- High temperatures may induce convection in and around the borehole and increase the sensitivity to ground water flow in general. Some norms for conducting TRTs, like ASHRAE's (ASHRAE 2007) test protocol, prescribe a high heat rate and advocate very turbulent flow conditions but the possible effects of high heat rates need to be carefully considered. Also, prescribing very turbulent conditions means the test will not be carried out at the normal operating conditions of the system.
- With a low heat rate, longer test duration may be needed to ensure sufficient radial propagation of the temperature pulse into the ground. However, if the heat rate is too low, test duration may be too long to be practical. Also, the uncertainty of the resulting estimate will be lower, and sensitivity to changes in, for example, ambient conditions is greater.
- A low fluid flow rate allows the experimenter to select a high temperature difference, but may induce very laminar flow (high borehole resistance) that may not be representative of the final installation. Also, instabilities in flow may occur.
- A high fluid flow rate will result in a small temperature difference and may also execute the test at unrealistic high Reynolds numbers. One should keep in mind that high pressure loss in the pipes at high Reynolds is in fact a change from kinetic to thermal energy which is not measured by the temperature sensors (as the heat is generated and partly dissipated in the pipes) and therefore introduces error in the test.

If one has a reasonable idea about the thermal conductivity of the location one can use the line source solution (or other method that is going to be used for the evaluation) to calculate the expected temperature response and decide if the selected parameters need to be adjusted.

In many cases the TRT will be done for a project where the intention is to install a borehole heat exchanger system. In those cases test conditions should be selected in the range of actual operating conditions of the intended system.

Table 4.1 Reynolds number and heat rate for different flow rates (Q) and temperature differences (DT) for a 130 m deep U-tube heat exchanger (26 mm inner diameter) calculated with monopropylene glycol (MPG) (concentration 25% at a bulk temperature of 5°C) or water (bulk temperature 20°C) as a heat transfer medium

		25% MPG @ 0°C		Water @ 20°C	
Q (m^3/h)	DT (K)	Re	Heat rate (W)	Re	Heat rate (W)
0.3	3	1050	1000	4060	1045
0.3	6	1050	2000	4060	2090
0.6	3	2100	2000	8100	2090
0.6	6	2100	4000	8100	4180
0.9	3	3150	3000	12,185	3130
0.9	6	3150	6000	12,185	6260
1.2	3	4200	4000	16,250	4180
1.2	6	4200	8000	16,250	8360

In general the test should be set up to use a fluid flow rate with a Reynolds between about 1500 and 3000 and a temperature difference between 3 and 6 K (values that are in the typical operating range of most borehole heat exchanger systems). If an antifreeze mix is used (as it will be in a heat extraction test) the viscosity of the mix needs to be taken into account, and changes from laminar to turbulent flow during the constant-heat rate phase of the test should be avoided. Some indicative values for Reynolds number and heat rate are given in Table 4.1.

4.9 Test evaluation

We will present an example of a TRT and its evaluation based on the line source solution, which is easily reproduced in any spreadsheet.

Based on the desktop study for this specific locality, the anticipated thermal conductivity was 1.9 W/m K, the heat capacity 2.6 MJ/m^3 K and the subsurface temperature 9.5°C. On the test location a standard 40 mm U-loop (inner diameter 0.032 m) was installed to a depth of 130 m. The borehole diameter was 0.13 m, backfilled with thermally enhanced grout. The full test comprised a heating and cooling pulse, therefore a 25% monopropylene glycol mix was used (properties at bulk temperature of 15°C: density 1025.9 kg/m^3, heat capacity 2915 J/(kg K), thermal conductivity 0.467 W/m K, and viscosity 2.78 mPa s).

Test conditions were selected based on some calculations with the line source solution. A total heat rate of the first heat pulse was set at about 5500 W, the

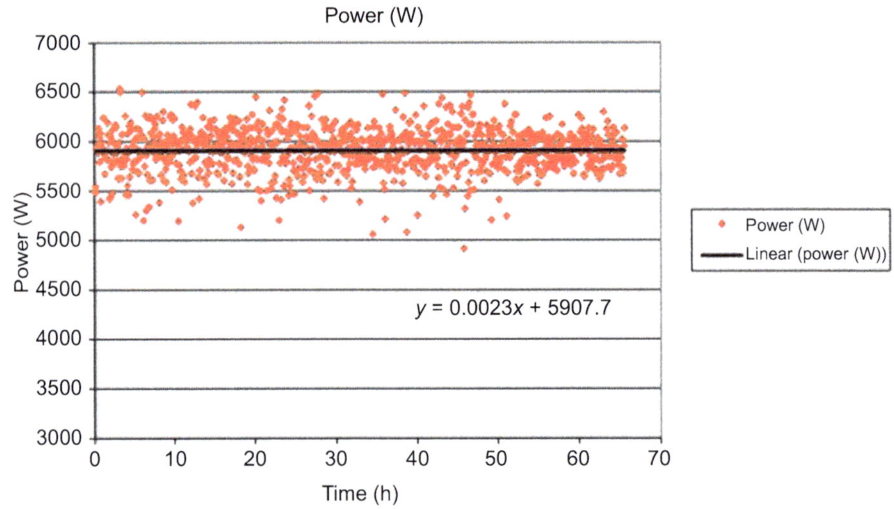

Figure 4.2 Vertical temperature profile measured before thermal response test.

anticipated temperature increase in the first 50 h is then about 13K. A flow rate of around 0.85 m^3/h and a temperature difference of 6K was selected. Flow conditions are turbulent throughout the test.

The actual vertical temperature profile on the location (Fig. 4.2) yielded an average ground temperature of 11.4°C over the complete interval: below 25 m the temperature was found to be 10.3°C. There was evidence of a slight geothermal gradient of about −0.025 K/m. It was noteworthy that the thermal gradient was inverted; because the test location was in an urban area we are observing the fact that the ground is gradually heating up.

From the general test statistics (Table 4.2) it can be observed that the measured test parameters (fluid flow and temperature difference) have a small variation throughout the test. The standard deviation on the heat rate is only 3.7%. More important, however, is to verify that the heat rate is constant throughout the test. In Fig. 4.3 the heat rate is plotted as a function of time. If a trend line is plotted (Fig. 4.3) the relation between heat rate (W) and time (s) is $0.00022 \times t + 5676.5$: the slope coefficient is not significantly different from zero ($p > 0.05$).

Plotting the average fluid temperature with log time (Fig. 4.4) gives a linear relationship: $T_f(t) = 1.0 \times \ln(t) + 1.855$.

Substituting the coefficients into Eq. [4.7] yields a thermal conductivity of 1.78 W/m K. With Eq. [4.8] we can subsequently estimate the borehole resistance, which is found to be 0.154 (m K/W). However, these values are based on using the complete dataset and this introduces an error. Using the criterium $\alpha t/r^2$ and selecting a 10% accuracy as acceptable, the first 10 h (8.7 to be more precise) of data need to be discarded from the evaluation. The result then becomes:

$$T_f(t) = 1.900 \times \ln(t) + 2.481$$

Table 4.2 Experiment measured main parameters and standard deviation of measured parameters

Experiment parameter	Pulse #1
Pulse type	Injection
Duration (h:min)	66.00
Flow rate (m^3/h)	0.85 ± 0.015
Average fluid temperature (°C)	24.06 ± 1.928
Reynolds	4444
ΔT (°C)	6.00 ± 0.199
Heat flux (W)	5679.58 ± 207.94
Specific heat flux (W/m)	43.67 ± 1.60
Log interval (s)	240

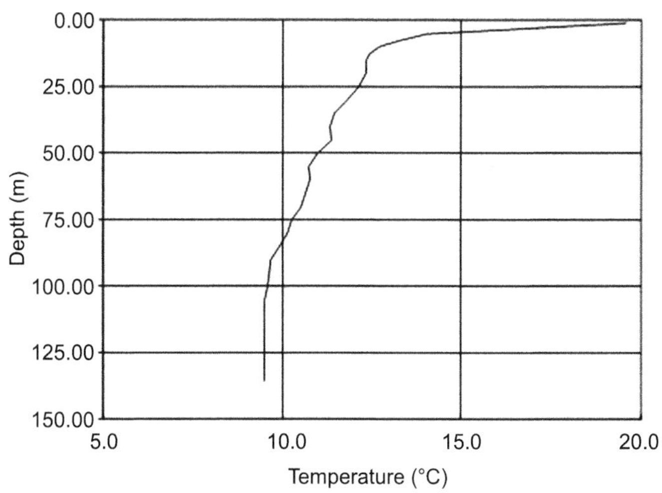

Figure 4.5 Plot of heat rate with time, showing constancy of heat rate. A trendline with linear regression coefficient is shown; the slope coefficient is not significantly different from 0.

Yielding a final estimate of the equivalent thermal conductivity of 1.83 W/m K and a borehole resistance of 0.154 m K/W.

4.10 Quality of the test result

In the interpretation of the test data we have to consider the standard deviations of the measured parameters as well as the constancy of the heat rate. In order to do this we

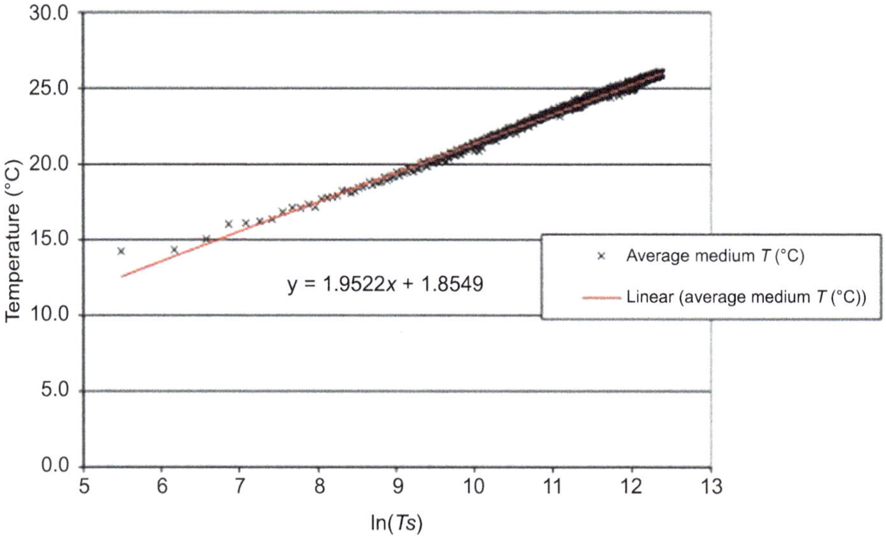

Figure 4.4 Plot of average fluid temperature (arithmetical average between entering and leaving temperature) and log time. A linear regression line and equation are shown.

have to be able to assign an uncertainty to the estimate of the thermal conductivity or borehole resistance. Witte (2013) has examined the uncertainty of the TRT based on the line source solution and presents equations for the calculation of the uncertainty of the thermal conductivity, borehole resistance, and undisturbed ground temperature, among others. The final equations for the uncertainty of thermal conductivity (Witte, 2013) include:

$$\frac{\delta\lambda_{\mathrm{TRT}}}{\lambda_{\mathrm{TRT}}} = \sqrt{\left(\frac{\delta q_{\mathrm{v}}}{q_{\mathrm{v}}}\right)^2 + \left(\frac{\delta\rho}{\rho}\right)^2 + \left(\frac{\delta c}{c}\right)^2 + \left(\frac{\delta\Delta T}{\Delta T}\right)^2 + \left(\frac{\delta H}{H}\right)^2 + \left(\frac{\delta k}{k}\right)^2} \quad [4.14]$$

And for the borehole resistance:

$$\partial R_{\mathrm{b}} = \sqrt{\left(\frac{\Delta R_{\mathrm{b}}}{\Delta H}\delta H\right)^2 + \left(\frac{\Delta R_{\mathrm{b}}}{\Delta Q}\delta Q\right)^2 + \left(\frac{\Delta R_{\mathrm{b}}}{\Delta m}\delta m\right)^2 + \left(\frac{\Delta R_{\mathrm{b}}}{\Delta T_{\mathrm{g}}}\delta T_{\mathrm{g}}\right)^2 + \left(\frac{\Delta R_{\mathrm{b}}}{\Delta\lambda}\delta\lambda\right)^2 + \left(\frac{\Delta R_{\mathrm{b}}}{\Delta C}\delta C\right)^2 + \left(\frac{\Delta R_{\mathrm{b}}}{\Delta r_{\mathrm{o}}}\partial r_{\mathrm{o}}\right)^2} \quad [4.15]$$

The uncertainty of the estimates, based on the example test data presented above (Table 4.2), can then be calculated (making some reasonable assumptions about the different uncertainties in constant such as borehole diameter and borehole length):

Estimated equivalent ground thermal conductivity: 1.83 ± 0.084 W/m K.

Estimated borehole thermal resistance: 0.154 ± 0.020 K/(W/m)

A final important quality indicator is if the test result converges to a single estimate of the parameters of interest. This, the constancy of the result, can be evaluated by a

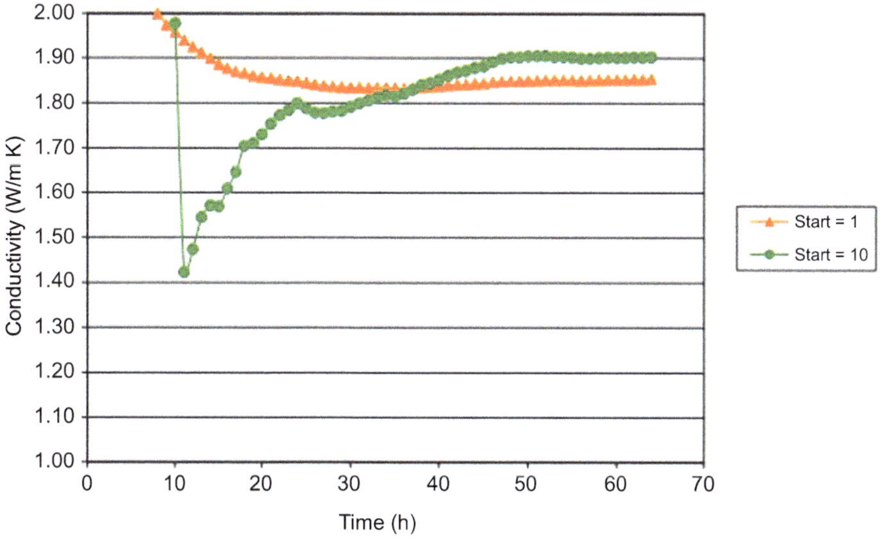

Figure 4.5 Constancy of regression solution, forward cumulative windowing. Convergence on the estimated equivalent thermal conductivity is obtained after about 50 h.

cumulative sum control chart (CUSUM) test (Brown et al., 1975) where the parameter of interest (thermal conductivity) is plotted as a function of the amount of data used in the evaluation. This can be done in a forward or backward way; Fig. 4.5 gives an example of a forward calculation and shows that in the present experiment convergence is reached after about 50 h.

4.11 Dealing with problems

The main problem with the line source solution is that the heat rate is assumed to be constant. Nonconstant heat rates in practice can consequently introduce errors. This can, first of all, be avoided by using an active control on the heat rate by using a proportional-integral-derivative (PID) control with a modulating three port valve controlled according to the temperature difference between the water entering and returning from the borehole. This solution was first implemented in the field test developed by Groenholland (van Gelder et al., 1999). To have a quick response to small perturbations an additional control on the pumping rate can be coupled to the control on temperature difference.

Other solutions to a varying heat rate are found in postprocessing. Several methods have been proposed, using numerical models that explicitly include varying heat rate (Yavuzturk, 1999), deconvolution (Beier & Smith, 2003), or stepwise analysis.

Another problem often encountered is the effect of changing ambient temperature conditions. The field test equipment originally used by Austin (1998) even included an air conditioning unit to keep conditions in the machine as constant as possible. Effects of changing ambient temperature conditions can be negated by installing the temperature sensors on the flow and return pipes of the borehole heat exchanger slightly subsided in the borehole, and then covering the sensors with a layer of earth and shielding to protect against radiation and possible infiltration of water. Although insulation of the connecting piping helps, it is not nearly as effective as installing the actual temperature sensors in the borehole proper.

4.12 Summary

An introduction to the TRTs of borehole heat exchangers has been presented. The purpose of the test, which is to measure as accurately as possible the thermal conductivity of the ground or find the main borehole heat exchanger design parameters (thermal conductivity, borehole thermal resistance, and undisturbed ground temperature profile), has been discussed. Different solutions (line source, cylinder source, numerical models) on which the test design can be based and that are used in the evaluation of the results have been presented and the equivalence with hydraulic pumping tests discussed. The practical application of the test and some suggestions concerning selecting appropriate test parameters has been given as well as procedures for assessment of test quality.

Recent research into TRTs has covered a wide range of subjects. The following is a selection of some recent publications.

Witte (2006a,b) and also Gustafson (2010) present tests based on a sequence of heating and cooling pulses at different heat rates (including heat injection and heat extraction), showing that such a procedure is more sensitive to groundwater flow and convection effects. Marcotte and Pasquier (2008) discuss the effects of different ways of calculating the bulk fluid temperature and show that this affects the results of the TRT analysis significantly. Fujii et al. (2009) and also Acuña et al. (2011) have published improved TRTs using fiber optic temperature measurements to evaluate differences in vertical temperature evolution. Bandos et al. (2009) have published a method to incorporate vertical temperature variations while Bandos et al. (2011) developed methods to account for effects of ambient temperature. Lamarche and Beauchamps (2007) present contributions to the finite line source for special geometries. Bauer et al. (2011) developed thermal resistance and capacitance models for borehole heat exchangers. Raymond et al. (2011b) discuss the temperature evolution during TRTs. Bruno et al. (2011, 2013a,b) present an evaluation of TRTs based on geostatistical modeling. Bruno et al. (2013b) found that a mildly modulating heat rate made the test more sensitive to groundwater movement, as measured in a controlled ground water extraction experiment. Pasquier (2015) published a stochastic interpretation method for TRT data. Loveridge et al. (2015) has published on TRTs for energy piles.

A number of alternative methods have been developed, for example, by accurately logging the temperature profile in a borehole heat exchanger with a miniature temperature probe (Geowatt; Martos et al., 2011) or using heating cables (Raymond et al., 2010).

Symbols

$\Delta T(r,t)$	Temperature change at radial distance r (m) and time t (s)
Q	Heat injection rate per unit of borehole length (W/m)
λ	Thermal conductivity (W/m K)
α	Thermal diffusivity (m^2/s)
E_1	Exponential integral
γ	Eulers constant (0.5772...)
r	Borehole radius (m)
ρ	Density (kg/m^3)
C_p	Specific heat capacity of the ground (J/(kg K))
$T_f(t)$	Average fluid temperature at time t (s)
R_b	Borehole thermal resistance (W/(m/K))
T_g	Undisturbed ground temperature (°C)
y	The change in temperature ΔT (K)
x	$\ln(t)$ (s)
c	Constants in the general linear equation, at $t = 0$
m	Intercept with y-axis of the linear regression (K)
k	Slope coefficient of the regression T_f with $\ln(t)$ (K/ln(s))

References

Acuña, J., Mogensen, P., Palm, B., 2011. Distributed thermal response tests on a multi-pipe coacial borehole heat exchanger. HVAC&R Research 17 (6), 1012–1029.

ASHRAE., 2007. ASHRAE Handbook – HVAC Applications. Atlanta: American Society of Heating, Refrigerating and Air-Conditioning Engineers, Inc.

Assael, M.J., Antoniadis, K.D., Wakeham, W.A., 2010. Historical evolution of the transient hot wire technique. International Journal of Thermophysics 31, 1051–1072.

Austin III, W.A., 1998. Development of an In Situ System for Measurement for Ground Thermal Properties (M.Sc. thesis). Oklahoma State University.

Bandos, T.V., Montero, Á., Fernándeza, E., Santandera, J.L.G., Isidroa, J.M., Péreza, J., Fernández de Córdoba, P.J., Urchueguía, J.F., 2009. Finite line-source model for borehole heat exchangers: effect of vertical temperature variations. Geothermics 38, 263–270.

Bandos, T.V., Montero, A., Fernández de Córdoba, P., Urchueguía, J.F., 2011. Improving parameter estimates obtained from thermal response tests: effect of ambient air temperature variations. Geothermics 40, 136–143.

Bauer, D., Heidemann, W., Müller-Steinhagen, H., Diersch, H.-J.G., 2011. Thermal resistance and capacity models for borehole heat exchangers. International Journal of Energy Research 35, 312–320.

Beier, R.A., Smith, M.D., 2003. Removing variable heat rate effects from borehole tests. ASHARAE Transactions 109 (2), 463–474.

Brown, R.L., Durbin, J., Evans, J.M., 1975. Techniques for testing the constancy of regression relationships over time. Journal of the Royal Statistical Society - Series B 149–192.

Bruno, R., Focaccia, S., Tinti, F., 2011. Geostatistical modeling of a shallow geothermal reservoir for air conditioning of buildings. In: Proceeding of IAMG 2011 Salzburg, "Mathematical Geosciences at the Crossroads of Theory and Practice, pp. 146–163.

Bruno, R., Focaccia, S., Tinti, F., 2013a. A Software Tool for Geostatistical Analysis of Thermal Response Test Data: GA-TRT. Computers and Geosciences.

Bruno, R., Salvatore, M., Tinti, F., Witte, H.J.L., 2013b. Probabilistic Approach to TRT Analysis: Evaluation of Groundwater Flow Effects and Machine - Borehole Interaction. European Geothermal Congress 2013, Pisa, Italy.

Carslaw, H.S., Jaeger, J.C., 1959. Conduction of Heat in Solids, second ed. Clarendon Press, Oxford.

Chiasson, A.D., Rees, S.J., Spitler, J.D., 2000. A preliminary assessment of the effects of groundwater flow on closed-loop ground-source heat pump systems. ASHRAE Transactions 106 (1), 380–393.

Claesson, J., 1985. Markvärme, en handbok om termiska analyser. Statens råd för Byggforskningsrådet, Stockholm.

De Vries, D.A., Peck, A.J., 1958. On the cylindrical probe method of measuring thermal conductivity with special reference to soils, part I: extention of theory and discussion of probe characteristics. Australian Journal of Physics 11, 255–271.

Deerman, J.D., Kavenaugh, S.P., 1991. Simulation of vertical U-tube ground-coupled heat pump systems using the cylindrical heat source solution. ASHRAE Transactions 97 (1), 287–295.

Eklöf, C., Gehlin, S., 1996. TED – a Mobile Equipment for Thermal Response Tests (Master's thesis 1996: 198E). Lulea University of Technology, Sweden.

Eskilson, P., 1986. Superposition Borehole Model, Manual for Computer Code.

Eskilson, P., 1987. Thermal analysis of Heat Extraction Boreholes (thesis). Dep. of Mathematical Phyisics, University of Lund, Sweden.

Fujii, H., Okubo, H., Nishi, K., Itoi, R., Ohyama, K., Shibata, K., 2009. An improved thermal response test for U-tube ground heat exchanger based on optical fiber thermometers. Geothermics 38 (4), 399–406.

Gehlin, S., 1998. Thermal Response Test, In-situ Measurements of Thermal Properties in Hard Rock (Licentiate thesis). LULEÅ University of Technology.

Geowatt A.G. http://www.geowatt.ch/index.php?pid=22.

Gustafson, A.M., 2010. Thermal Response Tests, Influence of Convective Flow in Groundwater Filled Borehole Heat Exchangers (Ph.D. thesis). Lulea University of Technology, Sweden.

Hellström, G., 1991. Ground Heat Storage, Thermal Analysis of Duct Storage Systems I: Theory (thesis). University of Lund, Sweden.

IEA Annex 21, 2013. Thermal Response Test (TRT). Final Report. http://www.iea-eces.org/files/a4.1_iea_eces_annex_21_final_report_1.pdf.

Ingersoll, L.R., Plass, H.J., 1948. Theory of the ground pipe heat source for the heat pump. ASHVE Transactions 54.

Ingersoll, L.R., Zobel, O.J., Ingersoll, A.C., 1954. Heat Conduction with Engineering, Geological and Other Applications. University of Wisconsin Press.

Kavenaugh, S.P., Rafferyty, K., 1997. Ground Source Heat Pumps: Design of Geothermal Systems for Commercial and Institutional Buildings. ASHRAE.

Kelvin, Thomson, W., 1882. Mathematical and Physical Papers. II, p. 41 ff.

Lamarche, L., Beauchamp, B., 2007. A new contribution to the finite line-source model for geothermal boreholes. Energy and Buildings 39, 188–198.

Loveridge, F., Olgun, C.G., Brettmann, T., Powrie, W., 2015. Group thermal response testing for energy piles. In: XVI European Conference for Soil Mechanics and Geotechnical Engineering, Edinburgh, GB, 13–17 Sep 2015, 6 pp.

Marcotte, D., Pasquier, P., 2008. On the estimation of thermal resistance in borehole thermal conductivity test. Renewable Energy 33 (11), 2407–2415.

Martos, J., Montero, A., Torres, J., Soret, J., Marinez, G., Garcia-Olicna, R., 2011. Novel wireless sensor system for dynamic characterization of borehole heat exchangers. Sensors 11, 7082–7094.

Mogenson, P., 1983. Fluid to duct wall heat transfer in duct system heat storages. In: Proceedings of the International Conference on Subsurface Heat Storage in Theory and Practice. Swedish Council for Building Research.

Monzó, P.M., 2011. Comparison of Different Line Source Model Approaches for Analysis of Thermal Response Test in a U-pipe Borehole Heat Exchanger (M.Sc. thesis). KTU, Stockholm.

Pasquier, P., 2015. Stochastic interpretation of thermal response test with TRT-Sinterp. Computers and Geosciences 75, 73–87.

Poppei, J., Schwarz, R., Peron, H., Silvani, C., Steinmann, G., Laloui, L., Wagner, R., Lochbühler, T., Rohner, E., 2008. Innovative Improvements of Thermal Response Tests. Swiss Federal Office of Energy. Project 101'690.

Priestley, J., 1781. Experiments and Observations Relating to Various Branches of Natural Philosophy; with a Continuation of the Observations on Air, vol. II. Pearson and Rollason, London.

Raymond, J., Robert, G., Therrien, R., Gosselin, L., 2010. A Novel Thermal Response Test Using Heating Cables. In: Proceedings World Geothermal Congress, Bali, Indonesia.

Raymond, J., Therrien, R., Gosselin, L., Lefebre, R., 2011a. A review of thermal response test analysis using pumping test concepts. Ground Water 49 (6), 932–945.

Raymond, J., Therrien, R., Gosselin, L., 2011b. Borehole temperature evolution during thermal response test. Geothermics 40, 69–78.

Schleiermacher, A., 1888. Ueber die wärmeleitung der Gase. Annalen der Physik 70 (8A), 623–646.

Shonder, J.A., 1999. Determining effective soil formation properties from field data using a parameter estimation technique. ASHRAE Transactions 105 (1), 458–466.

Signorelli, S., Basetti, S., Pahud, D., Kohl, T., 2006. Numerical evaluation of thermal response tests. Geothermics 36, 141–166.

Stålhane, B., Pyk, S., 1931. Ny metod för bestämning av värmeledningskoefficienter. Teknisk Tidskrift 28 (Svenska Teknologföreningen).

Theis, C.V., 1935. The relation between lowering of piezometric surface and rate and duration of discharge of a well using groundwater storage. Transactions of the American Geophysical Union 16, 519–524.

Van Gelder, A.J., Witte, H.J.L., Kalma, S., Snijders, A., Wennekes, R.G.A., 1999. In-situ Messungen der thermische Eigenschaften des Untergrunds durch Wärmeentzug. In: Hitziger, T. (Ed.), OPET Seminar Erdgekoppelte Wärmepumpen zum heizen und Klimatisieren von Gebäuden.

Wagner, R., Clauser, C., 2005. Evaluating thermal response tests using parameter estimation for thermal conductivity and thermal capacity. Journal of Geophysics and Engineering 2 (4), 349–356.

Witte, H.J.L., van Gelder, A.J., Spitler, J.D., 2002. In-situ measurement of ground thermal conductivity: the Dutch perspective. ASHRAE Transactions 108 (1).

Witte, H.J.L., 2002. Ground thermal conductivity testing: effects of groundwater on the estimate. In: Wärmetransport in der Kruste - Beiträge zur allgemeinen und angewandten 3. Kolloquium des AK Geothermik der DGG 3–4 October 2002, Aachen, Germany.

Witte, H.J.L., 2006a. Advances in geothermal response testing. In: Paksoy (Ed.), Nato Advanced Study Institute on Thermal Energy Storage for Sustainable Energy Consumption (TESSEC), Fundamentals Case Studies and Design, June 6–17, 2005, Izmir-Cesme, Turkey.

Witte, H.J.L., 2006b. Geothermal response tests using controlled multi-power level heating and cooling pulses (MPL-HCP): qauntifying ground water effects on heat transport around a borehole heat exchanger. In: Stiles (Ed.), The Tenth International Conference on Thermal Energy Storage, Ecostock 2006 Proceedings, May 31–June 2, 2006. Stockton College, New Jersey, USA.

Witte, H.J.L., 2013. Error analysis of thermal response tests. Applied Energy 109, 302–311.

Yavuzturk, C., 1999. Modeling of Vertical Ground Loop Heat Exchangers for Ground Source Heat Pump Systems (Ph.D. thesis). Oklahoma State University.

Horizontal and compact ground heat exchangers

5

S.J. Rees
University of Leeds, Leeds, United Kingdom

Abstract

This chapter describes horizontal ground heat exchanger (GHE) technology that is commonly applied to residential buildings and small non-residential buildings along with examples of innovative compact GHE designs. The essential physical properties and processes that are involved in shallow GHE behaviour are described – some of which are arguably more complex than have to be considered in other forms of GHE. Sources of design information and associated uncertainties are discussed along some of the developments in design models and methodologies. Two recently developed models that can be used to simulate GHE behaviour and capture the geometric complexities of certain forms of horizontal GHE are described in detail towards the end of this chapter.

5.1 Introduction

Vertical borehole heat exchangers (BHEs) continue to be the predominant form of ground heat exchanger (GHE) utilised in both residential and non-residential ground-source heat pump (GSHP) applications. Such heat exchangers have been the subject of academic study since the 1980s and many models for design and simulation purposes exist and some of these have been commented on in chapters 'Vertical borehole ground heat exchanger design methods', 'Calculation of borehole thermal resistance', and 'In situ estimation of ground thermal properties'. Shallow horizontal forms of GHE offer lower complexity and potential economic advantages compared to vertical designs and, although more land area may be required to meet a given heat exchange requirement, they continue to be an appealing technology for residential and smaller non-residential applications. Although the technology is essentially very simple, there are some significant challenges in modelling the physical processes that determine their performance, perhaps more so than in the case of vertical BHE. Some of the physical complexities of the behaviour of shallow soil zones are discussed in this chapter.

Although there is clearly a need to be able to calculate horizontal GHE performance there has been, we suggest, less progress in developing validated models and consequently design methods remain relatively simplified – design data tables and charts often being used in practice. The capabilities of a number of models and the

Advances in Ground-Source Heat Pump Systems. http://dx.doi.org/10.1016/B978-0-08-100311-4.00005-4
Copyright © 2016 Elsevier Ltd. All rights reserved.

difficulties in developing and applying models of horizontal heat exchangers will be discussed. Two recently developed modelling methods that can be applied to such heat exchangers and other forms of compact shallow heat exchanger will be presented later in this chapter.

5.1.1 Horizontal heat exchanger forms

Horizontal GHEs avoid the need for specialist drilling operations. They are consequently intended to be installed in trenches that can be efficiently created with trenching or excavation machinery commonly used for utility service or foundation construction. The heat exchangers are often sufficiently light and flexible that they can be installed in the trenches without other machinery. Installation depths are typically in the range 1–2 m. Although greater depths are found this often requires larger scale pits to be excavated. The simplest forms of horizontal heat exchanger consist of pairs of straight pipes with a loop or bend fittings at the end of the trench. More than one pair of pipes can be installed at different depths in the same trench. This is illustrated in Fig. 5.1. Careful attention has to be paid to the use of backfilling materials and installation procedures to ensure there is good contact between the pipe and the ground in the finished heat exchanger.

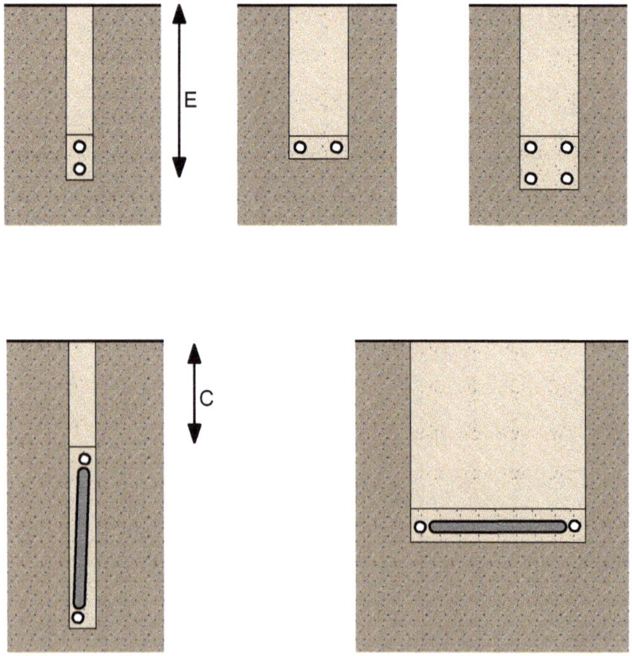

Figure 5.1 Common forms of horizontal ground heat exchanger trench arrangements. Trench depths (E) are typically in the range 1.2–2 m and minimum cover (C) of approximately 0.75 m depending on frost conditions.

Figure 5.2 Slinky ground heat exchangers being installed in a vertical orientation in a narrow trench (left) and being assembled prior to installation (right). Relatively fine backfill material is used near the pipes to ensure good thermal contact and protect against damage.
With permission of GeoScience.

As small diameter plastic pipe (25–32 mm outside diameter pipes are common) are manufactured in coiled form, it is relatively simple to form a 'Slinky'[1] arrangement where loops are tied to each other and are laid out – often with some overlap – in wide trenches (of the order of 1 m wide). Slinky heat exchangers can also be installed in vertical orientation in narrow trenches as illustrated in Fig. 5.2. Slinky heat exchangers have some advantage over parallel pipe designs in that it is possible to increase the pipe heat transfer area per unit length of trench (and hence overlying ground surface area). Increasing the effective heat exchanger area between the fluid and the ground in this way can be advantageous, but the value of doing so can be negated by increasing the pressure drop and hence pump energy. Some care in optimizing the design is consequently required.

Directional drilling methods can also be used to install horizontal GHEs (Fig. 5.4) although this is less common and requires specialist machinery. Directional drilling allows a curved path to be drilled (approximately 100 mm diameter) to avoid any below-ground obstacles or to avoid disturbing the ground surface. A pair of straight pipes with a U-tube fitting is drawn into the end of the bore, where it appears above ground, in reverse. The bore naturally closes around the pair of pipes which become closely spaced. Installation depth can be controlled so as to be of similar depth to a conventional trench (1–2 m) or can deviate by as much as 10 m on a curved path over distances of 100 m if required.

[1] Slinky is a registered trademark of Poof-Slinky Inc., 45400 Helm St., Plymouth, MI, USA.

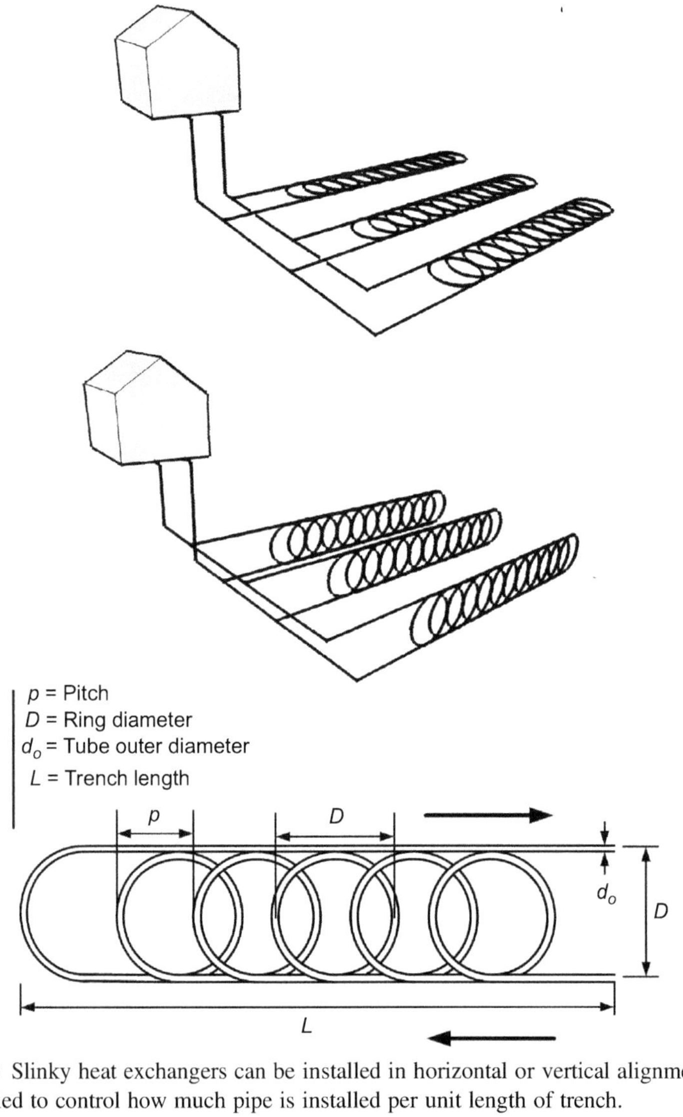

Figure 5.3 Slinky heat exchangers can be installed in horizontal or vertical alignments. Pitch can be varied to control how much pipe is installed per unit length of trench.
After Xiong, Z., Fisher, D.E., Spitler, J.D., 2015. Development and validation of a Slinky ground heat exchanger model. Applied Energy, 141, 57–69.

5.1.2 Advanced shallow ground heat exchangers

A number of other forms of GHE designed to be installed at relatively shallow depths that have economic advantages or require less land area to install, have been developed commercially or demonstrated in research projects. The first type of design – a number of similar products or site-assembled versions are in use – consists of small diameter pipes formed into a closely pitched spiral. These are usually intended to be installed

Figure 5.4 A directional drilling machine being used to install a horizontal ground heat exchanger.
With permission of GeoTheX b.v.

vertically in auger-drilled holes of the order of 0.5 m diameter and 4–8 m depth. They are intended for residential buildings but can be used in small non-residential applications. Typically between four and eight heat exchangers are necessary for residential applications and such heat exchanger arrangements have the advantage of requiring less land footprint than required for an equivalent conventional horizontal GHE and also avoiding the need for borehole drilling. There is some similarity with spiral heat exchangers used in energy pile designs (Zarrella et al., 2013), except that in this case the heat exchangers are simply backfilled with soil material. An example of a spiral device is shown in Fig. 5.5.

A form of horizontal GHE for low energy residential buildings termed a foundation heat exchanger (FHX) has been developed and monitored by researchers at Oak Ridge National Laboratory (Im et al., 2012). In this design, digging trenches for the heat exchanger is avoided by maximising the use of the excavations around the house basement and utility service trenches approaching the house. In this form of FHX the pipes are fixed to the wall of the excavation and backfilled with soil material and not cast into the concrete structure. The basement wall is insulated on the outside to limit the effect of the heat exchanger on the basement heating or cooling loads. The circuit arrangement is equivalent to three pairs of pipe loops in parallel. Such a heat exchanger is illustrated in Fig. 5.6 and is also discussed in one of the case studies in chapter 'Borehole thermal energy storage'. One of the modelling methods used to analyse this GHE design, the dynamic thermal network (DTN) approach, is discussed in detail later in Section 5.4.2.

5.1.3 Shallow ground conditions and heat transfer

Where vertical BHEs are installed, at depths of the order of 100 m or more, hydrostatic pressures ensure that the ground is fully saturated in most geologies and thermal

Figure 5.5 Spiral ground heat exchangers being installed following auger drilling (above) and backfilling (below).
With permission of REHAU Ltd.

properties can be safely assumed not to be subject to seasonal variations. This is reflected in the fact that models of vertical BHEs generally ignore heat transfer at the exposed ground surface or assume a constant temperature.

Conditions at the depths at which horizontal heat exchangers are installed are quite different to those around deep boreholes. At depths up to a few metres – and depending on water table depth – the ground is only partially saturated and moisture content varies both spatially and with time. In this ‘vadose zone’ the ground is composed of matter in three phases – solid porous material, liquid water and air/water vapour. In some conditions ice can form. Furthermore, behaviour is complicated by non-linearities in the relationships between water potential and water content (Philip and De Vries, 1957). Significant hysteresis in the water saturation (and hence thermal properties) has been demonstrated so that there can be significant differences between the drying and wetting

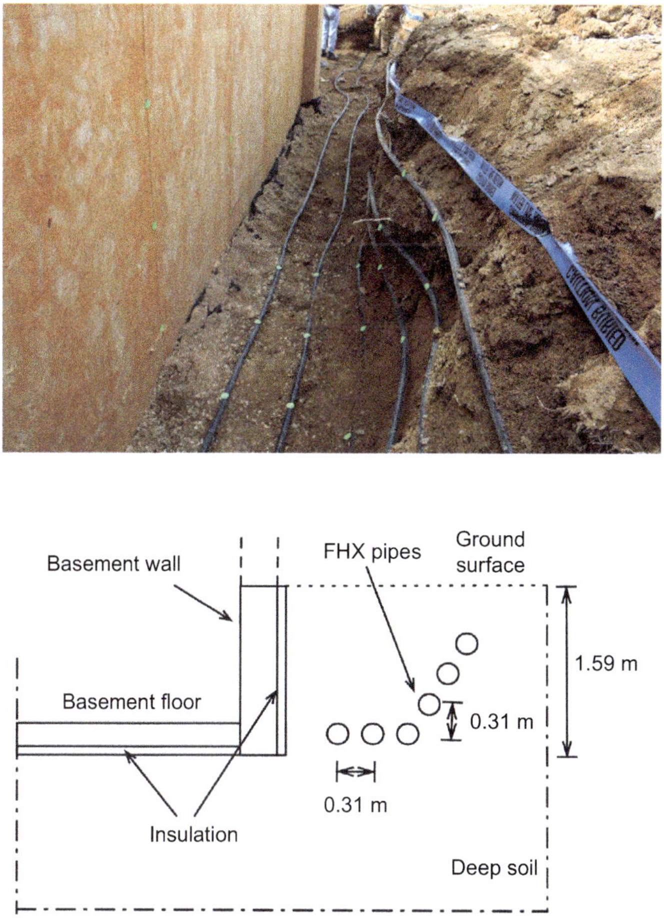

Figure 5.6 A foundation heat exchanger system installed alongside a basement. Shown before backfilling (above) and diagrammatically (below).
With permission of Oak Ridge National Lab.

phases. This is partly due to capillary forces being significant and highly dependent on the soil pore size and structure and hence, broadly speaking, soil type (Richards, 1931). The physics of the vadose zone is naturally important in study of plant growth and irrigation and has, accordingly, been studied by soil scientists for more than 100 years (Buckingham, 1907). Although soil behaviour is relatively well understood in these contexts, many of these complications are ignored or treated in a very simplified manner in studies of horizontal GHE performance.

At the ground surface thermal equilibrium is maintained with the environment by a number of mechanisms. The surface temperature is strongly driven by weather conditions and the conduction and moisture fluxes just beneath so that surface temperatures can vary significantly during the diurnal cycle. The thermal capacity of the ground is sufficient that at only a short distance below ground diurnal swings in temperature are reduced. At depths greater than a few 100 mm the variations in temperature are sufficiently damped that fluctuations predominantly follow a seasonal rather than a diurnal pattern. These trends are demonstrated in the ground temperature measurements that are usually taken at depths up to 2 m at meteorological observation sites.

In the early 1960s, Kusuda and Achenbach (1965) analysed a large number of these data sets and concluded that the near-surface ground temperatures could be characterised by an analytical solution to Fourier's equation for a semi-infinite solid subject to cyclic convective boundary conditions. This solution characterises temperature fluctuations varying sinusoidally on a seasonal basis with an amplitude that decays exponentially with depth and also demonstrates a characteristic phase lag. The mean ground temperature is found to be close to the mean annual air temperature. At depths of approximately 5 m there is often no observable seasonal variation in temperature, ie, ground temperatures are isolated from monthly climatic variations. At greater depths the natural geothermal gradient (approximately 1−2 K/m) becomes evident. This is illustrated in Fig. 5.7 where the analytical solution with typical parameter values is superimposed on the geothermal gradient.

The near-surface ground temperature characteristics indicated by meteorological data and Kusuda and Achenbach's model (1965) has notable implications for horizontal GHE design and modelling. In general, when considering typical horizontal GHE installation, sensitivity to surface conditions is strongly dependent on pipe depth. In many climates pipe installation depths much less than 1 m mean that there would be risk of freezing towards the end of the winter. Above the typical installation depths of 1−2 m there is an evident summer recharging of the heat stored in the ground

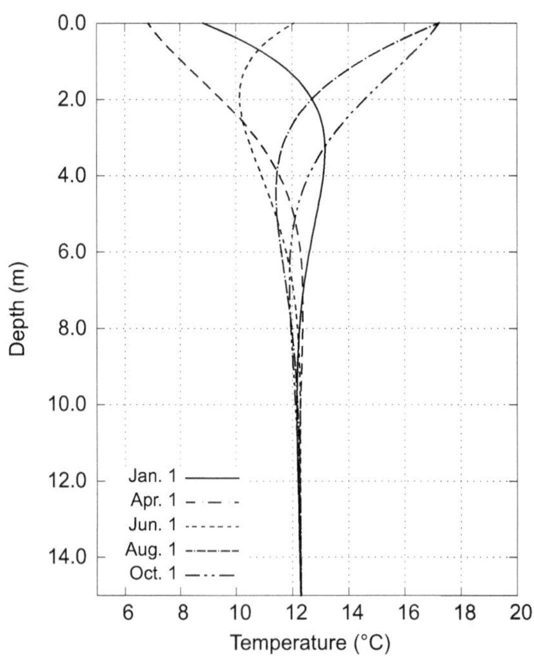

Figure 5.7 Typical shallow ground temperatures at different times of year. Ground temperatures lag the air temperature by several weeks and amplitude reduces quickly with depth.

and a discharge during winter – with some noticeable time lag. At depths of more than approximately 3 m heat transfer with a horizontal pipe could be expected to be dominated, over short and medium timescales, by conduction to/from the surrounding ground by conduction and largely isolated from surface conditions – much as it is in a vertical BHE. The type of compact spiral heat exchangers noted above, although oriented vertically, are short enough that they could also be expected to be subject to these near-surface seasonal temperature fluctuations.

It should be noted that there are some situations where the type of temperature profile described by the Kusuda and Achenbach model, and illustrated in Fig. 5.7, would be modified. First, the temperature profile can be modified by the presence of ground water if the water table is high and where there is groundwater movement. Second, where the climate is such that the upper surface of the ground is subject to prolonged frost, surface ice or snow cover. This was demonstrated in a study by Xu and Spitler (2014) who used a one-dimensional numerical solution of Richard's equation (Richards, 1931) that took account of both moisture and surface freezing, which showed that ground temperature profiles were not well described by Kusuda and Achenbach's model and that freezing needed to be taken into account in cold regions. A further situation where the temperature profile may be modified is in dense urban environments, in which case it is possible for long-term build up of heat through buildings in contact with the ground.

In very cold climates, such as those found in parts of Canada and Sweden, ground conditions can be dominated by permafrost and considerations of operating temperature and freezing around the GHE pipes are rather different than those in mild climates. Where naturally occurring ground temperatures are in the freezing range there is generally no detrimental effect of operating the heat pump ground loop at sub-zero temperatures as any increased freezing is unlikely to disturb the ground. Fontaine et al. (2011) suggest that such freezing may, in fact, be of benefit in stabilising ground conditions. In milder climates without permafrost, it may be detrimental to operate the system at sub-zero temperatures as localised freezing around the pipes may induce 'frost heave' which can be damaging near structures or where pipes approach the surface (eg, on entry to the building). Another primary consideration is that heat source operating temperature is related to thermodynamic efficiency. In mild climates designers often aim to achieve minimum fluid temperatures in the ground loop a few degrees Kelvin above freezing to maintain good efficiency: otherwise there would be little advantage in ground coupling compared to an air-source heat pump system. In cold climates, fluid temperatures a few Kelvin below freezing still correspond to a significantly improved operating efficiency when compared to air source systems. It has furthermore been shown that the insulating effect of snow cover and the latent heat storage by virtue of ice surrounding the pipes in such situations is beneficial (Tarnawski, 1989).

Being able to take proper account of ground surface heat transfer (boundary conditions) is important in proper design methods and simulation models. Heat transfer is clearly subject to convection driven by surface and atmospheric air temperature differences. It is also naturally influenced by solar insolation. Indeed, horizontal GHEs are sometimes thought of as solar collectors but this analogy is not accurate as other processes also strongly influence net surface heat transfer (Garcia Gonzalez et al., 2012). (The presence of the heat exchanger has no influence over how much solar insolation is absorbed by the ground surface above, after all.) The ground surface is subject to long wave heat transfer with the environment and with the hemisphere of the sky in

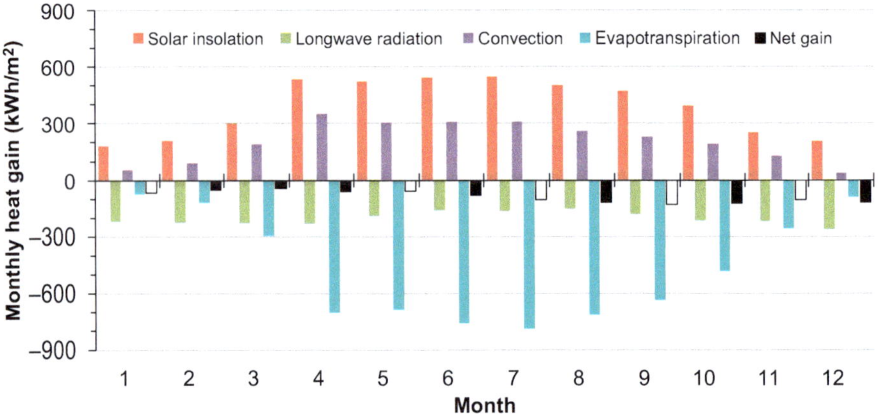

Figure 5.8 Ground surface monthly heat flux budgets and the net flux calculated for a demonstration site in Oak Ridge, Tennessee.
After Fan, D., Rees, S., Spitler, J., 2013. A dynamic thermal network approach to the modelling of foundation heat exchangers. Journal of Building Performance Simulation, 6 (2), 81−97.

particular. As the sky has an effective temperature at night time that is very low when the sky is clear, the long wave heat flux can amount to a significant loss of heat from the ground. The ground surface moisture balance also has a significant effect on the net heat transfer with the environment. Moisture is added to the ground vadose zone during rainfall and snow/ice melting and is evaporated from the ground, and furthermore, moisture is drawn out by any plant growth and subsequently evaporated. The phase change involved in the evaporation process means that the net evapotranspiration flux can be a significant heat loss from the surface.

The net effect of the heat fluxes at the ground surface governs the change in shallow subsurface ground temperatures. The net effect is dependent on large opposing fluxes − solar insolation being a significant gain and both long-wave radiation and evapotranspiration being large losses, typically. These fluxes are balanced by the heat that is exchanged with the subsurface material by conduction. The net effect is consequently rather sensitive to differences between opposing fluxes so that the net flux can swing from being a net gain to a net loss according to rainfall and irrigation, shading and plant growth (Blight, 2003). These heat fluxes have been calculated on a monthly basis and are shown in Fig. 5.8 (Fan et al., 2013). The magnitude of the net flux can be seen to be rather less than some of the heat flux components. In modelling exercises this means that uncertainties in surface parameters or environmental conditions can have a significant effect on the predicted heat exchanger performance (Xing et al., 2011; Fan et al., 2013).

5.2 Shallow ground thermal properties

Ground thermal conductivity (or perhaps we should say effective thermal conductivity) is a critical design parameter in shallow horizontal GHEs just as it is in vertical borehole systems. Although thermal diffusivity is also significant in determining timescales for long-term temperature variations, thermal conductivity is the main parameter that relates system fluid temperatures to heat transfer rates. Accordingly, this is one of the main parameters in all the design methods discussed in the following section. In order to develop a heat exchanger that is optimised with respect to some blend of efficiency and cost, one would like to use a thermal conductivity value with a good level of certainty. However, this is just one of the parameters that is quite uncertain in horizontal GHE design.

In situ thermal conductivity testing methods and data analysis methods for vertical borehole systems are now well established and well understood – as chapter 'In situ estimation of ground thermal properties' has shown. Uncertainty can be as good as 10% using well calibrated in situ test equipment and modern analysis methods. The scale of many non-residential vertical BHE projects means that in situ tests can be afforded in many cases and achieving a more optimal design can justify the expense. The situation is very different where horizontal GHEs are concerned and systems are small-scale (Remund, 1988). In situ test methods are not widely used and a wider variety of data sources are relied upon (Busby, 2015).

5.2.1 Available property data

The saturated ground conditions that are experienced over the depths typical of BHEs means that properties such as thermal conductivity are not found to vary over time (seasons) or vary very much from one location on a particular site to another. Shallow ground conditions at the depths of horizontal GHEs, on the other hand, have been found to vary noticeably across sites and according to season, with low values occurring in dry summer periods (Naylor et al., 2015; King et al., 2012). King et al. (2012) present a number of measured conductivity data sets (using the probe method discussed below) which show considerable variation in readings across particular sites. For example, a site with a mean value of 1.28 W/mK but minimum and maximum values of 0.43 and 1.93 W/mK respectively and standard deviation of 0.35 W/mK is reported, and such variations are said to be expected.

5.2.1.1 Soil type identification

Variations from one site to another with different soil types can be significant according to basic soil type but also local moisture conditions. First-order estimates of soil thermal conductivity are often made on the basis of identifying or classifying the predominant soil type and referring to tabulated data sources (Marlowe et al., 1989; Farouki, 1981; IGSHPA, 2009). Soil types are often identified according to the degrees of clay, sand and silt content and particle size. Naylor et al. (2015) compared such approaches with in situ measurements of a number of types at sites in Indiana, USA. They found that estimates based on soil identification methods and tabulated values led to conservative horizontal GHE lengths by as much as a factor of two. This may be somewhat reassuring (in that tabulated values are conservative) but also suggests that many systems designed on the basis of such data are not as economic as they could be.

5.2.1.2 *Geomapping of thermal property data*

A good deal of geological data is available in digital map form and is, in principle, a convenient way to access ground property data. However, until recently GIS data sets relating to thermal properties at shallow depths relevant to horizontal GHE design have been lacking (Busby, 2015). The European Union research project ThermoMap (www.thermomap-project.edu) sought to address this by firstly harmonising national digital data sets such as climate and underlying soil type. Combinations of such data were used to derive thermal conductivity values at depths up to 10 m (Thermomap, 2013). In this approach, soil type GIS data and climate data were used to deduce soil characteristics such as bulk density and pore size distribution so that Kersten's empirical model (Kersten, 1949) could be applied to derive calculated values of soil thermal conductivity.

An example view of the ThermoMap visualisation and calculation tool is shown in Fig. 5.9. This data set does not include values within many dense urban areas. In these locations soil type can deviate from the underlying type due to disturbance during construction and backfilling with non-native soils, rock and aggregates, and so mapped values would be particularly uncertain. The use of GIS data sets derived from modelling studies has also been demonstrated by García-Gil et al. (2015).

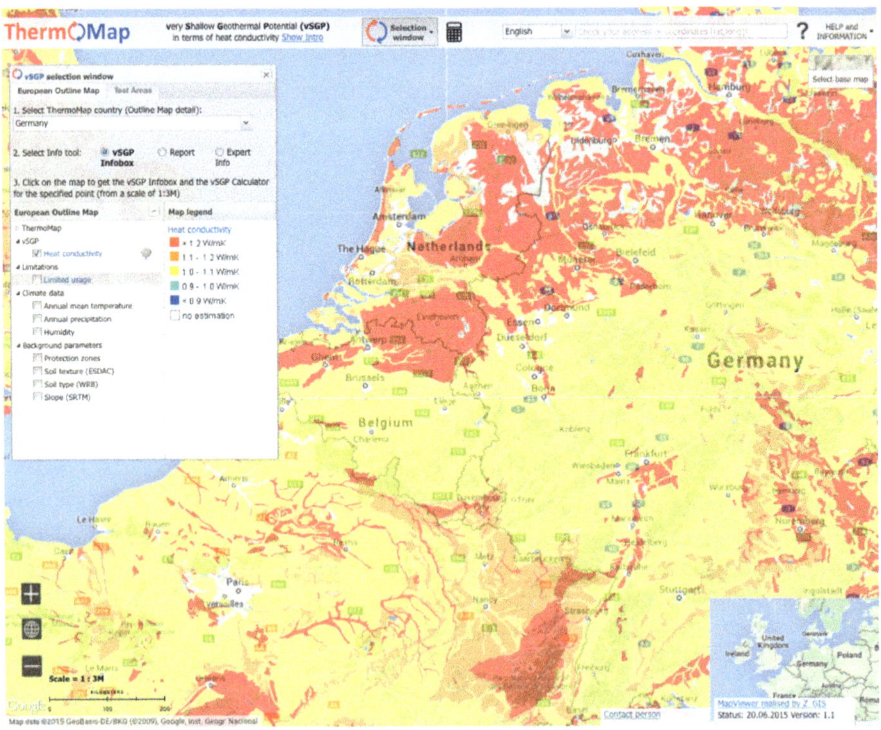

Figure 5.9 A geomapping tool for interrogation of soil thermal properties in Europe. Screen shot of the software available from www.thermomap-project.eu.

5.2.1.3 In situ conductivity testing

Soil scientists have been using inverse methods based around line-source analytical solutions to Fourier's equation – the same approach as used in BHE thermal response tests (TRTs) – to make site investigations of soil properties for many years (Bristow et al., 1994; Campbell et al., 1991) and recognised test standards have been established (ASTM, 2008). Such test methods use small needle probes that incorporate an electric heating element and temperature sensors. The probes are pressed into the soil and a constant heating flux produces a thermal response close to that of an ideal line source. Inverse methods are used to infer the local thermal conductivity.

King et al. (2012) give a detailed description of test procedures and data analysis using one form of commercially available needle probe (also shown in Fig. 5.10). The probe in this device is 170 mm long and 6.35 mm diameter (Hukseflux, 2010). Data is collected and simultaneously analysed by a handheld data logging computer and controller. Several thermal conductivity values are derived as data is collected over several minutes until an acceptably repeatable value is obtained (eg, standard deviation <10%). King et al. (2012) suggest that 12–16 measurement points may be necessary over a typical site to derive statistically meaningful values of bulk thermal conductivity and ground temperatures that can be used in design

Figure 5.10 Site investigation of soil thermal conductivity using a probe test instrument. With permission of Carbon Zero Consulting.

calculations. Some differences in results were found with higher heating rates and from summer to winter seasons, ie, varying moisture levels.

An alternative approach to experimental measurement would be to take samples and use laboratory procedures as has sometimes been done for energy pile applications (Low et al., 2014). These laboratory tests can be carried out with small needle probes or steady-state test cell procedures (Nicholson et al., 2008). The advantage of in situ probe testing compared to removing samples for testing is that compression and moisture levels are representative and not subject to change when transferred to a laboratory. These forms of off-site testing have not been reported to date for horizontal GHE applications.

Although probe testing does not appear to have become common practice, the published results and observations do serve to highlight the variability and uncertainty in ground thermal properties in shallow ground conditions.

5.3 Horizontal heat exchanger design methods

5.3.1 The design task

Design calculations for vertical BHE arrays commonly rely on computer design tool results that are, in turn, derived from analytical and numerical methods. In the design of horizontal GHEs for residential building applications a wider variety of tools and methods are in use:

1. numerical models of ground heat (and possibly moisture) transfer;
2. computer implementations of analytical heat transfer models;
3. manual methods based on design chart and tabular data.

There are historical reasons why there has been a preference for simpler models and manual approaches – lack of accessible computing power in earlier decades being just one reason. Relatively sophisticated numerical models can now be run with the computing resources available to most practitioners and recently developed models are able to deal with the full range of heat exchanger configurations and geometries – as will be shown in Section 5.4. Although the higher fidelity treatment of the physical processes available with such models is technically appealing, there are good reasons why simpler approaches are still relevant and appropriate for residential building applications.

The overall aim when designing a horizontal GHE is to decide for given ground conditions, climate and imposed thermal loads, a heat exchanger ground surface area and tube arrangement that allow long-term variations in fluid temperatures to be maintained within the desired limits. There are other considerations and constraints in the design problem but this is the technical task we would want a design method or tool to assist with. This requires consideration of both instantaneous heat transfer rates (peak loads) but also seasonal net heat balances (energy demands) to arrive at a sustainable design. In systems providing heating, only the length of the heating season is significant as reliance is placed on natural recharging during the summer to maintain a long-term heat balance. In heating and cooling systems then the balance depends on the relative significance of the heating and cooling loads and is highly dependent on climate conditions in residential applications.

The challenge in deriving a design methodology is to find an approach that is robust – that is, produces designs with acceptably low risk of poor performance – given the uncertainties in the design information. The significant uncertainties in the design inputs and parameters are what make horizontal GHE design for residential applications particularly challenging. The two main types of information that are project specific and that the designer must provide are the ground properties and the building heating and cooling loads and these can both be highly uncertain. In the case of non-residential buildings the building design and knowledge of the buildings' usage give a reasonable basis for calculating both peak loads and annual energy demands. Any calculation of heating and cooling loads is uncertain because of the limitations of the calculation methods (eg, ground heat transfer, wall corners and junctions are not well treated in most methods) and variability in construction (eg, air tightness) but also local conditions such as shading and wind exposure varying from one building plot to another. In the case of residential buildings the prediction of annual heating and cooling demands is made more uncertain because occupancy levels (family sizes) patterns of occupancy and room temperature preferences and other behaviours mean that demands can vary significantly from one house to another. There is little that can be done to reduce this design uncertainty.

The uncertainty in determining ground thermal properties has already been commented upon. The predominant method of determining ground properties in residential projects is to make a choice based on identification of soil type and observation of ground moisture conditions. Surveys of soil thermal properties and site investigations have always shown considerable variations in thermal properties for given identifiable soil types (King et al., 2012). The other uncertainties that should be acknowledged in methods that assume heat transfer is by conduction alone, are that the effects of moisture transport and the complexities of evapotranspiration at the ground surface are being ignored (see discussion of Fig. 5.8) as are seasonal variations in ground properties. Although more sophisticated models of ground heat transfer exist, and have been available since the first wave of research in such systems, they do not necessarily lend themselves to practical design methodologies. The difficulty in using more sophisticated models is often that a greater number of parameter values, along with initial condition information that are difficult to determine, need to be supplied by the designer. From a pragmatic design point of view simplified methods may be quite adequate at arriving at acceptable designs if one considers the overall uncertainties in the design process.

There are different ways to manage or estimate risk in the design process. In simple design procedures (such as the manual methods noted later in this section) where the designer has to choose certain parameter values from tables, one approach to deal with the various uncertainties and limit risk is to ensure the values in the tables are conservative: for example, by using thermal conductivity values that are in the lower quartile of those reported rather than the mean value reported. In the case of residential building heating and cooling applications, values of annual demand (kWh) are often arrived at by calculating the peak loads by a standard procedure and then assuming the system effectively runs for a standard number of hours at that rate of heat extraction or rejection – so-called 'full-load equivalent run hours'. (This value is highly dependent on user behaviour and usage patterns in reality.) Again, to limit risk one could use a conservative value. The difficulty in taking conservative values is, of course, that one may produce designs that are more expensive than they need to be in some cases. The topic of risk and uncertainty in the design process is discussed again at the end of this chapter.

5.3.2 Background

The first collective effort to investigate the performance of ground-coupled heat pump systems that is reflected in scientific literature was in North America starting in the 1940s and developing through the 1950s and was initiated with the support of a number of electricity utility companies (Svec, 1987). Much of this work was experimental in nature and results are reported from a variety of horizontal heat exchanger designs with a range of pipe sizes, loop lengths and spacings. Much of this work is reported simply in terms of heat transfer rates achieved (over a range of timescales) and operating temperatures without offering generic conclusions (eg, Coogan, 1949; Penrod et al., 1950; Kidder and Neher, 1952; Harlow and Klapper, 1953; Hapley, 1949). These experimental studies were of systems that used metal pipes and sought to both heat and cool residential buildings.

The first significant development in horizontal GHE design methods was the publication of an analytical solution by Ingersoll and Plass (1948) based on Kelvin's line source theory. There was already some understanding of soil heat transfer and the importance of heat capacities, thermal conductivity and moisture by virtue of electrical supply companies existing interests in buried cable design, although design methods were limited to steady-state conditions. Efforts to develop and market heat pumps in the 1950s in the USA were not that successful and there was something of a lapse in research effort and Ingersoll's method does not seem to have received much attention. Ambrose's text book on heat pumps written in the 1960s only refers to experimental data (he casts doubt on the viability of GSHP systems for cooling in view of poor thermal contact at the pipes in summer conditions) and similarly Heap only offers the reader a limited amount of empirical data on ground loop heat transfer rates (Ambrose, 1966; Heap, 1979).

International interest in GSHP applications did not gather pace until the energy crisis of the 1970s (Svec, 1987). In the later 1970s and the 1980s research efforts grew both in North America and in Europe – mostly the Nordic and Germanic countries, eg, Rosenblad (1979) and Schär (1977). This period also coincides with the availability of computing facilities (at least in research organisations and large corporations) and plastic pipe suitable for ground loops. In Europe development of design methods was focused on vertical BHEs whilst in the United States, although there was work on vertical BHEs, there was an enduring interest in horizontal systems and significant work was undertaken at certain US National Laboratories (Metz, 1979; Mei, 1986a) and Universities (Bose, 1982).

Writing his review in 1987, Svec cites one dozen North American computer models that were developed in the 1970 and 1980s either specifically for ground-coupled heat pump applications or adapted from other soil heat and moisture analysis tools. A similar number of European computer models are also noted. These were mostly finite difference models and many treated both heat and moisture transfer. Writing at the end of the last century, Rawlings and Sykulski (1999) tabulated 13 computer codes avail able and in regular use, several of which are in use at the time of writing. Only a few of these are traceable to the earlier codes cited by Svec (1987).

5.3.3 Design methodologies and standards

When ASHRAE commissioned its first project to develop design guidance for GSHP systems in 1982 the approach taken by the contractors Bose and Parker (1984) was to rely heavily on the analytical methods set out by Ingersoll and Plass (1948). In this approach Kelvin's line source solution is used by superimposing line sources/sinks corresponding to the sets of horizontal pipes of interest. Virtual sources and sinks are added at mirrored positions to impose lines of symmetry and ground surface boundary conditions. Temperature responses can be added to those due to the natural seasonal variation in surface conditions to find the overall response. Temporal superposition is used to find maximum and minimum temperatures occurring during the system life-cycle given applied heat pulses equivalent to the heat pump ground loop loads (Bose and Parker, 1984). This approach seems to follow that developed in Europe about the same time by Claesson and Dunand (1983).

A computer implementation of this approach was used to derive design charts and tables for ASHRAE (there was a preference for manual methods rather than software at the time). In this design methodology the user provides load and thermal property data along with desired minimum and maximum temperature limits. Heat exchanger sizes are the output given a range of pipe arrangements, depths and spacings the designer can select from. Guidance on the choice of ground thermal properties was also provided by drawing on a systematic review of the soil property data available at the time (Bose and Parker, 1984). The model was incorporated a little later into the CLGS software (Bose, 1989) supplied by the International Ground Source Heat Pump Association (IGSHPA) that remains in use today – more sophisticated numerical models being beyond the capabilities of personal computers at the time. This approach was also used to produce data in IGSHPA's own design guides for horizontal heat exchangers (IGSHPA, 2009).

One of the most widely known GHE standards in Europe is the German standard VDI 4640 (VDI, 2001). This four-part standard discusses the fundamental considerations – regulatory, environmental, technical and practical – relating to using the ground for thermal energy exchange (VDI, 2010). It does not prescribe a particular horizontal GHE design method but it does state limits to allowable heat transfer rates and seasonal heat extraction rates for various soil types along with allowable minimum and maximum seasonal temperatures (some freezing being allowable). It does not deal with GHE designs like Slinky loops and leaves the option open for the designer to use other design tools. Reuss et al. (1998) describes the development of the standard through a collaborative effort of industry and academic stakeholders (including some from countries neighbouring Germany) to arrive at the final standard. The limiting heat transfer rates are consequently a consensus view drawn from more than one source of evidence – models and laboratory testing – and selected with some view to acceptable design risk.

Although the VDI 4640 standard has been used as a source of design data in countries other than Germany and has been incorporated into tools provided by some heat pump manufacturers, there is some question as to whether the data is universal and should be viewed in the context of central European climate and geological conditions (Curtis et al., 2013). Certainly where systems that provide both heating and cooling, summer operating conditions and ground temperatures are important and may drive the design process more than heating considerations.

In drawing up design standards for UK climate conditions (with its temperate maritime climate) VDI 4640 was used as a source of information but, in the case of horizontal GHEs, the CLGS software (Bose, 1989) was also used to develop limiting heat extraction rate (W/m) data (DECC, 2011; Curtis et al., 2013) suited to UK climate conditions and efficiency targets (minimum fluid temperatures designed to be above freezing). The CLGS model uses a superimposed line source approach and so pipe spacings giving the pipe length per metre of trench equivalent to Slinky heat exchangers of common pitch were used to derive Slinky design data. Slightly conservative values of ground thermal conductivity are recommended in this standard. The 'equivalent full-load run hours' idea was also used to relate peak heating loads to annual energy extraction. In the interests of designs with reasonable risk, peak loads are to be calculated with a standard method (EN21831) and assuming the heat pump normally has capacity to supply the whole of the peak load (DECC, 2012). This conservative approach was adopted to avoid problems with high usage of electrical resistance supplementary heat found in early UK installations (Dunbabin and Wickins, 2012).

5.4 Advances in modelling

A wide range of computer models of horizontal and compact GHEs have been developed since the 1970s (Svec, 1987) and these have utilised a range of numerical methods and analytical solutions. Whilst analytical models are limited to simplified geometries and purely conduction heat transfer in homogeneous media, numerical models have great potential to capture more realistic physics, complex geometries and a wider variety of environmental and operating conditions and are becoming more practical to use given growing desktop computing power. Accordingly, research into new models and improvements to earlier modelling methods continues. Two modelling approaches of particular merit are presented later in this section to illustrate recent advances.

Models of horizontal GHEs differ chiefly according to:

- numerical method: finite difference, finite volume or finite element methods
- dimensionality: one, two or three-dimensional meshes

- geometric complexity
- treatment of the pipe and fluid flow
- treatment of moisture
- treatment of freezing
- treatment of snow cover

Some of the earliest noted numerical models applied the finite difference method and modelled conduction heat transfer in two-dimensions (Mei, 1986a,b). Tarnawski and Leong (1993) went further and produced a two dimensional finite difference model that modelled both heat and moisture transfer along with complex surface boundary conditions. This model was proposed as a design tool but was not incorporated into any commercial tools such as those noted later by Rawlings and Sykulski (1999).

Piechowski (1998) developed a similar model but aimed at improved computational efficiency by resolving the temperature gradients around the pipe with a fine cylindrical finite difference mesh and coupling this to a coarse rectilinear grid of the surrounding soil and ground surface (Piechowski, 1998). The model included both moisture and conduction effects so that some shrinkage effects at high ground temperatures could be diagnosed but was not modelled explicitly. Modelling of freezing was not included as the work was done in the context of an experimental installation in Melbourne, Australia (Piechowski, 1999), where interests were focussed on cooling applications.

A very similar approach incorporating coupled fine and coarse meshes was implemented more recently by Lee et al. (2013) in the EnergyPlus simulation environment (Crawley et al., 2001). This model does not include any equation for moisture transport but does deal with freezing by modifying the heat capacity over a narrow temperature range around the freezing point – as suggested by Lamberg et al. (2004). The model is quasi three-dimensional in that the pipe circuit is discretised into several pieces so that the variation of fluid temperature around the circuit is simulated. The circuit arrangement can be configured by the user and so there is some flexibility to represent a range of horizontal GHE types including those near basement walls such as FHXs (Fig. 5.6) but not Slinky designs.

In colder climates, freezing of the region surrounding the heat exchanger pipes is commonly intended in peak winter conditions and can make a good deal of difference to the behaviour of the system (Tarnawski and Yuet, 1988; Xu and Spitler, 2014). Tarnawski (1989) suggested that the insulating effects of snow cover could be taken advantage of by burying the heat exchanger at shallower depths and developed a numerical model that paid particular attention to freezing that was validated using Canadian data.

Fontaine et al. (2011) revisited the question of freezing and compared a model composed of a series of finite line sources with a three-dimensional finite element model. The line-source model was tested with thermal conductivities and heat capacities corresponding to frozen and non-frozen conditions. They found that using thermal property values in the analytical model representative of frozen ground gave reasonable predictions of temperatures for winter heating conditions but poor results for summer. On the whole, the reference finite element model that included phase change gave rather different results at certain points in the annual cycle. The numerical method was, nevertheless, too computationally demanding for common design tasks. Using frozen ground property values may therefore be a pragmatic approach to dealing with heating applications in cold climates. There seems to be, then, a continuing need for models that can deal with freezing in a computationally efficient manner. Recent advances in related applications (eg, artificial ground freezing for tunnel stabilisation) may give some insights into modelling freezing effects (Vitel et al., 2015).

The two models explained in some detail below can both be classed as hybrid weighting factor approaches. Both use temporal superposition to efficiently calculate responses to time-varying boundary conditions according to discrete weighting factors. In this sense both track the thermal history of the heat exchanger and aggregate long timescale responses. The model by Xiong et al. (2015) is closely related to models of vertical BHEs that use weighting factors defined as g-functions described in earlier chapters, eg, Yavuzturk and Spitler (1999). In this model the g-functions are derived from an analytical model of a Slinky pipe geometry. This model is promising in that it allows accurate representation of the pipe coils and also uses a computationally efficient time-stepping procedure that makes it suitable for integration in building and system simulations.

The GHE model by Fan et al. (2013) is a hybrid weighting factor model in a different sense. In this case the weighting factors are derived from application of a numerical model. This DTN model has the particular advantage of being able to deal with arbitrary complex geometries and interaction between two or three surfaces. It has been applied to FHXs (in which case there are three active surfaces) but can be applied to many other forms of compact heat exchanger such as Slinky and helical heat exchangers as well as energy pile systems.

5.4.1 A hybrid analytical approach

Kelvin's infinite line source solution has been used in the analysis of GHEs in several ways since Ingersoll and Plass (1948) first applied it. An alternative basis for developing analytical solutions is to use a point source solution. Superposition principles allow pipe arrangements to be represented by integrating the point source response along the path of the pipe. The point source solution defines the spatial and temporal variation of temperature in an infinite solid due to a continuous point heat source. The temperature response for such a point source can be expressed in the following form (Marcotte and Pasquier, 2009):

$$\Delta T(d,t) = \frac{q}{4\pi kd} erfc\left(\frac{d}{2\sqrt{\alpha t}}\right) \qquad [5.1]$$

The point source solution has been used to develop analytical solutions for finite line sources, inclined boreholes and pile applications (Marcotte and Pasquier, 2009; Zeng et al., 2002; Cui et al., 2011). Integrating around a circle allows the representation of a ring source and this can be useful in representing Slinky and spiral pipe geometries. Use of the approach to model a Slinky heat exchanger was demonstrated by Li et al. (2012) and the model presented below by Xiong et al. (2015) uses a similar approach as its starting point. The model has the following notable features:

- The model seeks to resolve fluid temperatures with time-varying boundary conditions and is intended to be used in integrated system simulation applications.
- The pipe wall is resolved in the integration to improve fluid temperature predictions.
- The ring source solution is applied to derive g-function temperature-response functions.
- The response to ground surface and climatic conditions is superimposed with the fluid temperature response in order to find the overall GHE response.
- The model has been validated with the experimental data of Fujii et al. (2012).

The model makes similar assumptions to other analytical models of conduction heat transfer: the ground is homogeneous and moisture and freezing are ignored. It is also assumed that the heat fluxes in each ring used to represent the Slinky heat exchanger (Fig. 5.3) are the same in every ring along the heat exchanger length.

The response of the whole heat exchanger is built up by first considering he response of a single ring. The point source integration process for a particular ring is illustrated in Fig. 5.11. In this figure point P_i is a fictions point on ring i, and the function $d\ (P_i,\ P_j)$ is the distance between the point i and point j on the other ring. The distance is calculated in terms of the angular integration variables ω and φ of the respective rings and the design parameters of the Slinky such as pitch and diameter (p and D in Fig. 5.3). The temperature perturbation at point P_i caused by ring source j is then found by integrating the point sources around the rings as follows:

$$\Delta T(P_i, t) = \frac{qlR}{4\pi k} \int_0^{2\pi} \frac{erfc\left(d(P_i, P_j)/2\sqrt{\alpha t}\right)}{d(Pi, Pj)} d\omega \qquad [5.2]$$

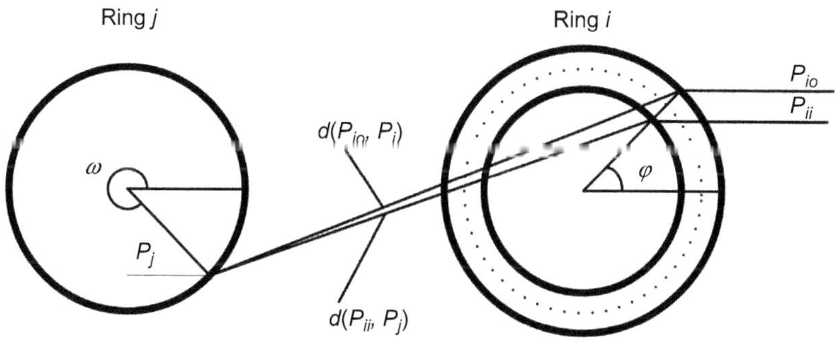

Figure 5.11 Distances between points P_i and P_j on a ring source are calculated from the average of positions on the pipe wall for particular angles ω and φ.
After Xiong, Z., Fisher, D.E., Spitler, J.D., 2015. Development and validation of a Slinky ground heat exchanger model. Applied Energy, 141, 57–69.

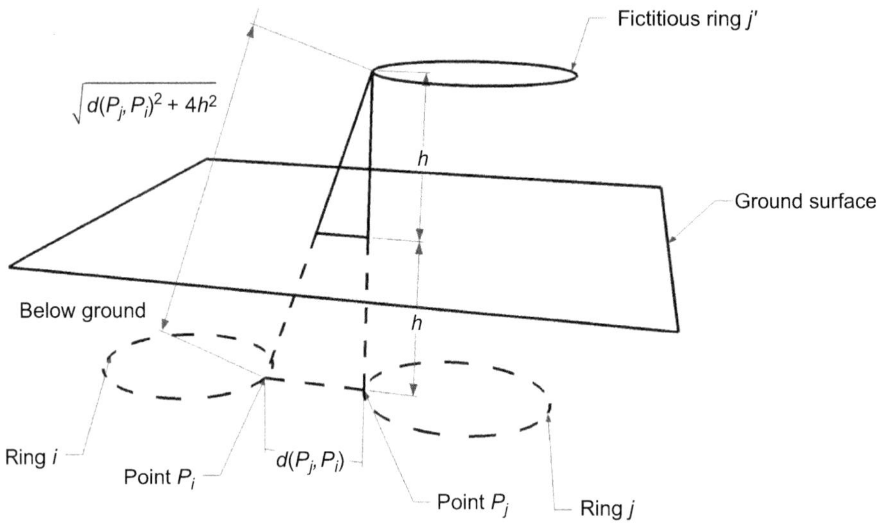

Figure 5.12 A perspective view of the heat exchanger ring and the imaginary ring source reflected in the ground plane.
After Xiong, Z., Fisher, D.E., Spitler, J.D., 2015. Development and validation of a Slinky ground heat exchanger model. Applied Energy, 141, 57–69.

In order to represent a ring that is buried below the ground surface rather than in an infinite medium, it is necessary to use the method of images and introduce an imaginary ring displaced by twice the depth (h) and to subtract this response. The positioning of the imaginary ring is illustrated in Fig. 5.12 and the distance between the rings can be expressed as $\sqrt{d(P_j, P_i)^2 + 4h^2}$. To calculate the effect of a perturbation at one ring on another it is also necessary to integrate with respect to φ. The average temperature perturbation then includes a double integral as follows:

$$\Delta T_{j,i}(t) = \frac{qlR}{8\pi^2 k} \int_0^{2\pi} \int_0^{2\pi} \left\{ \frac{erfc\left(d(P_i, P_j)/\left(2\sqrt{\alpha t}\right)\right)}{d(P_i, P_j)} - \frac{erfc\left(\sqrt{d(P_j, P_i)^2 + 4h^2} \Big/ \left(2\sqrt{\alpha t}\right)\right)}{\sqrt{d(P_j, P_i)^2 + 4h^2}} \right\} d\omega d\varphi \qquad [5.3]$$

A variation of this equation was used to find the response of rings aligned vertically.

The response of the whole heat exchanger is found by summing these perturbation for all the combinations of ring sources, so that:

$$\Delta \overline{T}(t) = \frac{1}{N_{\text{ring}}} \sum_{i=1}^{N_{\text{ring}}} \sum_{j=1}^{N_{\text{ring}}} \Delta T_{j,i}(t) \quad (j \neq i) \qquad [5.4]$$

The responses are not calculated from the analytical solution during a simulation. Rather, the approach taken by Xiong et al. (2015) was to use the *g*-function concept developed by Eskilson (1987) so that time-varying temperatures can be calculated by superposition of heat pulses from one time-step to the next – in a similar manner to the model for vertical BHEs by Yavuzturk and Spitler (1999). Following the standard definition of the *g*-function and normalising by $2\pi k/ql$ gives

$$\Delta\overline{T}(t) = \frac{2\pi k}{ql}\frac{1}{N_{\text{ring}}}\sum_{i=1}^{N_{\text{ring}}}\sum_{j=1}^{N_{\text{ring}}}\Delta T_{j,i}(t) \quad (j \neq i) \qquad [5.5]$$

where $\Delta\overline{T}(t)$ is found from Eq. [5.3]. Examples of *g*-functions calculated in this way and using the heat exchanger designs used by Fujii et al. (2012) are shown in Fig. 5.13.

The double integrals in Eq. [5.3] and the double summation in Eq. [5.5] mean that the *g*-functions in this model are not trivial to calculate. Consequently, procedures were developed to reduce the number of integrations by making use of symmetry in

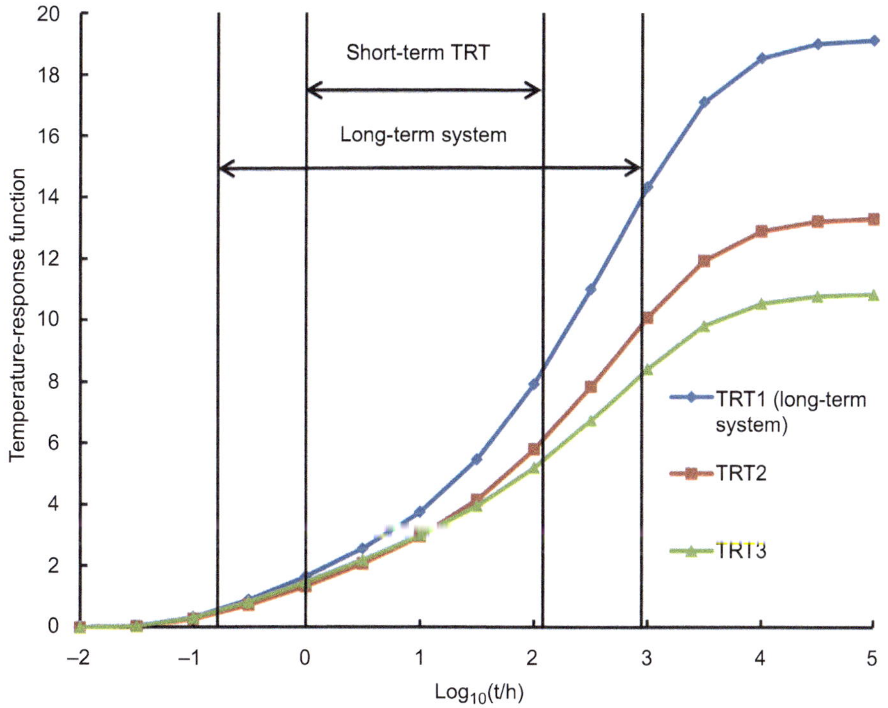

Figure 5.13 Temperature-response *g*-functions calculated with the experimental parameters used by Fujii et al. (2012).
After Xiong, Z., Fisher, D.E., Spitler, J.D., 2015. Development and validation of a Slinky ground heat exchanger model. Applied Energy, 141, 57–69.

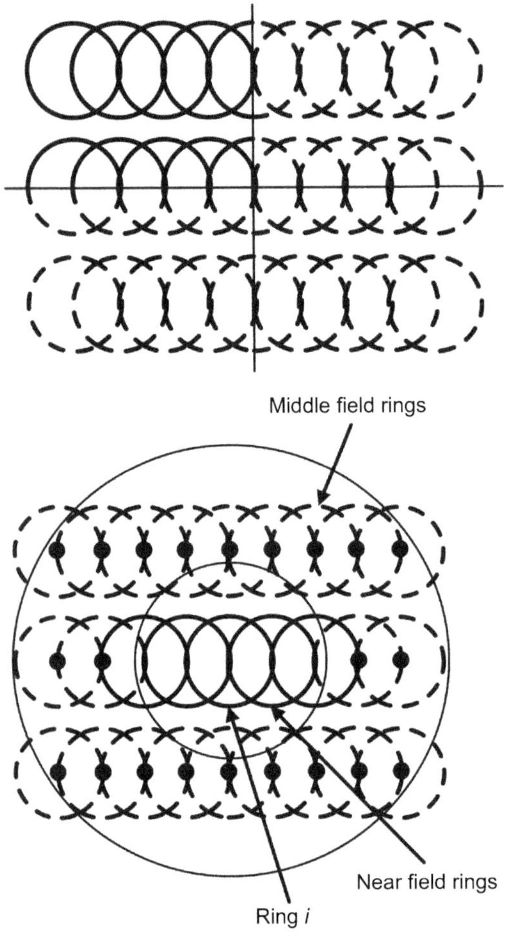

Figure 5.14 Grouping and symmetry can be taken advantage of to simplified the calculation of *g*-functions for Slinky heat exchangers.
After Xiong, Z., Fisher, D.E., Spitler, J.D., 2015. Development and validation of a Slinky ground heat exchanger model. Applied Energy, 141, 57–69.

rectangular Slinky configurations and grouping rings such that rings sufficiently far away could be ignored in the summations. Interaction with the middle rings is approximated by assuming equivalent point sources. Example symmetry and grouping features are illustrated in Fig. 5.14. In this model the *g*-functions are used to calculate the response of the ground to fluctuations in conditions at the pipes. In order to include the response of the heat exchanger to the climatic boundary conditions at the ground surface, a further model of time-varying undisturbed ground temperature is used and a further superposition of temperature perturbation introduced. Xiong et al. (2015) used an existing numerical model of ground heat exchange (Lee et al., 2013), in one-dimensional form, for the sake of convenient implementation in the EnergyPlus environment (Crawley et al., 2001). This model uses a ground surface boundary condition that accounts for the combined effect of convection, evapotranspiration and both long and shortwave radiation. The model also accounts for snow cover at the ground surface. Although this introduces a non-linearity that, in theory, negates the principle of superposition, this feature was thought to be better included rather than ignoring snow and freezing effects altogether.

Data from four tests conducted in 2008 and 2009 by Fujii et al. (2012) have been used in a validation exercise. The data consisted of three short TRTs with slightly different heat exchanger configurations and one longer test of 40 days duration. The boundary conditions in these tests were the measured climatic variables and the heat exchanger inlet fluid temperatures. Outlet temperatures and heat transfer rates were calculated. Results from one of the shorter tests are shown in Fig. 5.15 and the longer test in Fig. 5.16.

In this validation study the root mean square error (RMSE) of the predicted temperatures was less than 0.3 K and the RMSE of the predicted heat transfer rates no more than 3% in the case of the short term tests. In the case of the longer test, RMSEs of the predicted outlet temperatures were 0.3 K but with slightly larger divergence at the end of the test than the shorter test cases.

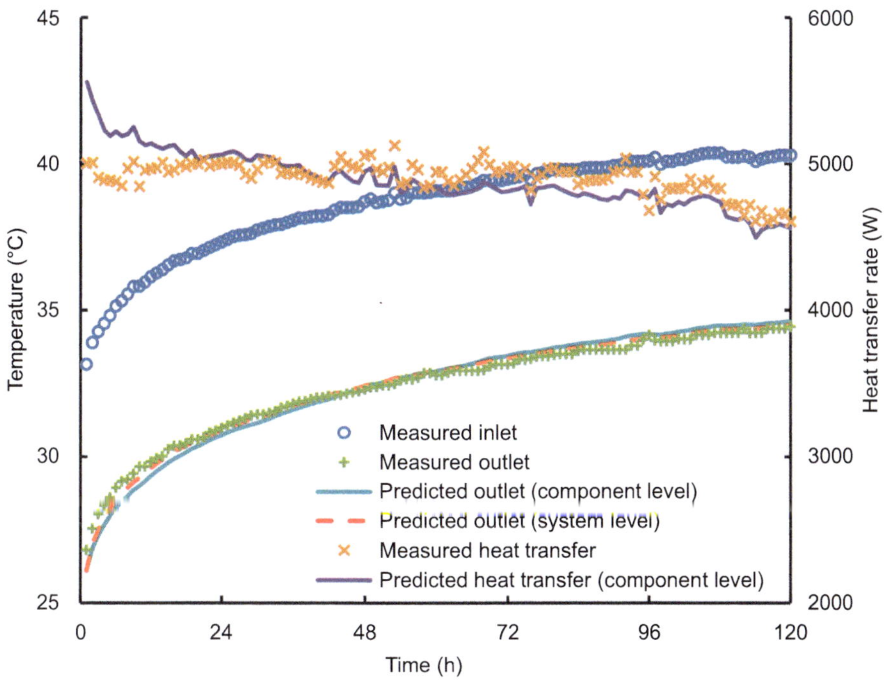

Figure 5.15 Comparisons of predicted and measured outlet temperatures and heat transfer rates for a Slinky heat exchanger (test TRT2).
After Xiong, Z., Fisher, D.E., Spitler, J.D., 2015. Development and validation of a Slinky ground heat exchanger model. Applied Energy, 141, 57–69.

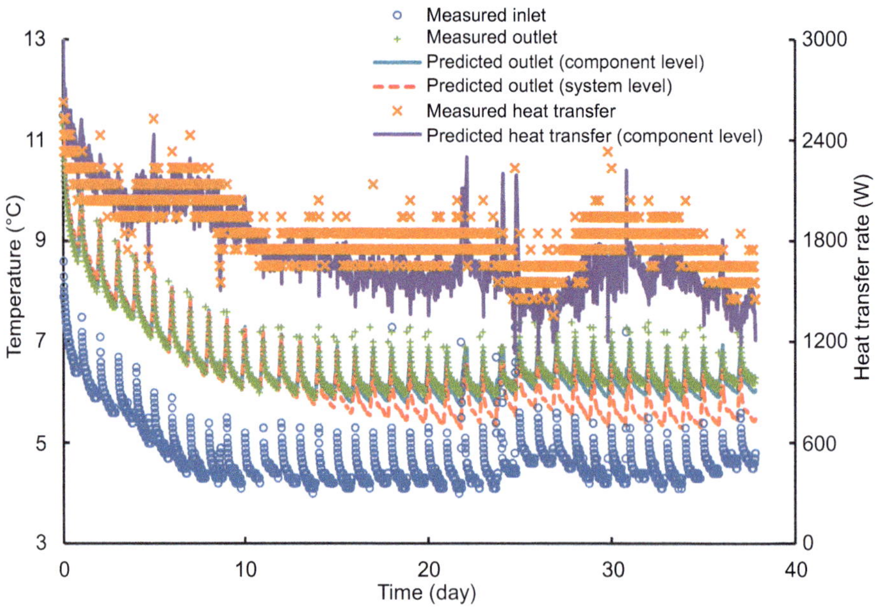

Figure 5.16 Comparisons of 10-minutely temperatures and heat transfer rates in a 40 day test of Slinky heat exchanger response.
After Xiong, Z., Fisher, D.E., Spitler, J.D., 2015. Development and validation of a Slinky ground heat exchanger model. Applied Energy, 141, 57–69.

The advantages of this model are primarily that it is able to deal with variations in Slinky geometry explicitly and that it is computationally efficient enough to be used in integrated system simulation studies. The authors point out that some of the assumptions such as homogeneity of the ground, equal heat exchange rates at each ring and also the assumption of the effective fluid temperature being the arithmetic mean of the inlet and outlet temperatures, requires further investigation – preferably with longer time series data sets.

5.4.2 A hybrid numerical approach

The DTN approach to the modelling of transient conduction was conceived by Claesson as an extension of the network representation of steady-state conduction processes (Claesson, 2002b, 2003). Details of the development of the method, along with proofs of the essential mathematical principles, can be found in the report by Claesson (2002a) and also in the thesis by Wentzel (2005). The DTN approach is applicable to any conduction problem driven by boundary temperatures and fluxes. It can, in principle, be applied to a wide variety of GHE types and was applied to FHXs by Fan et al. (2013) with the required weighting factor data being derived from a numerical model. The model is summarised below and readers wanting further details can find these in Rees and Fan (2013).

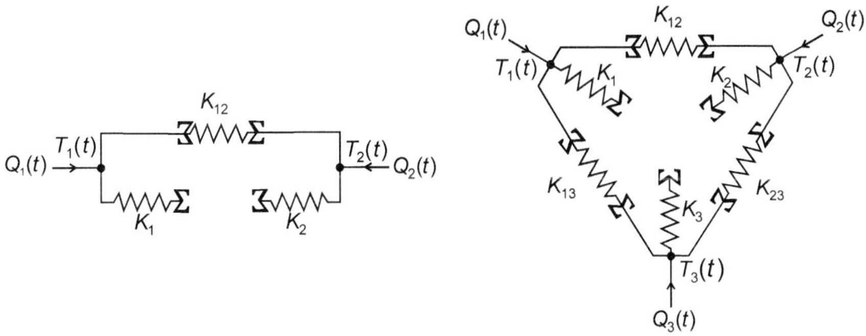

Figure 5.17 Dynamic thermal network representations of a two- (left) and three-surface problem (right). The sigma symbols indicate driving temperatures that are weighted averages.

The DTN model differs from conventional steady-state nodal network models in that heat fluxes at each surface are separated into admittive and transmittive components. This is reflected in the network model in that a single-ended heat transfer path is associated with each surface node to represent the admittive heat transfer path. Representations of two- and three-surface problems are shown in dynamic network form in Fig. 5.17.

The dynamic form of thermal network (Fig. 5.17) includes constant conductances in a similar manner to the steady-state form. The reversed summation symbols (Σ) adjacent these conductances indicate that the driving temperatures are averages of the current and previous temperatures rather than the instantaneous value. In the single-ended admittive path the single summation sign indicates the driving temperature is a function of the average temperature at that boundary alone. The FHX investigated by Fan et al. (2013) used a three-surface representation where the pipe, ground and basement wall formed the three surfaces at which time-varying boundary conditions were applied. In the case of a shallow GHE a two-surface representation is sufficient – one being the pipe and the other being the exposed ground surface.

The temperatures, $T_i(t)$, and fluxes, $Q_i(t)$, of the dynamic network are defined at environmental temperature nodes rather than at the surfaces themselves. The conductances in the admittive path (shown with a single subscript K_i) are equal to the surface area multiplied by a constant heat transfer coefficient. There are constant conductances between each pair of surfaces (shown with double subscripts K_{ij}) that are equivalent to the overall steady-state conductances between the boundaries and include the surface conductances K_i and K_j. The nodal heat balance equations for a two- and three-surface problems the heat balance equations are, respectively,

$$\begin{aligned} Q_1(t) &= Q_{1a}(t) + Q_{12}(t) \\ Q_2(t) &= Q_{2a}(t) + Q_{21}(t) \end{aligned} \qquad [5.6]$$

and

$$\begin{aligned} Q_1(t) &= Q_{1a}(t) + Q_{12}(t) + Q_{13}(t) \\ Q_2(t) &= Q_{2a}(t) + Q_{21}(t) + Q_{23}(t) \\ Q_3(t) &= Q_{3a}(t) + Q_{31}(t) + Q_{32}(t) \end{aligned} \qquad [5.7]$$

The transmittive fluxes $Q_{ij}(t)$ are from surface i towards surface j as indicated by the ordering of the subscripts. The admittive fluxes at surface i are denoted by $Q_{ia}(t)$.

Claesson (2002a) showed that the temperature differences driving the absorptive and transmittive fluxes can be defined in an exact manner by the current and weighted averages of temperatures $\overline{T}_{ij}(t)$ and $\overline{T}_{ia}(t)$ associated with each boundary. These can be defined exactly in terms of series of admittive and transmittive weighting factors κ_{ia} and κ_{ia} as follows:

$$\overline{T}_{ia}(t) = \int_0^\infty \kappa_{ia}(\tau) \cdot T_i(t-\tau) d\tau \qquad [5.8]$$

$$\overline{T}_{ij}(t) = \int_0^\infty \kappa_{ij}(\tau) \cdot T_i(t-\tau) d\tau \qquad [5.9]$$

The dynamic relations between boundary heat fluxes and temperatures for a two-surface problems are then, simply,

$$\begin{aligned} Q_1(t) &= K_1 \cdot \left[T_1(t) - \overline{T}_{1a}(t)\right] + K_{12} \cdot \left[\overline{T}_{12}(t) - \overline{T}_{21}(t)\right] \\ Q_2(t) &= K_2 \cdot \left[T_2(t) - \overline{T}_{2a}(t)\right] + K_{12} \cdot \left[\overline{T}_{21}(t) - \overline{T}_{12}(t)\right] \end{aligned} \qquad [5.10]$$

When extended to a three-surface problem the heat balance equations can be written as:

$$\begin{aligned} Q_1(t) &= K_1 \cdot \left[T_1(t) - \overline{T}_{1a}(t)\right] + K_{12} \cdot \left[\overline{T}_{12}(t) - \overline{T}_{21}(t)\right] + K_{13} \cdot \left[\overline{T}_{13}(t) - \overline{T}_{31}(t)\right] \\ Q_2(t) &= K_2 \cdot \left[T_2(t) - \overline{T}_{2a}(t)\right] + K_{12} \cdot \left[\overline{T}_{21}(t) - \overline{T}_{12}(t)\right] + K_{23} \cdot \left[\overline{T}_{23}(t) - \overline{T}_{32}(t)\right] \\ Q_3(t) &= K_3 \cdot \left[T_3(t) - \overline{T}_{3a}(t)\right] + K_{13} \cdot \left[\overline{T}_{31}(t) - \overline{T}_{13}(t)\right] + K_{23} \cdot \left[\overline{T}_{32}(t) - \overline{T}_{31}(t)\right] \end{aligned} \qquad [5.11]$$

As the steady-state is approached each average temperature approaches the related boundary temperature and the admittive fluxes become zero. It can be seen that, in the steady-state Eq. [5.10] reduces to the usual expression for flux in terms of overall conductance and boundary temperatures ($Q_1 = K_{12} \cdot [T_1 - T_2]$, etc.).

5.4.2.1 Discretisation and weighting factors

Claesson (2002a, 2003) showed that the DTN calculation method could be expressed in discrete form, in an exact way, for piecewise linear variations in boundary conditions. When the boundary temperatures are defined by a discrete time series, the average temperatures are calculated by the summation of weighting factor sequences multiplied by boundary temperature sequences that represent the state at previous time steps. The discrete form of Eqs [5.8] and [5.9] is, for time-step n,

$$\overline{T}_{ia,n} = \sum_{\rho=1}^{\infty} \kappa_{ia,\rho} \cdot T_{i,n-\rho} \quad [5.12]$$

$$\overline{T}_{ij,n} = \sum_{\rho=0}^{\infty} \kappa_{ij,\rho} \cdot T_{i,n-\rho} \quad [5.13]$$

A convenient way to derive suitable weighting functions is to consider the fluxes resulting from step changes in boundary temperatures. All the required weighting functions can be found by applying a step boundary condition at one of the surfaces and holding the other boundary temperatures at zero and repeating this for each surface. When the equations are expressed in discrete form the weighting factors can be obtained using the step-response admittive and transmittive fluxes averaged over each step (size Δt). The discrete weighting factors are then obtained from the differences in these average time-step fluxes as follows:

$$\kappa_{ia,\rho} = \frac{\overline{Q}_{ia}(\varphi) - \overline{Q}_{ia}(\omega)}{\overline{K}_i}$$
$$\kappa_{ij,\rho} = \frac{\overline{Q}_{ij}(\omega) - \overline{Q}_{ij}(\varphi)}{K_{ij}} \quad [5.14]$$

where the time differences are between $\varphi = (\rho\Delta t - \Delta t)$ and $\omega = \rho\Delta t$.

The relationship between the step-response data and the flux averages at each time step are illustrated in Fig. 5.18. Finally, the discrete form of the heat balance equations for a two-surface problem, Eq. [5.10], can be written as follows:

$$Q_{1,n} = \overline{K}_1 \cdot \left[T_{1,n} - \sum_{\rho=1}^{\rho_s} \kappa_{1a,\rho} T_{1,n-\rho} \right] + K_{12} \cdot \left[\sum_{\rho=0}^{\rho_s} \kappa_{12,\rho} \left(T_{1,n-\rho} - T_{2,n-\rho} \right) \right] \quad [5.15]$$

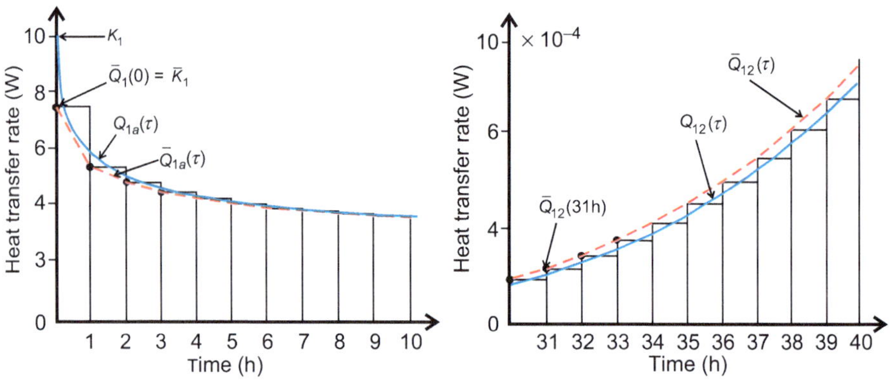

Figure 5.18 Step-response for (left) admittive, and (right) transmittive fluxes shown as *solid lines*. Flux averages are shown as *dotted lines*, with bars representing the average value over each time step.

$$Q_{2,n} = \overline{K}_2 \cdot \left[T_{2,n} - \sum_{\rho=1}^{\rho_s} \kappa_{2a,\rho} T_{2,n-\rho} \right] + K_{12} \cdot \left[\sum_{\rho=0}^{\rho_s} \kappa_{12,\rho} \left(T_{2,n-\rho} - T_{1,n-\rho} \right) \right] \tag{5.16}$$

The extension to three or more surfaces is straightforward.

In a simulation, these equations can be solved for the heat fluxes or the current temperature at each time step and the weighted temperature histories updated in preparation for the following step. Both the weighting factor data series and the temperature histories that need to be stored and processed can be very long, particularly for GHEs. To make the computation more efficient a weighting-factor reduction strategy that aggregates later values and that was developed by Wentzel (2005) and enhanced for thermally massive constructions by Fan et al. (2013). In this procedure, the weighting factor series is divided into several sub-series (levels) that have increasing time-step size. Examples of discrete admittive and transmittive weighting factor series and their reduced or aggregated form are shown in Figs. 5.19 and 5.20. In the few test cases reported Wentzel (2005) the error associated with the data reduction process was demonstrated to be less than 0.3%.

5.4.2.2 Dynamic thermal network model boundary conditions

The DTN is formulated, and the step-response data calculated, assuming that surface heat transfer coefficients (h) are constant. Simple convective boundary conditions could be used at pipe and ground surfaces but a more sophisticated treatment (sensitive to variable flow rates and several atmospheric variables respectively) is desirable in a GHE application of the method. This can be done by defining a boundary temperature that is an 'effective temperature' (T_e) that, when applied using the predefined constant heat transfer coefficient, gives the expected surface heat flux as applying a more complex boundary condition model. This effective temperature (or environmental

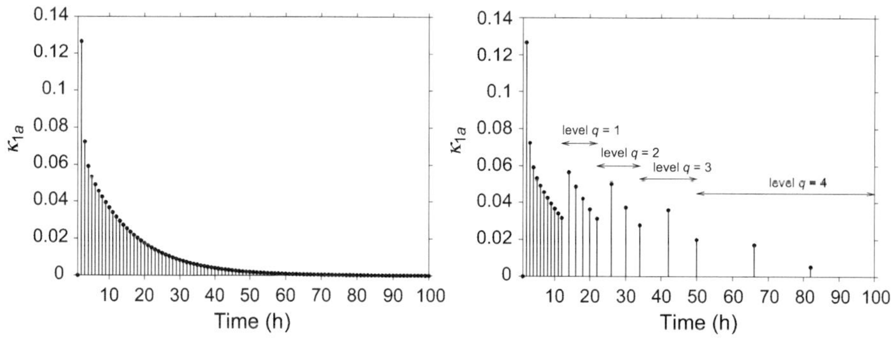

Figure 5.19 Discrete admittive weighting factor values and the corresponding reduced set. This shows the first 100 h from a series with a time-step size of 1 h.

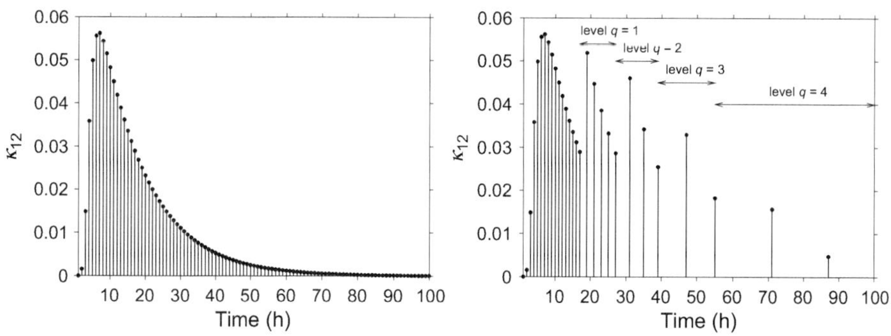

Figure 5.20 Discrete transmittive weighting factor values and the corresponding reduced set. This shows the first 100 h from a series with a time-step size of 1 h.

temperature) does not correspond directly to a physical boundary temperature but is applied in the DTN heat balance equations and when the weighted average temperature is updated (Rees and Fan, 2013).

At surfaces exposed to the external environment, convection processes act in combination with shortwave (R_{sw}) and long-wave radiant (R_{lw}) and evapotranspiration fluxes. A surface heat balance defining such a boundary condition is

$$\frac{Q_i}{S_i} = R_{sw} + R_{lw} + h_{ca}(T_a - T_{Si}) - \frac{\rho_w L_w}{3.6} ET_0 \qquad [5.17]$$

where S_i is the surface area, L_w is a vegetation lengthscale and the evapotranspiration rate (ET_0, mm/h) can be calculated by the Penman–Montieth equation (Penman, 1948). The effective boundary temperature is intended to give the equivalent heat flux and hence is defined by

$$\frac{Q_i}{S_i} = h_i(T_e - T_{Si}) \qquad [5.18]$$

Hence, the equivalent or environmental temperature must be

$$T_e = \left(R_{sw} + R_{lw} + h_{ca}T_a + (h_{ca} - h_i)T_{Si} - \frac{\rho_w L_w}{3.6} ET_0\right) \Big/ h_i \qquad [5.19]$$

This expression involves the surface temperature and so implies that some iteration may be required to calculate T_e. If the surface temperature changes slowly relative to the time step – as it does in ground heat transfer problems – then the surface temperature calculated at the previous time step can be used in the last term of Eq. [5.19] with little error and iteration avoided.

In geometries with embedded or buried pipes it is usually necessary to define the relationship between the boundary temperature and both the pipe fluid inlet and outlet temperatures. The approach taken by Fan et al. (2013) was to assume the pipe surface temperature does not vary along its length and make an analogy with an evaporating–condensing heat exchanger and so to define a characteristic normalised effectiveness parameter, ε. The pipe fluid heat balance can then be defined by the maximum possible temperature difference and the effectiveness as follows:

$$Q_p(t) = \varepsilon \dot{m} C(T_{\text{in}}(t) - T_p(t)) \qquad [5.20]$$

For such a heat exchanger, $\varepsilon = 1 - e^{-\text{NTU}}$, and this is related to the total pipe area and fluid heat transfer coefficient by the number of transfer units (NTU) according to

$$\text{NTU} = \frac{2\pi r H \cdot h_p}{\dot{m} C} \qquad [5.21]$$

The relationship between the pipe surface temperature and the boundary temperature ($T_i\,(t)$) that needs to be found for use in the DTN heat balance equations is defined by the surface heat flux relationship:

$$Q_i(t) = h_i \cdot S_i(T_i(t) - T_p(t)) \qquad [5.22]$$

Just a few steps of an iterative procedure are required to find the boundary temperature (T_i) consistent with the pipe surface heat balance. Finally, the outlet temperature can then be found from the fluid heat balance: $T_{\text{out}}(t) = T_{\text{in}}(t) - Q_P(t)/\dot{m}C$.

5.4.2.3 Application of the dynamic thermal network model

Wentzel (2005) used a combination of analytical solutions and numerical methods to generate the step-response data required to derive the weighting factors for DTN calculations. Fan et al. (2013) used an entirely numerical method. This numerical approach used a combination of a parametrically generated meshes, a block-structured finite volume method solver and variable time-step solution procedure (Rees and Fan, 2013). Using a numerical model in this way allows a wide variety of problems to be tackled – from plane walls to GHEs with complex 3D shapes. Hence

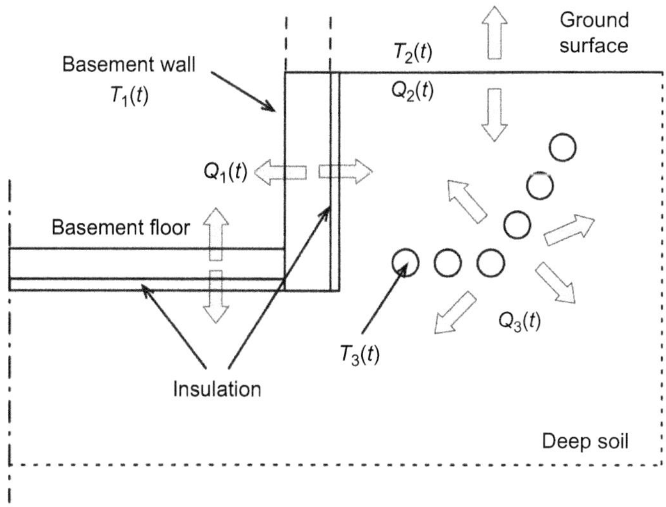

Figure 5.21 In a DTN model of a foundation heat exchanger the three surfaces in the network correspond to the basement, ground and pipe surfaces.

the model could be applied to Slinky and helical pipe arrangements provided that a suitable numerical mesh could be generated. In their study of FHXs, Fan et al. (2013) used a three-surface DTN where the three surfaces corresponded to the ground, basement wall and pipes as indicated in Fig. 5.21. An example of the type of numerical mesh used to calculate step responses for a DTN model of an FHX is shown in Fig. 5.22.

When the DTN model is used to simulate heat exchanger behaviour the input data are the climate data and the fluid inlet temperature. The sequence of operations required to derive the weighting factors and carry out a simulation can be summarised as follows:

1. parametric generation of a numerical mesh based on geometric design parameters
2. numerical step-response calculations for each surface using variable time steps
3. analysis of step-response data to derive discrete weighting factor series
4. application of a weighting factor reduction procedure
5. initialization of the discrete temperature data and calculation of the initial mean temperatures
6. calculation of the surface heat fluxes using time-varying boundary conditions (Eqs [5.15] and [5.16])
7. updating the mean temperature data series

The last two steps are repeated to the end of the boundary condition time series. The process can be separated between the weighting factor calculations and the simulation process at the end of step four if the reduced weighting factors are stored. The latter suggestion would be more suitable for geometrically complex and thermally massive structures. For planar and simple two-dimensional geometries, the process is efficient enough to calculate the weighting factors at the start of every simulation.

Figure 5.22 A numerical model is used to calculate the step-response data to derive the weighting factor data in the dynamic thermal network model. In this example the model is a three-dimensional representation of a foundation heat exchanger around the corner of a basement.

Rees and Fan (2013) demonstrated that the DTN method could produce heat transfer rates accurate to within a fraction of a percent compared to analytical and reference numerical results. Calculation speed was increased by four orders of magnitude compared to numerical models and overall simulation time amounted to a few minutes for a GHE problem. The calculation effort in the DTN approach is in the derivation of the weighting factors (as it is in all weighting factor methods). In the case of simple two-dimensional geometries applying a numerical model to derive the weighting factor data is still of the order of minutes. More complex three-dimensional geometries can require a considerably longer time to process. However, once calculated the weighting factors can be stored and used repeatedly in further simulations. Work to apply the method to pile heat exchangers, borehole thermal energy storage (BTES) arrays and district heating pipe systems is to be continued.

5.5 Summary

This chapter has provided an overview of the common horizontal GHE technologies and some recent innovations in compact GHE design. Although this technology is in many ways simple and economically attractive, there are many challenges in developing design methodologies, modelling methods and standards that are robust.

The physics of heat and mass transfer in the shallow soil and heat and mass exchange with the ambient environment is complex. In some ways horizontal GHEs are more challenging to analyse than vertical borehole systems and have furthermore received less attention from researchers until relatively recently.

Whether simple or sophisticated calculation methods are used or not, it is the large uncertainties in more than one type of design data that have to be accepted in designing for residential applications that make robust design difficult. Both building load data and ground thermal property data are major sources of uncertainty. In situ testing methods to evaluate thermal conductivity in shallow soil are certainly technically feasible and can be used to address some uncertainty; just as they can in vertical BHE designs. However, proper evaluation requires several tests and careful statistical analysis be made and so the cost of testing is often prohibitive on residential projects. Currently, there is a good deal of research into soil thermal property test methods being carried out in relation to energy pile applications, and it is possible that advances made in that context may produce benefits for the design of small-scale systems with horizontal GHEs.

Development of design methods and standards for horizontal GHEs has been discussed in this chapter. There has been industrial success using these methods and standards on the whole. Some of the uncertainties that are inevitable in residential system design can be guarded against by taking a conservative view of the parameter values used in such design methods and published data. There would be benefit in making systematic surveys of designs completed to these standards to evaluate whether designs are optimal, successful but overly expensive, or at risk of failure.

This chapter has highlighted some developments in modelling of shallow heat exchanger systems and has presented two promising models in some detail. Recent models such as these are able to deal with the complex geometries of Slinky heat exchangers and other compact designs. It is evident, however, that there is scope for further investigation of freezing conditions and how these might be modelled. Despite these advances it is hard to say these models are in regular use and it does not automatically follow that better models will lead to better designs. This is partly because improved models do not address some of the underlying uncertainties in the design data. It also has to be kept in mind that more sophisticated models need more parameters and data to initialise them and such data is not always available – otherwise the user has to introduce their own uncertainties by making assumptions about such data.

One way to make good use of some of the more sophisticated models of horizontal GHEs and other system component models would be to carry out large sets of simulations with a view to quantifying the effects of the various uncertainties in the input data. Such probabilistic approaches that evaluate designs on a risk basis and life-cycle performance are now feasible given recent improvements in desktop computing power and software tools. Such approaches have been demonstrated (Zhu et al., 2012; Garber et al., 2013) in studies of larger systems and in other spheres of building simulation. Applying such approaches may require something of a shift in thinking regarding what is viewed as a compliant or successful design. However, a risk evaluation approach may give a more realistic view of the trade-offs being made between economies, efficiency and ultimately risks of technical failure.

References

Ambrose, E.R., 1966. Heat Pumps and Electric Heating. John Wiley and Sons, New York.

ASTM, 2008. D 5334-08 Standard Test Method for Determination of Thermal Conductivity of Soil and Soft Rock by Thermal Needle Probe Procedure.

Blight, G.E., 2003. The vadose zone soil-water balance and transpiration rates of vegetation. Géotechnique 53 (1), 55–64.

Bose, J.E., 1982. Earth coil/heat pump research at Oklahoma State University. In: Proceedings of the Sixth Heat Pump Technology Conference. Oklahoma State University, Tulsa.

Bose, J.E., 1989. CLGS Ground Heat Exchanger Design Program. International Ground Source Heat Pump Association, Stillwater, Oklahoma.

Bose, J.E., Parker, J.D., 1984. A report on the ASHRAE project to develop a design data manual for ground-coupled heat pumps. In: Proceedings of the 7th Heat Pump Technology Conference October 15–16, pp. 159–168 (Tulsa, Oklahoma).

Bristow, K.L., Kluitenberg, G.J., Horton, R., 1994. Measurement of soil thermal properties with a dual-probe heat-pulse technique. Soil Science Society of America Journal 58 (5), 1288.

Buckingham, E., 1907. Studies on the Movement of Soil Moisture. Bulletin No. 38. U.S. Department of Agriculture, Bureau of Soils, Washington, DC.

Busby, J., 2015. Determination of thermal properties for horizontal ground collector loops. In: Proceedings of World Geothermal Congress 2015, 19–25 April, pp. 19–25 (Melbourne, Australia).

Campbell, G.S., Calissendorff, C., Williams, J.H., 1991. Probe for measuring soil specific heat using a heat-pulse method. Soil Science Society of America Journal 55 (1), 291.

Claesson, J., 2002a. Dynamic Thermal Networks. Background Studies I: Elements of a Mathematical Theory of Thermal Responses. Chalmers University of Technology, Departmental Report, Göteborg, Sweden.

Claesson, J., 2002b. Dynamic thermal networks. Outlines of a general theory. In: Proceedings of the 6th Symposium on Building Physics in the {Nordic} Countries, pp. 47–54 (Trondheim, Norway).

Claesson, J., 2003. Dynamic thermal networks: a methodology to account for time-dependent heat conduction. In: Proceedings of the 2nd International Conference on Research in Building Physics, Leuven, Belgium, pp. 407–415.

Claesson, J., Dunand, A., 1983. Heat Extraction from the Ground by Horizontal Pipes—a Mathematical Analysis. Technical report, (Document D1:1983). Swedish Council for Building Research, Stockholm, Sweden.

Coogan, C.H., 1949. Summary of Heat Absorption Rates for an Experimental Ground Coil System. Technical report. University of Conneticut Engineering Experimental Station.

Crawley, D.B., Lawrie, L.K., Pedersen, C.O., Strand, R.K., Liesen, R.J., Winkelmann, F.C., Buhk, W.F., Huang, Y., Fisher, D.E., Witte, M.J., Glazer, J., 2001. EnergyPlus: creating a new-generation building energy simulation program. Energy & Buildings 33 (4), 319–331.

Cui, P., Li, X., Man, Y., Fang, Z., 2011. Heat transfer analysis of pile geothermal heat exchangers with spiral coils. Applied Energy 88 (11), 4113–4119.

Curtis, R., Pine, T., Wickins, C., 2013. Development of new ground loop sizing tools for domestic GSHP installations in the UK. In: European Geothermal Congress 2013, pp. 1–10 (Pisa, Italy).

DECC, 2011. MCS 022: Ground Heat Exchanger Look-up Tables, Supplementary Material to MIS 3005 (Issue 1.0). Department of Energy and Climate Change, London, UK.

DECC, 2012. Microgeneration Installation Standard : MIS 3005 Requirements for Contractors Undertaking the Supply, Design, Installation, Set-to-work, Commissioning and Handover of Microgeneration Heat Pump Systems (Issue 3.1a). Department of Energy and Climate Change, London, UK.

Dunbabin, P., Wickins, C., 2012. Detailed Analysis from the First Phase of the Energy Saving Trust's Heat Pump Field Trial. Technical Report 3. Department of Energy and Climate Change, London, UK.

Eskilson, P., 1987. Thermal Analysis of Heat Extraction Boreholes. Doctoral thesis. University of Lund.

Fan, D., Rees, S., Spitler, J., 2013. A dynamic thermal network approach to the modelling of foundation heat exchangers. Journal of Building Performance Simulation 6 (2), 81–97.

Farouki, O.T., 1981. Thermal Properties of Soils (CRREL Monograph 81-1). Technical report. U.S. Army Cold Regions Research and Engineering Laboratory, Hanover, New Hampshire, USA.

Fontaine, P.O., Marcotte, D., Pasquier, P., Thibodeau, D., 2011. Modeling of horizontal geo-exchange systems for building heating and permafrost stabilization. Geothermics 40 (3), 211–220.

Fujii, H., Nishi, K., Komaniwa, Y., Chou, N., 2012. Numerical modeling of slinky-coil horizontal ground heat exchangers. Geothermics 41, 55–62.

Garber, D., Choudhary, R., Soga, K., 2013. Risk based lifetime costs assessment of a ground source heat pump (GSHP) system design: methodology and case study. Building and Environment 60, 66–80.

García-Gil, A., Vázquez-Suñe, E., Alcaraz, M.M., Juan, A.S., Sánchez-Navarro, J.A., Montlleó, M., Rodríguez, G., Lao, J., 2015. GIS-supported mapping of low- temperature geothermal potential taking groundwater flow into account. Renewable Energy 77, 268–278.

Garcia Gonzalez, R., Verhoef, A., Vidale, P.L., Main, B., Gan, G., Wu, Y., 2012. Interactions between the physical soil environment and a horizontal ground coupled heat pump, for a domestic site in the UK. Renewable Energy 44, 141–153.

Hapley, W.A., 1949. Operating characteristics of heat pump ground coils. Edison Electric Institute Bulletin.

Harlow, J.H., Klapper, G.E., 1953. Residential heat pump experiments in Philadelphia—Installation and operating experience. Transactions of the American Institute of Electrical Engineers, Part II: Applications and Industry 71 (6), 366–375.

Heap, R.D., 1979. Heat Pumps, first ed. F N Spon, London.

Hukseflux, 2010. FTN01 Field Thermal Needle System for Thermal Resistivity/Conductivity Measurement.

IGSHPA, 2009. Ground Source Heat Pump Residential and Light Commercial: Design and Installation Guide. International Ground Source Heat Pump Association, Oklahoma State University, Stillwater, Oklahoma.

Im, P., Hughes, P., Liu, X., Ridge, O., 2012. Demonstration and performance monitoring of foundation heat exchangers (FHX) in ultra-high energy efficient research Homes Field test of the foundation heat exchanger concept. In: Proceedings of the 17th ACEEE Summer Study on Energy Efficiency in Buildings, pp. 114–126. Pacific Grove, CA. August 12–17.

Ingersoll, L.R., Plass, H.J., 1948. Theory of the ground pipe heat source for the heat pump. ASHVE Transactions 47 (7), 339–348.

Kersten, M.S., 1949. Thermal Properties of Soils. University of Minneapolis Engineering Experimental Station, Minnesota.

Kidder, A.H., Neher, J.H., 1952. Residential heat pump experiments in Philadelphia—Earth as a heat source. Transactions of the American Institute of Electrical Engineers, Part II: Applications and Industry 71 (5), 343–350.
King, W., Banks, D., Findlay, J., 2012. Field determination of shallow soil thermal conductivity using a short-duration needle probe test. Quarterly Journal of Engineering Geology and Hydrogeology 45 (4), 497–504.
Kusuda, T., Achenbach, P.K., 1965. Earth Temperature and Thermal Diffusivity at Selected Stations in the United States (NBS 8972). Technical report. National Bureua of Standards, Building Research Division, Washington, DC.
Lamberg, P., Lehtiniemi, R., Henell, A.-M., 2004. Numerical and experimental investigation of melting and freezing processes in phase change material storage. International Journal of Thermal Sciences 43 (3), 277–287.
Lee, E.S., Fisher, D.E., Spitler, J.D., 2013. Efficient horizontal ground heat exchanger simulation with zone heat balance integration. HVAC&R Research 19 (3), 307–323.
Li, H., Nagano, K., Lai, Y., 2012. Heat transfer of a horizontal spiral heat exchanger under groundwater advection. International Journal of Heat and Mass Transfer 55 (23–24), 6819–6831.
Low, J.E., Loveridge, F.A., Powrie, W., 2014. Thermal conductivity of soils by the needle probe method, for energy foundation applications. In: 32nd International Thermal Conductivity Conference, p. 11 (West Laffayette, Indiana).
Marcotte, D., Pasquier, P., 2009. The effect of borehole inclination on fluid and ground temperature for GLHE systems. Geothermics 38 (4), 392–398.
Marlowe, J.I., Salomone, L.A., Bose, J.E., 1989. Soil and Rock Classification for the Design of Ground-coupled Heat Pump Systems: Field Manual. International Ground Source Heat Pump Association, Stillwater, Oklahoma.
Mei, V.C., 1986a. Horizontal Ground-coil Heat Exchanger Theoretical and Experimental Analysis (ORNL/CON-193). Technical report. Oakridge National Laboratory.
Mei, V.C., 1986b. Theoretical heat pump ground coil analysis with variable ground farfield boundary conditions. AIChE Journal 32 (7), 1211–1215.
Metz, P.D., 1979. Design, construction, and operation of the solar assisted heat pump ground coupled storage experiments at Brookhaven National Laboratory. In: 4th Annual Heat Pump Technology Conference, Stillwater, Oklahoma, 9 April, vol. 1 (Stillwater, Oklahoma).
Naylor, S., Ellett, K.M., Gustin, A.R., 2015. Spatiotemporal variability of ground thermal properties in glacial sediments and implications for horizontal ground heat exchanger design. Renewable Energy 81, 21–30.
Nicholson, D., Agab, A., Clarke, B.G., 2008. Model specification to determine thermal conductivity of soils. Proceedings of the ICE - Geotechnical Engineering 161 (3), 161–168.
Penman, H.L., 1948. Natural evaporation from open water, bare soil and grass. Proceedings of the Royal Society of London. Series A: Mathematical and Physical Sciences 193 (1032), 120–145.
Penrod, E., Knight, R., Baker, M., 1950. Earth heat pump research—Part II. Engineering Experimental Station Bulletin, University of Kentucky 5 (18).
Philip, J.R., De Vries, D.A., 1957. Moisture movement in porous materials under temperature gradients. Transactions American Geophysical Union 38 (2), 222–232.
Piechowski, M., 1998. Heat and mass transfer model of a ground heat exchanger: validation and sensitivity analysis. International Journal of Energy Research 22 (11), 965–979.
Piechowski, M., 1999. Heat and mass transfer model of a ground heat exchanger: theoretical development. International Journal of Energy Research 23 (7), 571–588.

Rawlings, R.H.D., Sykulski, J.R., 1999. Ground source heat pumps: a technology review. Building Services Engineering Research and Technology 20 (3), 119–129.

Rees, S.J., Fan, D., 2013. A numerical implementation of the Dynamic Thermal Network method for long time series simulation of conduction in multi-dimensional non-homogeneous solids. International Journal of Heat and Mass Transfer 61, 475–489.

Remund, C.P., 1988. Predicted Effects of Soil Dry Density and Initial Water Content on Performance of a Heat Exchanger Buried in an Unsaturated Loess Parent Material (Ph.D. thesis). University of Nebraska–Lincoln.

Reuss, M., Konstantinidou, E., Sanner, B., 1998. 10 Years VDI 4640—German guidelines for ground coupled heat pumps, UTES and direct thermal use of the underground. In: Proceedings of ECOSTOCK, Stockton College, New Jersey, USA, p. 8.

Richards, L.A., 1931. Capillary conduction of liquids through porous mediums. Journal of Applied Physics 1 (5), 318–333.

Rosenblad, G., 1979. Earth heat pump system with vertical pipes for heat extraction and storage. In: Proceedings of Nordic Symposium of Earth Heat Pump Systems, Göteborg, pp. 102–110.

Schär, O., 1977. Di Anwendung der Wämepumpe zur Heizung von Wohnbuaten mit Wämentnahme aus dem Erdreight (The use of the heat pump for heating dwelllings with heat drawn from the soil). Bulletin of the Association of Suisse Electical Engineers 68 (4), 182–184.

Svec, O.J., 1987. Potential of ground heat source systems. International Journal of Energy Research 11 (4), 573–581.

Tarnawski, V., 1989. Effect of snow cover on ground heat pump performance and soil moisture freezing. International Journal of Refrigeration 12 (2), 71–76.

Tarnawski, V., Yuet, P., 1988. Winter performance of residential heat pump. Heat Recovery Systems and CHP 8 (3), 271–278.

Tarnawski, V.R., Leong, W.H., 1993. Computer analysis, design and simulation of horizontal ground heat exchangers. International Journal of Energy Research 17 (6), 467–477.

Thermomap, 2013. ThermoMap Project Newsletter.

VDI, 2001. VDI 4640 Part 2. Thermal Use of the Underground: Ground Source Heat Pump Systems. Verein Deutscher Ingenieure, Düsseldorf, Germany.

VDI, 2010. VDI 4640 Part 1. Thermal Use of the Underground: Fundamentals, Approvals, Environmental Aspects. Verein Deutscher Ingenieure, Düsseldorf, Germany.

Vitel, M., Rouabhi, A., Tijani, M., Guérin, F., 2015. Modeling heat transfer between a freeze pipe and the surrounding ground during artificial ground freezing activities. Computers and Geotechnics 63, 99–111.

Wentzel, E.-L., 2005. Thermal Modeling of Walls, Foundations and Whole Buildings Using Dynamic Thermal Networks (Ph.D. thesis). Chalmers University of Technology, Göteborg, Sweden.

Xing, L., Cullin, J., Spitler, J., Im, P., Fisher, D., 2011. Foundation heat exchangers for residential ground source heat pump systems—Numerical modeling and experimental validation. HVAC&R Research 17 (6), 1059–1074.

Xiong, Z., Fisher, D.E., Spitler, J.D., 2015. Development and validation of a Slinky ground heat exchanger model. Applied Energy 141, 57–69.

Xu, H., Spitler, J.D., 2014. The relative importance of moisture transfer, soil freezing and snow cover on ground temperature predictions. Renewable Energy 72, 1–11.

Yavuzturk, C., Spitler, J., 1999. A short time step response factor model for vertical ground loop heat exchangers. Ashrae Transactions 105 (2), 475–485.

Zarrella, A., De Carli, M., Galgaro, A., 2013. Thermal performance of two types of energy foundation pile: helical pipe and triple U-tube. Applied Thermal Engineering 61 (2), 301–310.

Zeng, H.Y., Diao, N.R., Fang, Z.H., 2002. A finite line-source model for boreholes in geothermal heat exchangers. Heat Transfer—Asian Research 31 (7), 558–567.

Zhu, Y., Tao, Y., Rayegan, R., 2012. A comparison of deterministic and probabilistic life cycle cost analyses of ground source heat pump (GSHP) applications in hot and humid climate. Energy and Buildings 55, 312–321.

Analytical methods for thermal analysis of vertical ground heat exchangers

6

M. Li[1], *K. Zhu*[2], *Z. Fang*[2]
[1]School of Energy Science and Engineering, Central South University, Changsha, China;
[2]School of Thermal Engineering, Shandong Jianzhu University, Jinan, China

Abstract

Analysis of heat transfer in ground (or geothermal) heat exchangers is essential to their design and integration within ground-source heat pump (GSHP) systems. This chapter provides a critical review of analytical models for heat transfer through borehole and foundation pile ground heat exchangers (GHEs). The analytical models are classified in terms of the underlying thermal processes and are evaluated in a time-scale framework because of the diversity of time and space scales involved in their operation. We summarize, discuss, and evaluate the major advances in this field, including conventional heat source models, conduction–advection models, short-term models, solutions for diverse time scales, and analytical solutions for energy piles. An approach to integration of several analytical models for the unit-step response functions (ie, *G*-function) to form a multi-timescale representation of a U-tube GHE is presented.

6.1 Introduction

A vertical ground heat exchanger (GHE) is essentially a channel (eg, U-, W-, or helix-shaped) buried in a borehole or a foundation pile of a building (Fig. 6.1), in which a circulating heat-carrier fluid absorbs (or discharges) heat from (or to) the ground. Analysis of heat transfer by GHEs imposes a great challenge to geothermal engineers, which involves a complex set of influencing factors and uncertainties, including complex geological conditions, unknown thermal properties of the ground, as well as the diversity of the involved time and length scales. To meet this challenge, a vast number of studies have been performed and great progress has been made. The reports of these achievements, however, are scattered in various sources over decades, being responsible for a major obstacle to the understanding of this topic and the development of this field. In order to remove the obstacle, this chapter provides a thorough review of the analytical models for the heat transfer of vertical GHEs.

Advances in Ground-Source Heat Pump Systems. http://dx.doi.org/10.1016/B978-0-08-100311-4.00006-6
Copyright © 2016 Elsevier Ltd. All rights reserved.

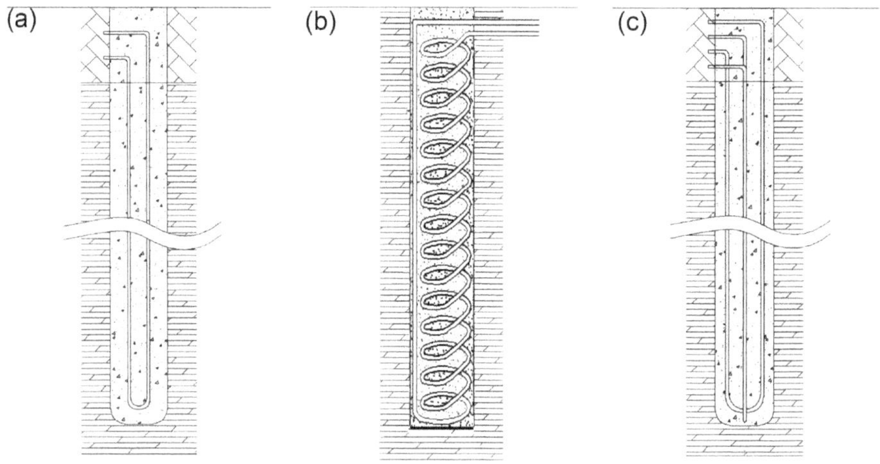

Figure 6.1 Schematic diagram of ground heat exchangers with (a) U-tube, (b) helix-shaped, and (c) double U-tube pipes.

Both numerical and analytical methods can be used in the thermal analyses of GHEs. Analytical approaches seem to be more practical than numerical methods for engineering applications. Numerical methods, such as finite-difference, finite-volume, and finite-element methods are elaborate enough to deal with all the thermal processes, but they are time-consuming for year-round and life-cycle simulations of GHEs, where diverse time and length scales need to be addressed (Li et al., 2014; Spitler and Bernier, 2011). Despite the complicated mathematical forms, analytical methods are much more desirable for academic studies and engineering applications. This is because they can facilitate encoding and computation, yield explicit relations between dependent and independent variables, and provide valuable insights into the underlying physical mechanisms.

The widely used analytical approaches to calculating thermal responses of GHEs are Kelvin's theory of heat sources and the Laplace transform method (Ingersoll et al., 1954; Li and Lai, 2015). Kelvin's theory of heat sources has been proven to be extremely successful for the thermal analyses of GHEs, because many configurations of ground loops, which are intractable by other means, can be well treated by line or surface heat sources. The Laplace transform method is often used to derive short-term solutions for borehole GHEs or to obtain fundamental solutions for the heat-source method. So far, a large number of analytical solutions have been developed for pile and borehole GHEs. We present a possible classification of these solutions in Fig. 6.2.

Most of the analytical models presume the heat transfer of GHEs to be pure heat conduction (referred as to pure heat conduction models), ie, other mechanisms of heat transfer, such as moisture transfer, groundwater movement, and ice formation of groundwater, are ignored. By contrast, the analytical models involving these effects are called nonpure heat conduction models in this chapter (Fig. 6.2).

By reviewing the progress in the analytical models, the development of this field can be roughly divided into three stages. The earliest efforts to analyze the thermal problems of GHEs can be traced back to the theoretical work of Ingersoll et al. during

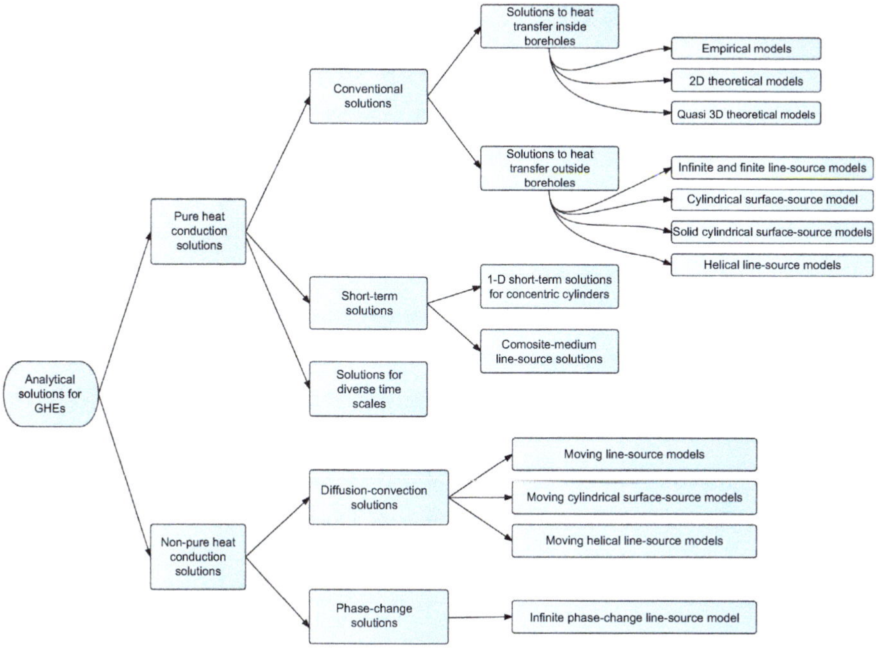

Figure 6.2 A classification of existing analytical solutions for vertical ground heat exchangers.

1940–1950s. Ingersoll et al. (1954) summarized in their book the theoretical models developed at this stage. Although the GHEs described in their book differ from the forms of the GHEs used today, their work has provided a solid foundation for subsequent developments in this field. Another significant breakthrough in the thermal analyses of GHEs was achieved in the 1980s (the second stage). A research group at University of Lund built a series of analytical models for the U-tube GHEs (Eskilson, 1987; Claesson and Eskilson, 1988; Hellström, 1991). One of their key ideas is to decompose the thermal process in the ground into two parts: the processes inside and outside the borehole. All the analytical solutions based on this idea are referred to as conventional solutions in this chapter (Fig. 6.2). The conventional solutions have been widely accepted and used in the design and simulation of ground-source heat pump (GSHP) systems and borehole thermal energy storage (BTES). Since 1990s, the application of GHEs in GSHPs and BTES has become an area of great practical and scientific interests on account of their potential for energy conservation. At this stage, considerable research has been devoted to extend and improve the conventional models from various aspects. Good examples of this are the analytical short-term solutions developed during the recent decade (reviewed in Section 6.3.2).

According to the classification (Fig. 6.2), this chapter offers a critical review of analytical models, and is organized as follows. Section 6.2 introduces a framework for the analysis of thermal problems of GHEs. Section 6.3 summarizes the recent progress in pure heat conduction models; Section 6.3.1 provides a review of developments in conventional solutions; Section 6.3.2 summarizes the newly developed short-term solutions; and solutions suitable for diverse time scales are evaluated in Section 6.3.3, which is followed by a comparison study of seven response functions in Section 6.3.4. Section 6.4 presents a review on nonpure heat conduction solutions, including heat source solutions addressing the effects of groundwater flow and freezing/thawing. Finally, this chapter closes in Section 6.5 with some conclusions.

6.2 A framework for analysis of heat transfer in ground heat exchangers

6.2.1 Scale analysis

A great challenge in modeling the heat transfer via GHEs is the involved diverse time and space scales (Fig. 6.3). It is highly necessary to identify these time-length scales because most analytical heat-transfer models make assumptions about the time

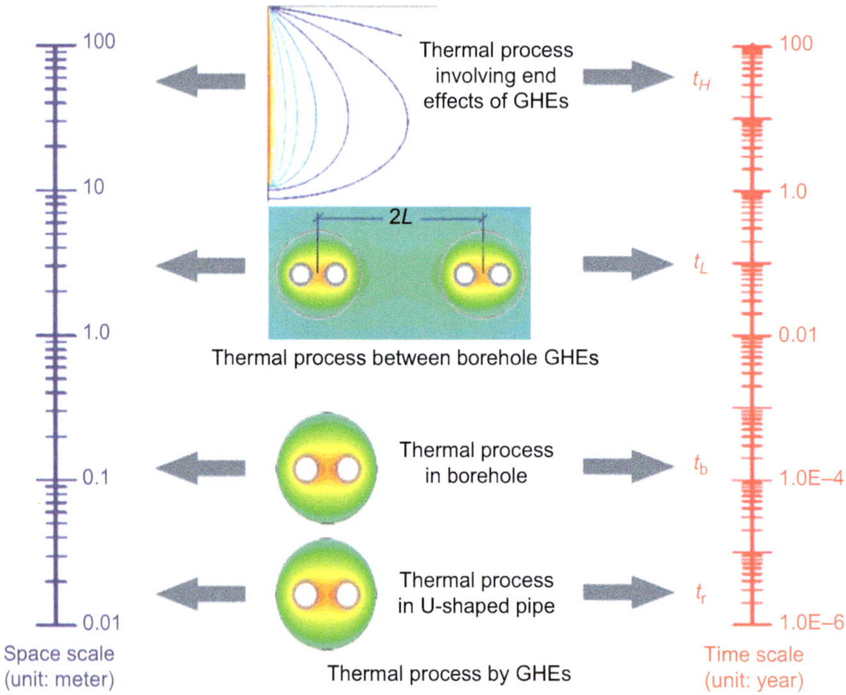

Figure 6.3 Time and space scales involved in thermal process of borehole ground heat exchangers. The thermal process can span four space scales and eight time scales. Modified after Li, M., Li, P., Chan, V., Lai, A.C.K., 2014. Full-scale temperature response function (G-function) for heat transfer by borehole ground heat exchangers (GHEs) from sub-hour to decades. Applied Energy 136, 197–205.

characteristics of the thermal process. We can evaluate the time scales important for heat transfer through GHEs by analyzing the differential equation of heat conduction. According to Bejan (2013), estimates of the orders of magnitude of the terms in Eq. [6.1a] are given by Eq. [6.1b] for a given radial space range, Δr:

$$\frac{\partial T}{\partial t} = \frac{\alpha}{r}\frac{\partial}{\partial r}\left(r\frac{\partial T}{\partial r}\right) \qquad [6.1a]$$

$$\frac{\Delta T}{t} \sim \frac{\alpha}{\Delta r}\frac{1}{\Delta r}\left(\Delta r\frac{\Delta T}{\Delta r}\right) \qquad [6.1b]$$

where t is the estimated time and α is the thermal diffusivity of the heat conduction medium. The symbol "$\sim$" used here means "is of the same order of magnitude as" (Bejan, 2013). When the transient term on the left is comparable to the heat conduction term on the right, the time t can be estimated by rearranging the scaling relation Eq. [6.1b]:

$$t \sim \Delta r^2/\alpha. \qquad [6.2]$$

The transient term can be ignored when t is about equal to $10\Delta r^2/\alpha$. From Eq. [6.2], it is easy to identify several time scales that are of practical importance, provided that the orders of magnitude of the thermal diffusivities of the grouting material, α_b, and the ground, α_s, are known (eg, 10^{-6} m^2/s).

Four length scales and eight time scales can be involved in the thermal process of GHEs (Fig. 6.3). The first important space scale is the radius of the borehole, r_b, ($\sim$0.05 m), and the associated time $\left(t_b \sim r_b^2/\alpha_b\right)$ is of the order of 1 h, during which the effect of the heat capacity of the backfilling material is significant (Eskilson, 1987; Claesson and Eskilson, 1988; Hellström, 1991). The second space dimension having practical importance is the half distance between two adjacent boreholes, L, which is of the order of several meters. The corresponding time $t_L \sim L^2/\alpha_s$ is of the order of 1 month, during which the thermal interaction between adjacent boreholes becomes important (Li et al., 2014). The largest space scale can be tens of meters or more, such as the half-length of a borehole, $H/2$, and the horizontal scale of a GHE cluster. The time scale involved, $t_H \sim H^2/4\alpha_s$, is as long as the lifetime of a GHE (decades) (Zeng et al., 2002). There is another time, $t_r \sim H/u$, (where u denotes the velocity of the fluid in the GHEs) called the residence time (Hellström, 1991). It is generally several minutes long. It is the time during which the heat capacity of the heat-carrier fluid needs to be addressed.

Quantifying short-term hourly temperature response of GHEs is vital for hourly energy analysis and the optimum control of GSHP and hybrid GSHP systems. In contrast, the long-term response determines the overall feasibility of a system from the life-cycle standpoint. Addressing the complete spectrum of time scales can require vast computational resources, particularly in large-scale applications.

6.2.2 General assumptions

Some assumptions must be made when deriving an analytical solution. The most common ones are as follows:

Assumption 1	The ground is assumed to be infinite or semiinfinite in extent, depending on whether or not the influence of the surface is considered.
Assumption 2	The ground has a uniform initial temperature (effective undisturbed ground temperature). If the surface is considered, this initial temperature can be used as a constant-temperature boundary condition (BC) for the surface.
Assumption 3	The BC for the wall of borehole or heat transfer pipe is either a constant flux or a constant temperature, with the constant-flux BC being more convenient.
Assumption 4	If the effect of the seepage of groundwater cannot be ignored, the flow is generally assumed to be homogeneous and parallel to the surface.
Assumption 5	Although the ground is usually layered and inhomogeneous, the ground can be treated as a medium with an equivalent thermal conductivity, $k_{s,e}$. There are several theoretical expressions for $k_{s,e}$ (Bear, 1972). For example, the equivalent conductivities are $$k_{s,e} = \frac{1}{\sum_{i=1}^{N} b_i} \sum_{i=1}^{N} k_{s,i} b_i \quad [6.3a]$$ for a horizontally layered ground medium and $$\frac{1}{k_{s,e}} = \frac{1}{\sum_{i=1}^{N} b_i} \sum_{i=1}^{N} \frac{b_i}{k_{s,i}} \quad [6.3b]$$ for a vertically layered ground medium, where $k_{s,i}$ and b_i are the thermal conductivity and thickness of the ith layer of the ground, respectively, and N is the total number of layers. An expression for ground stratified in an arbitrary direction is given by Bear (1972).

Assumptions 1–5 are widely used in analytical models of GHEs, and they have been verified to some extent by numerical simulations (Eskilson, 1987; Claesson and Eskilson, 1988; Lee, 2011).

6.2.3 Formalization of key problem

In the language of heat transfer, the problem of heat transfer by GHEs can probably be expressed as

$$q_l = \frac{T_f(t) - T_{s,0}}{R(t)} = \frac{\Delta T}{R(t)} \quad [6.4]$$

where T_f is the average temperature of the circulating fluid, $T_{s,0}$ is the effective, undisturbed temperature of the ground, q_l is the heat transfer rate of the GHE per unit time per unit length (W/m), and R is the total thermal resistance (m K/W).

Despite $R(t)$ being a function of time, analytical models exclusively decompose it into a time-independent part and a time-dependent part to simplify the analysis. The time-dependent part of R is often denoted by $G(t)$ (often referred to as G-function). There are minor differences between the G-functions proposed by different researchers (Li and Lai, 2012a; Claesson and Eskilson, 1988). The G-function was first suggested in a dimensionless form by Ingersoll et al. (1954), and was subsequently used and further developed by Eskilson (1987) and Claesson and Eskilson (1988). The dimensionless form is widely used because it facilitates the analysis and summary of results.[1] By contrast, the G-function used in this chapter has a clear physical meaning: It has the same dimension as thermal resistance and is entirely consistent with the unit-step response function defined in the principle of superposition (Section 6.2.4). In other words, the unsteady thermal resistance, G, can be understood as the temperature response in the ground due to a unit-step change in the heat flux, q_l. Using the definition of G, Eq. [6.4] becomes

$$\Delta T = q_l R(t) = q_l [R_s + G(\mathbf{x}, t)] \qquad [6.5]$$

where $\mathbf{x}$ denotes the coordinates of the point under consideration; and R_s is the time-independent part of the thermal resistance R, which depends on the choice of $\mathbf{x}$.

The temperature, T_f, in Eqs. [6.4] and [6.5] is the average temperature of the ground loops. Another model or assumption is necessary to determine the temperatures of the inlet and outlet fluids. One commonly used assumption is that T_f approximately equals the average of the temperatures of the inlet and outlet fluids of the GHE (ASHRAE, 2011). Thus, the temperatures at the inlet and outlet, $T_{f,i}$ and $T_{f,o}$, respectively, are

$$\begin{cases} T_{f,i}(t) = T_f(t) + \dfrac{q_l H}{2\rho_f c_f V_f} \\ T_{f,o}(t) = T_f(t) - \dfrac{q_l H}{2\rho_f c_f V_f} \end{cases}, \qquad [6.6]$$

where subscript f denotes fluid; H is the length of the GHE; and ρ, c, and V are density, mass specific heat, and volume flow rate, respectively. Eq. [6.6] is equivalent to $\rho_f c_f V_f (T_{f,i} - T_{f,o}) = q_l H$. More accurate expressions for $T_{f,o}$ can be found by solving energy balance equations of the circulating fluid for a given $T_{f,i}$ and for a specific point $\mathbf{x}$. Good examples of this are the quasi-three-dimensional models for R_b reviewed in Section 6.3.1.

To conclude, the key to heat transfer analysis is first to determine G and R_s. Then, T_f, $T_{f,i}$, and $T_{f,o}$ can be determined. Existing models, as discussed in Sections 6.3

[1] The nondimensional form is used in Chapter 2 and is distinguished by use of a lower case "g" notation.

and 6.4, can be evaluated in terms of the extent to which the predictions are affected by the constant thermal resistance, R_s.

6.2.4 Duhamel's theorem

Duhamel's theorem, which is also known as the principle of superposition (Carslaw and Jaeger, 1959; Özisik, 1993), is another theoretical tool for developing solutions to heat conduction with time-dependent boundary conditions and/or time-varying energy source terms. In actual applications, the heat transfer rate of GHEs varies continuously due to variations in the cooling or heating loads of buildings. According to Duhamel's theorem, the problems of time-varying loads can easily be tackled by using the solution to the problem for a unit-step load (Carslaw and Jaeger, 1959; Özisik, 1993):

$$T_1(\mathbf{x},t) = T_{s,0} + \int_0^t q_l(\tau)\frac{\partial G(\mathbf{x},t-\tau)}{\partial t}d\tau, \quad [6.7]$$

where T_1 is the temperature at the point under consideration, $\mathbf{x}$; the time-dependent function $q_l(t)$ is the heat exchange rate of a GHE; and τ is an integral variable with the dimension of time. The function G is the response function for unit-step change in the load of a GHE; it is entirely identical to that defined in Eq. [6.5]. Duhamel's theorem indicates that how the preceding loads $q_l(\tau)$ influence the current temperature $T_1(\mathbf{x}, t)$.

Since the heating and cooling loads of a building are commonly available as stepwise constant values (Fig. 6.4), it is more convenient to write Eq. [6.7] as Eq. [6.8] (Özisik, 1993):

$$T_1(\mathbf{x},t) = T_{s,0} + \sum_{j=0}^{N-1} \Delta q_j G(\mathbf{x}, t - j\Delta t). \quad [6.8]$$

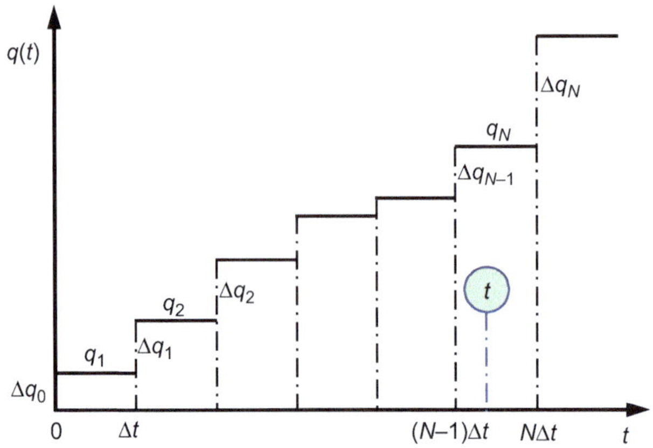

Figure 6.4 Definitions of variables used in Eq. [6.8] and stepwise varying load $q(t)$ of ground heat exchangers.
Modified after Li, M., Li, P., Chan, V., Lai, A.C.K., 2014. Full-scale temperature response function (G-function) for heat transfer by borehole ground heat exchangers (GHEs) from sub-hour to decades. Applied Energy 136, 197–205.

Here, $\Delta q_j = q_{j+1} - q_j$ is the stepwise change in q_l at the beginning of the $(j+1)$-th time interval as illustrated in Fig. 6.4. The derivation of Eq. [6.8] from Eq. [6.7] can be found in Özisik (1993). Eqs. [6.7] and [6.8] imply that the temperature response of any time-varying q_l is readily determined once the temperature response to the unit-step load, that is, the G-function, is available. Thus, a heat transfer analysis should focus on the development of the G-function.

6.3 Pure heat conduction solutions

6.3.1 Conventional solutions for separate regions

The conventional way in which the thermal processes of a borehole GHE are decomposed (ie, R) is in terms of the radius of the borehole (Hellström, 1991): The process in the borehole is assumed to be in a steady-flux state (or quasi−steady state) that can be described by the effective thermal resistance of the borehole, R_b. The process outside the borehole depends on time and is represented by a G-function (Ingersoll et al., 1954). For simplicity, all the solutions based on this assumption are called conventional solutions in this chapter. This approach was originally proposed to design GHEs for BTES (Hellström, 1991); it is only applicable to cases in which the rate of heat transfer varies slowly. But this approach is also widely accepted and used for the analysis and design of GSHPs (ASHRAE, 2011). It could be argued that the conventional solutions may result in discrepancies when used to model the high-frequency (short-term) thermal response of a GSHP.

6.3.1.1 Solutions to heat transfer inside boreholes

Heat transfer inside a borehole depends mainly on the configuration of the flow channels, the thermal properties of the grouting materials and the surrounding ground. This local thermal process consists of three components:

1. convective heat transfer between the circulating fluid and the inner surface of the U-tube pipes,
2. conductive heat transfer through the wall of the U-tube pipe, and
3. conductive heat transfer through the backfilling material.

In a strict sense, if t is greater than or equal to 5-10t_b, the heat transfer in the borehole approaches a steady-flux state (not a steady state), in which the temperature difference between the fluid and the wall of the borehole (not the temperatures themselves) is constant. Under this steady-flux condition, the three thermal processes can be characterized by three time-independent thermal resistances; and their sum yields the effective fluid-to-ground thermal resistance, R_b (Hellström, 1991).

The thermal resistances associated with the first two processes are referred to as the thermal resistance of the pipe, R_p (m K/W). It is easy to calculate from the formula

$$R_p = \frac{1}{2\pi k_p} \ln\frac{r_o}{r_i} + \frac{1}{2\pi r_i h}, \qquad [6.9]$$

where k_p is the thermal conductivity of the U-tube pipe (W/m K); r_o and r_i denote the outer and inner radii of the legs of the U-tube pipe, respectively; and h is the convective heat transfer coefficient (W/m^2 K). There are many empirical formulas for estimating h (Bejan, 2013). An empirical correlation only can yield a rough estimate of h due to the inherent complexity of convective heat transfer, but the influence of such a rough estimate on the calculation of R_b is negligible because convective thermal resistance accounts for only 2–3% of R_b in the cases of turbulent flows (Li and Lai, 2013a).

The thermal resistance of heat conduction through the backfilling material constitutes the majority of R_b. Models for this process (ie, for R_b) can be divided into empirical and theoretical groups (Table 6.1). Typical empirical models are one-dimensional models that take the U-tube pipe to be a pipe of "equivalent" diameter (Gu and O'Neal, 1998; Shonder and Beck, 1999). This equivalent-diameter assumption simplifies a two-dimensional geometric region into a concentric annular region, thereby reducing a complicated multidimensional problem to a simple one-dimensional one. Empirical models may contain several empirical constants, which may be determined by fitting experimental or computational data to the model for a particular geometric arrangement (Paul, 1996; Sharqawy et al., 2009). Empirical models are widely used because of their simplicity; but they provide little insight into the underlying heat transfer processes, which makes them difficult to generalize.

Theoretical models can be broken down into two-dimensional and quasi-three-dimensional models. Hellström (1991) presented two two-dimensional models for R_b, one derived from the steady-state line-source assumption and one based on a multipole method. The difference between them lies in the dimensionless variable η (see Table 6.1). For the configuration with a single U-tube pipe, either η is equal to 0, if the steady-state line-source assumption is used, or it is calculated from Eq. [6.10], which is derived from the first-order multipole expression (Hellström, 1991):

$$\eta = \frac{\left[\frac{r_o}{2D}\left(1 - \frac{4\sigma D^4}{r_b^4 - D^4}\right)\right]^2}{\frac{1+2\pi k_b R_p}{1-2\pi k_b R_p} + \frac{r_o^2}{4D^2}\left[1 + \frac{16\sigma D^4 r_b^4}{\left(r_b^4 - D^4\right)^2}\right]}, \qquad [6.10]$$

where D is the half spacing between the legs of the U-tube tube; the dimensionless variable σ is defined to be $(k_b - k_s)/(k_b + k_s)$; and k_s and k_b denote the thermal conductivities of the ground and backfilling materials, respectively. The difference between the line-source model and the first-order multipole model (ie, η) is usually less than 15% in most situations (Hellström, 1991).

It should be noted that k_s enters the two-dimensional models via the dimensionless variable σ, highlighting the fact that the thermal conductivities of not only the grouting material but also the surrounding ground can affect the thermal process inside the borehole. All empirical expressions incorrectly ignore the influence of k_s by arbitrarily imposing a constant-temperature BC on the wall of the borehole (Gu and O'Neal, 1998; Shonder and Beck, 1999; Paul, 1996; Sharqawy et al., 2009). Theoretical models address the influence of k_s by imposing a constant-temperature BC at a certain distance r_c from the wall (Hellström, 1991). Therefore, the predictions of these models

Table 6.1 Models of R_b for U-tube GHEs

		Expressions for R_b	Comments
Empirical models		$R_b = \frac{1}{2\pi k_b} \ln\left(\frac{r_b}{r_p}\sqrt{\frac{r_p}{D}}\right)$	This is for a GHE with a single U-tube pipe. It is also derived from the equivalent-diameter assumption (Gu and O'Neal, 1998).
		$R_b = \frac{1}{2\pi k_b} \ln \frac{r_b}{\sqrt{n} r_p}$	This is derived from the equivalent-diameter assumption; n denotes the number of pipes in a borehole (Shonder and Beck, 1999).
		$R_b = \frac{1}{\beta_0 k_b (r_b/r_p)^{\beta_1}}$	This expression uses the shape-factor concept. The empirical coefficients are obtained by fitting experimental data (Paul, 1996).
Theoretical models	Two-dimensional models	$R_b = \frac{1}{4\pi k_b}\left\{\ln\left[\frac{r_b^2}{2Dr_o}\left(\frac{r_b^4}{r_b^4 - D^4}\right)^{\sigma}\right] - \eta\right\} + \frac{R_p}{2}$	The influence of k_s is represented by the dimensionless ratio σ. η is equal to 0 or calculated from Eq. [6.10] (Hellström, 1991).
	Quasi-three-dimensional models	$R_b^* = R_b + \frac{1}{3R_a}\left(\frac{H}{C_f V_f}\right)^2$	This expression is derived by assuming a uniform heat flux on the borehole wall. C_f and V_f are volumetric specific heat and flow rate of the fluid. R_a is given by Hellström (1991).
		$R_b^* = R_b \eta \coth(\eta), \quad \eta = \frac{H}{C_f V_f} \frac{1}{2R_b}\sqrt{1 + \frac{4R_b}{R_{12}^{\Delta}}}$	This expression is derived by assuming a uniform temperature on the borehole wall. R_b is obtained by a two-dimensional model; the details of R_{12}^{Δ} can be found in Hellström (1991).
		$R_b^* = \frac{H}{C_f V_f}\left(\frac{T_{f,i} - T_b}{T_{f,i} - T_{f,o}} - \frac{1}{2}\right)$	$T_{f,o}$ is the fluid outlet temperature obtained by solving energy equations for U-tube channels (Zeng et al., 2003).

depend slightly on the choice of r_c. To completely eliminate the dependence on r_c, Claesson and Hellström (2011) have reformulated the multipole model and derived a physically and mathematically correct definition of R_b; Hermanns and Pérez (2014) also proposed a mathematically rigorous boundary condition at r_c based on a matched-asymptotic-expansion technique.

Empirical and two-dimensional theoretical models consider the local heat exchange at a given depth but ignore the variation in the temperature of the fluid in the downward and upward channels. To address this variation, Hellström (1991) proposed two quasi-three-dimensional models for R_b, one employing a uniform-flux BC for the borehole wall and the other one employing a uniform-temperature BC. The final expressions include additional correction factors for the two-dimensional models (Table 6.1). Based on Hellström's two-dimensional models, Zeng et al. (2003) also built quasi-three-dimensional models for GHEs with single and double U-tube pipes. They employed functions of the temperature of the fluid at the GHE outlet (Table 6.1). Recently, Ma et al. (2015) reported a new set of quasi-3D models for single and double U-tube GHEs based on a full-scale line source model.

Last but not least, all the models for R_b are applicable only for time $t > 5\text{-}10t_b$, which means that the steady-flux state is attained in the borehole. These models for R_b will be used together with a G-function for the thermal process outside the borehole (Section 6.3.1.2). Therefore, no matter which G-function is used, the conventional approach is unsuitable for $t < 5\text{-}10t_b$ when a rapidly-varying heat flux has to be addressed. To tackle the thermal response to a high-frequency heat flux, we must deal with the effect of the heat capacities of the materials in the borehole. Work on this is reviewed in Section 6.3.2.

6.3.1.2 Solutions to heat transfer outside boreholes

The models for heat transfer outside the borehole have been inspired by the seminal work of Ingersoll et al. (1954), who presented an infinite line source (ILS) model and an infinite cylindrical-source model for heat transfer through buried pipes. Even though they did not develop concrete expressions, they did propose ideas for dealing with additional complicated factors, including the method of images of heat sources used to account for the influence of the ground surface, the moving heat-source method for determining the effect of the movement of groundwater, and the principle of superposition for addressing the issue of variable heating rates. They also discussed relevant topics such as the effect of ice formation, moisture transfer, temperature recovery, and heat storage. Subsequently, the ideas suggested by Ingersoll et al. were extended to the analysis of borehole GHEs. Table 6.2 lists some typical G-functions, and all of them are given for the point $r = r_b$ on the borehole wall because the thermal resistance, R, is split depending on the radius of the borehole in the conventional approach.

An assumption adopted in a number of analytical solutions is that the ground can be considered as an infinite medium. Thus, heat transfer outside a borehole can be assumed to be heat conduction in an infinite region bounded internally by a borehole wall, which is subject to the boundary condition of a constant flux or temperature. The solution to this problem can be obtained by employing the Laplace transform method and was summarized in Carslaw and Jaeger (1959). Ingersoll et al. (1954) used that

Table 6.2 Response functions of heat conduction outside a borehole (given for the position $r = r_b$)

Expression for G-function	Comments
Infinite cylindrical surface source model (Ingersoll et al., 1954; Carslaw and Jaeger, 1959) $G(t; r_b) = \frac{1}{\pi^2 r_b k_s} \int_0^\infty (e^{-\alpha_s u^2 t} - 1) \frac{J_0(ur_b)Y_1(ur_b) - Y_0(ur_b)J_1(ur_b)}{u^2[J_1^2(ur_b) + Y_1^2(ur_b)]} du$	It is more complicated than the line-source model, and it is difficult to extend to handle other influences (eg, surface of ground, seepage flow, etc.)
Infinite line source model (Ingersoll et al., 1954; Carslaw and Jaeger, 1959) $G(t; r_b) = \frac{1}{4\pi k_s} \int_{r_b^2/4\alpha_s t}^\infty \frac{\exp(-u)}{u} du = \frac{1}{4\pi k_s} E_1\left(\frac{r_b^2}{4\alpha_s t}\right)$	It has the same performance as the infinite cylindrical surface model, and it is easy to extend to handle other relevant factors.
Finite line source model (Zeng et al., 2003) $G(t; r_b, z) = \frac{1}{4\pi k_s} \int_0^H \left[\frac{\operatorname{erfc}\left(\frac{\sqrt{r_b^2 + (z - z')^2}}{2\sqrt{\alpha_s t}}\right)}{\sqrt{r_b^2 + (z - z')^2}} - \frac{\operatorname{erfc}\left(\frac{\sqrt{r_b^2 + (z + z')^2}}{2\sqrt{\alpha_s t}}\right)}{\sqrt{r_b^2 + (z + z')^2}} \right] dz'$	It is suitable for predicting long-term responses because the influence of the ground surface is considered. This G depends on z, thus a double integration is required for determining the average temperature on the borehole wall. The value at $z = H/2$ could be used as an approximation of the average G-function to simplify the computation.
Infinite moving line source model (Diao et al., 2004) $G(t; r_b) = \frac{1}{4\pi k_s} I_0\left(\frac{Ur_b}{2\alpha_s}\right) \int_0^{r_b^2/4\alpha_s t} \frac{1}{\eta} \exp\left(-\frac{1}{\eta} - \frac{U^2 r_b^2 \eta}{16\alpha_s^2}\right) d\eta$	This model accounts for the effect of the characteristic velocity of groundwater, U, on the responses of borehole GHEs. An extension to finite moving line-source solution is given by Molina-Giraldo et al. (2011a).

Continued

Table 6.2 Continued

Expression for *G*-function	Comments
Infinite phase-change line source model (Ingersoll et al., 1954) $G(t;r_b)=\frac{T_1}{q_l}+\frac{1}{4\pi k_{fr}}\left[E_1(\beta^2)-E_1\left(\frac{r_b^2}{4\alpha_{fr}t}\right)\right]$	This is the only analytical solution that can address the issue of ice formation. T_1 is the melting-point of groundwater; a_{fr} and k_{fr} are the effective properties of frozen soil; and β is the root of Eq. [6.22].
Infinite solid cylindrical surface source (Man et al., 2010) $G(r,t)=-\frac{1}{4\pi^2 k_s}\int_0^{\pi}\mathrm{Ei}\left(-\frac{r^2+r'^2-2rr'\cos\varphi'}{4\alpha_s t}\right)d\varphi'$	r' denotes the radial position of the cylindrical surface-source. Extensions to the finite cylindrical surface-source models in homogeneous and anisotropic media are reported in Man et al. (2010) and Li and Lai (2012b).
Infinite helical line source model (Li and Lai, 2012b) $G(r,\varphi,z,t)=\frac{b}{4\pi k_s}\int_{-\infty}^{+\infty}\frac{1}{d}\mathrm{erfc}\left(\frac{d}{2\sqrt{\alpha_s t}}\right)d\beta$ $d=\sqrt{(r\cos\varphi-r'\cos\beta)^2+(r\sin\varphi-r'\sin\beta)^2+(z-b\beta)^2}$ $\begin{cases} x'=r'\cos\beta \\ y'=r'\sin\beta \quad \beta\in[-\infty,+\infty] \\ z'=b\cdot\beta \end{cases}$	The position of helical line source (x', y', z') is described by a helix equation. Extensions to the helical-shaped pipe of finite length is given by Li and Lai (2012b).

solution to model the thermal processes of buried pipes, and Deerman and Kavanaugh (1990) used it to model U-tube GHEs. This solution is often called the infinite cylindrical surface-source model. It is given mathematically by

$$\begin{cases} \rho_s c_s \dfrac{\partial T_s}{\partial t} = k_s \left(\dfrac{\partial^2 T_s}{\partial r^2} + \dfrac{1}{r} \dfrac{\partial T_s}{\partial r} \right) \\ r = r_b \quad -2\pi r_b k_s \dfrac{\partial T_s}{\partial r} = q_l \\ r \to \infty \quad T_s = T_{s,0} \\ t = 0 \quad T_s = T_{s,0} \end{cases} \qquad [6.11]$$

Here, r denotes radial distance. The solution of the model Eq. [6.11] is (Carslaw and Jaeger, 1959)

$$T_s(t, r_b) = T_{s,0} + q_l G(t, r_b) \qquad [6.12]$$

where the temperature response function $G(t, r_b)$ is given in Table 6.2. For large values of $\alpha_s t / r_b^2$, the G-function for a single borehole heat exchanger can also be written as

$$G(r,t) = \frac{1}{4\pi k_s} \left[\ln \frac{4\alpha_s t}{r^2} - \gamma + \frac{r^2}{2\alpha_s t} \left(\ln \frac{4\alpha_s t}{r^2} - \gamma + 1 \right) \right] \qquad [6.13]$$

where $\gamma = 0.5772$ is Euler's constant (see also Section 2.2.1).

The radius of the borehole of a GHE is usually several centimeters. It is very small compared to the length, which is tens of meters. Thus, this very thin borehole may be thought of as a line of infinite length with no volume, which releases (or absorbs) heat to (or from) the surrounding soil (Ingersoll et al., 1954; Hellström, 1991; Zeng et al., 2002; Lamarche and Beauchamp, 2007a). This approach is known as the ILS assumption and is given mathematically by

$$\begin{cases} \rho_s c_s \dfrac{\partial T_s}{\partial t} = k_s \left(\dfrac{\partial^2 T_s}{\partial r^2} + \dfrac{1}{r} \dfrac{\partial T_s}{\partial r} \right) \\ r \to 0 \quad -2\pi k_s \lim\limits_{r \to 0} r \dfrac{\partial T_s}{\partial r} = q_l \\ r \to \infty \quad T_s = T_{s,0} \\ t = 0 \quad T_s = T_{s,0} \end{cases} \qquad [6.14]$$

The difference between Eqs. [6.14] and [6.11] is the way in which the BC on the borehole wall ($r = r_b$) is dealt with. The solution to Eq. [6.14] is (Ingersoll et al., 1954; Carslaw and Jaeger, 1959)

$$T_s(t, r_b) = T_{s,0} + \frac{q_l}{4\pi k_s} \int_{r_b^2/4\alpha_s t}^{\infty} \frac{\exp(-u)}{u} du = T_{s,0} + q_l G(t, r_b). \quad [6.15]$$

For large values of $\alpha_s t/r_b^2$, the exponential integral E_1 occurred in Eq. [6.15] and Table 6.2 can be approximated by (Carslaw and Jaeger, 1959)

$$\frac{1}{4\pi k_s} E_1\left(\frac{r_b^2}{4\alpha_s t}\right) = \frac{1}{4\pi k_s}\left(\ln\frac{4\alpha_s t}{r_b^2} - \gamma\right), \quad [6.16]$$

with a maximum error of 2% for $t > 5t_b$. Eq. [6.16] is also equal to the first two terms on the right-hand side of Eq. [6.13]. The ILS solution is widely used in GHE simulation because of its simplicity. Nevertheless, it suffers from two limitations on the time scale.

First, Eq. [6.15] is unsuitable for small time because of the line-source assumption. This assumption can initially delay the temperature response at the borehole wall if a hollow borehole is employed, as in the infinite cylindrical surface source model (Eq. [6.12]). The relative error between these two models is less than about 10% when $t \approx 5t_b$ (Hellström, 1991) and 2% when $t \approx 20t_b$ (Ingersoll et al., 1954). An interesting observation is that $t > 5\text{-}10t_b$ is required when using any of these G-functions in conjunction with any of the models for R_b reviewed in Section 6.3.1.1 (ie, the conventional approach). In this sense, if we compare the complicated cylindrical surface source model with the ILS model, it appears that the former has little advantage because the difference between the two is less than 10% for $t > 5\text{-}10t_b$.

Second, the ILS model (and the infinite cylindrical surface source model) cannot be used to predict the response of GHEs for long times ($t \sim t_H$) because it ignores the influence of the ground surface. This is a common disadvantage of all the models that assume the ground to be an infinite medium.

The ground surface has an impact on long-term temperature variations in the ground, especially when the heating and cooling loads differ from each other. Excess heating or cooling loads can accumulate in the ground, causing the average ground temperature to gradually increase or decrease over time to such an extent that heat transfer through the ground surface becomes significant. If the time is long enough, a quasi–steady state can be approached whereby the excess heat is transferred to the air through the ground surface (Zeng et al., 2002; Li and Lai, 2012b).

As discussed in Section 6.2.2, the effect of the ground surface is generally simplified as a constant-temperature BC. Whereas the influence of the ground surface cannot be addressed by the infinite cylindrical surface source model, it can be treated analytically by combining Kelvin's theory of heat sources and the method of images: the ground surface is thought of as a reflecting plane, and by the process of taking images in the reflecting plane, we obtain a mirror-image line of heat sinks (strength: $-q_l$). The symmetrical distribution of line source and sink yields a temperature distribution that is constant on the ground surface. The addition of the temperature responses of the heat sources and sinks leads to the finite line source (FLS) models (Table 6.2). The FLS model requires a double integral to obtain the integral average temperature of the borehole wall, which leads to a marked increase in computational cost. Lamarche and Beauchamp (2007a) and Claesson and Javed (2011) proposed two reformulations of the FLS model so as to avoid the computation of the double integral, contributing to a sharp decrease in computation time.

In addition to U-tube pipes, GHEs can take other forms. A good example of this is the helix-shaped tubes used in foundation (energy) pile and compact GHEs (eg, Fig. 6.5). However, modeling GHEs with helical coils is somewhat indirect compared with GHEs with U-tube pipes. In these cases, analytical solutions can be developed by adapting line-source approaches or integrating point source solutions along curved paths. Three analytical models have been peculiarly derived for the helix-shaped tubes. Man et al. (2010) developed a solid cylindrical surface source model simplifying the spiral pipe to be a cylindrical surface that releases or absorbs heat in an infinite or semi-infinite homogenous medium (Table 6.2). Cui et al. (2011) developed a ring-coil-source model assuming the helix-shaped coil to be a series of separated rings. Li and Lai (2012b) derived helical line-source models approximating the spiral tube to be a helical line-source (Table 6.2). The helical line-source should be the most physically reasonable assumption for a helix-shaped GHE. But Li and Lai (2012b) found that the helical line-source model and the solid cylindrical surface source model yield almost the same average temperature for the wall of the helix-shaped pile. Moreover,

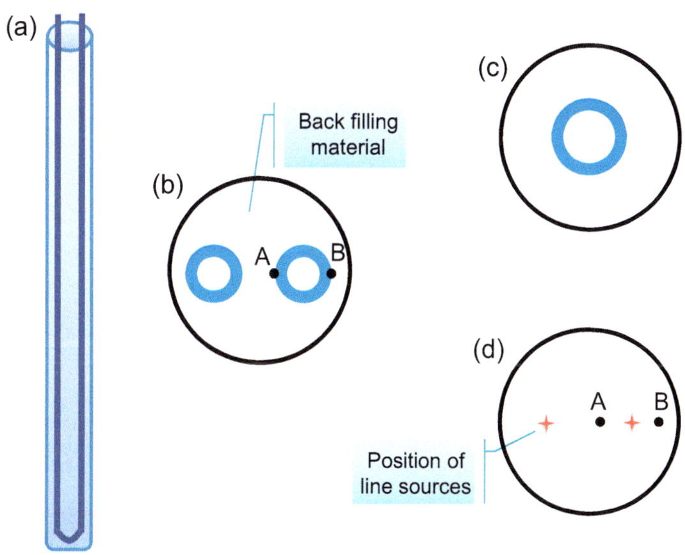

Figure 6.5 The assumption used in short-term models for a U-tube ground heat exchanger. (a) Borehole GHE. (b) Borehole GHE with a single U-shaped tube. (c) Equivalent-diameter assumption for U-shaped tube. (d) Composite-medium line source assumption for a sing U-tube.

the homogenous medium assumption used in these heat-source models may cause unacceptable errors when applying to energy piles because they ignore the difference in thermal properties between the ground and a concrete foundation pile.

6.3.2 Short-term solutions

Short-term and high-frequency temperature fluctuations are important in the design, optimization, and hourly energy analysis of GSHPs and hybrid GSHPs (Yavuzturk and Spitler, 1999). The heating and cooling loads of buildings vary continuously with time due to changes in the weather, the number of occupants, their activities, and so forth. In fact, cooling loads are often calculated on an hourly basis and sometimes sub-hourly. Such variations in loads can cause corresponding high-frequency fluctuations in the temperatures of a ground loop. According to Yavuzturk and Spitler (1999), the supply and return temperatures of a ground loop typically fluctuate up to 5.6−10°C over a given day. The scale analysis in Section 6.2.1 reveals that $t_b \sim 1$ h, implying that predicting the hourly high-frequency response requires handling the effect of heat capacity in a borehole. It becomes a complicated problem of transient heat conduction through a composite medium, in conjunction with various U-tube pipe installations (Li and Lai, 2012a). Conventional analytical models reviewed in Section 6.3.1 cannot meet this challenge.

One way to tackle this challenge is by numerical methods (Rees and He, 2013; Pasquier and Marcotte, 2012). Since this chapter focuses only on analytical models, numerical models are not discussed here. Another way to model the short-term response of a GHE is to simplify the geometrical arrangement in the borehole. Most short-time analytical models opt for the equivalent-diameter assumption (Gu and O'Neal, 1995; Beier and Smith, 2003; Lamarche and Beauchamp, 2007b; Bandyopadhyay et al., 2008a,b; Javed and Claesson, 2011; Li and Lai, 2013b; Shirazi and Bernier, 2013). As shown in Fig. 6.5(c), the equivalent-diameter assumption simplifies a problem involving complicated U-tube geometry to one in hollow concentric composite cylinders and thus reduces the original two-dimensional problem to a one-dimensional problem. The concentric composite cylinders can be considered to be infinite (Beier and Smith, 2003; Lamarche and Beauchamp, 2007b; Bandyopadhyay et al., 2008a,b; Javed and Claesson, 2011) or finite (Gu and O'Neal, 1995; Li and Lai, 2013b) in the radial direction because of the very short time involved in this problem. It is possible to obtain the simplified one-dimensional short-term solution by different analytical methods, such as the generalized orthogonal expansion technique (Gu and O'Neal, 1995) and the Laplace transform method (Lamarche and Beauchamp, 2007b; Li and Lai, 2013b). This simplification also makes it possible to tackle the transient effect of fluid circulating in the U-tube (Beier and Smith, 2003). But the equivalent-diameter approach is a pure empirical assumption. It cannot handle GHEs of other forms (eg, W- and helix-shaped GHEs) and also generates a mismatch between short- and long-term temperature responses as discussed in Section 6.3.3.

Recently, Li and Lai (2012a, 2013c) have proposed a theoretically complete approach to calculating the short-term response of a GHE based on Jaeger's infinite composite-medium line-source solution for a cylindrical medium. By this approach, one can assume that the legs of U-tube pipes (not the borehole) to be lines of heat source releasing heat into the surrounding composite media (Fig. 6.5(d)). The infinite composite-medium line-source model can accurately evaluate the influence of the heat capacity of grouting materials and the impact of the difference in thermal properties between the ground and the backfilling material while it is flexible enough to model various configurations of GHEs such as single and double U-tube, W-shaped pipes, helical-coils, etc. It has, however, a complicated form. Moreover, this model belongs to the group of ILS models and thus cannot be used to calculate long-term thermal processes if there is an imbalance between heat injection and extraction in the GHE.

6.3.3 Solutions for diverse time scales

As illustrated in Fig. 6.3, the heat transfer by GHEs can span four length scales and eight time scales. Accurately calculating the temperature response of GHEs is an intricate and time-consuming task because the solution requires time resolution ranging from minutes to decades (Claesson and Javed, 2011). Conventional FLS solutions appear to be appropriate models for predicting the long-term thermal process in the ground, but they are unsuitable for modeling the short-term thermal response. By contrast, the infinite composite-medium line-source model is valid for calculating the short-term response but invalid for the long-term process. An efficient analytical model that can address the entire time-space spectrum will be beneficial. Very few attempts have been made to develop such a model, however.

Claesson and Javed (2011) developed a combined response function applicable to time from minutes to decades for single- and multiborehole GHEs. According to Claesson and Javed (2011), a short-term model based on equivalent-diameter assumption is used up to a so-called breaking time t_{bt}; after that time, a long-term response function derived from the FLS solution is used. Claesson and Javed recommended, somewhat heuristically, that a breaking time of approximately 100 h could be used. It should be noted that the two solutions are not guaranteed to blend. The accuracy of the used short-term solution determines the goodness of match between the combined short- and long-term solutions.

Li et al. (2014) developed a temperature response function applicable to time from minutes to decades by combining the composite-medium line-source solution and the conventional ILS and FLS solutions. The three analytical solutions are theoretically complete; they can exactly match each other (Fig. 6.6). Based on this finding, Li et al. (2014) developed using the idea of matched asymptotic expansions the following response function suitable for diverse time scales (named full-scale G-function):

$$G_c(t) = G_i + G_o - G_m \qquad [6.17]$$

where G_m and G_o are the response functions for the borehole wall derived respectively from the conventional ILS and FLS solutions (Table 6.2). G_i is a short-term solution (eg, composite-medium line-source solution) for the average temperature on the wall of U-tube pipes. The short-term response functions for a single and double U-tube

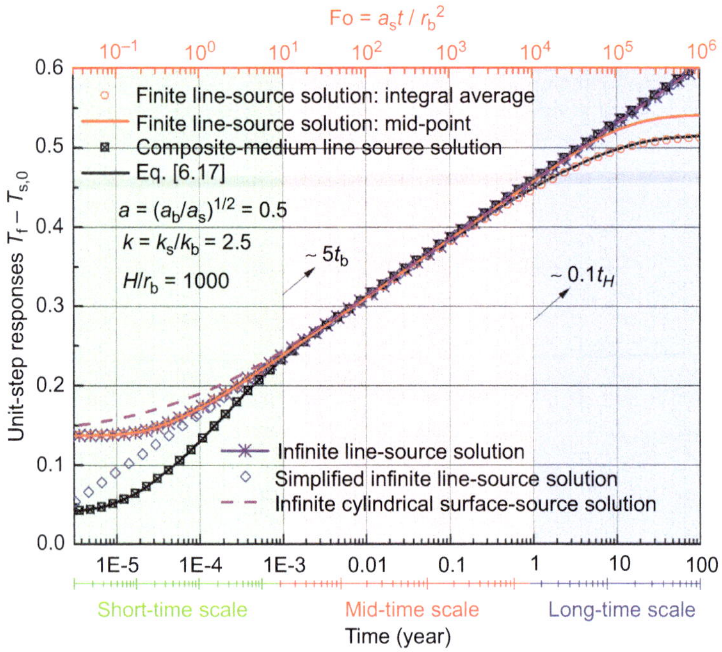

Figure 6.6 Comparison of seven analytical *G*-functions for a single U-tube ground heat exchanger.

GHEs were given by Li and Lai (2012a, 2013c). All the *G*-functions used in Eq. [6.17] should be theoretically correct and complete to ensure the entire match between the solutions.

It is important to remember that Eq. [6.17] is built on the idea of matched asymptotic expansions and is independent of the used short-, mid-, and long-term solutions. Therefore, any theoretically complete solutions can be used. For example, the computationally efficient expressions for the FLS solution developed by Claesson and Javed (2011) or Lamarche and Beauchamp (2007a) can be used for G_o instead of the original double-integration formula.

6.3.4 A comparison study of response functions

In this section, seven analytical G-functions are calculated and compared for a single U-tube GHE. Among them, five response functions are based on the conventional solutions: the infinite cylindrical surface source solution Eq. [6.12], the ILS model Eq. [6.15], the simplified ILS solution Eq. [6.16], and two FLS solutions. According to the conventional approach, the five *G*-functions must be used with R_b to calculate the response of the circulating fluid:

$$T_f(t) - T_{s,0} = R_b + G(t; r_b), \quad [6.18]$$

where the effective borehole thermal resistance, R_b, is calculated using the two-dimensional model derived from the line-source assumption (Table 6.1). The FLS model (Table 6.2) is used in two ways to determine the average temperature of the borehole wall: The first one is the computationally intensive integral average and the second one uses the temperature at the midpoint $z = H/2$ as an approximation.

The other two G-functions compared in Fig. 6.6 are the full-scale G-function Eq. [6.17] and the G-function based on the composite-medium line-source solution. For these two G-functions, the average fluid temperature is given by

$$T_f(t) - T_{s,0} = \frac{R_p}{2} + G(t; r_p), \quad [6.19]$$

where the thermal resistance R_p is defined in Eq. [6.9]. The expression for the G-functions can be found in Li et al. (2014) and Li and Lai (2012a, 2013c).

Fig. 6.6 shows the calculated fluid temperatures and the input parameters that were used. The time range of the calculation is from several minutes to 100 years and is divided into three subintervals: short range ($0 < t < 5t_b$), medium range ($5t_b < t < 0.1t_H$), and long range ($>0.1t_H$).

The calculated fluid temperatures are in complete accord with one another in the medium range ($5t_b < t < 0.1t_H$), which is roughly from several hours to one year. But the agreement is poor for the short and long ranges. For times larger than $0.1t_H$, all the models assuming that a borehole is infinitely long yield temperature responses that increase to infinity, whereas the FLS models show temperatures approaching a steady state. The results of the finite-length models are thus more logical and can be explained by the following physical insight: Heat transferred through the GHE is balanced by that through the ground surface. The comparison indicates that the use of the infinitely long models leads to an overestimation of the temperature variation for the long range, resulting in a more conservative design. There is also a discrepancy between the two FLS models: The prediction of the model using the midpoint temperature is higher than that using the integral average value. The impact of this difference on GHE design, particularly in large-scale applications, deserves to be explored.

Moreover, Fig. 6.6 clearly reveals the marked difference between the calculated short-term temperature responses. The conventional models using the steady-state thermal resistance, R_b, (infinite cylindrical-surface model, and conventional ILS and FLS models) give temperatures noticeably larger than that yielded by the composite-medium line-source model. The difference arises as a result of the neglected heat capacity of the grouting material in the borehole. Comparison of the composite-medium model with a set of reference sandbox data and an elaborate finite-volume model has demonstrated that the composite-medium line-source model is suitable for times as short as several minutes (Yang and Li, 2014). Thus, it seems clear that the discrepancies shown herein validate that all the conventional models fail to accurately predict the short-term (high frequency) temperature variations of GHEs.

Finally, Fig. 6.6 shows that the full-scale response function, Eq. [6.17], yield very fair prediction for the entire time range. It is in agreement with the composite-medium line-source solution in the short time range and agrees completely with the FLS model (integral average) in the long range. This result confirms the validity of the full-scale G-function.

To conclude, all the models examined here can be used to predict the medium-term temperature responses; the composite-medium line-source model is suitable for predicting the short-term responses; the FLS models can be used to calculate the long-term responses; and the full-scale model is applicable to the entire range. Although all the G functions examined here are developed for a single GHE, it is very direct to extend these G functions to include thermal interaction between boreholes by the principle of superposition. Since the thermal interaction between boreholes occurs in a time scale of weeks, the superimposed temperature should be from long-term solutions (eg, Claesson and Javed, 2011).

6.4 Nonpure heat conduction solutions

In general, heat conduction through GHEs is often accompanied by other heat transfer processes, including convection due to groundwater (seepage) flow, moisture migration, or phase change. Exact calculations of these processes are very difficult, but some idea of the order of magnitude for groundwater flow and ice formation can be reached in analytical ways. Unfortunately, to the best knowledge of the authors, no analytical solution has been reported yet for calculating the effect of coupled heat and moisture transfer through GHEs.

6.4.1 Diffusion–convection solutions

The movement of groundwater affects the heat transfer by involving gross heat convection (advection in some texts), which is significant for the long-term temperature response of GHEs (Pavlov and Olesen, 2012). While a conservative design of GSHPs can assume no benefit from this flow (Kavanaugh and Rafferty, 1997), it is nevertheless desirable to estimate its influence. Most analytical models solve this diffusion–convection problem by using the moving heat-source method, which was first proposed by Ingersoll et al. (1954) and was further explored or extended by Diao et al. (2004), Sutton et al. (2003), Molina-Giraldo et al. (2011a), Chiasson and O'Connell (2011), Tye-Gingras and Gosselin (2014), and Zhang et al. (2015). All these models are theoretically identical to each other, and that of Diao et al. (2004) is given in Table 6.2.

The theoretical basis of applying the moving heat source method to the diffusion–convection problem is the equivalence between them. The moving heat source problems may be regarded either as problems in which heat sources move through a fixed medium or as cases in which a uniformly moving medium flows through fixed heat sources (Carslaw and Jaeger, 1959; Eckert and Drake, 1987). Suppose that x, y, and

z denote the fixed coordinates and the velocity of the moving medium U is parallel to the axis of x. In the solution of moving heat source problems, it is convenient to define a coordinate system, say (ξ, η, ζ), moving with the medium. In the moving coordinate system, heat conduction in the moving medium is governed by

$$\frac{\partial T}{\partial t} = \alpha\left(\frac{\partial^2 T}{\partial \xi^2} + \frac{\partial^2 T}{\partial \eta^2} + \frac{\partial^2 T}{\partial \varsigma^2}\right) \qquad [6.20]$$

Since the relations between the fixed and the moving coordinates is $x = \xi + Ut$, $y = \eta$, and $z = \zeta$, substituting these relations into Eq. [6.20] yields:

$$\frac{\partial T}{\partial t} + U\frac{\partial T}{\partial x} = \alpha\left(\frac{\partial^2 T}{\partial x^2} + \frac{\partial^2 T}{\partial y^2} + \frac{\partial^2 T}{\partial z^2}\right) \qquad [6.21]$$

Clearly, the form of Eq. [6.21] is identical to the governing equation for diffusion–convection in a porous aquifer (Bear, 1972; Diao et al., 2004).

The flow of groundwater may be very complex, being vertical, horizontal, or both. All the moving heat source solutions, however, use Assumption 4 in Section 6.2.2. Therefore, the accuracy of these solutions depends heavily on how closely this assumption represents the actual groundwater flow conditions. A further extension of analytical solutions to consider more complicated groundwater flows is extremely difficult. The movement of groundwater in a porous aquifer may also cause heat transfer by thermal dispersion (Bear, 1972). Several authors have addressed the effect of thermal dispersion through the concept of modified thermal conductivity (Molina-Giraldo et al., 2011b; Erol et al., 2015). Since the modified thermal conductivities generally depend on the direction, heat source solutions for an anisotropic medium must be used to develop the response function (Li and Lai, 2012b; Molina-Giraldo et al., 2011b; Erol et al., 2015).

6.4.2 Phase-change solutions

Only one analytical expression is available for estimating the influence of freezing. It combines the ILS solution with Neumann's treatment for the phase-change interface (Ingersoll et al., 1954; Carslaw and Jaeger, 1959). This expression may be applicable to a porous medium if proper modifications are made. Table 6.2 gives a G-function derived from this solution, in which parameter β is the root of Eq. [6.22]:

$$\frac{q_l}{4\pi}\exp\left(-\beta^2\right) - \frac{k_s\theta_1 \exp\left(-\beta^2\alpha_{fz}/\alpha_s\right)}{E_1\left(\beta^2\alpha_{fz}/\alpha_s\right)} = \beta^2\alpha_{fz} n\rho_f Q_L, \qquad [6.22]$$

where E_1 is the exponential integral defined in Table 6.2; $\theta = T_{s,0} - T_s$ is the excess temperature; the subscript fz denotes frozen ground; the subscript 1 denotes the freezing-point of water; n is the porosity of a porous aquifer; Q_L is the latent heat of freezing; and ρ_f is the density of the groundwater. Ingersoll et al. (1954) conducted a simple case study using this solution to explore the effect of groundwater freezing. Their preliminary conclusion is that, among the factors affecting heat transfer, freezing is much less important than moisture transfer.

Even though heat transfer due to moisture transfer may be a minor contribution, the accompanying variations in the thermal properties of the ground can be large, thereby significantly affecting the thermal process in the ground at shallow depths. Moisture transfer coupled with heat transfer may involve the evaporation (or condensation) of water and diffusion driven by temperature (Soret/Luikov effect) or pressure differences (Luikov, 1975). These processes make the analysis of the zone more complicated. Reuss et al. (1997) used the finite-difference method to investigate coupled heat and moisture transfer in high-temperature ground heat storage and found that a deep moisture gradient was detected, and drying out and cracking out occurred in some ground media. However, very little investigative work has been done on a coupled heat and moisture process associated with the temperature levels of a GSHP system.

6.5 Conclusions

The analysis of heat transfer by vertical GHEs involves geothermal engineers tackling several difficult tasks. The first one is the diverse time and length scales involved in the thermal process. In spite of the rapid development of computers, the numerical methods are still time consuming and less efficient. Recent advancement of analytical models offers a plausible approach to meeting this challenge.

The second difficulty is the uncertainties of the thermal properties of the ground. The estimation of thermal properties is an inverse problem of heat transfer (a subject lying beyond the scope of this chapter) (Witte, 2013; Witte et al., 2002). Estimation procedures based on analytical models are simpler, faster, and more robust than those based on numerical methods; the new progress in analytical solutions may lay solid groundwork for establishing new algorithms of high accuracy and reliability.

Another difficulty is the involved complex geometric factors and geologic conditions. The geometric characteristics of GHEs make Kelvin's theory of heat source very attractive and efficient at solving related heat transfer problems. This theory, in some situations, can be extended to deal with the effects of movement and ice formation of groundwater. However, it seems to be extremely difficult to develop analytical solutions for more complex factors, eg, multidimensional groundwater flows and coupled heat and moisture transfer.

Finally, the authors would like to draw attention to the need for the validation of heat transfer models. The reference data set reported by Beier et al. (2011), Beier (2014) is suitable only for validating the short-term performance of the models. Reliable experimental data for validating the heat transfer models in the medium range is becoming available (Cullin et al., 2015) but remains lacking for long timescales.

References

ASHRAE, 2011. ASHRAE Handbook: HVAC Applications. ASHRAE, Inc., Atlanta, GA.

Bejan, A., 2013. Convection Heat Transfer, fourth ed. John Wiley & Sons, Inc., Hoboken.

Bear, J., 1972. Dynamics of Fluids in Porous Media. American Elsevier Publishing Company Inc., New York.

Beier, R.A., Smith, M.D., 2003. Minimum duration of in-situ tests on vertical boreholes. ASHRAE Transactions 109, 475–486.

Bandyopadhyay, G., Gosnold, W., Mann, M., 2008a. Analytical and semi-analytical solutions for short-time transient response ground heat exchangers. Energy and Buildings 40, 1816–1824.

Bandyopadhyay, G., Kulkarni, M., Mann, M., 2008b. A new approach to modeling ground heat exchangers in the initial phase of heat-flux build up. ASHRAE Transactions 114, 428–439.

Beier, R.A., Smith, M.D., Spitler, J.D., 2011. Reference data sets for vertical borehole ground heat exchanger models and thermal response test analysis. Geothermics 40, 79–85.

Beier, R.A., 2014. Transient heat transfer in a U-tube borehole heat exchanger. Applied Thermal Engineering 62 (1), 256–266.

Claesson, J., Eskilson, P., 1988. Conductive heat extraction to a deep borehole – thermal analyses and dimensioning rules. Energy 13, 509–527.

Carslaw, H.S., Jaeger, J.C., 1959. Conduction of Heat in Solids, second ed. Claremore Press, Oxford.

Claesson, J., Hellström, G., 2011. Multipole method to calculate borehole thermal resistances in a borehole heat exchanger. HVAC&R Research 17 (6), 895–911.

Claesson, J., Javed, S., 2011. An analytical method to calculate borehole fluid temperatures for time-scales from minutes to decades. ASHRAE Transactions 117, 279–288.

Cui, P., Li, X., Man, Y., Fang, Z.H., 2011. Heat transfer analysis of pile geothermal heat exchangers with spiral coils. Applied Energy 88, 4113–4119.

Chiasson, A., O'Connell, A., 2011. New analytical solution for sizing vertical borehole ground heat exchangers in environments with significant groundwater flow: parameter estimation from thermal response test data. HVAC&R Research 17 (6), 1000–1011.

Cullin, J.R., Spitler, J.D., Montagud, C., Ruiz-Calvo, F., Rees, S.J., Naicker, S.S., Konečný, P., Southard, L.E., 2015. Validation of vertical ground heat exchanger design methodologies. Science and Technology for the Built Environment 21 (2), 137–149.

Deerman, J.D., Kavanaugh, S.P., 1990. Simulation of vertical U-tube ground source heat pump systems using the cylindrical heat source solution. ASHRAE Transactions 97, 287–295.

Diao, N.R., Li, Q.Y., Fang, Z.H., 2004. Heat transfer in ground heat exchangers with groundwater advection. International Journal of Thermal Sciences 43, 1203–1211.

Eskilson, P., 1987. Thermal analysis of Heat Extraction Boreholes (Ph.D. thesis). University of Lund, Sweden.

Eckert, E.R.G., Drake, R.M., 1987. Analysis of Heat and Mass Transfer. Hemisphere Publishing Corporation, Washington.

Erol, S., Hashemi, M.A., François, B., 2015. Analytical solution of discontinuous heat extraction for sustainability and recovery aspects of borehole heat exchangers. International Journal of Thermal Sciences 88, 47–58.

Gu, Y., O'Neal, D.L., 1998. Development of an equivalent diameter expression for vertical U-tubes used in ground source heat pumps. ASHRAE Transactions 104, 347–355.

Gu, Y., O'Neal, D.L., 1995. An analytical solution to transient heat-conduction in a composite region with a cylindrical heat-Source. Journal of Solar Energy Engineering – Transactions ASME 117, 242–248.

Hellström, G., 1991. Ground Heat Storage – Thermal Analysis of Duct Storage Systems I. Theory. University of Lund, Lund.
Hermanns, M., Pérez, J.M., 2014. Asymptotic analysis of vertical geothermal boreholes in the limit of slowly varying heat injection rates. SIAM Journal on Applied Mathematics 74 (1), 60–82.
Ingersoll, L.R., Zobel, O.J., Ingersoll, A.C., 1954. Heat Conduction with Engineering, Geological, and Other Applications, Revised ed. The University of Wisconsin Press, Madison WI.
Javed, S., Claesson, J., 2011. New analytical and numerical solutions for the short-term analysis of vertical ground heat exchangers. ASHRAE Transactions 117, 3–12.
Kavanaugh, S.K., Rafferty, K., 1997. Ground-source Heat Pumps: Design of Geothermal Systems for Commercial and Institutional Buildings. American Society of Heating, Refrigerating and Air-Conditioning Engineers, Inc., Atlanta, GA.
Li, M., Lai, A.C.K., 2015. Review of analytical models for heat transfer by vertical ground heat exchangers (GHEs): a perspective of time and space scales. Applied Energy 151, 178–191.
Li, M., Lai, A.C.K., 2012a. New temperature response functions (G functions) for pile and borehole ground heat exchangers based on composite-medium line-source theory. Energy 38, 255–263.
Li, M., Lai, A.C.K., 2012b. Heat-source solutions to heat conduction in anisotropic media with application to pile and borehole ground heat exchangers. Applied Energy 96, 451–458.
Li, M., Lai, A.C.K., 2013a. Thermodynamic optimization of ground heat exchangers with single U-tube by entropy generation minimization method. Energy Conversion and Management 65, 133–139.
Li, M., Lai, A.C.K., 2013b. Analytical solution to heat conduction in finite hollow composite cylinders with a general boundary condition. International Journal of Heat and Mass Transfer 60, 549–556.
Li, M., Lai, A.C.K., 2013c. Analytical model for short-time responses of ground heat exchangers with U-shaped tubes: model development and validation. Applied Energy 104, 510–516.
Li, M., Li, P., Chan, V., Lai, A.C.K., 2014. Full-scale temperature response function (G-function) for heat transfer by borehole ground heat exchangers (GHEs) from sub-hour to decades. Applied Energy 136, 197–205.
Lee, C.K., 2011. Effects of multiple ground layers on thermal response test analysis and ground-source heat pump simulation. Applied Energy 88 (12), 4405–4410.
Lamarche, L., Beauchamp, B., 2007a. A new contribution to the finite line-source model for geothermal boreholes. Energy and Buildings 39, 188–198.
Lamarche, L., Beauchamp, B., 2007b. New solutions for the short-time analysis of geothermal vertical boreholes. International Journal of Heat and Mass Transfer 50, 1408–1419.
Luikov, A.V., 1975. Systems of differential equations of heat and mass transfer in capillary-porous bodies. International Journal of Heat and Mass Transfer 18 (1), 1–14.
Ma, W., Li, M., Li, P., Lai, A.C.K., 2015. New quasi-3D model for heat transfer in U-shaped GHEs (ground heat exchangers): Effective overall thermal resistance. Energy 90, 578–587.
Man, Y., Yang, H., Diao, N.R., Liu, J., Fang, Z.H., 2010. A new model and analytical solutions for borehole and pile ground heat exchangers. International Journal of Heat and Mass Transfer 53, 2593–2601.
Molina-Giraldo, N., Blum, P., Zhu, K., Bayer, P., Fang, Z.H., 2011a. A moving finite line source model to simulate borehole heat exchangers with groundwater advection. International Journal of Thermal Sciences 50, 2506–2513.
Molina-Giraldo, N., Bayer, P., Blum, P., 2011b. Evaluating the influence of thermal dispersion on temperature plumes from geothermal systems using analytical solutions. International Journal of Thermal Sciences 50, 1223–1231.

Özisik, M.N., 1993. Heat Conduction, second ed. John Wiley &Sons, Inc., New York.
Paul, N.D., 1996. The Effect of Grout Thermal Conductivity on Vertical Geothermal Heat Exchanger Design and Performance (M.Sc. thesis). South Dakota University, USA.
Pasquier, P., Marcotte, D., 2012. Short-term simulation of ground heat exchanger with an improved TRCM. Renewable Energy 46, 92–99.
Pavlov, G.K., Olesen, B.W., 2012. Thermal energy storage—a review of concepts and systems for heating and cooling applications in buildings: Part 1—Seasonal storage in the ground. HVAC&R Research 18 (3), 515–538.
Rees, S.J., He, M., 2013. A three-dimensional numerical model of borehole heat exchanger heat transfer and fluid flow. Geothermics 46, 1–13.
Reuss, M., Beck, M., Müller, J.P., 1997. Design of a seasonal thermal energy storage in the ground. Solar Energy 59 (4), 247–257.
Spitler, J., Bernier, M., 2011. Ground-source heat pump systems: the first century and beyond. HVAC&R Research 17 (6), 891–894.
Shonder, J.A., Beck, J.V., 1999. Field test of a new method for determining soil formation thermal conductivity and borehole resistance. ASHRAE Transactions 106, 843–850.
Sharqawy, M.H., Mokheimer, E.M., Badr, H.M., 2009. Effective pipe-to-borehole thermal resistance for vertical ground heat exchangers. Geothermics 38, 271–277.
Shirazi, A.S., Bernier, M., 2013. Thermal capacity effects in borehole ground heat exchangers. Energy and Buildings 67, 352–364.
Sutton, M.G., Nutter, D.W., Couvillion, R.J., 2003. A ground resistance for vertical bore heat exchangers with groundwater flow. Journal of Energy Resources Technology – Transactions ASME 125 (3), 183–189.
Tye-Gingras, M., Gosselin, L., 2014. Generic ground response functions for ground exchangers in the presence of groundwater flow. Renewable Energy 72, 354–366.
Witte, H.J.L., 2013. Error analysis of thermal response tests. Applied Energy 109, 302–311.
Witte, H.J.L., van Gelder, A.J., Spitler, J.D., 2002. In situ measurement of ground thermal conductivity: the Dutch perspective. ASHRAE Transactions 108 (1), 263–272.
Yavuzturk, C., Spitler, J.D., 1999. A short time step response factor model for vertical ground loop heat exchangers. ASHRAE Transactions 105 (2), 475–485.
Yang, Y., Li, M., 2014. Short-time performance of composite-medium line-source model for predicting responses of ground heat exchangers with single U-shaped tube. International Journal of Thermal Sciences 82, 130–137.
Zeng, H.Y., Diao, N.R., Fang, Z.H., 2002. A finite line-source model for boreholes in geothermal heat exchangers. Heat Transfer-Asian Research 31, 558–567.
Zeng, H.Y., Diao, N.R., Fang, Z.H., 2003. Heat transfer analysis of boreholes in vertical ground heat exchangers. International Journal of Heat and Mass Transfer 46, 4467–4481.
Zhang, W.K., Yang, H.X., Lu, L., Diao, N.R., Fang, Z.H., 2015. Study on spiral source models revealing groundwater transfusion effects on pile foundation ground heat exchangers. International Journal of Heat and Mass Transfer 84, 119–129.

Energy geostructures

7

K. Soga, Y. Rui
University of Cambridge, Cambridge, United Kingdom

Abstract

Energy geostructures (eg, energy piles, walls and tunnels) incorporate the primary heat exchangers of ground-source heat pump (GSHP) systems through the foundation elements (eg, piles and basement walls) or into the tunnel linings. Concrete used for such structures is an ideal medium as the heat absorber in the ground due to its good thermal conductivity and thermal storage capacity. Thus, an energy geostructure functions not only to carry mechanical load but also as a heat exchanger, circulating cooler or warmer fluid liquid to extract stored geothermal energy from the ground. When designing such dual purpose structures, it is necessary to investigate the thermo-hydromechanical interactions between the heat exchanger–coupled foundation elements and the surrounding ground. In this chapter, the basic concepts of energy piles, walls and tunnels are introduced. The current understanding on the performance of such structures are summarised and some design considerations are discussed. More work is needed to build confidence in the use of energy geostructures. Once clear design guidance has been developed, the uptake of energy geostructures will increase from the current rate.

7.1 Introduction

Ground-source heat pumps (GSHPs) provide sustainable heating and cooling energy for housing, offices and retail spaces. For new building and underground developments, it is possible to incorporate the primary heat exchangers through the foundation elements (eg, piles and basement walls) or into the tunnel linings. They are called energy (or thermal) piles/walls/tunnels, or collectively, energy geostructures. For example, in an energy pile, heat exchanger pipes are attached to the reinforcement cage of the pile before concrete is poured to create the pile. As concrete has excellent thermal conductivity and good heat storage properties, foundation piles are an ideal medium for heat transfer into the surrounding ground.

The first well-documented energy piles were installed in Austria in 1984 (Brandl, 2006). Since the mid-1980s, energy piles have found application across Northern Europe (eg, Suckling and Smith, 2002; Brandl, 2006; Adam and Markiewicz, 2009). In the United Kingdom, there have been a number of energy pile schemes constructed such as the 'One New Change' project (Garber et al., 2013), as well as many smaller projects (Keble college, Suckling and Smith, 2002; Lambeth college, Bourne-Webb et al., 2009). Brandl (2006) describes the first well-documented

Advances in Ground-Source Heat Pump Systems. http://dx.doi.org/10.1016/B978-0-08-100311-4.00007-8
Copyright © 2016 Elsevier Ltd. All rights reserved.

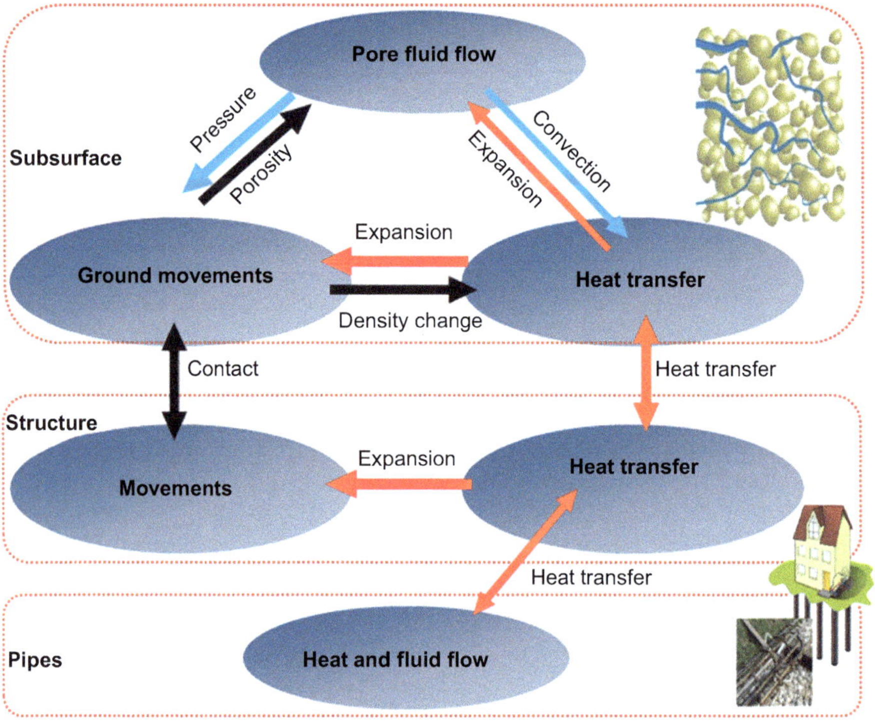

Figure 7.1 Coupling process in energy geostructure problems.

energy wall installation in Switzerland in 1996. In the United Kingdom, the first energy wall was installed at the Bulgari Hotel in London in 2010 (Amis et al., 2010), triggering the implementation of such systems in other high-profile projects such as the Crossrail tunnel project (Nicholson et al., 2013).

There are three important coupling processes relevant to energy geostructures, as shown in Fig. 7.1. The first process is the heat and fluid coupling inside heat exchanger pipes. The second is the thermomechanical coupling in the substructure, which refers the impact of temperature change on the mechanical state of concrete structures. The third is the thermo-hydromechanical (THM) coupling of the soil, which refers to the interactions between the temperature field, the displacement field and the pore water pressure distribution in the soil. The full understanding of these three coupling processes is essential in order to assess the performance of energy geostructures.

7.2 Energy piles

7.2.1 Concept

In an energy pile system, a heat exchanger pipe network is installed in the structural piles of a building as shown in Fig. 7.2. In their early years, energy piles were mainly

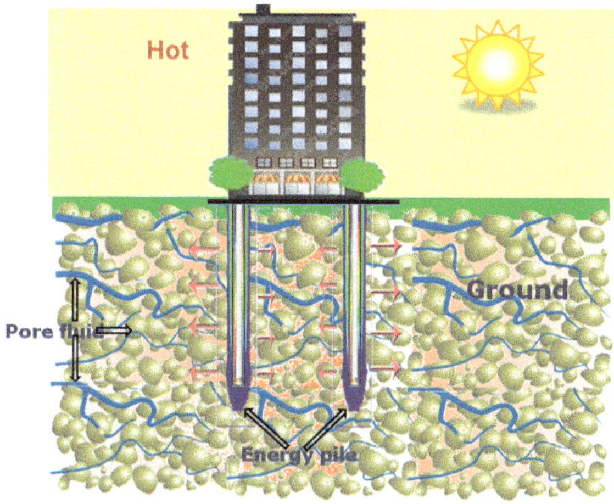

Figure 7.2 Heat transfer of energy pile.

built as driven precast concrete piles with integrated heat exchange tubes. Now the technology has been extended to large-diameter deep-bored piles, which allow for placing multiple U-shaped loops of pipes used to circulate the carrier fluid. In bored energy piles, the pipes are made of high-density polyethylene and are 20–25 mm in diameter and about 2-mm thick. They are fixed to the reinforcement cage, as shown in Figs. 7.3 and 7.4. A heat carrier fluid used in the primary circuit is water, saline solutions or, most commonly, water–glycol mixtures (the choice of fluid is discussed further in chapter: Borehole thermal energy storage). As water–glycol fluids are relatively high viscosity, there is a distinct possibility that fluid flow can be laminar. Some

Figure 7.3 Installation of geothermal loops into an energy pile.

Figure 7.4 Geothermal loops inside an energy pile before concreting.

care, therefore, has to be taken in the choice of fluid flow rates and pipe diameters to achieve turbulent flow conditions so that heat transfer can be effective. Due consideration also has to be given to the system pressure drop to ensure circulating pump energy demands are not burdensome.

The use of energy piles increased sharply in the 1990s and 2000s. Brandl (2006) report that the total number of energy piles installed in Austria increased by almost an order of magnitude in the period from 1990 to 2004. Energy piles have been installed in many countries including the Netherlands (eg, Koene and Charles, 2000), Switzerland (eg, Laloui et al., 2006; Pahud and Hubbuch, 2007), Belgium (eg, Desmedt and Hoes, 2007), China (eg, Gao et al., 2008), Japan (eg, Sekine et al., 2007; Hamada et al., 2007) and the United States (McCartney et al., 2010). In the United Kingdom, the first energy piles were installed at Keble College in Oxford in 2001 (Suckling and Smith, 2002). Since then, installation of energy foundations has been increasing rapidly. Approximately 150 energy piles were installed per year in the United Kingdom by 2005 and the total installed had risen to nearly 1600 energy piles by 2012 (Amis and Loveridge, 2014). However, after 2010 the rate of increase reduced in line with the economic slowdown, but a strong interest in this relatively new technology remains.

In bored piles, heat exchanger pipes are fixed to the structural pile steel reinforcement cage before placing the cage in the pile bore and concreting (Fig. 7.5(a)). For contiguous flight auger piles, heat exchanger pipes may be plunged into the centre of the pile bore after placing the concrete (Fig. 7.5(b)). The overall thermal performance of an energy pile depends on the position of the pipes within the pile as well as the size of the concrete part of the pile in relation to the size of the pipes. When designing energy piles, such differences between energy piles and GSHP borehole exchangers as well as variation in pipe positions and numbers need to be considered.

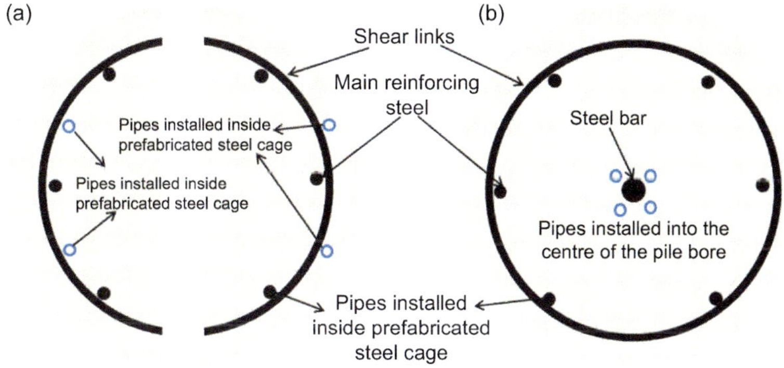

Figure 7.5 Typical energy pile construction details. (a) Pipework fixed to a rotary bored pile cage; (b) pipework installed in the centre of a pile (Loveridge, 2012).

A comprehensive framework towards energy pile design has been produced by the Ground Source Heat Pump Association (GSHPA) in the United Kingdom (GSHP Association, 2012). However, further improvements are needed to consider the issues discussed in this chapter.

7.2.2 Thermal behaviour of energy piles

When designing an energy pile, the thermal performance in relation to the expected heating/cooling load profiles is a primary concern. In broad terms, the thermal design problem is to maintain fluid temperatures within reasonable limits (depending on efficiency targets) in response to the imposed heating and cooling demands imposed on the pile heat exchange system. A simple approach to analyse the relationship between temperatures and heat transfer rates is to use a line source model (Ingersoll et al., 1954), which simulates the thermal performance of an energy pile as shown in Fig. 7.6. In this model, a heat exchanger is treated as a line heat source with constant heat flux in an infinite medium. The medium is assumed to be homogeneous and at a uniform initial temperature. For energy piles, however, it may be more appropriate to use a cylinder source model (Carslaw and Jaeger, 1959) because they often have small length-to-diameter ratios. The thermal performance of energy piles may be different from that of borehole heat exchangers (BHEs) and some key issues to consider are described in the next section.

7.2.2.1 Energy piles have small length–diameter ratios

The conventional BHE design often utilises temperature response functions, such as *g*-functions (Eskilson, 1987) to make the design tractable (see chapter: Vertical borehole ground heat exchanger design methods). However, it has been argued that it is not appropriate to apply Eskilson's *g*-function data into the energy pile design since the original functions were based on small radius heat sources with large depth/spacing

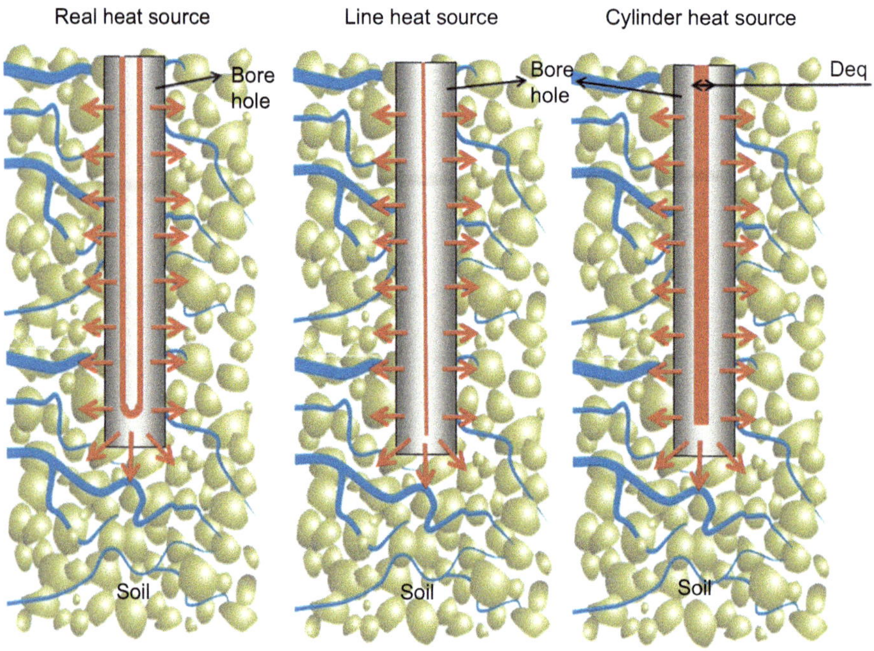

Figure 7.6 Mechanisms for line source model and cylinder source model.

ratio (Loveridge and Powrie, 2013). In contrast, the length−diameter ratio of energy piles is relatively small compared that of conventional BHEs. Therefore, when they are applied to large diameter energy piles, the solution may underestimate the temperature change at smaller time scales.

As most of the pipe systems in energy piles are multiple U-shaped pipes, some approximation is needed to use the conventional heat transfer models as shown in Fig. 7.7. For example, two pipes need to be approximated to a single pipe, and modelled as a cylindrical heat source. The key problem for the cylinder source model is to determine the diameter of the artificial pipe D_{eq}. It should be noted that the diameter of energy piles is much greater than that of a conventional GSHP borehole and the pipe geometry and layout can vary as illustrated in Fig. 7.7.

7.2.2.2 Heat transfer takes time to reach to the soil medium in large diameter energy piles

Simple heat source models assume that the temperature field within the pile body is at a steady state when used to predict the temperature changes in the ground. This assumption allows a constant resistance to be used when calculating the heat fluxes around the pipes. Although this method is successfully applied in the design of BHEs, large piles can have a diameter of more than 1 m and therefore have a much larger area of concrete in their cross-section. The analysis of heat transfer at shorter time scales, therefore, needs to account for the thermal mass of the concrete and the differences in thermal

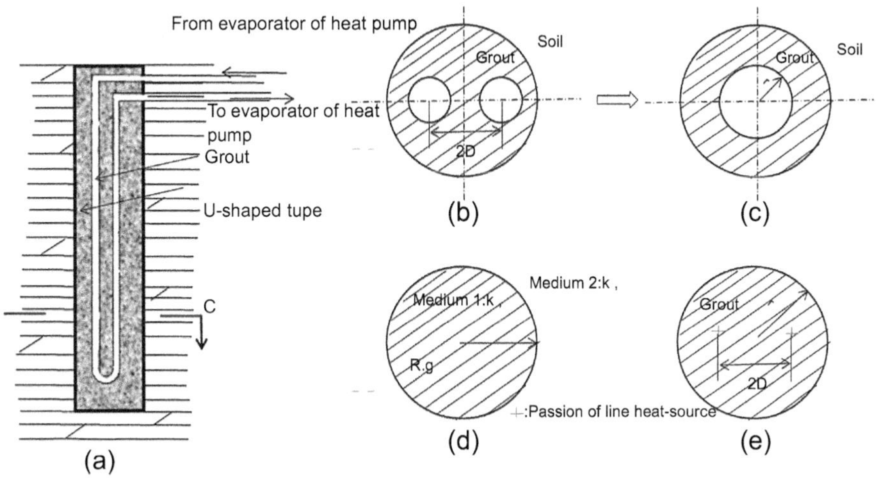

Figure 7.7 Schematic layout of an energy pile with a single U-shaped tube. (a) Borehole GHE, (b) C–C profile for single U-shaped tube, (c) equivalent diameter assumption for borehole GHE with single U-shaped tube, (d) line source positioned in an infinite composite medium, (e) line source assumption for borehole GHE with single U-shaped tube (Li and Lai, 2013).

properties between the pile and the surrounding ground. The constituents of concrete have widely differing thermal conductivities and overall thermal conductivity depends mainly on the aggregate lithology, aggregate volume ratio and water content.

Loveridge (2012) states that a transient approach to the heat transfer in both the concrete and the ground should be considered. Li and Lai (2012) address this by improving the *g*-functions. They applied infinite line sources in composite media to produce a two-region analytical model. The *g*-functions are normalised by the concrete conductivity and the temperature response function depends on the ground conductivity. Loveridge and Powie (2013) made a 3D numerical model of conduction heat transfer to consider a wide range of pile heat exchanger geometries. They developed correlations to improve the *g*-functions for a range of pile heat exchangers of different sizes and pipe arrangements. The advantages of their models are accounting for transient heat storage within the pile and applicable over a full range of timescales. This is achieved by using separate *g*-functions for the temperature responses of the ground surrounding the pile (pile *g*-functions) and of the pile itself (concrete *g*-functions).

7.2.2.3 Energy piles have many loops

As an energy pile has a large diameter, it is possible to add more pipes within the pile cross-section and the performance can be more efficient if more pipes are installed inside the borehole. For small diameter energy piles, Brandl (2006) characterised thermal output in terms of Watts per metre length of the pile. For larger diameter piles ($\geq$600 mm), he used Watts per square metre, reflecting the expected beneficial effects of larger surface areas of bigger piles.

Figure 7.8 Helix loop in an energy pile.
From Rehau.

It is usually beneficial to take consideration of thermal interactions between different parts of the U-loops. Installing two U-loops rather than one or separating the up and down parts of the U-loops with spacers to prevent thermal interactions have been considered for BHEs (Banks, 2012). Recently, there has been an interest in using spiral coil or helix-type heat transfer pipes rather than vertical pipes installed as U-loops (Park et al., 2012; Zarrella et al., 2013) (see Fig. 7.8). Man et al. (2010) derived a solid cylindrical source model to approximate the performance of a spiral coil, whereas Cui et al. (2011) modified the solid cylindrical source model to a ring–coil source model, which took into account the effect of the coil pitch.

Because of the complexities of pile heat exchanger geometries, the energy pile problem is well suited to numerical simulations using the finite element method, for example. The recent rapid increase in computational power allows this to be done even for design purpose. Fig. 7.9 shows a finite element model that was generated to represent the geometry of the energy piles installed at One New Change, London (Garber et al., 2013). There are a total of 192 foundation piles beneath the building footprint. The piles are 24–28 m long with a diameter of up to 2.5 m. Fig. 7.9(a) shows a plan view of the model. The dimension of the soil block is 450 × 450 m and 150 m deep. Close to the building footprint the mesh is finer, and increases in size further away. As shown in Fig. 7.9(b), the soil block is divided into four layers: Made Ground, London Clay, Lambeth Group and Chalk Group. To simulate the performance of the system accurately, modelling the fluid flow though the pipes embedded in the concrete piles is very important. Fig. 7.9(c) shows the 1D finite element model of heat exchanger pipes. Each pile contains four embedded loops in this case. The vertical spacing between adjacent pipe element nodes is 1 m, and each pipe has four U-loops connected in series (ie, each pile has one flow inlet and one flow outlet). The pipe elements are

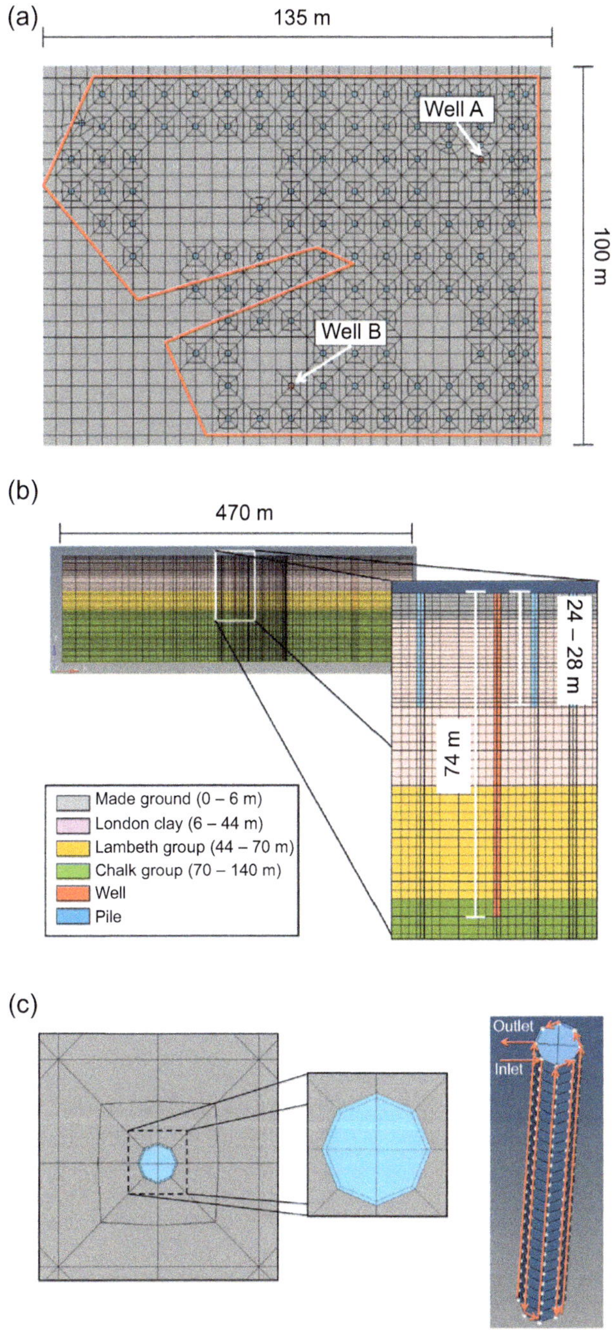

Figure 7.9 Finite element model: (a) plan view; (b) side view; and (c) pile model embedded in concrete piles (the red line indicates the pipe and the direction of the fluid flow) (Garber, 2013; Rui, 2015).

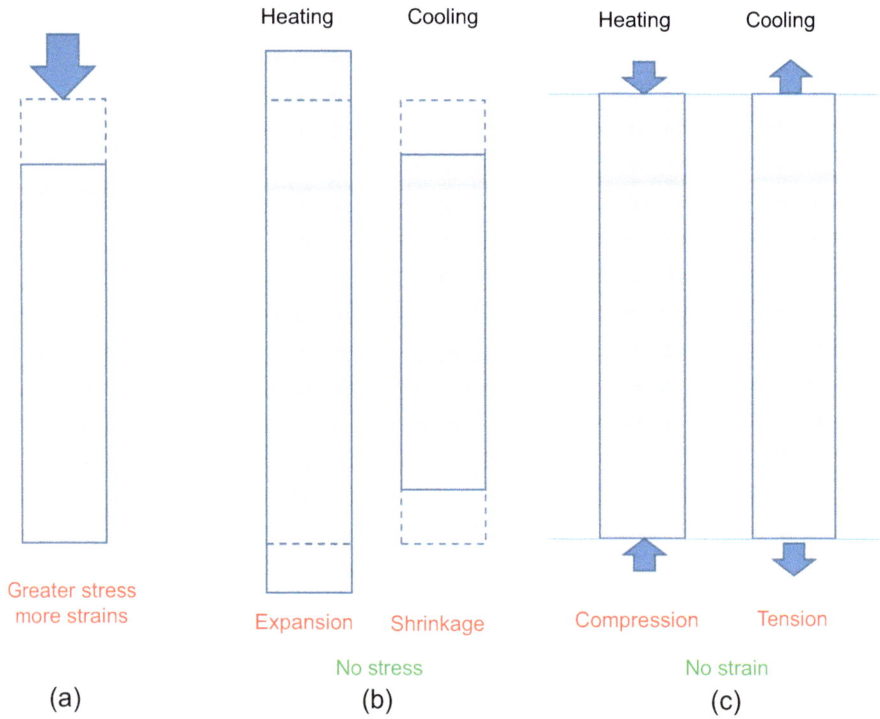

Figure 7.10 Pile with external mechanical load and internal thermal load. (a) External loading (eg, building load), (b) internally generated thermal load, free expansion (soft base, pile groups), (c) internally generated thermal load, end constraints (hard base, stiff structure).

located inside the pile, 75 mm below the soil−structure interface. Further details of the simulations and the results are given in Rui (2015).

7.2.3 Mechanical behaviour of energy piles

7.2.3.1 Basics: free expansion versus full restraint

When an external load (such as building load) is applied to a long structure (like pile), it compresses and stress is generated in the structure as shown in Fig. 7.10(a). Strain increases with applied stress.

When a structure is heated or cooled, it expands or shrinks, as shown in Fig. 7.10(b). The magnitude of the expansion or contraction depends on the change in temperature (ΔT) and the linear thermal expansion coefficient of the material α:

$$\varepsilon_T = \alpha \Delta T \quad [7.1]$$

where ε_T is the thermally induced strain (tension positive).

This means that if there is no soil resistance the pile expands both vertically and radially as shown in Fig. 7.10(b). The vertical elongation is $\Delta L = L \times \varepsilon_T = \alpha \Delta T L$ when the pile freely expands. There is no thermally induced stress.

Alternatively, if the two ends of the pile are fully restrained so that they would not move, the thermally induced strain/displacement becomes zero, as shown in Fig. 7.10(c). Instead, thermally induced stress is generated within the pile. The amount of stress is computed using the following equation:

$$\sigma_T = E(\varepsilon_{T\text{-real}} - \alpha \Delta T) = E\varepsilon_{T\text{-}Rstr} \qquad [7.2]$$

where E is Young's modulus, and $\varepsilon_{T\text{-real}}$ is the actual strain that develops within the structure by thermal changes. In the fully restrained case (ie, two ends fixed), the strain is zero ($\varepsilon_{T\text{-real}} = 0$) and hence the thermally induced stress is $\sigma_T = -E\alpha\Delta T$, which is compression (negative) when the pile is heated (ΔT).

$\varepsilon_{T\text{-}Rstr}$ ($= \varepsilon_{T\text{-real}} - \alpha\Delta T$) in Eq. [7.2] is the virtual strain value to compute the thermally induced stress from the actual strain value $\varepsilon_{T\text{-real}}$. In free expansion, $\varepsilon_{T\text{-real}} = \alpha\Delta T$ and hence $\varepsilon_{T\text{-}Rstr} = 0$; the thermally induced stress is therefore zero ($\sigma_T = 0$). If $\varepsilon_{T\text{-real}}$ is measured for a given change in ΔT (say in the field), then thermal induced stress σ_T can be estimated using Eq. [7.2]. It is important to note that some 'temperature compensated' monitoring instruments report $\varepsilon_{T\text{-}Rstr}$ instead of $\varepsilon_{T\text{-real}}$; hence, checking the outputs of the instruments when conducting field measurements is necessary.

7.2.3.2 Soil resistance

It is unlikely that an energy pile expands or shrinks freely as shown in Fig. 7.10(b) because the soil surrounding the pile resists the movements. The load applied to the pile is usually resisted by the shaft friction and/or the end-bearing. Herein two energy pile cases are considered: (1) friction pile (a building load is carried by side friction only; Fig. 7.11(a)) and (2) end-bearing pile (a building load is carried by end-bearing mainly but some is resisted by side friction; Fig. 7.11(b)). The strain and stress distribution profiles within the pile by the applied building load (ε_M and σ_M) are plotted in Fig. 7.11.

When the energy pile is heated or cooled, thermally induced strain ($\varepsilon_{T\text{-real}}$) and stress ($\sigma_T$) are added to the building load induced strain and stress (ε_M, σ_M) using the following equations:

$$\varepsilon = \varepsilon_M + \varepsilon_{T\text{-real}} \qquad [7.3]$$

$$\sigma = \sigma_M + \sigma_T = \sigma_M + E(\varepsilon_{T\text{-real}} - \alpha\Delta T) = E(\varepsilon_M - \varepsilon_{T\text{-}Rstr}) \qquad [7.4]$$

The profile of $\varepsilon_{T\text{-real}}$ depends on the deformation mechanism of the pile during heating and cooling, which is described below for the two cases.

(1) Friction pile: When the pile is heated, the top half section moves upwards, whereas the bottom half section moves downwards as shown in Fig. 7.12. If the two ends are free to move (ie, the bottom end is soft and the building founded by energy pile group is free to move as a

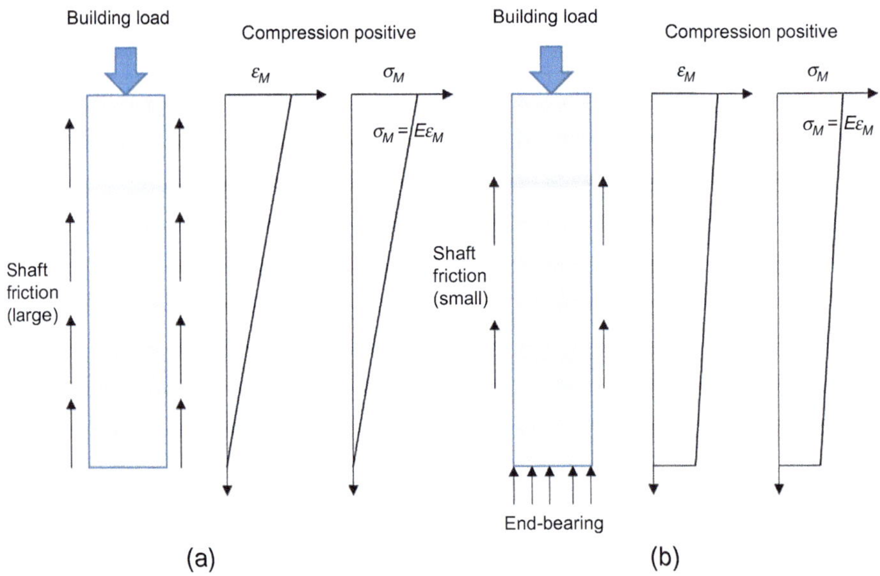

Figure 7.11 Friction pile (a) versus end-bearing pile (b).

rigid body), the strains at these two locations are equal to $\alpha\Delta T$. The largest restraint is located at the middle section of the pile and the expansive strain is at minimum. The expected strain and stress profiles are computed based on Eqs [7.3] and [7.4]. In general, compressive stress increases by heating. The gradient of the stress profile is proportional to the friction generated at the shaft. At the bottom section, the thermally induced movement gives friction in the same direction to the friction generated by the mechanical load. Hence, the magnitude of friction increases. At the top section, on the other hand, the friction reduces as the section moves upwards in the opposite direction of the friction generated by the mechanical load. When the pile is cooled, the opposite trend is expected to occur as shown in Fig. 7.12(b). The stresses in the pile reduce by cooling. Some part of the pile may go into tension, which needs to be checked during the design stage.

(2) End-bearing pile: When heated, limited movement occurs at the bottom of the pile due to the presence of the end-bearing hard soil. The whole pile tends to move upwards, as shown in Fig. 7.13(a). If the top end is free to move, the thermally induced strain increases with decreasing depth. On the other hand, the thermally induced stress is compressive and increases with depth, according to Eq. [7.4] with a given $\varepsilon_{T\text{-real}}$ profile. The shaft friction reduces and hence more load is transferred to the bottom. This increased stress needs to be checked against the concrete strength, for example. When cooled, an opposite trend is observed as shown in Fig. 7.13(b).

7.2.3.3 Design considerations

The conditions of end constraints at the top and bottom of an energy pile are very important to evaluate the thermomechanical response of the pile during heating and cooling. The mechanisms described above are based on a simple assumption of

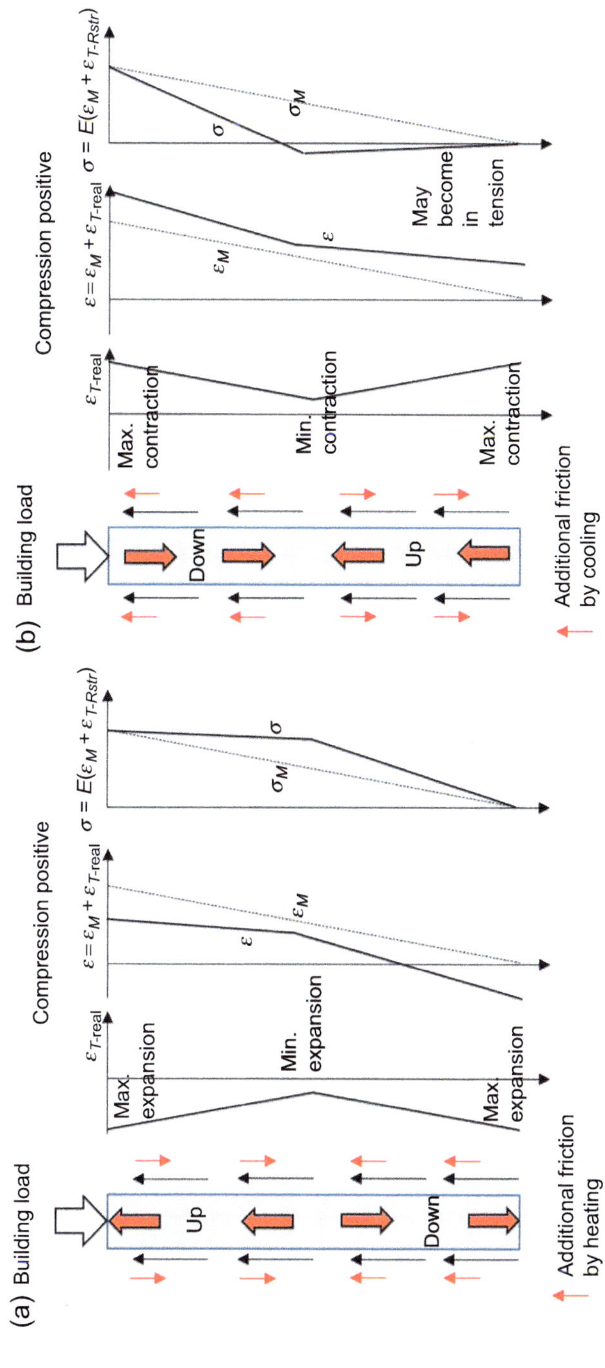

Figure 7.12 Thermomechanical performance of friction pile. (a) Heating and (b) cooling.

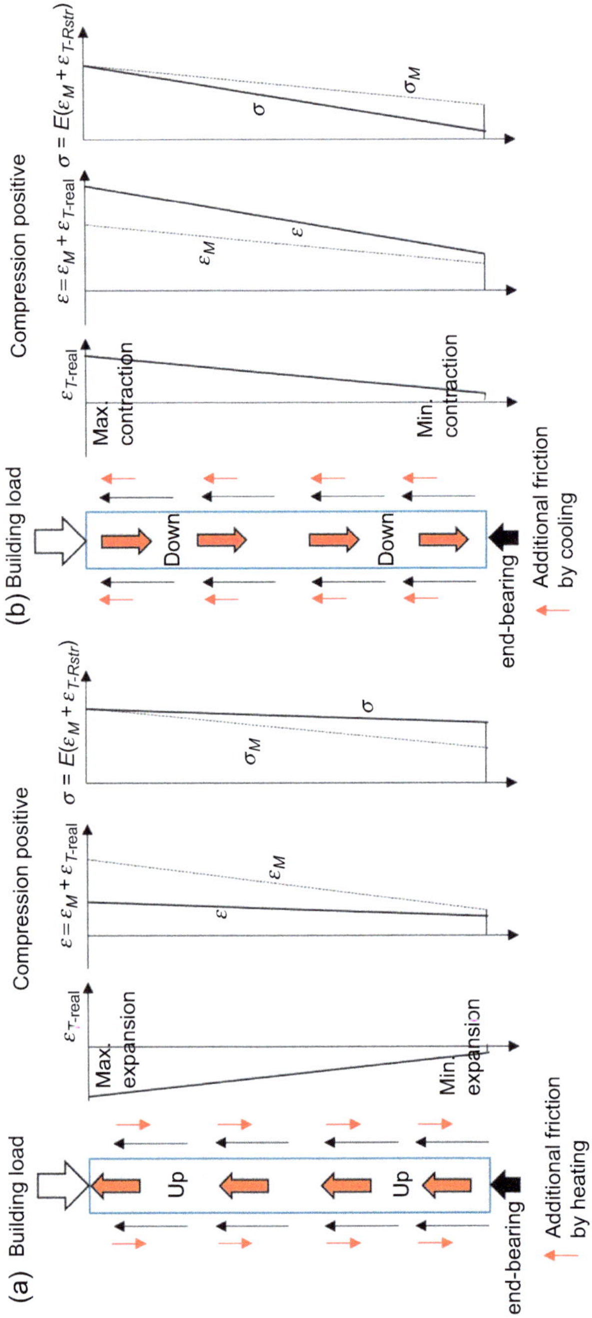

Figure 7.13 Thermomechanical performance of end-bearing pile. (a) Heating and (b) cooling.

homogenous soil; the actual pattern of soil resistance when an energy pile is heated or cooled can be complex because of the heterogeneous and nonlinear nature of soil–pile interaction. However, two extreme conditions can be considered: (1) the maximum additional heave or settlement of the pile is $\alpha\Delta TL$, where L is the pile length, and (2) the maximum thermally induced stress is $E\alpha\Delta T$. Hence, the first design screening can be performed simply by ensuring that the additional pile movement $\alpha\Delta TL$ is acceptable for building performance and checking that the generated stress is less than compressive or tensile strengths of pile material (eg, concrete). Since thermal loading is an internally generated one, it is not possible to have both situations at the same time. Therefore, satisfying the two criteria is a very conservative evaluation. If one of these criteria is missed, more complicated soil–pile interaction analysis can be performed to obtain a more realistic strain and stress profile of an energy pile by cooling and heating. It is also important to check whether the soil–pile interface behaviour is influenced by temperature or not. For example, the freezing at the pile–soil interface should be avoided. Further details of designing energy piles can be found in GSHPA's Thermal Pile Design, Installation & Materials Standards (2012), Bourne-Webb et al. (2009), Bourne-Webb (2013), Amatya et al. (2012), Laloui et al. (2006), Laloui (2011), Olgun et al. (2012, 2014), McCartney and Murphy (2012), Akrouch et al. (2014), Colls (2013), and Sutman et al. (2014).

7.3 Energy walls

7.3.1 Concept

GSHP systems can be coupled with underground basement structures or railway station boxes, as shown in Fig. 7.14. By installing heat exchanger pipes within the walls of an underground structure (Fig. 7.15), they can provide energy for aboveground buildings at the same time as being structurally load-bearing. These are called energy (or thermal) walls. In comparison to an energy pile system, potentially larger ground volumes can be used for geothermal heat transfer in energy wall system because of its large structure.

Similar to an energy pile system, heat exchanger pipes are fixed to the structural steel reinforcement cage inserted in a diaphragm wall, as shown in Fig. 7.16. The position of the pipes within the concrete is usually close to the soil side (see Fig. 7.17), which can reduce the thermal resistance of the concrete and improve the overall thermal performance of the energy wall.

There are two possible GSHP operating modes: (1) both heating and cooling (Fig. 7.18(a)) and (2) heating only (Fig. 7.18(b)). If an energy wall is used for a basement, then the interior side of the wall is insulated to ensure that the heat from the exchangers transfers into the soil rather than into the basement. By doing so, it can cater for both heating and cooling of the aboveground structures. For underground railway stations, there is a possibility to extract heat in winter time from both sides of the wall (station and soil) because stations often have excessive heat generated by train operations. In summer time, the GSHPs are not used and the excessive heat from the station

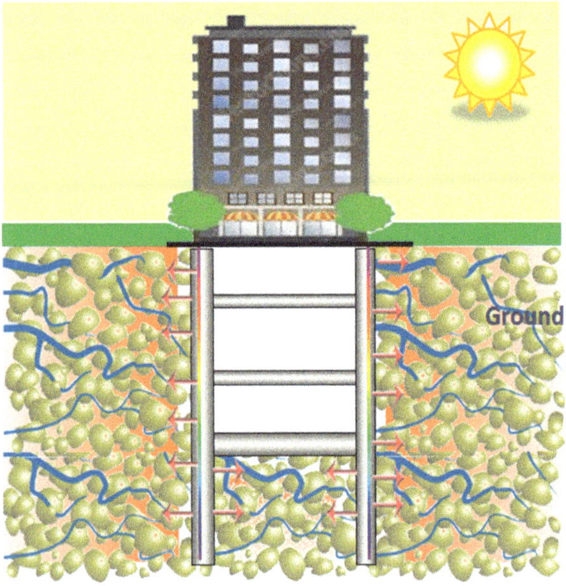

Figure 7.14 A description of an energy wall.

is transferred into the soil. The heat stored in summer then can be used in winter for heating the aboveground structures.

There have been many studies on energy piles, but limited research has been conducted on energy walls. Brandl (2006) describes a case in Vienna, in which geothermal loops were installed in metro station U2/2-Taborstraße as shown in Fig. 7.19. The loops were located in the diaphragm (D-) wall, piles, base slab and part of the tunnel linings. It was predicted that the whole system could provide 81 kW heating while the heat demand of the station was 95 kW at the peak hours. This project encouraged the Vienna underground to further utilise GSHP coupled foundations such as in the stations of Schottenring (base slab and piles), Messe (base slab) and Praterstern U2/3 (base slab and D-wall).

Another application of energy wall is in the Uniqa Tower, Vienna (Adam and Markiewicz, 2009). Heat exchangers were installed in diaphragm walls which reached down to 35 m below surface. A surface area of 7800 m^2 was used and the system produced a heating capacity of 420 kW and a cooling capacity of 240 kW. The annual heating output reached up to 818 MWh, whereas the annual cooling output was up to 646 MWh.

Several energy walls have been installed in the United Kingdom. GI Energy was the first to install geothermal loops into a diaphragm retaining wall at Bulgari hotel in Knightsbridge (Amis and Loveridge, 2014). Amis et al. (2010) described the installation of this thermal wall system and discussed the potential effects of thermal changes during operation. It was found that, once operational, daily loop temperature fluctuations were considerably less than those of exposed concrete during a winter's day. The thermal resistance and thermal conductivity detected from thermal response test in the

Figure 7.15 GSHP system in a diaphragm wall. (a) Diaphragm wall and (b) geothermal loop.

construction period would change when the station box is finished due to the removal of soil. Amis (2011) reports a rise of 20% in thermal resistance as well as reduction in thermal conductivity value by 13%.

Crossrail is a major infrastructure project in London. Various underground stations are currently being constructed. Ten new station boxes include energy foundations as listed in Table 7.1 (Amis, 2014). Dean Street station box is one of them and includes thermal walls as well as thermal piles, as shown in Fig. 7.20. The station box was constructed using the 'bottom up' method. A one-metre thick diaphragm wall (up to 40-m depth) was first installed by excavating a trench at a required depth. Heat exchanger pipes were attached to the reinforcement cage and lowered into the trench. Concrete was then poured in to cast the diaphragm wall. The soil inside the diaphragm box

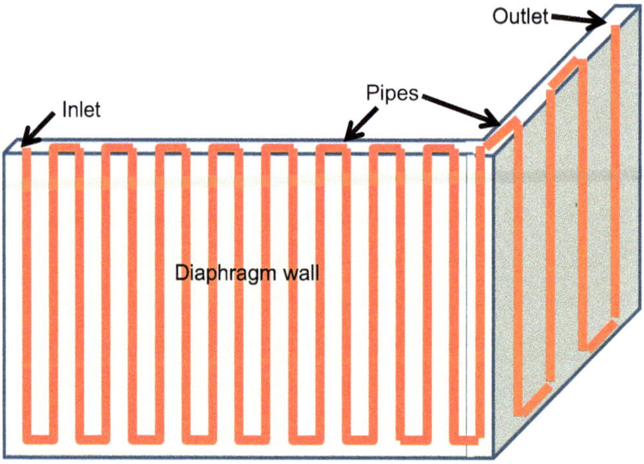

Figure 7.16 A typical heat exchanger configuration in an energy wall.

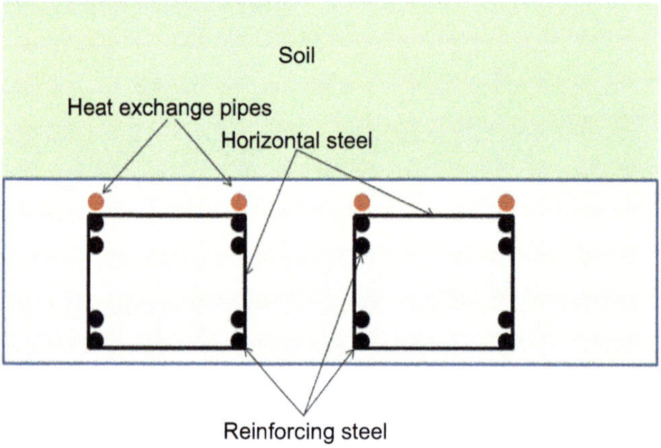

Figure 7.17 Cross-section of an energy wall showing the location of heat exchangers.

was excavated 25-m deep and temporary props were added to support the excavation. Slabs were cast from the bottom and work proceeded upwards, replacing the props with slabs to form five levels for the station box. The geothermal capacity of this particular wall is estimated to be 160 kW (Amis, 2014).

7.3.2 Thermal behaviour of energy walls

A BHE is a single system with one or two loops of heat exchanger pipes, whereas an energy wall contains several groups of loops inside the wall panel. An energy wall has one side covered by soil and the other side exposed into air, whereas a BHE is surrounded by soil only. All these special conditions make an energy wall different from a BHE system and the difference should be reflected in the design.

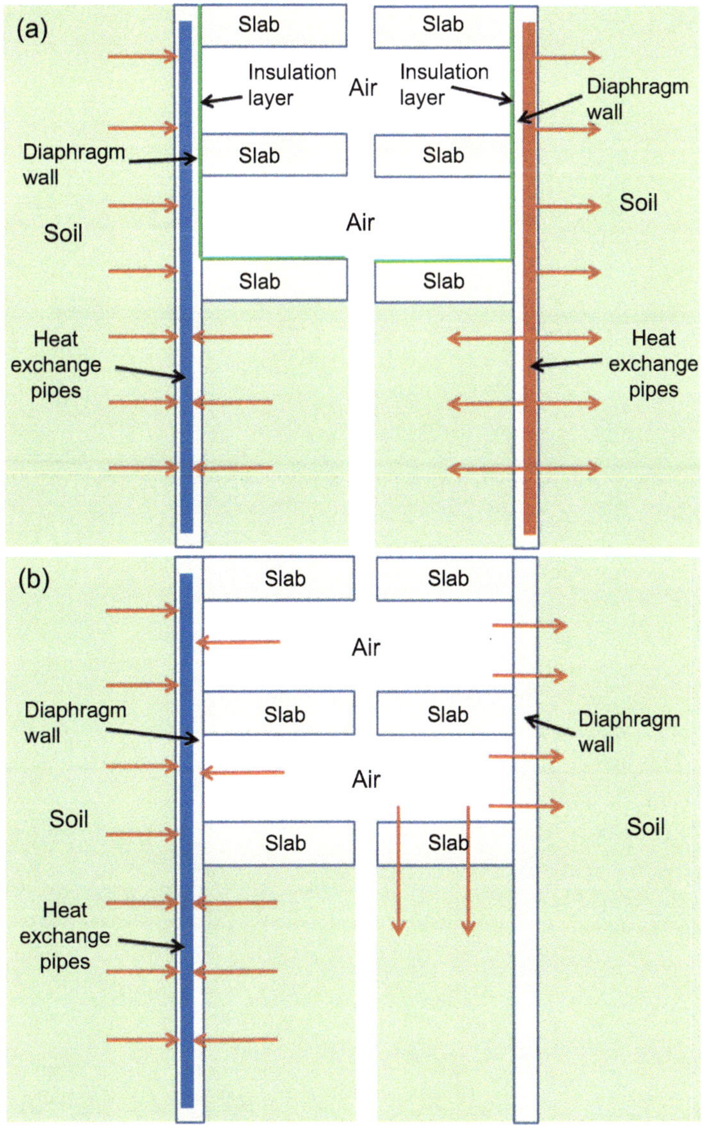

Figure 7.18 Ground-source heat pump operating modes of energy wall. (a) Heating and cooling mode and (b) heating only mode.

The thermal response of an energy wall can be conceived of as a plane heat source (see Fig. 7.21) or a point heat source (see Fig. 7.22). The former figure considers the cross section perpendicular to the wall, whereas the latter considers the cross section in plan view.

The plane heat source model (Fig. 7.21) has a semi-infinite domain extending over the region $x > 0$ and with a constant heat source boundary at $x = 0$; all the heat flow

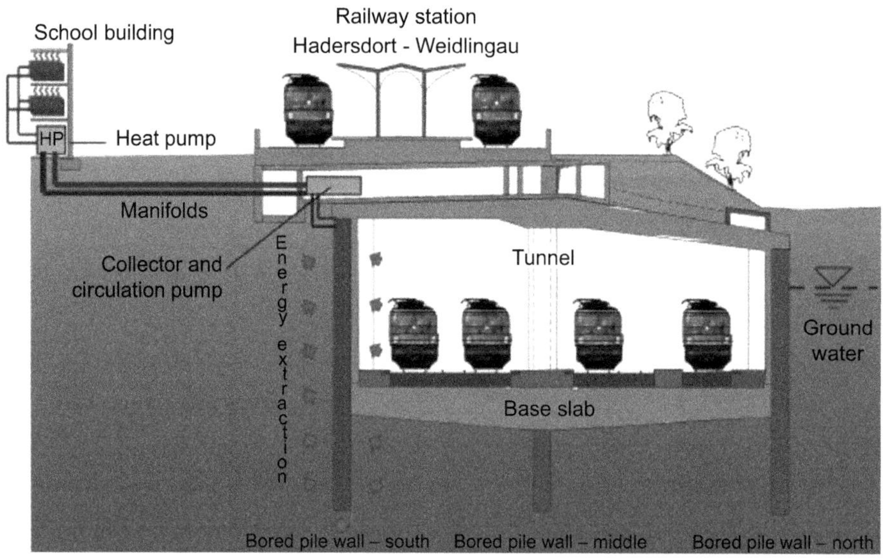

Figure 7.19 Energy walls in Vienna Metro (Brandl, 2006).

Table 7.1 **Design heating and Colling peak power from Crossrail's energy walls (Amis, 2014)**

Contract	Heating peak power (kW)	Cooling peak power (kW)	CO_2 savings (T/annum)
Paddington integrated	420	240	200
Bond St WTH	186	186	28
Bond St ETH	340	340	49
TCR West	160	160	21
TCR East	75	75	9
Fisher St	60	60	10
Liverpool St – Blomfield	150	80	11
Liverpool St – Moorgate	70	25	14
Farringdon West	120	120	16
Farringdon East	90	90	12
Totals	**1671**	**1376**	**370**

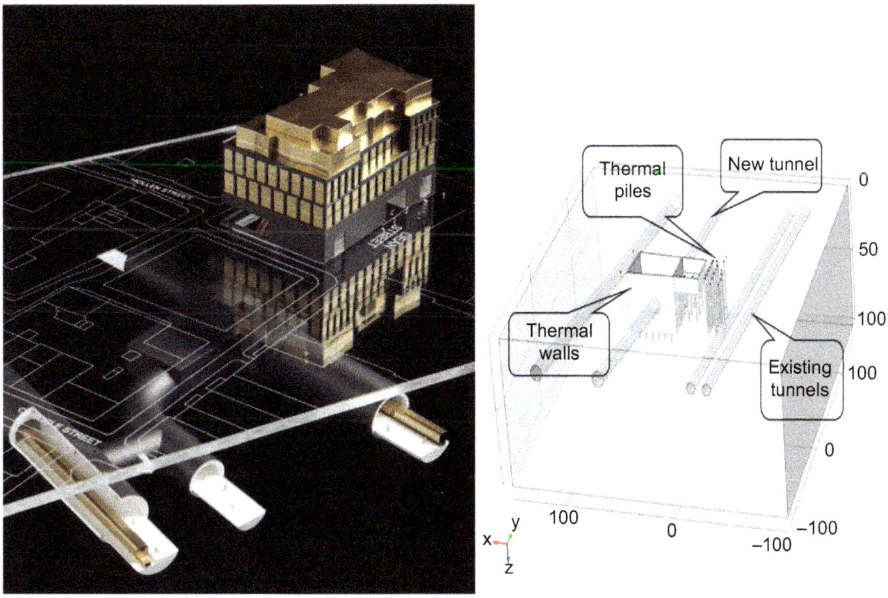

Figure 7.20 Location of Dean Street station box.
From Crossrail, Tottenham Court Road Station Website.

occurs in the direction of positive x. It is assumed that the non-soil side of the wall is insulated. If the central section of the wall is far away from the top and bottom boundaries, a 1D diffusion model such as by Carslaw and Jaeger (1959) can be used. However, it assumes that the heat source is an infinitely long plane source.

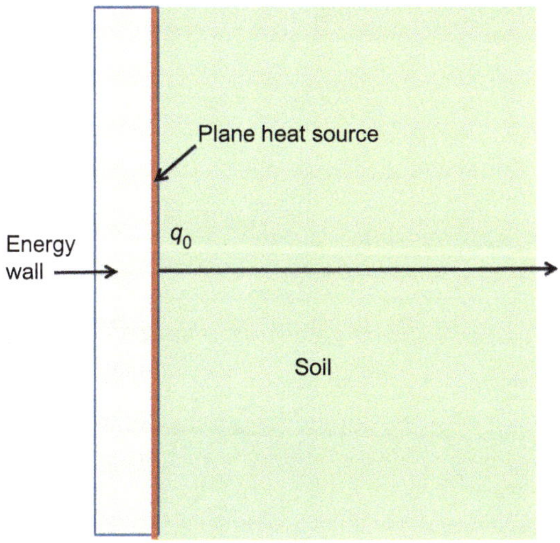

Figure 7.21 Heat sources in a semi-infinite medium.

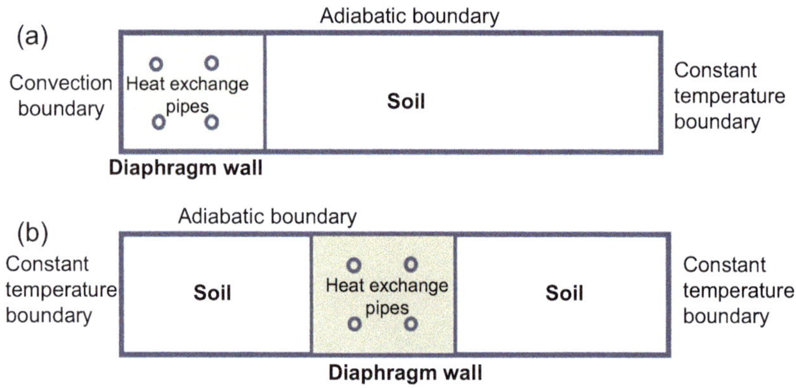

Figure 7.22 Heat transfer model for energy wall: (a) over excavation surface; (b) below excavation surface.
After Sun, M., Xia, C., Zhang, G., 2013. Heat transfer model and design method for geothermal heat exchange tubes in diaphragm walls. Energy and Buildings 61, 250–259.

If the geometry of the heat exchanger pipes embedded in the wall is of concern, it is possible to examine the heat transfer using a periodical boundary as shown in Fig. 7.22. Sun et al. (2013) developed analytical solutions of this problem using Green's function method and the separation of variables method, assuming that the heat exchanger tube is a point heat source. The heat exchange rate per metre is assumed to be constant and the heat flow in the vertical direction is ignored to simplify the vertical thermal gradient in the ground and the effect of the ground surface boundary condition. They developed two heat transfer models: one for the section over the excavation line, and the other for the section under the excavation line, as shown in Fig. 7.22.

Compared to analytical solutions, a numerical method like the finite element method is a more sophisticated and geometrically accurate way to analyse the heat transfer problem of energy walls. As computer technology improves, numerical methods are becoming cost-effective and widely used mathematical tools for solving the differential governing equations for various boundary value problems. In a similar way to energy pile problems, a finite line heat source model can be used to simulate the pipe system in energy walls, as shown in Fig. 7.23.

7.3.3 Mechanical behaviour of energy walls

The temperature changes that occur within an energy wall may affect the structural performance of the wall as well as the surrounding ground. Fig. 7.24 shows some possible effects of energy wall operation on the geotechnical performance of an energy wall.

If heat exchangers are embedded inside a wall, the temperature variation within the wall would induce additional bending movement inside the wall. The increased bending needs to be within the design bending moment envelope to ensure that the

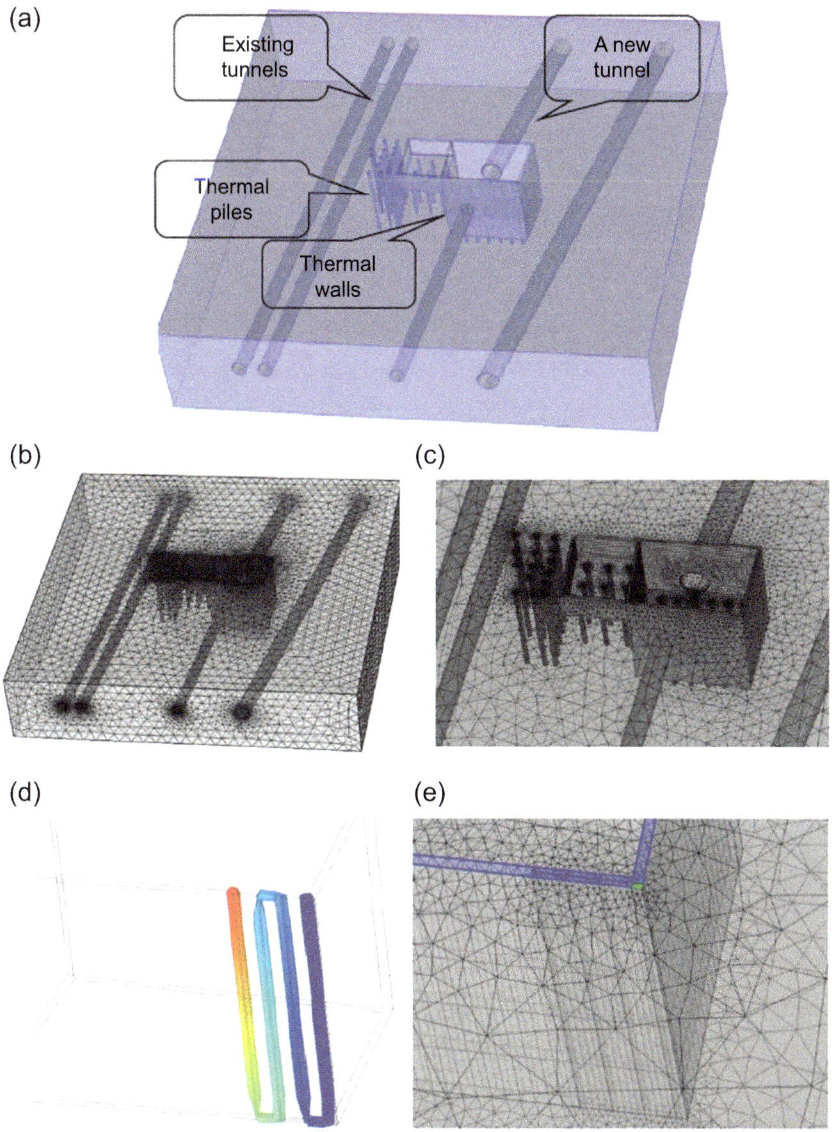

Figure 7.23 3D finite element model for thermal analysis of energy wall. (a) 3D model, (b) meshing of 3D model, (c) meshing of and around the station box, (d) model of the D-wall TRT and (e) meshing of the D-wall TRT.

concrete structures do not deteriorate during the operation. Also, possible thermally induced movements of the structures need to be assessed.

A pile is surrounded by soil, whereas only one side of a wall is exposed to soil. Therefore, there are differential stresses across the wall. The soil (both pore fluid and soil skeleton) would expand and contract due to temperature change. The

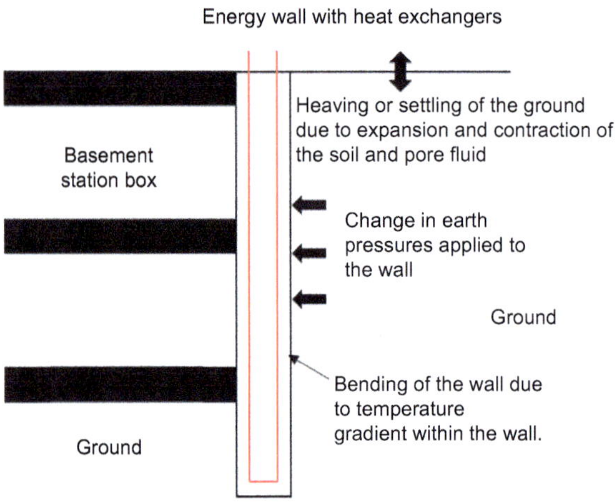

Figure 7.24 Effects of energy wall operation on geotechnical performance.

expansion/shrinkage of soil may change the earth pressures acting on the wall, inducing additional displacements and bending moments in the wall. Such thermally induced mechanical behaviour of the wall needs to be assessed during the design stage. The shrinkage or expansion of the soil may also induce heave or displacement of the ground nearby, which may affect the buildings adjacent to the wall. The long-term ground deformation due to cyclic thermal changes needs to be evaluated.

7.4 Energy tunnels

7.4.1 Concept

Underground metro systems provide a large amount of heat, which can be utilised. The motors in a London Crossrail train emit about 1 MW of heat and one train passes at 2.5 min intervals during peak times (Nicholson et al., 2013). The air conditioning adds an extra 0.1 MW of heat. Furthermore, braking adds to the heat generation. It is well known that many old underground tunnels such as the London Underground have made the underground hotter than the surface temperature (Waboso and Gilbey, 2007). The concept of an energy tunnel system is shown in Figs. 7.25 and 7.26. Typically, heat exchanger pipes are embedded in the tunnel linings. This enables the tunnels to extract heat from the train operations for the surrounding commercial and domestic needs.

For shield-driven tunnels, tunnel linings are pre-casted in a factory and then installed on site. Hence, segmental lining with heat exchanger pipes can be prepared in advance. The integrated heat-exchange pipes are connected to the reinforcement cage of tunnel linings, as shown in Fig. 7.27. If fibre-reinforced concrete is used for

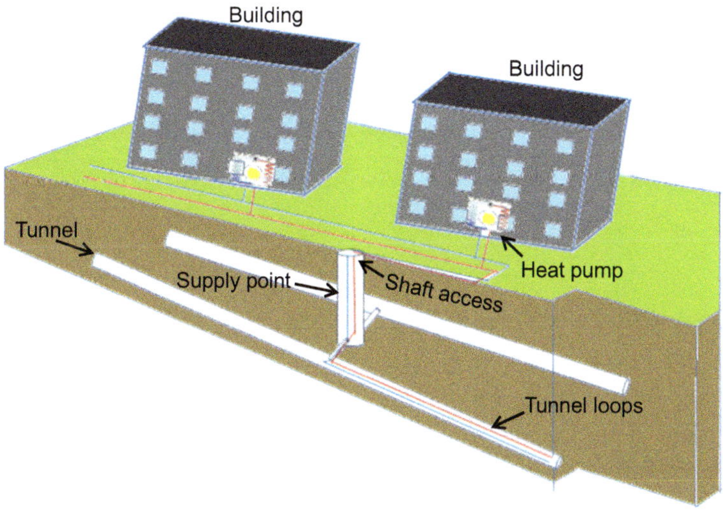

Figure 7.25 Energy tunnel concept.

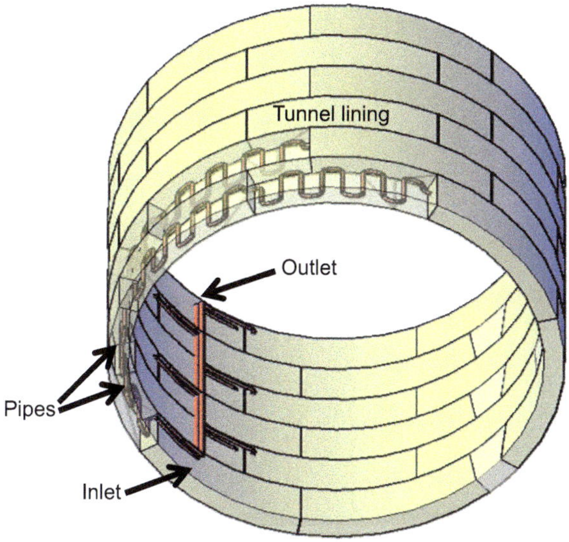

Figure 7.26 Typical energy tunnel.

segment material, a simple frame structure that supports the pipes is needed when concrete is cast. Ideally, the pipes (20−30 m for each segment) are connected at the inside of the reinforcement cage or the frame to ensure the integrity of concrete cover. The piles are placed into each segment in a meandering fashion. At a construction site, prefabricated segments are installed to create a series of tunnel lining ring, as shown in Fig. 7.26. The ends of the pipes in each segment are then connected to the ends of

Figure 7.27 Pipes attached to a reinforcement cage.
From Rehau and Zueblin.

the other segments as shown in Fig. 7.28. The connecting work to the heat pump systems located above the ground is performed after the tunnel is constructed so that it does not affect the tunnel construction process. An example of the completed installation is shown in Fig. 7.29.

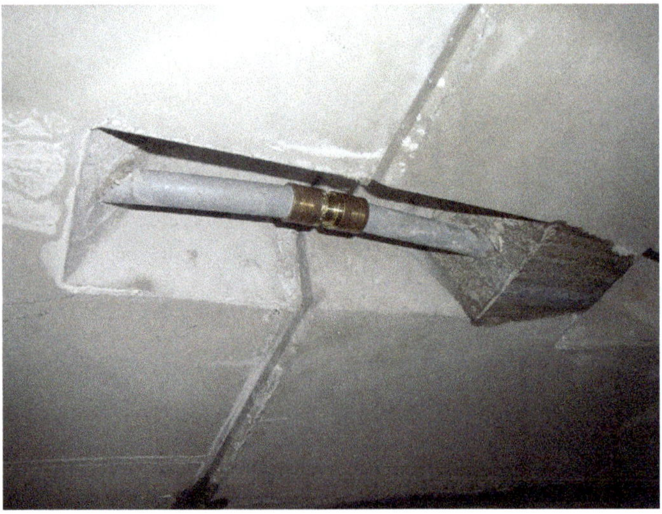

Figure 7.28 Pipe connections at the segmental joints.
From Rehau and Zueblin.

Figure 7.29 Pipe installation inside the tunnel.
From Rehau and Zueblin.

The efficiency of an energy tunnel system can be influenced not only by the surrounding ground temperature but also the air temperature inside the tunnel. In a short-running tunnel, air temperatures are often controlled by the surface air temperature and hence they are called 'cold' tunnels (see Fig. 7.30, left). In a long metro and cable tunnel, the internal air temperatures are often higher than the surface air temperature. They are called 'hot' tunnels (see Fig. 7.30, right).

There are various operational modes for an energy tunnel system as shown in Fig. 7.31. If the ground surrounding the tunnel is used as heat storage between summer and winter seasons, the system becomes more efficient if the inside section of the tunnel is insulated. If not, the temperature inside the tunnel may increase or decrease during summer or winter, creating problems for passenger comfort. For hot tunnels, energy tunnels can be used for mainly heat extraction. In summer, heat is extracted

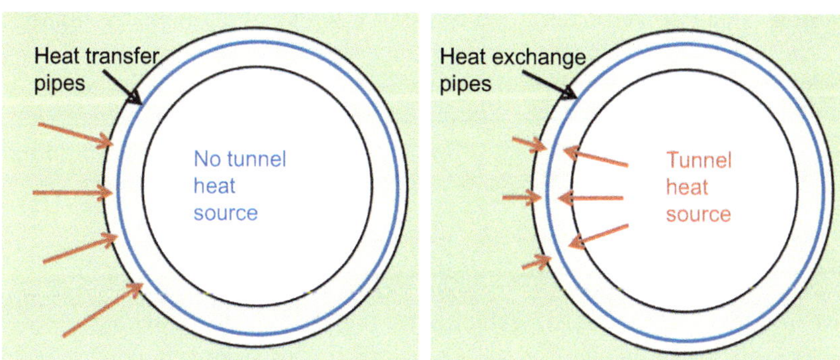

Figure 7.30 Cold and hot tunnels.

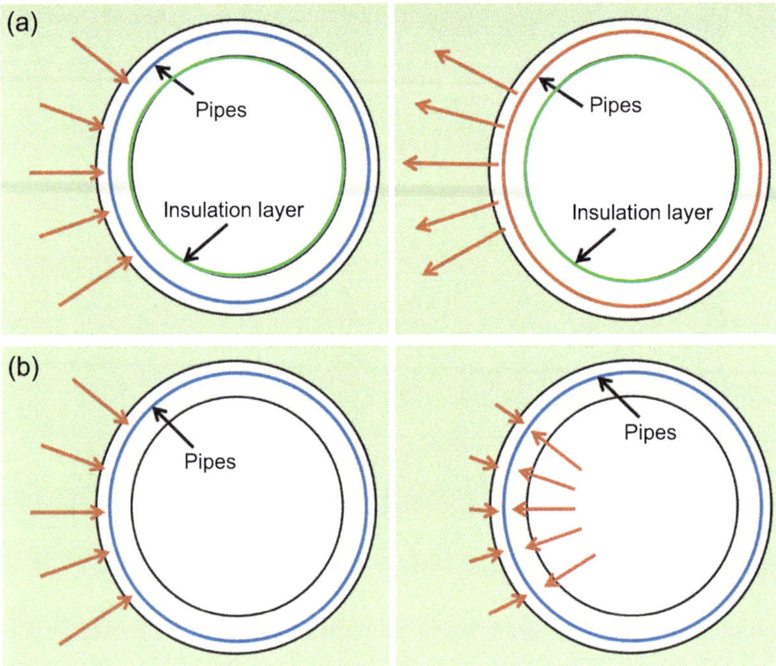

Figure 7.31 Operation modes of energy tunnels. (a) Heating and cooling and (b) heating only.

from both tunnel and soil sides to cool down for passenger comfort or improve the transport system efficiency (for example, increase the efficiency of an air conditioning unit). The extracted heat can be used as hot water supply for aboveground structures. In winter, heat is extracted to cool down the soil side, which is useful for cooling the tunnel in the following summer.

Applications of energy tunnels have been reported in the literature (Brandl, 2006; Franzius and Pralle, 2011; Nicholson et al., 2013; Schneider and Moormann, 2010; Zhang et al., 2013). Franzius and Pralle (2011) introduced a demonstration project in Austria. In this project, a 54-m length of a twin-track high-speed railway tunnel was equipped with geothermal heat pump system, as shown in Fig. 7.32. This system could supply around 40-kW heat for a municipal building in the town of Jenbach nearby. Adam and Markiewicz (2009) introduced an energy tunnel system to the new 15-km long Lainzer tunnel in Austria. Heat exchanger pipes were first attached to non-woven geosynthetic material off site, and then placed between the primary and secondary lining of the tunnel as shown in Figs. 7.33 and 7.34.

7.4.2 Thermal behaviour of energy tunnels

The line heat source model and the cylinder heat source model used for borehole ground heat exchangers are not suitable to assess the thermal performance of an energy tunnel system. One suitable method is the 'hollow' cylinder source model, as shown in Fig. 7.35.

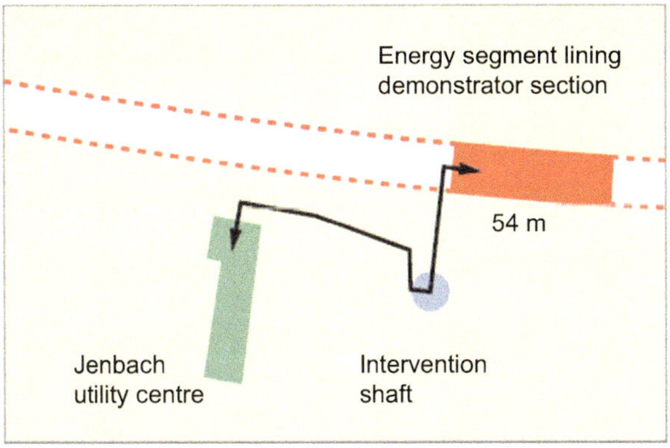

Figure 7.32 Layout of demonstrator project in Jenbach, Austria (Franzius and Pralle, 2011).

Figure 7.33 Detail of the four absorber loops with collection pipe (Adam and Markiewicz, 2009).

In this model, the domain $r_b < r < \infty$ is considered, and the heating flux is in the radial direction only. On the inner surface ($r = r_b$), the boundary condition is assumed as a time-dependent heat flux. Hellstrom (1991) proposed a solution for this model.

Zhang et al. (2013) proposed a new line heat source model as shown in Fig. 7.36. The tunnel lining and the surrounding rock/soil are assumed as homogenous and their thermal properties are independent of temperature. The absorber pipes are modelled as a collection of line heat sources in the longitudinal direction of the tunnel and the heat conduction of the pipes wall and carrier liquid is ignored. Heat transfer in both radial direction and circumferential direction is computed; that is, the interaction between the neighbouring pipes can be considered. The superposition principle and finite integral transform method are adopted to obtain this analytical solution.

The thermal behaviour of an energy tunnel can also be simulated by the finite element method. Heat transfer elements can be used to model the heat convection in

Figure 7.34 Energy geotextile at lot LT22 of Lainzer tunnel (Adam and Markiewicz, 2009).

the pipe and heat conduction between heat exchange tunnel and soil. A typical finite element model of an energy tunnel is shown in Fig. 7.37 that includes the soil, tunnel lining, track bed, pipe wall and fluid.

Usually, only heat conduction is considered in the tunnel lining and surrounding soil, and hence heat transfer solid element can be used. There are two methods to simulate the interaction of fluid and heat exchanger pipe. One method ignores the heat convection in the fluid and both the pipe wall and fluid can be modelled with solid

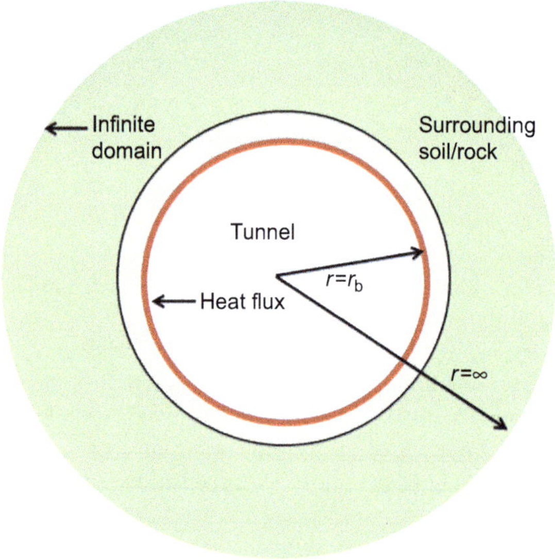

Figure 7.35 'Hollow' cylindrical model.

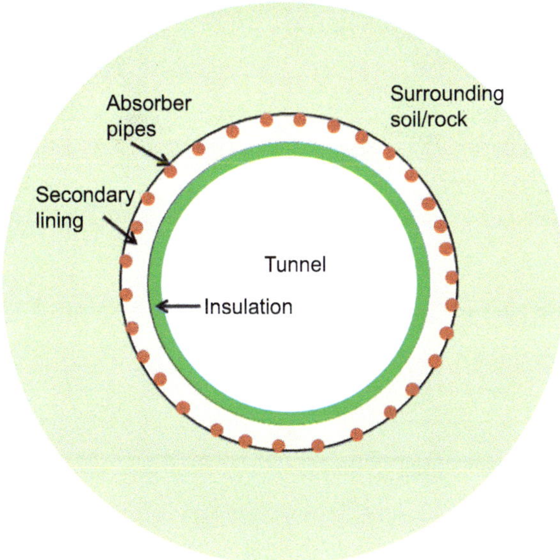

Figure 7.36 Schematic view of energy tunnel.
After Zhang, G., Xia, C., Sun, M., Zou, Y., Xiao, S., 2013. A new model and analytical solution for the heat conduction of tunnel lining ground heat exchangers. Cold Region Science Technology 88, 59–66.

elements. In this model, a thermal boundary condition like temperature boundary or heat flux is applied to the fluid element. The second method considers heat convection, so the thermo-hydro coupling element can be used to simulate the heat transfer in the pipe fluid.

7.4.3 Mechanical behaviour of energy tunnels

The temperature changes that occur within an energy tunnel may affect the structural performance of the tunnel. Fig. 7.38 shows some possible mechanical behaviour of an energy tunnel by its operation.

One of the important design aspects to consider for energy tunnels is the effect of thermal expansion/contraction of the segments on the overall engineering performance of the tunnel. The temperature distribution within a segment can be complex and they need to be assessed accurately so that the additional thermally induced stresses are within the range of the design axial load and bending moment envelope of the tunnel lining. The local stresses at the joint sections need to be assessed because it may influence the water tightness of the lining as well as induce additional joint movements. The locations of heat exchange pipes within the segments therefore need to be decided carefully to ensure that the structural integrity is not lost. The expansion or shrinkage of the soil around the tunnel by temperature change (drying or freezing in extreme cases) may induce changes in the earth pressures applied to the tunnel. The effect of

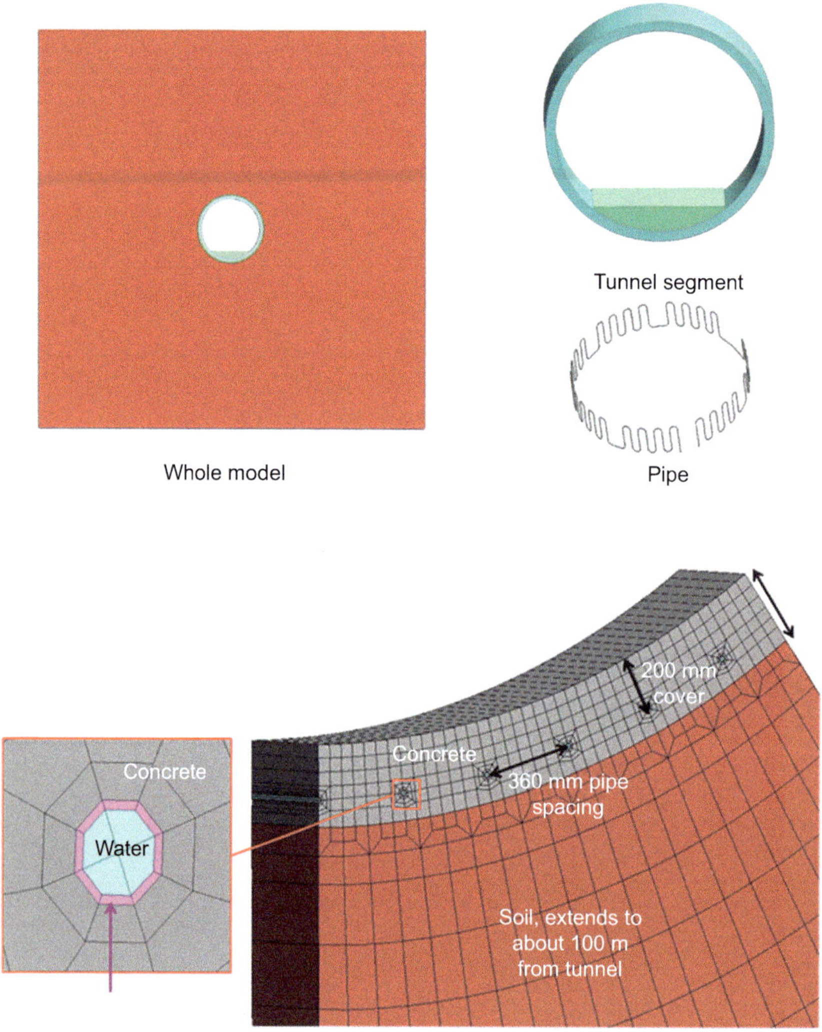

Figure 7.37 Finite-element model of energy tunnel.

this on the tunnel performance as well as neighbouring underground structures needs to be assessed as well.

Because of complex pipe geometries within the tunnel segments as well as soil–tunnel interaction, numerical methods such as the finite element method need to be adopted to evaluate the thermomechanical performance of energy tunnels. Nicholson et al. (2013) conducted a THM coupling analysis on the Crossrail tunnel model (Fig. 7.37). The principle stress in the tunnel lining is shown in Fig. 7.39. It is shown that the hoops stress increased by about 7% due to the thermal expansion of the lining and the ground during the heating process. In the cooling process, local tensile stress is observed.

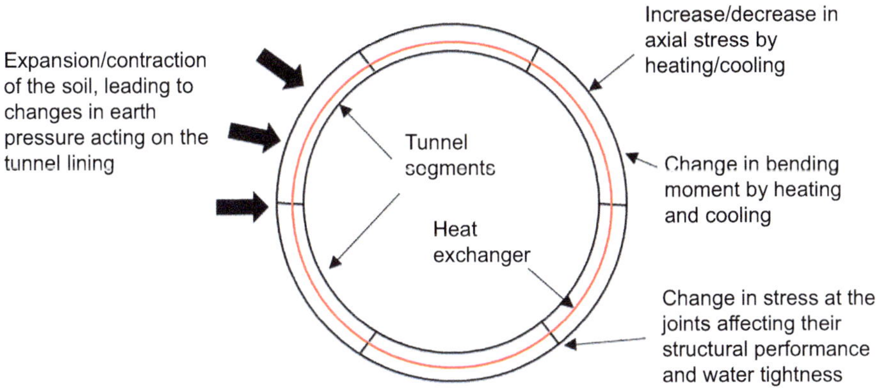

Figure 7.38 Effects of energy tunnel operation on the geotechnical performance.

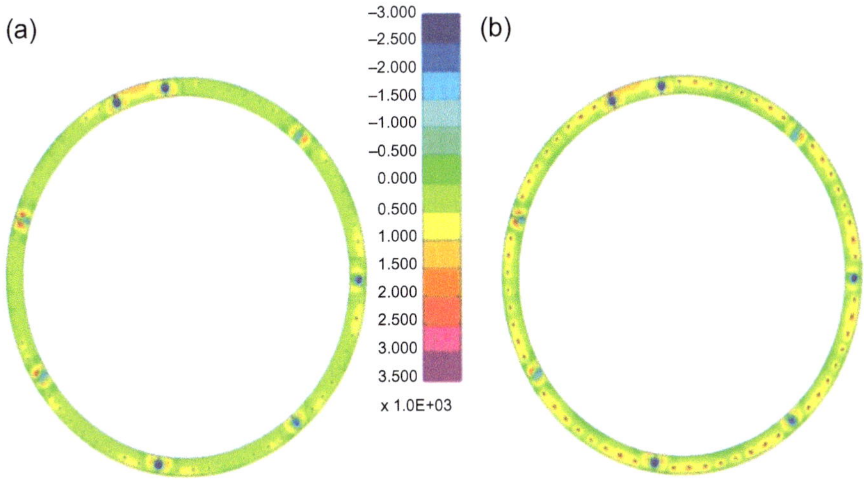

Figure 7.39 Comparison of maximum principal stress distribution in segments during summer. (a) No heat extracted by TES and (b) 30 W/m^2 heat extracted (Nicholson et al., 2013).

7.5 Conclusions

Energy geostructures (eg, energy piles, walls and tunnels) incorporate the primary heat exchangers of GSHP systems through the foundation elements (eg, piles and basement walls) or into the tunnel linings. Concrete, which is used for structural material, is an ideal medium as the heat absorber in the ground due to its good thermal conductivity and thermal storage capacity. Therefore, an energy geostructure functions not only to carry mechanical load but also as a heat exchanger, circulating cooler or warmer fluid liquid to extract stored geothermal energy from the ground. This dual purpose potentially makes them economically attractive as well as an environmentally sustainable solution.

This chapter describes the concepts and applications of energy geostructures. When designing such structures, it is necessary to investigate the THM interactions between the heat exchanger–coupled foundation elements and the surrounding ground. Some considerations and tools for analysing and designing energy geostructures are introduced. However, there are still large knowledge gaps to design, construct and operate them. Most of the field data available up to now are related to the short-term operation of energy geostructures. Data showing the long-term ground and structure response under actual operational processes of energy geostructures are very much needed in order to build confidence with geotechnical engineers to examine the effects of temperature on structural performance. Once clear design guidance is developed based on field data and the associated data analysis, the uptake of energy geostructures will increase from the current rate. Such development will help remove existing blocks to the use of energy geostructures within the construction market, and thus enhance the potential for CO_2 emission reductions in line with government targets.

Because the spatial scale of energy structures is often much larger than the conventional GSHP systems adopted for residential buildings, there is opportunity to design a cost-effective energy geostructure by careful consideration of energy flow in and out of the ground, building energy demand and heat distribution through the building. A common problem in the design of energy geostructures is the building energy demand profile is often given as an input without consideration of the coupling between the two. Ideally, the efficiency of an energy geostructure, which is a function of size, depth, layout and soil thermal conductivity, should be considered together with the building energy efficiency, which depends on a complex set of interrelated parameters including urban microclimate, the building's heat capacity, seasonal energy demand, as well as any variability in the building's heating and cooling requirements. Architects and engineers should work together throughout the design process of energy geostructures.

References

Adam, D., Markiewicz, R., 2009. Energy from earth-coupled structures, foundations, tunnels and sewers. Géotechnique 59 (3), 229–236.

Akrouch, G.A., Sánchez, M., Briaud, J.L., 2014. Thermo-mechanical behavior of energy piles in high plasticity clays. Acta Geotechnica 9 (3), 399–412.

Amatya, B.L., Soga, K., Bourne-Webb, P.J., Amis, T., Laloui, L., 2012. Thermo-mechanical behaviour of energy piles. Géotechnique 62 (6), 503–519.

Amis, T., Loveridge, F., 2014. Energy piles and other thermal foundations for GSHP–developments in UK practice and research. REHVA Journal 2014 (1), 32–35.

Amis, T., Robinson, C., Wong, S., 2010. Integrating geothermal loops into the diaphragm walls of the Knightsbridge Palace Hotel project. In: EMAP-Basements and Underground Structures 2010.

Amis, T., 2011. Energy Foundations in the UK [online] Available from: http://www.gshp.org.uk/GroundSourceLive2011/TonyAmis_Piles_gsl.pdf.

Amis, T., 2014. Renewable Energy Opportunities with Infrastructure Projects [online] Available from: http://www.rehau.com/download/1347278/gi-energy-renewable-energy-opportunities-with-infrastructure.pdf.

Banks, D., 2012. An Introduction to Thermogeology: Ground Source Heating and Cooling, second ed. Wiley-Blackwell, Oxford.

Bourne-Webb, P.J., Amatya, B., Soga, K., Amis, T., Davidson, C., Payne, P., 2009. Energy pile test at Lambeth College, London: geotechnical and thermodynamic aspects of pile response to heat cycles. Géotechnique 59 (3), 237–248.

Bourne-Webb, P., 2013. Observed response of energy geostructures. In: Laloui, L., Di Donna, A. (Eds.), Energy Geostructures: Innovation in Underground Engineering, pp. 45–77.

Brandl, H., 2006. Energy foundations and other thermo-active ground structures. Géotechnique 56 (2), 81–122.

Carslaw, H.S., Jaeger, J.C., 1959. Conduction of Heat in Solids, second ed. Oxford University Press, Oxford, p. 84.

Colls, S., 2013. Ground Heat Exchanger Design for Direct Geothermal Energy Systems (Ph.D. dissertation). Lund University.

Crossrail. Tottenham Court Road Station web site. [online] Available from: http://www.crossrail.co.uk/route/stations/tottenham-court-road/.

Cui, P., Li, X., Man, Y., Fang, Z., 2011. Heat transfer analysis of pile geothermal heat exchangers with spiral coils. Applied Energy 88 (11), 4113–4119.

Desmedt, J., Hoes, H., 2007. Case study of a BTES and energy piles application for a Belgian hospital. In: Proceedings ECOSTOCK, 2006, May 31–June 2, New Jersey, US.

Eskilson, P., 1987. Thermal Analysis of Heat Extraction Boreholes. Department of Mathematical Physics, University of Lund.

Franzius, J.,N., Pralle, N., 2011. Turning segmental tunnels into sources of renewable energy. Proceedings of ICE – Civil Engineering 164 (1), 35–40.

Gao, J., Zhang, X., Liu, J., Li, K., Yang, J., 2008. Numerical and experimental assessment of thermal performance of vertical energy piles: an application. Applied Energy 85 (10), 901–910.

Garber, D., Soga, K., Choudhary, R., 2013. Post-occupancy assessment of Thermal-pile and open-well ground source heat pump (GSHP) system-case study. ASHRAE Transactions 119 (2), 1–9.

Garber, D., 2013. Ground Source Heat Pump System Models in an Integrated Building and Ground Energy Simulation Environment. *Doctoral Thesis*. University of Cambridge.

Ground Source Heat Pump Association, 2012. Thermal Pile Design, Installation & Materials Standards. National Energy Centre. http://www.gshp.org.uk/pdf/GSHPA_Thermal_Pile_Standard.pdf.

Hamada, Y., Nakamura, M., Saitoh, H., Kubota, H., Ochifuji, K., 2007. Improved underground heat exchanger by using no-dig method for space heating and cooling. Renewable Energy 32 (3), 480–495.

Hellström, G., 1991. Ground Heat Storage: Thermal Analyses of Duct Storage Systems. Part I: Theory (thesis). Lund University.

Ingersoll, L.R., Zobel, O.J., Ingersoll, A.C., 1954. Heat Conduction with Engineering, Geological and Other Applications. Madison University of Wisconsin Press, Madison.

Koene, F., Charles, G., 2000. Energy piles as an efficient way to store heat. CADDET Energy Efficiency 4–6.

Laloui, L., Nuth, M., Vulliet, L., 2006. Experimental and numerical investigations of the behaviour of a heat exchanger pile. International Journal for Numerical and Analytical Methods in Geomechanics 30 (8), 763–781.

Laloui, L., 2011. Understanding the thermo-mechanical behaviour of energy piles. Civil Engineering 164, 184–191. EPFL-ARTICLE-167204.

Li, M., Lai, A.C.K., 2013. Analytical model for short-time responses of ground heat exchangers with U-shaped tubes: model development and validation. Applied Energy 104 (C), 510–516.

Li, M., Lai, A.C.K., 2012. New temperature response functions (G functions) for pile and borehole ground heat exchangers based on composite-medium line-source theory. Energy 38 (1), 255–263.

Loveridge, F., 2012. The Thermal Performance of Foundation Piles Used as Heat Exchangers in Ground Energy Systems (Ph.D. dissertation). University of Southampton.

Loveridge, F., Powrie, W., 2013. Temperature response functions (G-functions) for single pile heat exchangers. Energy 57, 554–564.

Man, Y., Yang, H., Diao, N., Liu, J., Fang, Z., 2010. A new model and analytical solutions for borehole and pile ground heat exchangers. International Journal of Heat and Mass Transfer 53 (13), 2593–2601.

McCartney, J.S., Rosenberg, J., Sultanova, A., 2010. Engineering performance of thermo-active foundations. In: Proceedings GeoTrends 2010, West Palm Beach, Florida, USA, pp. 27–42.

McCartney, J.S., Murphy, K.D., 2012. Strain distributions in full-scale energy foundations (DFI Young Professor Paper Competition 2012). DFI Journal: The Journal of the Deep Foundations Institute 6 (2), 26–38.

Nicholson, D.P., Chen, Q., Pillai, A., Chendorain, M., 2013. Developments in thermal pile and thermal tunnel linings for city scale GSHP systems. In: Proceedings of the 38th Workshop on Geothermal Reservoir Engineering.

Olgun, C.G., Martin, J.R., Bowers, G.A., 2012. Energy piles: using deep foundations as heat exchangers. Geo-Strata—Geo Institute of ASCE 13 (2), 46–51.

Olgun, C.G., Ozudogru, T.Y., Arson, C., 2014. Thermo-mechanical radial expansion of heat exchanger piles and possible effects on contact pressures at pile-soil interface. Geotech Letters (July–September), 170–178.

Pahud, D., Hubbuch, M., 2007. Measured thermal performances of the energy pile system of the dock midfield at Zürich Airport. In: European Geothermal Congress, Germany, pp. 1–7.

Park, H., Lee, S.R., Yoon, S., Shin, H., Lee, D.S., 2012. Case study of heat transfer behavior of helical ground heat exchanger. Energy and Buildings 53, 137–144.

Rui, Y., 2015. Finite Element Modelling of Thermal Piles and Walls (Ph.D. dissertation). University of Cambridge.

Schneider, M., Moormann, C., 2010. GeoTU6 – a geothermal research project for tunnels. Tunnel 2, 14–21.

Sekine, K., Ooka, R., Shiba, Y., Yokoi, M., Hwang, D.S., 2007. Development of a ground-source heat pump system with ground heat exchanger utilizing the cast-in-place concrete pile foundations of buildings. ASHRAE Transactions 113 (Part 1), 558–566.

Suckling, T., Smith, P., 2002. Environmentally friendly geothermal piles at Keble College, Oxford, UK. In: Proceedings of Deep Foundations Institute Conference. Nice, France.

Sun, M., Xia, C., Zhang, G., 2013. Heat transfer model and design method for geothermal heat exchange tubes in diaphragm walls. Energy and Buildings 61, 250–259.

Sutman, M., Brettmann, T., Olgun, C.G., 2014. Thermo-mechanical behavior of energy piles: full-scale field test verification. In: DFI 39th Annual Conference on Deep Foundations, Atlanta, GA, pp. 1–11.

Waboso, D., Gilbey, M., July 2007. Cooling the tube. T & T International 36–39.

Zarrella, A., De Carli, M., Galgaro, A., 2013. Thermal performance of two types of energy foundation pile: helical pipe and triple U-tube. Applied Thermal Engineering 61 (2), 301–310.

Zhang, G., Xia, C., Sun, M., Zou, Y., Xiao, S., 2013. A new model and analytical solution for the heat conduction of tunnel lining ground heat exchangers. Cold Region Science Technology 88, 59–66.

Part Two

Open-loop systems and energy storage

to isolate fouling-prone seawater or lakewater from the building's heating, ventilating, and air conditioning (HVAC) systems. These systems will not be covered in this chapter but are reviewed by Mitchell and Spitler (2013).

b. SWHP systems: systems that use heat pumps or chillers to provide heating and/or cooling, with their heat source or sink, surface water. Depending on location and application, the systems may provide heating only, cooling only, or heating and cooling.

c. Hybrid surface water heat pump (HSWHP) systems: systems that use heat pumps or chillers to provide heating and/or cooling; however, they can also use seawater or lakewater directly to provide cooling when water temperatures allow this to be done.

Despite SWHP systems having been used for more than 65 years, there is a paucity of design data and procedures. Therefore, this chapter gives an overview of the current state of the art, with an emphasis on system design. Design data are covered in Section 8.2. Since statistically significant design data are relatively scarce, simulation of lakes and other surface water bodies may be utilized, as discussed in Section 8.3. Then open-loop and closed-loop systems will be discussed in Sections 8.4 and 8.5. This chapter draws heavily on two research projects (Spitler et al., 2012, 2014) and a review paper by Mitchell and Spitler (2013).

8.2 Design data

For purposes of designing SWHP systems it would be ideal to have statistically characterized minimum and maximum lake (or other surface water body) temperatures over the year in much the same way that air temperatures and humidity data are available for design of building heating and cooling systems. Unfortunately, design data in the form of statistically characterized minimum and maximum lake temperatures based on measured data are essentially nonexistent. In fact, there are very few measured data sets of lake temperatures and almost all of those rely on spot measurements. The most extensive collection of measured lake water temperature data is that presented by Hattemer and Kavanaugh (2005) and available online (EIS, 2015). However, this source only covers 41 lakes in 14 states in the continental United States. By some estimates (Lehner and Döll, 2004) there are more than 11,000 lakes between 10 and 100 ha in the United States, so figuratively speaking, even the most extensive data set is only a drop in the bucket. Furthermore, lake temperature profiles depend on local weather conditions and the bathymetric profile of the lake. For example, comparing two lakes of equal surface area: one that is much deeper may remain stratified over much or all of the year while the shallower lake may become completely unstratified, even during the summer. Therefore, extrapolation of data from one lake to another lake is problematic, at best.

An alternative approach, described by Spitler et al. (2014) is to use simulation models to estimate lake temperatures from typical meteorological data. A validated simulation model, accompanied by an uncertainty analysis was used to derive design temperatures for three lakes with different sizes and bathymetries, placed in 926 locations for which typical meteorological year-type weather data are available. The approach can be used for any combination of specific lake bathymetry and weather

Part Two

Open-loop systems and energy storage

file. It is acknowledged that there are limitations to such an approach, but when statistically characterized measured temperature data are simply not available, this seems to be the next-best approach.

8.3 Physics and modeling of surface water bodies

This section briefly describes the physics of lakes, reservoirs, streams, and rivers along with modeling approaches used to predict water temperatures.

8.3.1 *Physics of lakes and reservoirs*

The spatial and temporal distribution of temperatures in lakes and reservoirs depends both on heat inputs and losses and on mixing mechanisms. In unfrozen lakes, solar irradiation usually represents the dominant heat input. Some of the incident solar radiation is absorbed at the water body surface; the remaining energy is transmitted through the water body, where it is absorbed in the water or at the bottom. Penetration of the solar radiation through the depth depends on the transmissivity of the water, which in turn depends on its turbidity. Evaporation is the main cooling mechanism for ponds and lakes. Surface convection heat transfer and longwave radiation may heat or cool the water body depending on the relative temperatures of the water body surface and the surroundings. Heat transfer from the ground below the water body (referred to as sediment heat transfer) can become highly significant in winters especially during the times when the water body freezes at the surface (Fang and Stefan, 1996a,b). Though often negligible, heat transfer to/from surface water heat exchangers (SWHE) could be important if the SWHP system size is large relative to the size of the lake.

Mixing in lakes can be due to advection or diffusion. Advection can be driven by wind shear, inflows, outflows, buoyancy, and internal waves. Strictly speaking, diffusion is a molecular level process, though we often treat turbulent mixing as if it were a diffusion process, referring to it as turbulent diffusion or eddy diffusion. Turbulent diffusion is characterized by irregular, random fluctuations due to turbulent eddies in the water system. Other mechanisms for mixing include surface wind stress, internal waves, shear waves, inflows, and outflows. The formation and cause of these different mixing mechanisms on the surface water body and their effects on the water temperatures are discussed in detail by Imberger et al. (1978). Fig. 8.1 illustrates some of the heat transfer and mixing mechanisms for a typical surface water body with a heat exchanger installation.

For an unfrozen lake, the formation of stratification is controlled by the balance between two phenomena: heat input at the surface leads to naturally stabilizing temperature and density gradients; and surface winds generate turbulent energy that is destabilizing. Surface winds over a shallow water body may completely overwhelm the naturally occurring temperature-density gradients in the water column and cause it to mix completely. Hence, shallow ponds and lakes are often

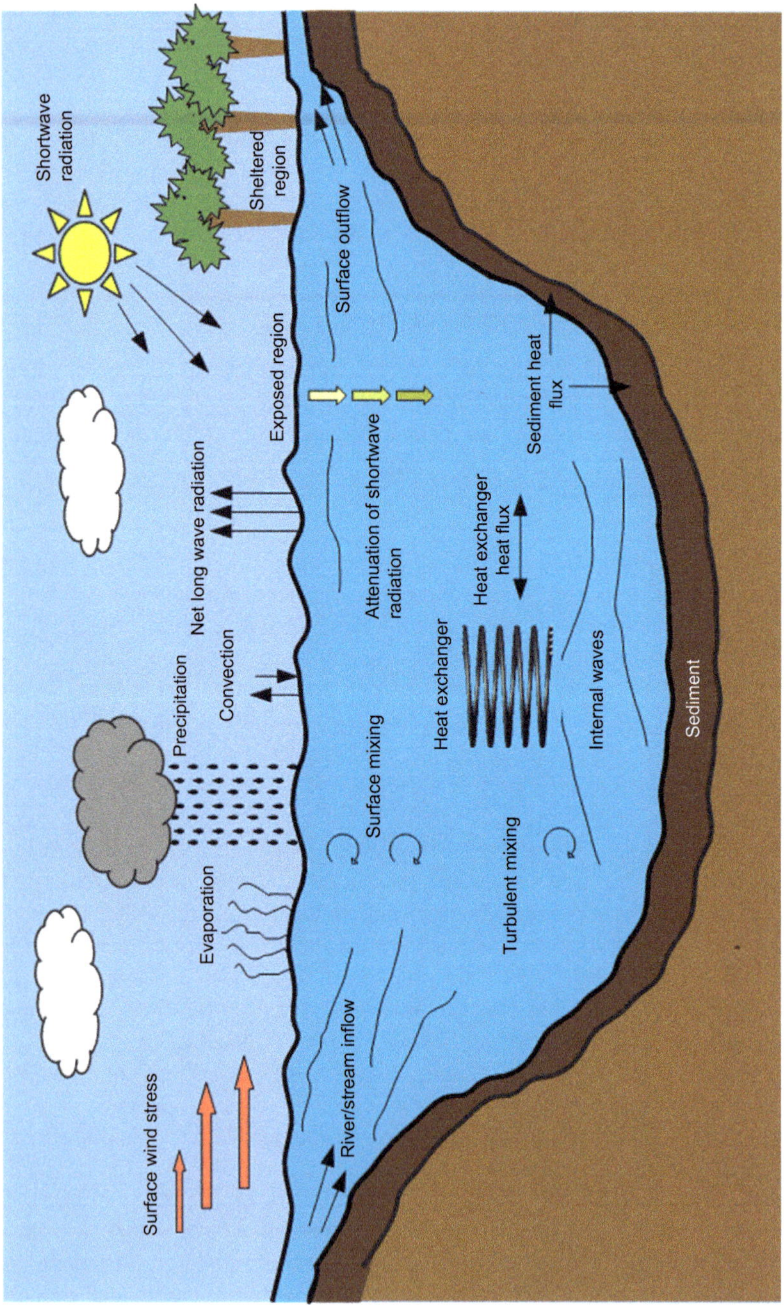

Figure 8.1 Heat transfer and mixing mechanisms.

unstratified for much of the year. Stratification is most commonly observed in deeper lakes where the wind-induced mixing is restricted to a certain depth near the surface. This gives rise to a temperature profile with three distinct regions as illustrated with experimental data for a 16.6 ha lake with maximum depth of 16 m in Fig. 8.2:

- The epilimnion or upper mixed region characterized by well-mixed and relatively high water temperatures.
- Metalimnion or thermocline region: This transition region between the warm surface water and cooler bottom water is characterized by large temperature and density gradients.
- Hypolimnion or lower mixed region: This relatively cold, well-mixed region below the thermocline extends to the bottom of the surface water body. It tends to be undisturbed by the surface winds.

Most deeper lakes exhibit these distinct temperature stratification regions throughout the summer and most of the fall season. By late fall, the lake surface temperature cools down, and the density gradient in the water column slowly diminishes. At this time, the turbulent energy from the surface wind is sufficient to completely destabilize or mix the entire water column. This phenomenon is called turnover, and during the turnover period the temperature of the lake is approximately uniform from the surface to the bottom. Lakes that develop surface ice cover exhibit the same phenomenon in the spring when the surface ice melts and as the temperature of the epilimnion approaches 3.8°C (the maximum density point of water).

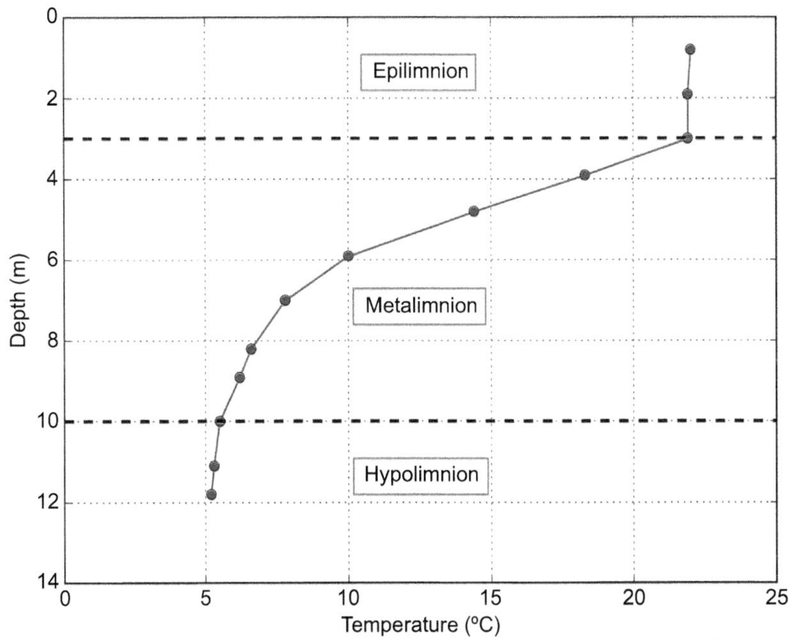

Figure 8.2 Temperature profile: Ice Lake, Minnesota, June 24, 2003.

Most but not all (Hakala, 2004) stratified lakes exhibit such turnover conditions at least once or twice annually.

8.3.2 Modeling of lakes and reservoirs

Modeling of lakes and reservoirs to predict their temperature variations and water quality for environmental purposes has been a subject of research for many years. Numerous one-dimensional (Tucker and Green, 1977; Imberger et al., 1978; Ford and Stefan, 1980; McCormick and Scavia, 1981; Saloranta and Andersen, 2007), two-dimensional (Hayter et al., 1998; Ji et al., 2001), and three-dimensional (Blumberg and Mellor, 1987; Hamrick, 1992) numerical models have been developed over the years to predict the water temperature of lakes and ponds. While two- and three-dimensional lake models have the potential to more accurately determine temperature distribution in a lake, real-world constraints on engineer time suggest that one-dimensional models are probably the most practical approach for design of SWHP systems. Likely the most comprehensive work on the subject is that of Stefan and his co-researchers at the University of Minnesota (Ford and Stefan, 1980; Riley and Stefan, 1988; Gu and Stefan, 1990, 1995; Ellis et al., 1991; Hondzo et al., 1991; Hondzo and Stefan, 1993; Fang et al., 1996; Fang and Stefan, 1996a,b). Saloranta and Andersen (2004, 2007) have also provided a helpful summary of methods used for one-dimensional modeling of lakes.

With a few exceptions, all of the lake models discussed in this section discretize the lake to numerically calculate the temperature profile and compute the vertical temperature transport using eddy diffusion. Exceptions include Ford and Stefan (1980) who only accounted for molecular diffusion; Pezent and Kavanaugh (1990) who estimated the thicknesses of the epilimnion, metaliminion and hypoliminion based on experimental measurements and Chiasson et al. (2000) who modeled unstratified shallow ponds.

Ford and Stefan (1980) initially developed a model to simulate temperatures and stratification in small lakes using an integral energy approach. This approach finds the depth where the turbulent kinetic energy from the wind balances the potential energy available from the natural stratification. Above this depth is the epilimnion region. Below this depth are the stratified metalimnion and hypolimnion regions. The model allowed only molecular diffusion in the thermocline and hypolimnion regions, but was able to predict the temperatures reasonably well when validated against three lakes with surface areas varying from 29 to 181 ha and depths from 9 to 27 m.

Riley and Stefan (1988) developed the MINLAKE program which extended the previous model by considering the effect of eddy diffusion for temperature transport. Separate eddy diffusion coefficients are determined for the epilimnion and hypolimnion layers. The model considers the effect of mixing, stratification, inflow, and outflow in lakes but does not consider the formation of ice/snow cover on the lake surface. Gu and Stefan (1990) added models for sediment heat transfer and the surface heat balance for frozen surface conditions. Ellis et al. (1991) calculated the eddy diffusion coefficients for lakes under ice and snow cover based on their measurements from a small lake in Minnesota. Fang and Stefan (1996b) developed a correlation to predict the eddy diffusion coefficients for lakes under ice and snow cover based on the results from Ellis et al. (1991).

A general limitation of these models is the nongenerality of the models for determining eddy diffusion coefficients. For very small lakes, it is recommended that some of the model parameters be estimated for the lake in question. For larger lakes, there are a range of models that can be used for determining eddy diffusion coefficients. Spitler et al. (2014) make some recommendations based on comparisons to experimental data.

Several less comprehensive lake models have treated SWHEs. Pezent and Kavanaugh (1990) developed a model to simulate lakes as heat sources/sinks for heat exchangers with water source heat pump systems. The model determines seasonal variation in stratification based on experimental observations of certain lakes in Alabama and does not consider the effects of wind stress on stratification or temperature transport by turbulent diffusion.

Chiasson et al. (2000) developed a shallow pond model with an SWHE intended to serve as a supplemental heat rejecter for ground-source heat pump (GSHP) systems. The model treats the pond as being well mixed and includes a full surface heat balance and heat transfer from the SWHE. The SWHE were horizontal and vertical "Slinky" configurations of high-density polyethylene (HDPE). The model was validated against two different specially constructed experimental "ponds" of 0.6 and 1.1 m depths for a 9-day summer period and a 25-day fall period.

8.3.3 Rivers

Rivers are also sometimes used as sources for SWHP systems. Compared to lakes and reservoirs, rivers tend to be less stratified and more closely follow air temperatures. However, as might be expected, the actual situation is complicated. River temperatures are affected by atmospheric conditions, shading, the nature of the flow, groundwater inflows, and conduction heat transfer to/from the stream bed. Heat transfer phenomena are reviewed by Caissie (2006) and Webb et al. (2008). A range of models are also available, but simpler models based on the equilibrium temperature concept (Edinger et al., 1968) are probably the most feasible for engineering use. Herb and Stefan (2011) present an improved version of the equilibrium temperature model.

8.4 Open-loop systems: surface water heat pumps and direct surface water cooling

Open-loop systems may be used in almost any application including residential buildings, commercial and institutional buildings, and district heating and cooling systems. However, because of the economics of scale, very large systems such as district heating and cooling systems often utilize open-loop SWHPs and DSWC systems. This section is largely drawn from a review paper written by Mitchell and Spitler (2013), which covers open-loop systems in more detail.

8.4.1 Open-loop surface water heat pump systems

Open-loop SWHP systems are "indirect"—that is, they utilize heat pumps to provide cooling and/or heating. Included within this category are systems that use heat pumps that provide cooling only, more commonly referred to as "chillers." Other authors have referred to these systems as "Indirect Seawater Utilization" (Akashi and Watanabe, 2003; Song et al., 2007), "Seawater Condenser Cooling" (Chien et al., 1986), "Open-loop Lakewater Heat Pump System" (Kavanaugh and Pezent, 1990; Chen et al., 2006), "pumped water—indirect system" (in Norwegian, "*Oppumpet vann—indirekte løsning*"; Smebye et al., 2011).

Eggen and Slettahjell (2005) surveyed 19 open-loop SWHP systems installed in Norway between 1981 and 1995. All of the systems provided heating and had capacities between 35 kW and 6.8 MW. Of the 19 systems, 15 were still in operation as of 2005. During the time of operation, systems operated with an average seasonal heat pump heating COP of 3.1 and average seasonal system heating COP of 2.7. Seasonal heat pump heating COPs ranged from 2.1 to 4; seasonal system heating COPs ranged from 2.0 to 4.

Chien et al. (1986) discussed systems with an opposite purpose: providing directly for condenser cooling in Hong Kong. The authors estimated that in 1985, total installed cooling capacity of systems utilizing seawater condenser cooling in Hong Kong exceeded 1 GW. The authors give a figure of 10–15% reduction in energy consumption compared to systems using indirect cooling. The paper also discusses installation details such as filtration, intake location, heat exchanger materials, and so forth.

Smebye et al. (2011) reports on a system installed at the University of Bergen in Norway that utilizes a central ammonia heat pump to supply water at 15–25°C to 14 secondary heat pumps located in different buildings. The secondary heat pumps have a total capacity of 4 MW heating or 2 MW cooling.

Akashi and Watanabe (2003) and Song et al. (2007) describe a cooling-only SWHP system serving a retail complex in Nagasaki, Japan. The system was monitored for 22 months. During the first 10 months of operation, the system COP (2.6) was adversely affected by the pumping scheme, which ran the pumps continuously due to fouling concerns. Changing the pumping scheme so that the pumps only ran when the chillers ran gave an overall system COP of 2.9 for the following 11 months of operation.

Chen et al. (2006) reported on an open-loop SWHP system in south China, used to provide both cooling and heating to a city center complex of four buildings: city hall, TV studio, theater, and shopping center. One year of experimental measurements were made in 2003 and 2004. In cooling mode, the heat pump COP varied between about 4.8 and 4.3 at lake temperatures between 25°C and 35°C; over the same lake temperatures, the system COP varied between 4.2 and 3.8. In heating mode the heat pump COP varied between about 3.8 and 4.3 at lake temperatures between 7°C and 15°C over these lake temperatures; the system COP varied between 3.4 and 4.

Lind (2010) and Baardsen (2011) describe a 60 MW district heating system in Drammen, Norway, that uses an SWHP system and gas-fired boilers to supply heat to a heating district. With ammonia as the refrigerant, the three-stage heat pump supplies 90°C water for a heat output of 15 MW and a heat pump COP of 3, using seawater from the Drammen Fjord as its source.

A district heating system (Friotherm, 2015) in Stockholm, Sweden, utilizes six seawater heat pumps at the Värtan Ropsten plant with total heating capacity of 180 MW. The heat pumps originally used refrigerant R22 when installed in the mid-1980s; they were converted to R134a beginning in 2003. The heat pumps supply hot water at 80°C. During the winter, water is taken from the Baltic at 3°C and returned at 0.5°C. Under these conditions a heat pump COP of 3.75 is claimed. "Falling film" evaporators are used—plate evaporators with the sea water sprayed on to the outside of the evaporator.

8.4.2 Hybrid surface water heat pump systems

HSWHP are very robust systems, in that they can cool directly when the water temperatures allow, and make up the difference with supplemental cooling when water temperatures will not allow direct cooling. They can also provide heating provided the heat pumps are reversible and designed appropriately. There appears to be no distinction between SWHP systems and HSWHP systems in the literature; however, we have adopted the term "HSWHP systems" to distinguish between the two system types.

Fermbäck (1995) outlines a 60 MW HSWHP cooling system in Stockholm, Sweden, that has four chillers with a total supplemental cooling capacity of 25 MW. Chillers are used in late summer for approximately 5 months to provide supplemental cooling when surface water intake temperatures rise above required district supply temperatures.

Viquerat et al. (2008, 2012) describe the HSWHP system that provides cooling to the United Nations campus in Geneva, Switzerland. The system draws 0.75 m^3/s from its source, Lake Léman, at a depth of 35 m and is capable of providing 23.5 MW at peak capacity. The system receives supplemental cooling from four chillers with a combined capacity of 6.5 MW. The chillers reject heat to Lake Léman via the HSWHP system.

The Toronto HSWHP system draws 4°C water from Lake Ontario at a depth of 83 m. From there it is filtered and then passed through the isolation heat exchangers at the heat exchange facility before being provided to the city's potable water supply. The direct cooling portion of this system alone is capable of producing 141 MW of cooling, while bottoming chillers—chillers which enhance system performance by reducing chilled water supply temperature to the design lower limit—provide an additional 58 MW for a rated capacity of 185 MW. Distribution water leaves the heat exchange facility at 5°C and is then reduced to 3.3°C by the bottoming chillers where it is then sent to the customer for space conditioning (Heffernan, 2001; Eliadis, 2003).

8.5 Major system components

8.5.1 Intake piping and screening

Proper intake sizing and design depends on several important factors—namely, surface water bathymetry and temperature, peak cooling and/or heating load, and screening requirements.

Different varieties of steel and fiberglass pipes have been used for underwater applications, but at present HDPE is the most widely used underwater pipeline material (Hirshman et al., 1975) because it is chemically inert, fusible, flexible, strong, and will float in water, which aids in pipeline installation. In addition, HDPE has a high thermal conduction resistance relative to steel pipe, which aids in reducing heat gain as the pipe passes through warmer water layers.

To prevent the entrainment of biological organisms or other submerged debris, some form of screening or filtration is implemented at the water intake. A typical intake structure and primary screening device are shown in Fig. 8.3 (Iso, 1977; Peer and Joyce, 2002; Viquerat et al., 2008; Zogg et al., 2008). This type of radial primary screen is intended to draw water from one specific horizontal stratum deep underwater and is located far enough above the sea or lake bed to avoid entraining sediment particles. The intake for the Cornell DSWC system is situated 3 m above the floor of Lake Cayuga (Cornell University, 2005).

Chien et al. (1986) discussed the use of a coarse stationary primary screen, followed by one or more stages of traveling band screens equipped with backwashing spray systems for SWHP system intakes in Hong Kong. This is necessary due to the shallow SWHP intake locations, which were all located just below sea level down to 4 m depth. Because

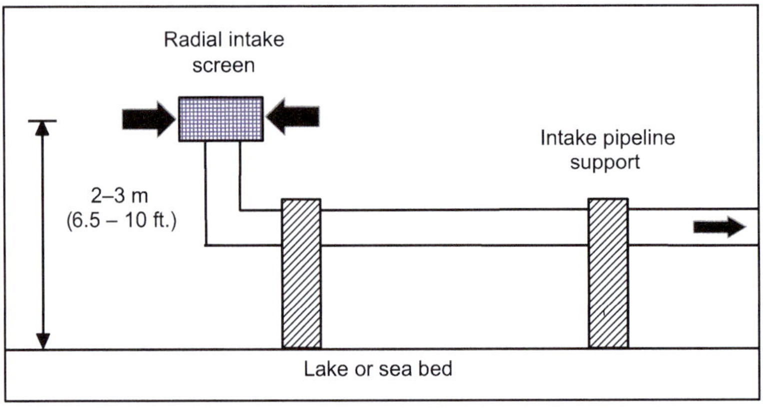

Figure 8.3 Typical surface water intake.

plastic garbage, seaweed, and other floating debris will impact the performance of screening equipment, traveling band screens with backwash systems were implemented.

The screened surface area should be large enough to maintain face water velocity at an acceptable level. By limiting the maximum face water velocity, biological organisms such as fish or fish larvae are not drawn to the intake which will minimize death by impingement and entrainment. Specific system designs are discussed by Stefan et al. (1986), Cornell University (2005), and Ng et al. (2005).

Another approach (Fariñas and López, 2006; Peters and Pintó, 2008) is using the sea or lake bed itself as the filter by horizontally drilling beneath the lake or sea floor and installing a perforated drill casing. Similarly, for residential open-loop SWHP systems, a sand filter (Cusack, 2012) has been used. The sand filter functions by passing the intake water through a fine medium, typically fine sand or some gradation of sand and other porous media. The filtered water is then pumped through the heat pump for space conditioning. In the system reported by Cusack (2012) once a predetermined time, maximum differential pressure, or filter water level has been exceeded, the sand filter units then reverse flow direction, backwashing the debris from the filter back to the water source.

8.5.2 Pumps and pumping configurations/dry sumps, wet sumps

For open-loop SWHP systems, the surface water pump is a primary energy consumer, so the piping system design and pump selection are very important to the overall success of the system. In the literature, there seems to be two main design strategies for configuring surface water pumps—namely, wet sump installations and dry sump installations. Kavanaugh and Pezent (1990) suggest that for small residential systems, a submersible pump may be placed at the pipe inlet. However, due to the depths and distances typically involved with large intake systems, maintenance and installation costs make this configuration impractical for large systems (Hirshman et al., 1975).

Wet sump pumping systems are configured so that the surface water pump is situated in a wet sump pumping pit. Fig. 8.4(a) shows a representation of the wet sump surface water pumping configuration. When the system is not operating, the water level in the wet well is the same as the water source. As water is pumped from the wet well, the water level in the wet well decreases. The resulting elevation difference between the water level of the source water body and the wet well is the driving force that causes water to flow into the wet well through the inlet and intake pipeline. This process is also reversible, meaning water could be pumped into the wet well to aid in removal of debris at the intake inlet screen.

The wet sump-pumping configuration is limited by the depth below the source water level that the intake pipe outlet can be installed. The wet sump–type pumping configuration may also require additional offshore dredging or trenching to install the pipeline inlet at a depth sufficiently below the source body water level (Hirshman et al., 1975). Wet sump pit designs must also consider the possibility of vortex formation (Osman et al., 1993; Tsou et al., 1994) at the pump inlet. Jones et al. (2008) is a comprehensive reference for designing pumping stations.

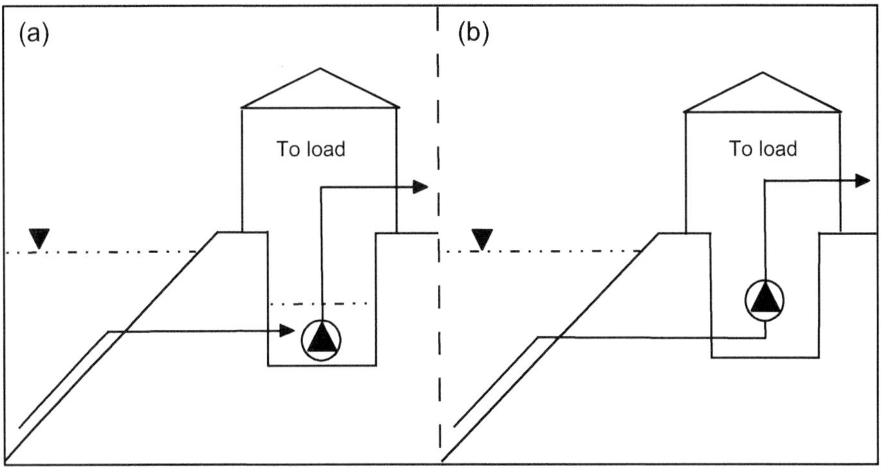

Figure 8.4 Wet (a) and dry (b) sump-pumping configurations.

The Cornell University DSWC system utilizes a wet sump-pumping configuration design. Water flows into the 8.5 m deep wet sump as water is pumped out of the wet sump pit and through the isolation heat exchangers. Three 261 kW variable frequency drive lakewater pumps are each designed to operate at 2952 m^3/h at 24 m of head at a speed of 1200 rpm (Cornell University, 2001, 2005).

Dry sump-pumping stations are configured so the pump is directly connected to the intake pipeline as is depicted in Fig. 8.4(b). The pump may be located at or below the source body water level as long as the net positive suction head available (NPSHA) is greater than the net positive suction head required (NPSHR). Placing the pump below sea level will decrease the likelihood of pump cavitation provided the NPSHA is always greater than the NPSHR.

The Genéve-Lac-Nations project in Geneva, Switzerland, utilizes a dry sump-pumping configuration for a DSWC system used to cool the United Nations buildings along the shore of Lake Léman. This system utilizes pumps located 6 m below the water surface with a capacity of 2700 m^3/h to circulate water through the district-cooling network (Viquerat et al., 2008).

Material selection for pumps is an important consideration, which depends on salt content and other water quality parameters such as suspended solids, water temperature, and pH. Salt content will cause corrosion of the pump components by electrochemical corrosion processes while higher water temperatures will accelerate these processes. Suspended solids content will accelerate the erosion of high velocity pump components, causing clearances to expand and pump performance to decrease. Material selection is discussed in more detail by Mitchell and Spitler (2013).

8.5.3 Isolation heat exchangers

A selection of isolation heat exchangers are influenced by a number of factors: required heat transfer rate, maintenance requirements, surface water characteristics, expected fouling resistance, pressure drop, space requirements, and, of course, cost. For DSWC or SWHP systems utilizing brackish or saltwater sources, electrochemical corrosion is a particular concern.

For seawater applications, titanium is the most widely used heat exchanger material in current DSWC and SWHP systems operation around the world (Fermbäck, 1995; Leraand and Van Ryzin, 1995; Newman and Herbert, 2009; War, 2011). Smebye et al. (2011) recommend titanium for seawater systems and high alloy steel for fresh water systems. Janikowski (2003) reviews materials for power generation heat exchangers. Pugh et al. (2005, 2009) also discuss materials for seawater cooling and freshwater cooling.

Another challenge for open-loop SWHP systems is performance degradation of the isolation heat exchangers due to biological fouling. If not addressed, biological film, algae, slime, and/or mollusks could form in the cooling system, which will increase the thermal resistance, as well as the required pumping power. To control the growth of biological organisms, biocide dosing schemes (cf. Mitchell and Spitler, 2013) have been implemented and reported for various freshwater and seawater cooling system applications.

Pugh et al. (2005, 2009) present a substantial amount of guidance regarding heat exchanger design, material selection, and fouling precautions. The two papers present an introduction to the relationships between flow velocity, surface temperature, heat exchanger construction materials, water quality, entering and exiting temperatures, and fouling resistance for freshwater and seawater heat exchangers.

Kavanaugh and Rafferty (1997) describe several other methods for coping with heat exchanger fouling including permanently installed brush systems. Disassembly and cleaning of bolted plate frame heat exchangers is also another maintenance option for fouling mitigation.

8.5.4 Heat pumps and chillers

For large SWHP or HSWHP systems, the heat pumps or chillers are often custom designed specifically for the site and application. This is due to the wide range of operating temperatures and conditions under which SWHP or HSWHP systems may operate. A number of the Scandinavian systems, for example, are providing hot water at 75–90°C for hydronic heating or district heating systems. Some of these systems (Friotherm, 2012) utilize two-stage compression. Because many of the companies providing district heating and cooling are privately held and in competition, heat pump/chiller performance information is often proprietary (Pietrucha, 2012). Most of the publicly available information is summarized by Mitchell and Spitler (2013).

8.5.5 Return piping and discharge

Open-loop SWHP outfall systems are used to return the warmer or cooler return water back to the water source without creating a concentrated pocket of water that is significantly warmer or cooler than the water source. These outfall systems are also intended to be placed at a depth that will not cause significant nutrient enhancement of the outfall area (Davidson, 2003; War, 2011). Water taken from deep underwater may contain higher amounts of nutrients than surface water, so the outfall should discharge at depth that will not promote growth of algae. The outfall system works by mixing the system water with the source water body over a large enough area so that temperature gradients are minimal. It is also typically discharged near the lake or sea bed instead of directly into the water column.

The Cornell University outfall system is composed of a 1.2 m pipe that extends 122 m from shore, which is then connected to a diffuser that is submerged 4.2 m underwater. The diffuser is 30 m long, situated approximately 0.3 m above the lake bed, and has 38 nozzles of diameter 15 cm pointed up at a 20° angle above horizontal (Peer and Joyce, 2002; Cornell University, 2005). The heat input into Lake Cayuga is roughly equivalent to two additional hours of sunlight each year (Peer and Joyce, 2002). No nutrient enhancement of the outfall area has been noted (Peer, 2012).

The HSWHP system in Toronto makes dual use of the outfall water by using it as raw water for the city's water treatment plant, where it is then treated as drinking water (Newman and Herbert, 2009) for the city. It has been claimed that the DSWC system in Toronto will have no environmental impacts on the surrounding ecosystem. The system currently draws about 5 m^3/s from Lake Ontario and the cold lakewater is considered renewable at extraction rates up to 10,000 m^3/s (Fotinos, 2003).

8.6 Closed-loop systems

Closed-loop systems exchange energy with surface water through heat exchangers that are submerged in the surface water body. Rather than circulating surface water directly, closed-loop systems typically circulate an antifreeze solution between the SWHE and the heat pumps or chillers used to meet the heating or cooling load. As with open-loop systems, closed-loop systems can be deployed for almost any application. These systems have the advantage of not requiring filtering or screening of the surface water but the disadvantage of having an SWHE submerged in the surface water body. Closed-loop SWHP systems have been installed for commercial purposes since the 1970s (Johansson, 1983) and continue to be popular today. This section reviews the different types of SWHE and briefly discusses additional design considerations.

8.6.1 Surface water heat exchanger types

8.6.1.1 Bottom sediment heat exchangers

Bottom sediment heat exchangers consist of pipes laid on or buried in the bottom sediment of the surface water body. Because these pipes can be in contact with the sediment layer, the heat exchangers are able to absorb or reject heat directly to/from the sediment layer as well as the surface water body. If the heat exchanger is expected to perform significant space heating operations, this can be an important advantage as heat transfer from the sediment becomes important when lakes are ice covered (Fang and Stefan, 1996a).

Researchers in Sweden pioneered bottom sediment heat exchangers in the early 1980s. Svensson and Sörman (1982, 1983a,b) experimentally investigated heat extraction from sediment heat exchangers in still and running water. They presented their results in terms of heat transfer rates that depended on temperature difference, temperature level, tube position, sediment type, and, for running water, flow velocity.

A novel sediment heat exchanger is described by Backlund (1982). Eleven pieces of 50 m straight pipe were connected together in parallel, placed on the lake bottom and weighted with stones. Inside each of these pipes were smaller diameter air-filled pipes. Buoyancy caused the inner tube to rise to the top of the cylinder, thus causing most of the flow to run along the bottom of the outer pipe in order to promote heat exchange with the sediment.

In Botngård, Norway, a small district heating network utilizes an SWHE consisting of approximately 20 km of piping embedded in the seabed of an intertidal area (Smebye et al., 2011). The SWHE is designed for 1 MW heating capacity. Calcium chloride antifreeze transports energy between the heat exchanger and the heat pumps.

8.6.1.2 High-density polyethylene piping bundles

HDPE piping is commonly used in spaced bundles. The spaced bundles might be called "stacked spiral" or "spiral-helical" heat exchangers as each horizontal layer is a spiral connected either at the outside or inside to the layer above. A spiral-helical heat exchanger is shown in Fig. 8.5. Loose bundles are also used, which consist of coils of HDPE piping bound together by loose bands and which therefore have uncontrolled tube-to-tube spacing. HDPE piping bundles are relatively inexpensive since they consist of only HDPE piping. They also can be fabricated on site as needed.

Figure 8.5 Spiral-helical heat exchanger made from 152 m of nominal 2.5 cm diameter high-density polyethylene (HDPE) pipe.

The bundles must be anchored since they have density near to that of the surrounding water. When icing conditions can occur, anchoring becomes more critical.

Hansen (2011) and Mitchell (2014) conducted experimental tests in a 1.2 ha (3 acre) pond, measuring overall thermal resistance and then deriving exterior convection correlations for heat exchanger types which include spiral-helical coils, flat spiral coils, vertical and horizontal Slinky coils, loose coils, and metal flat plate heat exchangers. In order to estimate the outside convective resistance, the inside convective resistance for the coils was estimated with convection correlations (Rogers and Mayhew, 1964) that considered the effect of tubing curvature.

Lloyd (2001) and Gyory (2008) describe large SWHP systems serving health care facilities. Both systems use HDPE bundles. The first system utilizes 171 km of HDPE piping in a 6 ha lake and has a total nominal cooling capacity in the building of 5.3 MW. The second system utilizes 296 km of HDPE piping in a 6 ha lake and has a total nominal cooling capacity of 8.4 MW.

8.6.1.3 Flat plate heat exchangers

Flat plate heat exchangers used as SWHE (Fig. 8.6) are commonly constructed from spot-welding two metal (stainless steel for freshwater; titanium for saltwater) panels together, welding the edges and expanding the gap between the plates. The resulting gap between the panels forms one or more flow paths or noncircular and irregular cross-section. For the flat plate heat exchangers, the hydraulic diameter of the flat plate is estimated and the inside Nusselt number is calculated based on the Dittus–Boelter correlation for straight pipes, using the calculated hydraulic diameter. Hansen (2011) correlated external convection for vertical flat plate heat exchangers but stated that the widely used correlation by Churchill and Chu (1975) also provides accurate results.

Figure 8.6 Flat plate heat exchangers installed in Big Sky, Montana.
Photo reproduced by permission of Terry Proffer, Major Geothermal.

Flat plate heat exchangers may be the most widely adopted type of SWHE, based on limited survey data report by Spitler et al. (2014). Compared to HDPE bundles, flat plate heat exchangers are more compact, are often easier to install, and have design expertise available from the manufacturer; but tend to be more expensive.

An SWHP system provides King's Mill Hospital in Mansfield, United Kingdom, with 5.4 MW of cooling capacity and 5 MW of heating capacity (Skanska, 2008). One hundred and forty stainless steel flat plate heat exchangers with total surface area of 1560 m^2 are submerged in the King's Mill Reservoir, which has a surface area of 16.5 ha (41 acre) and a maximum depth of 5 m (Banks, 2012).

8.6.1.4 Other surface water heat exchanger types

Several other types of SWHEs have also been utilized. These include Slinky-type heat exchangers (Chiasson et al., 2000). Hansen (2011) conducted multiple experiments on vertical and horizontal Slinky heat exchanger and developed correlations for the exterior natural convection.

Several SWHEs utilizing small-dimension HDPE or crosslinked polyethylene (PEX) tubing (Limnion, 2015; TEVAenergy, 2015) have been put on the North American market. Another SWHE developed in Sweden (Sjöberg, 2004) utilized a submersible motor and axial propeller to enhance the convective heat transfer over a helical HDPE coil.

8.7 Closed-loop design considerations

There are a few important considerations regarding the design of closed-loop systems. When used to provide heating, there is a risk of icing when the surface body water temperature is low and the antifreeze mixture inside the SWHE drops below the freezing point of water. A small amount of icing will increase the thermal resistance, thus causing a decrease in performance. But, if the heat extraction rate is high for a long enough period of time, and the ice layer builds up to the point of blocking the water flow between tubes in a bundle heat exchanger, the thermal resistance will increase dramatically and heat pump–entering fluid temperatures may drop low enough to shut off the heat pump. Therefore, a backup heat source may be required in some circumstances. In addition, ice formation on the SWHE creates a buoyancy force. If the anchoring is insufficient, the SWHE may float to the surface. There are anecdotal accounts of the SWHE being damaged under these circumstances; Spitler et al. (2014) documents one such case. However, there have also been anecdotal accounts of SWHE being damaged by the anchoring hardware, particularly where the SWHE is shallow enough to be affected by waves.

For HDPE heat exchangers, the conductive resistance of SDR-11 HDPE tube can account for 60–75% of the total thermal resistance of the heat exchanger (Mitchell, 2014). Use of thinner-walled and/or thermally enhanced HDPE tubing may be advantageous.

8.8 Conclusions

SWHP systems have been utilized since the late 1930s in a wide range of climates, from subarctic conditions in Scandinavia to subtropical conditions in Hawaii and Hong Kong. In general, there is a shortage of guidelines for system design, installation, and operation. Large commercial systems have received more treatment in the literature, and these systems tend to be open-loop, so there is a greater availability of design information for open-loop systems (Mitchell and Spitler, 2013) than for closed-loop systems. Some recent experimental research (Spitler et al., 2014) has improved the understanding of heat transfer performance for SWHEs.

There is also greater availability of measured performance data for open-loop systems; seasonal COPs for closed-loop systems have not been reported in the literature. For open-loop SWHP systems, seasonal heat pump COPs for heating in Scandinavian climates on the order of 2−4 have been reported for systems providing high-temperature water installed in the 1980s and early 1990s, with system COPs lower on average by about 0.4. In more temperate climates in Japan and China, heat-pump cooling COPs of 4.3−4.8 have been reported, with system cooling COPs lower by 0.5−0.6 for one system in China. As with GSHP systems, the overall efficiency of the system depends heavily on the design of the pumping and piping systems. The system in China also gave heat pump heating COPs of 3.8−4.3 with system heating COPs lower by 0.3−0.4.

For both types of systems, statistically characterized, experimentally measured surface water body temperatures are desirable, but not available. Simulations of lakes using typical meteorological year−type weather data can provide necessarily imperfect estimates of temperatures versus depth and time of year.

Further work in field monitoring of SWHP systems, particularly closed-loop systems serving residential and light commercial buildings, would be useful for establishing performance, design guidelines, and validation of models. Additional measurement programs of lake temperatures and further development of eddy diffusivity models would be useful for all types of SWHP systems.

References

Akashi, Y., Watanabe, T., 2003. Energy and cost performance of a cooling plant system with indirect seawater utilization for air-conditioning in a commercial building. Journal of Asian Architecture and Building Engineering 2 (1), 67−73.

Baardsen, B., 2011. The market for heat pumps in Norway. IEA Heat Pump Centre Newsletter 29 (1), 11−12.

Backlund, L., 1982. Resultant frän prov i Storsjön med värmeväxlare för utvinning av termisk energi ur vattendrag. In: Billfalk, L. (Ed.), Sjövärmesystem − BRF-seminarium. Statens Råd för Byggnadsforskning, Stackholm, pp. 14−28.

Banks, D., 2012. Pond and Lake Based Ground Source Heat Pump Systems. In: An Introduction to Thermogeology. John Wiley & Sons, Ltd, Chichester, West Sussex, UK, p. 366.

Blumberg, A.F., Mellor, G.L., 1987. A description of a three-dimensional coastal ocean circulation model. Coastal and Estuarine Sciences 4, 1–16.
Caissie, D., 2006. The thermal regime of rivers: a review. Freshwater Biology 51 (8), 1389–1406.
Chen, X., Zhang, G., Peng, J., Lin, X., Liu, T., 2006. The performance of an open-loop lake water heat pump system in south China. Applied Thermal Engineering 26 (17–18), 2255–2261.
Chiasson, A.D., Spitler, J.D., Rees, S.J., Smith, M.D., 2000. A model for simulating the performance of a shallow pond as a supplemental heat rejecter with closed-loop ground-source heat pump systems. ASHRAE Transactions 106 (2), 107–121.
Chien, P.Y.B., Tse, V.K.C., Yeung, D.P.L., 1986. Using seawater for cooling. ASHRAE Journal 28, 18–24. Atlanta, ASHRAE.
Churchill, S.W., Chu, H.H.S., 1975. Correlating equations for laminar and turbulent free convection from a vertical plate. International Journal of Heat and Mass Transfer 18 (11), 1323–1329.
Cornell University, 2001. Lake Source Cooling. Report LSC Project Description. Cornell University, Ithaca, NY.
Cornell University, 2005. How Lake Source Cooling Works. From: http://energyandsustainability.fs.cornell.edu/util/cooling/production/lsc/works.cfm.
Cusack, J., 2012. Telephone Conversation with Author. UMR Geothermal, Maple Plain, MN.
Davidson, J.R., 2003. California's Energy Future and Cold Ocean Water. Sea Technology 44 (7), 30–34.
Edinger, J.E., Duttweiler, D.W., Geyer, J.C., 1968. The response of water temperatures to meteorological conditions. Water Resources Research 4 (5), 1137–1143.
Eggen, G., Slettahjell, O., 2005. Varmepumpens bidrag til redusert energibruk i Norge. Norges Vassdragsog Energidirektorat, Oslo, Norway, 12/2005.
EIS, 2015. Energy Information Services – Temperature Profiles of Surface Water Bodies. Retrieved June 3, 2015, from: http://geokiss.com/surwatertemps.htm.
Eliadis, C., 2003. Deep Lake Water Cooling – A Renewable Technology. Electric Line 9 (3), 26–28.
Ellis, C.R., Stefan, H.G., Gu, R., 1991. Water temperature dynamics and heat transfer beneath the ice cover of a lake. Limnology and Oceanography 36 (2), 324–335.
Fang, X., Ellis, C.R., Stefan, H.G., 1996. Simulation and observation of ice formation (freeze-over) in a lake. Cold Regions Science and Technology 24 (2), 129–145.
Fang, X., Stefan, H.G., 1996a. Dynamics of heat exchange between sediment and water in a lake. Water Resources Research 32 (6), 1719–1727.
Fang, X., Stefan, H.G., 1996b. Long-term lake water temperature and ice cover simulations/measurements. Cold Regions Science and Technology 24, 289–304.
Fariñas, M., López, L.A., 2006. New and innovative sea water intake system for the desalination plant at San Pedro del Pinatar. Desalination 203, 199–217.
Fermbäck, G., 1995. Sea water plays key role in Stockholm cooling system. District Energy 81 (1), 41–43.
Ford, D.E., Stefan, H.G., 1980. Thermal predictions using integral energy model. Journal of the Hydraulics Division, ASCE 106 (1), 39–55.
Fotinos, D., 2003. Deep Lake Water Cooling. Power Engineering Society General Meeting, July 13–17, 2297–2300.
Friotherm, 2012. 5 Unitiop 50FY Heat Pump/Chiller Units Simultaneously Generate 90 MW Heat Energy and 60 MW Chilled Water. From: http://www.friotherm.com/downloads/katri_vala_e012_uk.pdf.
Friotherm, 2015. Värtan Ropsten—The Largest Sea Water Heat Pump Facility Worldwide, with 6 Unitop® 50FY and 180 MW Total Capacity. Retrieved June 3, 2015, from: http://www.friotherm.com/webautor-data/41/vaertan_e008_uk.pdf.

Gu, R., Stefan, H.G., 1990. Year-round temperature simulation of cold climate lakes. Cold Regions Science and Technology 18 (2), 147–160.
Gu, R., Stefan, H.G., 1995. Stratification dynamics in wastewater stabilization ponds. Water Research 29 (8), 1909–1923.
Gyory, J., 2008. Running hot and cold. Healthcare Design 8 (7), 70–75.
Hakala, A., 2004. Meromixis as a part of lake evolution – observations and a revised classification of true meromictic lakes in Finland. Boreal Environment Research 9 (1), 37–53.
Haldane, T.G.N., 1930. The heat pump. Journal of the Institution of Electrical Engineers 68, 666–675.
Hamrick, J.M., 1992. A Three-Dimensional Environmental Fluid Dynamics Computer Code: Theoretical and Computational Aspects. Virginia Institute of Marine Science, Gloucester Point, VA.
Hansen, G.M., 2011. Experimental Testing and Analysis of Spiral-Helical Surface Water Heat Exchanger Configurations (Masters thesis). Oklahoma State University.
Hattemer, B., Kavanaugh, S.P., 2005. Design temperature data for surface water heating and cooling systems. ASHRAE Transactions 111 (1), 695–701.
Hayter, E.J., Bergs, M.A., Gu, R., McCutcheon, S.C., Smith, S.J., Whiteley, H.J., 1998. HSCTM-2D, a Finite Element Model for Depth-Averaged Hydrodynamics, Sediment and Contaminant Transport. U. S. E. E. R. Laboratory, Athens, GA.
Heffernan, V., 2001. Toronto Waterfront: Toronto Cools Off Naturally – A Deep Lake Water Cooling System. Canadian Consulting Engineer 2001 (12), 22–24.
Herb, W.R., Stefan, H.G., 2011. Modified equilibrium temperature models for cold-water streams. Water Resources Research 47 (6), W06519.
Hirshman, J., Whithaus, D.A., Brooks, I.H., 1975. Feasibility of a District Cooling System Utilizing Cold Seawater. Phase I: Final Report. Report T/M727117-1. Energy Research and Development Administration, Washington, DC.
Hondzo, M., Ellis, C.R., Stefan, H.G., 1991. Vertical diffusion in small stratified lake: data and error analysis. Journal of Hydraulic Engineering 117 (10), 1352–1369.
Hondzo, M., Stefan, H.G., 1993. Lake water temperature simulation model. Journal of Hydraulic Engineering 119 (11), 1251–1273.
Imberger, J., Patterson, J., Hebbert, B., Loh, I., 1978. Dynamics of reservoir of medium size. Journal of Hydraulics Division 104 (5), 725–743.
Iso, S., 1977. Problems on sea water intake and discharge facilities in coastal regions. Desalination 22, 159–168.
Janikowski, D.S., 2003. Selecting Tubing Materials for Power Generation Heat Exchangers. Southwest Chemistry Workshop, Irving, TX. July 28–31.
Ji, Z.G., Morton, M.R., Hamrick, J.M., 2001. Wetting and drying simulation of estuarine processes. Estuarine, Coastal and Shelf Science 53 (5), 683–700.
Johansson, I., 1983. Comparison between soil, rock, groundwater and lakes as heat source for heat pumps – theory and practice. In: XVth International Congress of Refrigeration, Paris.
Jones, G.M., Bosserman, B.E., Sanks, R.L., Tchobanoglous, G., 2008. Pumping Station Design. Elsevier Inc, Oxford, UK.
Kavanaugh, S., Rafferty, K., 1997. Surface Water Heat Pumps. Ground-Source Heat Pumps: Design of Geothermal Systems for Commercial and Institutional Buildings. American Society of Heating, Refrigerating, and Air-Conditioning Engineers, pp. 114–131.
Kavanaugh, S.P., Pezent, M.C., 1990. Lakewater applications of water-to-air heat pumps. ASHRAE Transactions 96 (1), 813–820.
Lehner, B., Döll, P., 2004. Development and validation of a global database of lakes, reservoirs and wetlands. Journal of Hydrology 296 (1–4), 1–22.

Leraand, T.K., Van Ryzin, J.C., 1995. Air Conditioning with Deep Seawater: A Cost-Effective Alternative for West Beach, Oahu, Hawaii.. In: OCEANS '95. MTS/IEEE. Challenges of Our Changing Global Environment, San Diego, CA, pp. 1100–1109. Oct 9–12.

Limnion, 2015. Limnion Lima-1. Retrieved June 17, 2015, from: http://www.limnion.com/products.html.

Lind, O., 2010. Drammen først i verden med ammoniakk varmepumpe som leverer 90°C. Norsk Energi 86, 10–12.

Lloyd, W., 2001. A hospital on the lake. Consulting-specifying Engineer 29, 32–36.

McCormick, M.J., Scavia, D., 1981. Calculation of vertical profiles of lake-averaged temperature and diffusivity in Lakes Ontario and Washington. Water Resources Research 17 (2), 305–310.

Mitchell, M.S., 2014. Experimental Investigations and Design Tool Development for Surface Water Heat Pump Systems (Masters thesis). Oklahoma State University.

Mitchell, M.S., Spitler, J.D., 2013. Open-loop direct surface water cooling and surface water heat pump systems – a review. HVAC&R Research 19 (2), 125–140.

Newman, L., Herbert, Y., 2009. The use of deep water cooling systems: two Canadian examples. Renewable Energy 34, 727–730.

Ng, K.Y., Zheng, Y., Taylor, S.W., 2005. Recent developments in hydraulic design of power plant cooling water intake structures. In: Proceedings of World Water and Environmental Resources Congress, Frederick, MD, pp. 1–12.

Osman, H., Sultan, A.H., Radif, A., Aziz, S.A., 1993. Large pumps in power and desalination plants. Desalination 93, 181–206.

Peer, T., Joyce, W.S.L., 2002. Lake-source cooling. ASHRAE Journal 44 (4), 37–39.

Peer, T., 2012. Personal correspondence with Matt Mitchell, research assistant. Oklahoma State University. Aug 7.

Peters, T., Pintó, D., 2008. Seawater intake and pre-treatment/brine discharge – environmental issues. Desalination 221, 576–584.

Pezent, M.C., Kavanaugh, S.P., 1990. Development and verification of a thermal model of lakes used with water source heat pumps. ASHRAE Transactions 96 (1), 574–582.

Pietrucha, U., 2012. Email Conversation with Author. Friotherm, Winterthur, Switzerland.

Pugh, S.J., Hewitt, G.F., Müller-Steinhagen, H., 2005. Fouling during the use of seawater as coolant – the development of a user guide. Heat Transfer Engineering 26 (1), 35–43.

Pugh, S.J., Hewitt, G.F., Müller-Steinhagen, H., 2009. Fouling during the use of "Fresh" water as coolant – the development of a "User Guide". Heat Transfer Engineering 30 (10–11), 851–858.

Riley, M.J., Stefan, H.G., 1988. Minlake: a dynamic lake water quality simulation model. Ecological Modelling 43 (3–4), 155–182.

Rogers, G.F.C., Mayhew, Y.R., 1964. Heat transfer and pressure loss in helically coiled tubes with turbulent flow. International Journal of Heat and Mass Transfer 7 (11), 1207–1216.

Saloranta, T.M., Andersen, T., 2004. MyLake (v.1.1): Technical Model Documentation and User's Guide. NIVA-report 4838. Norwegian Institute for Water Research, Olso, Norway, p. 44.

Saloranta, T.M., Andersen, T., 2007. MyLake—A multi-year lake simulation model code suitable for uncertainty and sensitivity analysis simulations. Ecological Modelling 207 (1), 45–60.

Sjöberg, K., 2004. Sjövärme ger effekt. Energi & Miljö, pp. 30–32.

Skanska, 2008. Kings Mill Hospital. Skanska AB, UK.

Smebye, H., Midttømme, K., Stene, J., 2011. Energi fra overflatevann i Norge – kartlegging av økonomisk potensial. Norges vassdrags- og energidirektorat.

Song, Y.-H., Akashi, Y., Yee, J.-J., 2007. Effects of utilizing seawater as a cooling source system in a commercial complex. Energy and Buildings 39 (10), 1080–1087.

Spitler, J.D., Bashyam, K., Hansen, G., Mitchell, M., Selvakumar, M., 2014. Development of Design Tools for Surface Water Heat Pump Systems. Draft Final Report, ASHRAE 1385-RP.

Spitler, J.D., Cullin, J.R., Conjeevaram, K., Ramesh, M., Selvakumar, M., 2012. Improved Design Tools for Surface Water and Standing Column Well Heat Pump Systems (Final Report to the US DoE for project DE-EE0002961). Oklahoma State University.

Stefan, H.G., Dahlin, W.Q., Winterstein, T., Fournier, P., 1986. Passive screen water intake design studies. Journal of Energy Engineering 112 (2), 115–126.

Sumner, J.A., 1948. The Norwich heat pump. Journal of the Institution of Mechanical Engineers 158 (22–29), 39–51 (discussion).

Svensson, T., Sörman, L.-O., 1983a. Pipe heat exchangers on lake bottoms – results of laboratory and field tests. In: International Conference on Subsurface Heat Storage in Theory and Practice, Stockholm, pp. 679–684.

Svensson, T., Sörman, L., 1983b. Värmeupptagning med bottenförlagda slangar i rinnade vatten – Laboratorieförsök. R145:1983.

Svensson, T., Sörman, L., 1982. Värmeupptagning med bottenförlagda kylslangar i stillastående vatten – Laboratorieförsök. Institutionen för Vattenbyggnad, Chalmers Tekniska Högskola. B:36.

TEVAenergy, 2015. High-Efficiency Loop for Geothermal Heat Pumps. Retrieved June 17, 2015, from: http://tevaenergy.com/wp-content/uploads/2012/09/hyperloop_brochure_20120906_web.pdf.

Tsou, J.L., Melville, B.W., Ettema, R., Nakato, T., 1994. Review of flow problems at water intake pump sumps. In: International Joint Power Generation Conference, Phoenix, AZ, vol. 25, pp. 91–102.

Tucker, W.A., Green, A.W., 1977. A time-dependent model of the lake-averaged, vertical temperature distribution of lakes. Limnology and Oceanography 22 (4), 687–699.

Viquerat, P.-A., Lacal, B., Hollmuller, P. (Eds.), 2012. French – Connexion des bâtiments de l'ONU au réseau GLN Analyse de la première année de fonctionnement. Université de Genève, Geneva, Switzerland.

Viquerat, P.A., Lachal, B., Weber, W., Mermoud, A., Pampaloni, E., 2008. Utilization of a deep lake water direct cooling network for cooling a large administrative district, energy and environmental demonstration and follow-up. In: Improving Energy Efficiency in Commercial Buildings, Frankfurt, Germany, 2008, p. 13.

War, J.C., 2011. Seawater Air Conditioning (SWAC) – A Renewable Energy Alternative. Oceans 2011. Kailua-Kona, Hawaii, pp. 1–9. Sept 19–22.

Webb, B.W., Hannah, D.M., Moore, R.D., Brown, L.E., Nobilis, F., 2008. Recent advances in stream and river temperature research. Hydrological Processes 22 (7), 902–918.

Zogg, R., Roth, K., Brodrick, J., 2008. Lake-Source district cooling systems. ASHRAE Journal 50, 55–56.

Open-loop heat pump and thermal energy storage systems

9

A.L. Snijders[1], B.C. Drijver[2]
[1]IFTech International B.V., Arnhem, The Netherlands; [2]IF Technology B.V., Arnhem, The Netherlands

Abstract

This chapter describes the application of open-loop wells to ground-source heat pumps (GSHPs) and to seasonal thermal storage for heating and cooling systems. The application of groundwater wells requires detailed knowledge of the aquifer[1] characteristics. The introduction describes the methods used to obtain these characteristics and the numerical tools available to model the behaviour of the warm and cold groundwater in the aquifer.

The importance of properly integrating the open-loop heat pump and thermal energy storage system into a building's heating and cooling system in order to achieve a good overall energy efficiency is illustrated.

To guarantee trouble-free operation of the groundwater loop of an open-loop system for many years, criteria are given for the design of the wells and the well field, as well as for preventive maintenance.

9.1 Introduction

Ground-source heat pumps (GSHPs), also referred to as geothermal heat pumps, all utilize the subsurface to extract heat in the winter and/or inject heat in the summer. When aquifers are available, groundwater can be applied as a low temperature heat source and sink for the heat pumps.

The advantages of the application of groundwater are the stable temperature of the groundwater (as compared to surface water and outside air) and the effective transfer of heat (as compared to borehole heat exchangers and outside air). By avoiding the need for intermediate antifreeze fluids and the thermal resistance of the borehole heat exchangers found in closed-loop systems, open-loop systems are able to operate at temperatures more closely approaching the ambient ground temperature and consequently have thermodynamic efficiency advantages. Opportunities for direct cooling (ie, bypassing any heat pump) can also be greater in open-loop systems.

A GSHP using water from an aquifer for heat exchange is a form of open-loop GSHP system (surface-water open-loop systems are discussed in chapter 'Surface water heat pump systems'). The principle of operation of this type of system is illustrated

[1] An aquifer is a subsurface geologic feature that is capable to yield large quantities of water (eg, a layer of sand, gravel, sandstone or fractured rock).

Advances in Ground-Source Heat Pump Systems. http://dx.doi.org/10.1016/B978-0-08-100311-4.00009-1
Copyright © 2016 Elsevier Ltd. All rights reserved.

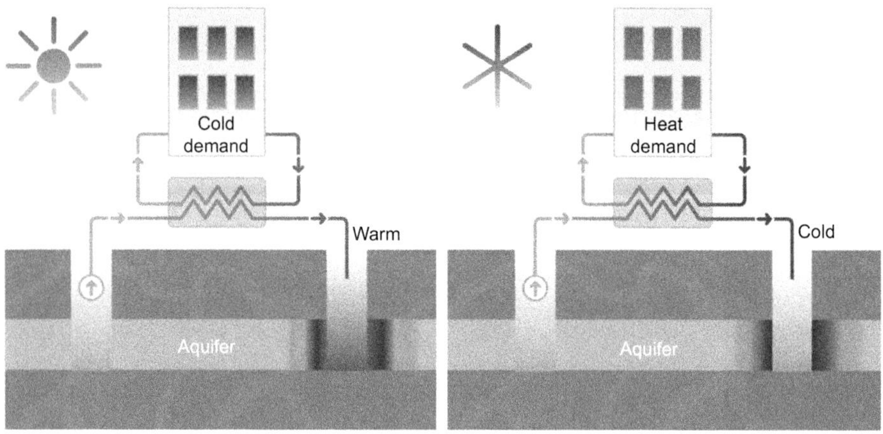

Figure 9.1 Principle of an open-loop ground source heat pump system for heating and cooling.

in Fig. 9.1. One or several groundwater wells supply the water necessary for the heat pump application; a similar number of wells would be used to inject the water. The heat pump(s) extract heat from the water in winter and/or inject heat into the water in summer. The water is typically returned to the same aquifer from which it was extracted. As can be seen from Fig. 9.1, the name 'open-loop' GSHP system is a bit misleading because the groundwater loop is separated from the building loop by a heat exchanger.

There are many single-family and small commercial building applications of open-loop well systems in Asia, Europe and North America. In the United States and Germany, for example, there were about 100,000 and 23,000 open-loop HP systems operational by the end of 2009 (Lund et al., 2010).

The thermal power that can be delivered by a well pair in an open-loop system is relatively high as compared to the thermal power that can be provided from a closed-loop borehole. A borehole heat exchanger typically provides 20–70 W/m (VDI, 2001), so a borehole to a depth of 100 m on average provides 5 kW. A well pair typically has a capacity in the range 10–100 m^3/h, so a well pair with a capacity of 40 m^3/h provides about 230 kW (assuming a temperature difference of 5°C between the extracted and the injected water). This implies that, for this example, one well pair has the same thermal capacity as 46 boreholes to 100 m. The investment to realize the two wells may be far less than the investment to realize the borehole field, resulting in the lowest first cost for the open-loop GSHP system. A smaller site footprint may also result. However, the investment for the wells (including the site investigation) will not decrease proportionally when a smaller thermal capacity is required, and so the scale of the installation is an important consideration.

A variation of the open-loop GSHP system is called aquifer thermal energy storage (ATES). An ATES system is an open-loop GSHP system optimized and operated to realize seasonal thermal storage, ie, by reversing extraction and injection wells seasonally. The basic ATES principle is shown in Fig. 9.2 for an application in which the system is used for both heating and cooling. In summer, groundwater is extracted

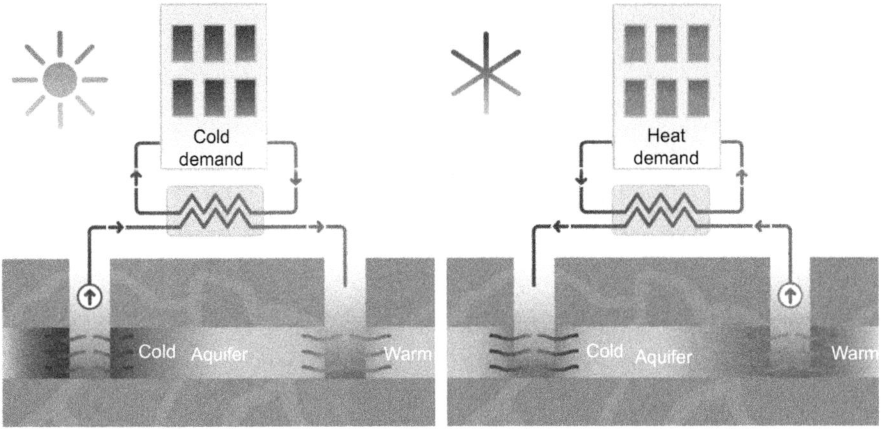

Figure 9.2 Principle of an aquifer thermal energy storage (ATES) system for heating and cooling.

from the cold well(s) and used for cooling purposes. The warmed water is injected into the warm well(s). In winter the process is reversed. Water is pumped from the warm well(s) and applied as low temperature heat source for a heat pump. After the exchange of heat the chilled groundwater is injected into the cold well(s).

It is common practice to size an ATES system to provide (part of) the cooling demand directly, ie, without running the heat pump in chiller mode. This cooling load would be provided with groundwater, which was injected during the winter into the aquifer at a temperature below the ambient aquifer temperature. ATES systems storing heat at higher temperatures, eg, solar heat, which enables heating without a heat pump, are not within the scope of this chapter (borehole thermal energy storage systems intended for higher temperature operation are discussed in chapter 'Borehole thermal energy storage').

In general, an ATES system should be a large-scale system to enable effective seasonal storage and is more dependent on site-specific hydrogeological conditions than closed-loop and other forms of open-loop GSHP systems.

By the end of 2013, the vast majority of the operating ATES systems can be found in Europe (about 3000 systems); mainly in the Netherlands (eg, Sommer, 2015) and Sweden (Andersson and Bjelm, 2013). These ATES systems are applied in larger commercial and institutional buildings as well as for apartment blocks and residential areas (CBS, 2014; Andersson and Bjelm, 2013).

9.2 Site information and modelling

Since open-loop GSHP systems use aquifers for the extraction and injection of groundwater, the main condition for the technical feasibility of an open-loop system is the presence of at least one aquifer. While aquifers are geographically limited in area around the world, they are typically located under large population areas such as the East Coasts of the United States and China and new river deltas such as the Netherlands.

Apart from possible legal restrictions, the suitability of an aquifer depends on the aquifer characteristics. When the information on the aquifer characteristics is insufficient, site investigation may be needed to be able to assess the feasibility of an open-loop system accurately. Subsequently, modelling tools can be used for improved insight in the feasibility, the design of the system and calculations of the environmental impact.

9.2.1 Aquifer characteristics

Aquifers that can be utilized by open-loop systems are limited to deposits of sands and gravels, sandstones and highly fractured or karstified carbonate rocks. For an economically feasible project the aquifer has a relatively large thickness and high hydraulic conductivity and the top of the aquifer is preferably within about 50 m (small scale projects) or 200 m (large-scale projects) from the surface. Furthermore the groundwater flow and groundwater composition are important factors. Other factors affecting ATES design are stratigraphy, grain size distribution, degree of consolidation, natural temperature, static head, groundwater flow direction, geological structures and fracture distribution, aquifer geometry, storativity and leakage factor of confining layers. Table 9.1 summarizes the range of parameters normally required for successful application of an open-loop GSHP system in an unconsolidated aquifer.

9.2.1.1 Groundwater flow

For ATES systems the regional groundwater flow velocity should be limited in order to prevent significant losses of heat and/or cold from the thermal store. However, for

Table 9.1 Criteria for open-loop ground-source heat pumps and aquifer thermal energy storage systems in unconsolidated aquifers

	Lower limit	Typical	Upper limit
Depth of the well(s) (m)	10 (injection pressure)	30–100	300 (economic feasibility)
Aquifer thickness (m)	2–5	20–50	None (partial use)
Hydraulic conductivity (m/s)	3×10^{-5}	1×10^{-4}–5×10^{-4}	1×10^{-3}/none[a]
Groundwater flow velocity (m/d)	0	0.03–0.15	0.3/none[a]
Static head (m below ground surface)	50	2–20	–5

[a]For aquifer thermal energy storage a high value usually results in a high groundwater flow velocity and a low storage efficiency as a consequence. For open-loop systems a high groundwater flow velocity is favourable.

open-loop GSHP systems that dispose more cold than heat (or vice versa) a low groundwater flow velocity increases the risk of thermal breakthrough: after a certain time (mainly depending on the distance between both wells, the pumped volumes per year and the aquifer thickness) the waste heat or waste cold can reach the extraction well, leading to a less favourable extraction temperature.

When thermal breakthrough occurs and the injection temperature cannot be (equally) adjusted, the temperature difference between the extraction and injection well will be reduced. As a consequence the amount of water that needs to be pumped to provide the same amount of energy increases, which causes the thermal shortcutting to increase even further. In the end this will lead to a loss of heating or cooling capacity (Banks, 2009).

One way to prevent disadvantageous effects is to restore the energy balance by injecting extra heat (or cold) in the summer (or winter) season. The energy balance can also be restored by providing less heat (or cold) than the total building demands and providing the remainder in another way. When the distance between both wells is increased and the injection well is placed downstream from the extraction well, the risk of thermal shortcutting is reduced and it will take more time before thermal shortcutting occurs. A high groundwater flow velocity is therefore favourable for open-loop systems.

9.2.1.2 Groundwater composition

Due to different geochemical processes in combination with groundwater flow, the groundwater composition in an aquifer can vary both in vertical and in horizontal direction. The most common zoning occurs in vertical direction and is caused by redox processes. The well screens of the open-loop systems extract water from a certain vertical depth interval and can also attract groundwater from other depths in the aquifer. When different types of groundwater are mixed, this can cause chemical reactions that lead to well clogging. This is a particular risk where mixing of groundwater that contains oxygen and/or nitrate with groundwater that contains dissolved iron can cause serious well clogging problems by iron hydroxide precipitates. This type of well clogging is well known in water wells used for extraction (eg, Houben and Treskatis, 2007). In comparison to extraction wells, injection wells are much more vulnerable to this type of well clogging because the injection of water with solid particles (consisting of the reaction products that are formed in the mixed water) is notoriously problematic.

9.2.2 Site investigation

Proper insight into the aquifer characteristics is needed for a realistic assessment of the feasibility, design and environmental impact of an open-loop system. Investigating a site usually starts with an inventory of existing information. When existing information is not sufficient, the missing information will have to be gathered on site. The method depends on the type of information that is missing. A number of methods are given in this section.

9.2.2.1 Test boring

A test boring is performed to the depth of interest and provides information on the hydrogeological sequence that is present. By installing piezometers at different depths in the borehole, the groundwater composition at different depths can be investigated. When the aquifer hydraulic properties are needed, a well screen can be installed that enables a single well test or a pumping test. In order to obtain reliable ground samples, reverse rotary, percussion and sonic drilling are good options. A borehole log is sometimes used to acquire more accurate information on the soil sequence (eg, in combination with rotary drilling) and on the depth of the freshwater–saltwater transition zone, if any.

9.2.2.2 Pumping test

A single well test or pumping test is standard procedure to acquire information on the average hydraulic conductivity of an aquifer (or part of an aquifer). A well test can be useful for the design of the system when the capacity of a well cannot be estimated accurately and the number of wells that are needed is uncertain because of that. For the interpretation of these tests, analytical solutions (Kruseman and de Ridder, 1994) and specialized software (eg, MLU for Windows or AQTESOLV) are available.

9.2.2.3 Cone penetration test

In unconsolidated formations, cone penetration tests can be used to get information on the depth of sand, silt and clay layers. With special cone penetration test equipment it is also possible to sample the groundwater at different depths. The depth that can be reached is usually no more than 30–50 m. Another limitation is that gravel layers cannot be penetrated.

9.2.2.4 Sampling existing piezometers

This can be useful when the information on the groundwater composition or groundwater flow rate/direction is insufficient provided that existing piezometers are present at the right depth and distance from the location.

9.2.3 Modelling tools

Modelling tools are used in various stages of an open-loop GSHP project to examine both energy performance and hydrogeological behaviour. A vital part is the estimation of the energy demand of the building—as it is in all GSHP applications. In combination with the energy concept chosen, the energy demand of the building determines the overall requirements for the open-loop system. The feasibility, design and environmental impact of the open-loop system will be based on this information.

Advanced building simulation software is available to make the best possible estimate of the heating and cooling demands (eg, TRNSYS). A common problem is that the results of the simulation may be unreliable when part of the input for the simulations is uncertain. In building simulations, this uncertainty may be concerned with the information on the operation of the building (eg, the internal heat load and the indoor climate requirements) or the behaviour of the user (eg, room temperature is set higher, windows are left open or sun shading is not used). This has been commented on elsewhere and is a recurring issue in GSHP system design (eg, chapter: Horizontal and compact ground heat exchangers and Hoes et al., 2009).

When the energy demand and the underground (aquifers and aquitards) have been characterized (Section 9.2.1), hydrogeological and hydrogeochemical modelling tools are used for the design of the system and for calculating the environmental impact. Such modelling tools enable assessment of the injection pressures (relevant for well design), distance between the wells and well field layout (well field design), and the hydraulic and thermal impact of the GSHP system on the underground (environmental impact).

For modelling of the hydraulic impact of open-loop systems, well-established numerical groundwater modelling software like Feflow, MicroFEM and MODFLOW can be used. Since open-loop systems extract and inject the same amount of water, there is no net extraction or injection. As a consequence, the hydraulic impact is relatively small in comparison to groundwater extraction without reinjection and the area of influence is limited. In many cases, assuming constant hydrogeological properties in the area of influence is a reasonable assumption. This means that relatively simple analytical groundwater modelling software can be used (eg, MLU for Windows).

A common limitation of groundwater modelling software is that the influence of temperature on the hydraulic conductivity (caused by the temperature dependency of the viscosity) is not included. For low-temperature open-loop systems (injection temperatures generally less than 10°C above or below the ambient groundwater temperature) the error is very small, but for high-temperature ATES systems the error can be significant.

Hydrothermal transport models do incorporate the temperature dependency of the hydraulic conductivity. These models are necessary to calculate the temperature and thermal energy of the water extracted, to ensure that thermal breakthrough does not occur and to assess the thermal impact on existing groundwater users. Examples of hydrothermal transport models are METROPOL, MOCDENSE, NAMMU, SUTRA (Konikow et al., 1997), FEFLOW (Diersch, 1998), SEAWAT and HST2D/3D (Kipp, 1997).

In summary, it is important that modelling tools are used to optimize the design of the open-loop system. In doing so, problems in later stages can be minimized. The reliability of the modelling results strongly depends on the reliability of the model input (in combination with possible limitations of the software).

9.3 Design and construction

9.3.1 Well design

Since open-loop systems are meant to operate over several decades, it is very important that the performance of the wells can be sustained over a long period of time.

Therefore, the design of the well should enable a long lifetime with limited maintenance. Much of the knowledge on well design has been derived from research in the drinking water industry.

9.3.1.1 Unconsolidated sediments

Well design in unconsolidated sediments is focused on the prevention of sand production and well clogging. For ATES systems, guidelines have been developed that limit the flow velocity in the aquifer around the well (NVOE, 2006). These guidelines apply also for open-loop GSHP systems. For extraction wells, these guidelines are meant to minimize the production of fine particles (fine sand, silt, clay) that could cause well clogging (eg, van Beek, 2010; de Zwart, 2007).

The guidelines for injection wells (Buik and Willemsen, 2002) enable prediction of the clogging rate of infiltration wells based on the membrane filtration index (MFI: indication of the clogging potential of the water), the well dimensions (screen length and diameter), the hydraulic conductivity of the aquifer and the well loads. The fines that move from the production well are collected in the gravel pack of the injection well. To maintain low injection resistance and low pumping energy in the operational phase, these fines have to be removed regularly by back-flushing the well (at least each half year). A smaller diameter well (higher flow velocity in the aquifer) will result in a higher clogging rate and thus higher maintenance cost. This implies that economic criteria, such as total cost of ownership, determine the design of infiltration wells. When a well is used for both extraction and injection, the design criterion that yields the lowest flow velocity determines the well capacity.

Apart from the maximum flow velocity during extraction and injection, the maximum injection pressure is the third criterion that has to be met. When the injection pressure is too high, hydraulic fracturing may occur and (part of) the water that is injected into the well may find a path to the surface instead of going into the aquifer. Since the maximum allowable injection pressure increases with the depth of the aquifer, this criterion is especially relevant for shallow aquifers. More detailed information on well design in unconsolidated aquifers can be found in Pyne (2005). The design criteria for open-loop systems result in much larger diameter wells than would normally be expected for a standard water supply well.

9.3.1.2 Rock aquifers

In rock aquifers (sandstones, carbonate rocks) the grains are cemented and can therefore not easily be mobilized. When the cementation is fully developed, the production of fines will be limited. In that case there is no need to limit the flow velocity in the formation like in unconsolidated aquifers. Furthermore, the rock will have extra strength in comparison to unconsolidated sediments, which means that the maximum allowable injection pressure increases significantly. In well-cemented rock aquifers, well design is therefore based on limitations to the drawdown in the production well or the energy efficiency of the groundwater loop, ie, the ratio between thermal energy transported and pump electricity required. In friable sandstone aquifers, prevention of the production of fines becomes more important and the design standards shift towards those of unconsolidated aquifers.

9.3.2 Well field design

The optimal well field layout depends on many issues, and in real life the well field layout is often determined by the possibilities the site offers. Well field design is meant to minimize (1) thermal interference between the production and infiltration wells and (2) undesirable hydraulic and thermal impact in the surrounding area.

9.3.2.1 Thermal interference

As discussed in Section 9.2.1 on aquifer characteristics, thermal breakthrough will lead to unfavourable impact on the extraction temperature of the extraction well. For ATES systems that store and produce equal amounts of heat and cold over an annual cycle (ie, the energy exchange is well balanced), the NVOE guidelines (2006) state that a minimum distance between a warm and a cold well of three times the thermal radius of the stored cold or heat is sufficient to minimize thermal interference (Fig. 9.3). The thermal radius r_{th} can be calculated from

$$r_{th} = \sqrt{\frac{c_W Q}{c_a H \pi}}$$

where: r_{th} = thermal radius of the stored cold or heat (m); c_W, c_a = heat capacity of water and aquifer (water and solid part) (J/(m^3 °C)]; Q = amount of pumped/injected water per season (m^3); H = Length of the well screen (m).

When the distance between the warm and the cold well is decreased, thermal losses due to thermal interference will increase (eg, Sommer et al., 2013). In case the temperatures of both the warm and cold wells are on one side of the natural groundwater temperature (both above or below the natural temperature), some short-circuiting will increase the thermal efficiency. In that case, the distance between the wells can be decreased to around 1–2 times the thermal radius (Pyne, 2005).

For ATES systems that do not have a thermal balance, either the warm or the cold bubble will increase in size year after year (Fig. 9.4). To prevent thermal breakthrough in the long run (eg, over a 20-year period), the distance between the warm and cold

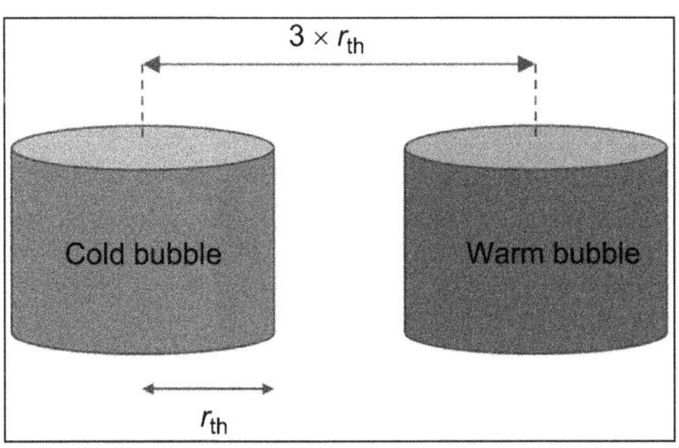

Figure 9.3 According to the guidelines of the NVOE (2006), the distance between the warm and the cold well of an aquifer thermal energy storage (ATES) system (thermally balanced) should be three times the thermal radius.

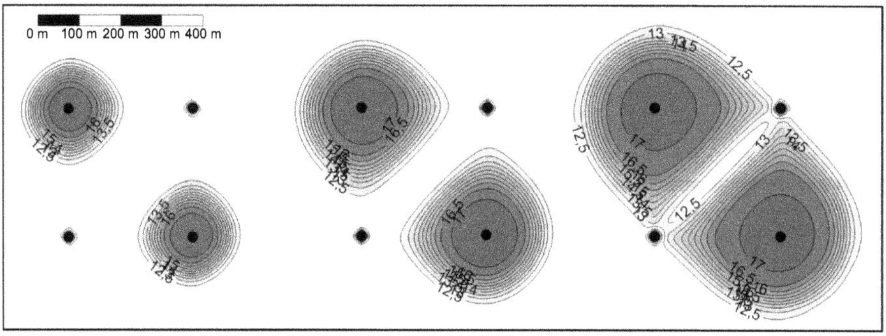

Figure 9.4 Calculated temperatures at the end of the winter season after (from left to right) 10, 20 and 30 years for an unbalanced aquifer thermal energy storage (ATES) system (underground heat surplus) in an aquifer without groundwater flow.

wells has to be increased. The extra distance that is needed will depend on the degree of energy imbalance, the groundwater flow velocity (Fig. 9.5) and the orientation of the wells with respect to the direction of groundwater flow.

For one-directional open-loop systems, the water from the injection well can reach the extraction well after some time. When the system injects both warm and cold water, there will be an alternation of warmer and colder zones close to the injection well (see Fig. 9.1). Further away from the injection well the temperature differences will fade away, resulting in the average injection temperature. For a thermally balanced system, the average injection temperature will be equal to the ambient temperature. This means that there is little or no impact on the extraction temperature when the injection water reaches the production well. In case of a thermally unbalanced system, the extraction temperature can be adversely affected. If, on an annual basis, more heat is extracted than injected, the extraction temperature will decrease when the injection water reaches the production well. This phenomenon may limit the lifetime of the

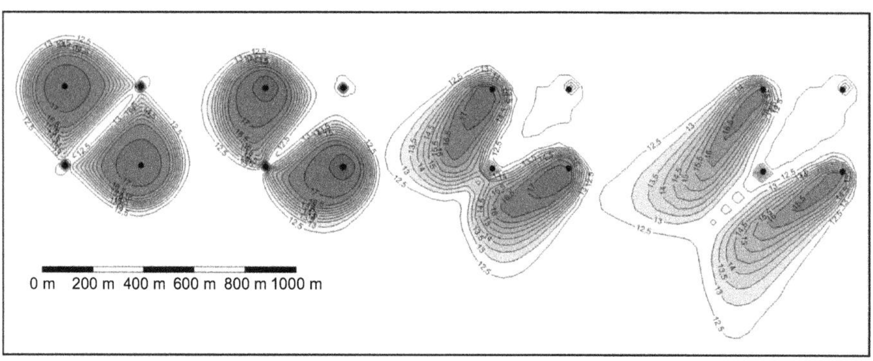

Figure 9.5 Calculated temperatures at the end of the winter season after 30 years for an unbalanced aquifer thermal energy storage (ATES) system (underground heat surplus) in an aquifer with a groundwater flow velocity of (from left to right) 0, 10, 20 and 50 m/y.

open-loop system, which means that these systems are not automatically sustainable (Banks, 2009). Possible measures to prevent (or reduce) thermal breakthrough are to increase the distance between the wells and placing the injection well downstream from the production well. Analytical solutions are available for a first estimate of the most optimal well positions (Banks, 2009). For a more detailed well field design, modelling tools are used.

9.3.2.2 Hydraulic impact

According to the superposition principle, the hydraulic impact of the production well (decrease in hydraulic head) and the injection well (increase in hydraulic head) will partly compensate each other. In the extreme case of a production well and an injection well at the same location and depth, the net effect will be zero. When the distance between the production well and the injection well is infinite, there is no compensating effect and the hydraulic impact will be equal to that of a single well. In reality the wells are at some distance apart and there will be a certain degree of compensation. The smaller the distance between the wells, the stronger the compensation will be. This principle can be used to minimize the area of hydraulic influence during well field design.

Fig. 9.6 shows the difference in the calculated hydraulic head changes between an extraction without reinjection and with reinjection of the extracted water. When an open-loop system consists of several doublets, there are several options for the well configuration. Fig. 9.7 shows the hydraulic impact for three well field layout options of a system consisting of four extraction and four injection wells. In Option A, all production wells are grouped in one cluster (right-hand side) and all injection wells in the other cluster. This option results in a relatively large hydraulic impact in the surrounding area in comparison to Option B (alternating single wells) and Option C (two clusters of two production wells and two clusters of two injection wells). Option A has the lowest investment cost due to the limited length of the piping that connects the wells. In Options B and C the length of the connecting piping increases (higher investment cost). These options can be used when the hydraulic impact of Option A is unacceptable or undesirable.

9.3.2.3 Thermal impact

The temperature changes that occur in the surrounding area can be relevant for existing groundwater users, especially for other open-loop systems. Minimizing thermal interference is not only relevant on the scale of one project, but also on the scale of an area with several projects. In many cities in The Netherlands there are areas where a large number of ATES systems are present. Fig. 9.8 shows an example of the upper aquifer in the city centre of The Hague, where 16 ATES systems are located in an area of 3 km^2. On an area scale, clusters of (up to four) warm and cold wells are alternated. A modelling study on the interference between the warm and cold wells in this area (Bakr et al., 2013) showed that the storage efficiency of the ATES wells can both increase and decrease due to interference with other ATES wells. Positive thermal

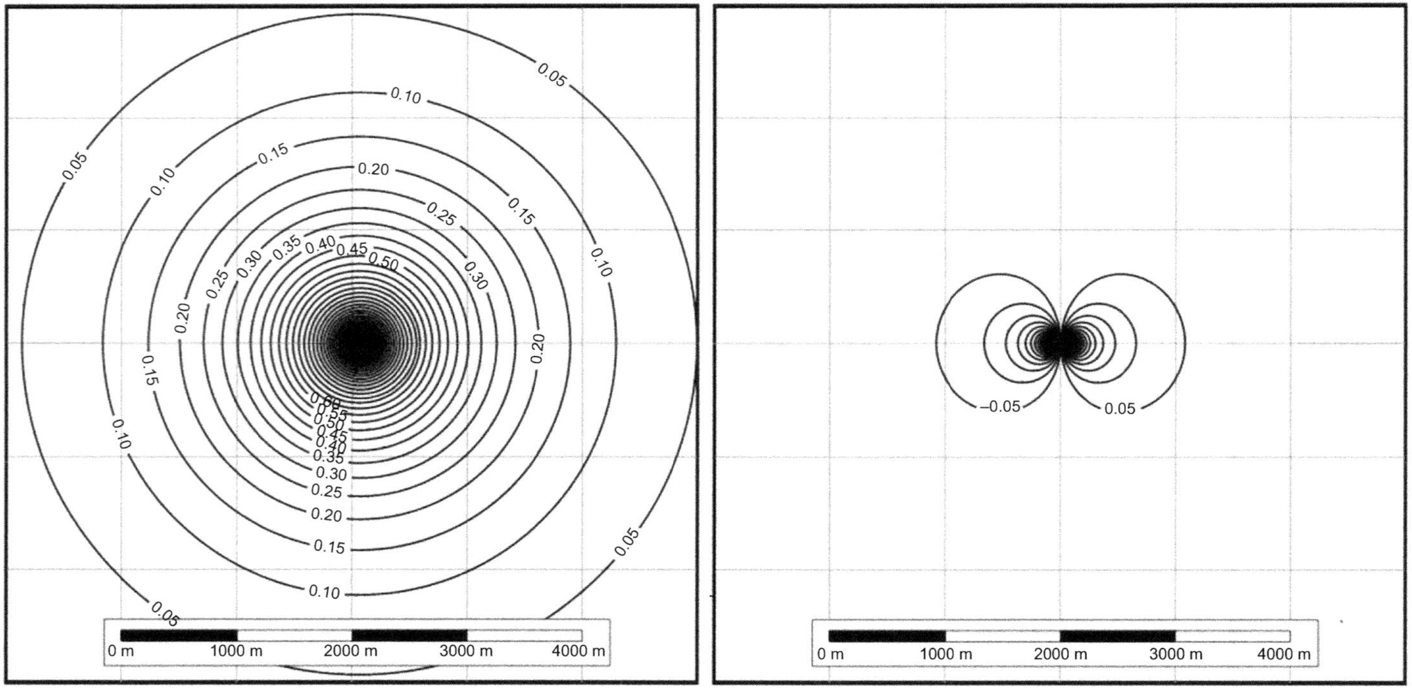

Figure 9.6 Calculated drawdown in the aquifer used for groundwater extraction (0.05 m contours) without reinjection of the extracted water (left) and with reinjection (right). The well configuration on the right side is equal to Option A from Fig. 9.7. Groundwater extraction 330 m^3/h.

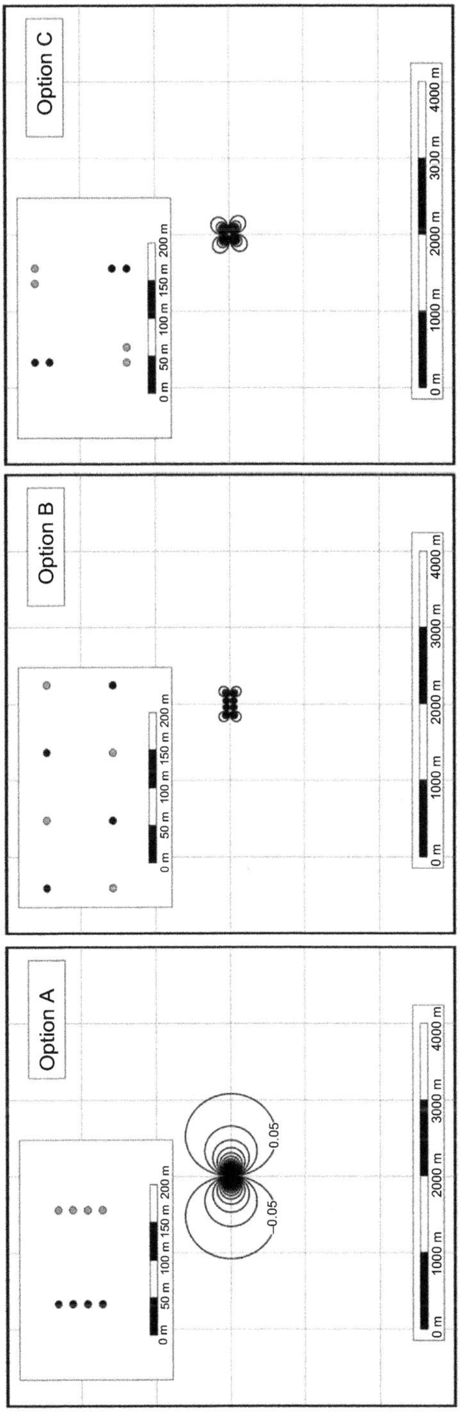

Figure 9.7 Calculated drawdown in the aquifer used by the open-loop system (0.05 m contours) for three different options of the well field layout. Groundwater extraction 330 m^3/h.

Figure 9.8 Calculated temperatures (°C) after 20 years of use for the aquifer thermal energy storage (ATES) systems in the upper aquifer in the city centre of The Hague, The Netherlands.

interference occurs when a cold well is placed near another cold well and negative interference occurs when a cold well is placed close to, or downstream from, a warm well. In the area shown in Fig. 9.8 the wells have been placed in such a way that the average effect on the storage efficiency of the wells is positive. Because wells in clusters of several 'equal wells' will automatically experience a certain positive influence of the other wells, this conclusion does not necessarily mean that the average thermal impact between the ATES systems is also positive. However, since all systems have been granted a permit, undesirable interference was apparently not expected.

In areas where many open-loop systems are installed the availability of underground aquifer space can be a limiting factor. When this is expected in advance, it can be considered to reduce the distance between the production and injection wells (Sommer et al., 2015). This will increase the amount of thermal interference but will also increase the amount of systems that can be implemented in the same area. Disadvantages of reducing the well distances might be that this leaves little or no room for increasing the pumped volumes (if that appears to be necessary later on), and extra interference might occur when the thermal balance shifts from year to year and the impact of aquifer heterogeneity on thermal performance will be larger.

In the city centre of The Hague there are areas where it is not possible to add another ATES system in the upper aquifer without adversely affecting an existing system. However, in The Hague a second and a third deeper aquifer are available. Due to a change in legislation, use of the first aquifer is currently not allowed anymore and therefore new ATES systems use the combined second and third aquifer.

9.3.3 Well drilling, completion and development

9.3.3.1 Well drilling

The aim of well drilling, completion and development is creating a well with a good specific capacity that produces a minimum amount of fines and is not prone to clogging. The first step is to use a drilling technique that minimizes the intrusion of drilling mud into the aquifer and enables obtaining reliable soil samples needed for careful selection of the depth interval to use. In the Netherlands, reverse rotary drilling is the industry standard to meet this requirement. Since the drilling phase determines the chance of success and the effort needed for well development, it is important to minimize the amount of drilling mud used and avoid the use of a type of mud that is hard to remove afterwards (for that reason, the use of bentonite as a drilling mud additive for open-loop systems should be avoided). Similarly in rock drilling, care is required to prevent closing off of the pores or fractures.

9.3.3.2 Well completion

In most wells, a well screen and well riser is installed and the well screen is gravel packed. In stable rock aquifers, a casing can be installed that reaches the top of the aquifer and the rest of the borehole can be left open. The thickness of the gravel pack varies between a minimum of 100 mm to a maximum of 300 mm. The grain size of the gravel and the size of the screen slots has to be adjusted to the grain size of the screened aquifer, or (the other way around) the layers that are selected for screening should not be too fine considering the grain size of the gravel pack (Sterrett, 2007). To prevent mixing of different types of groundwater and to be sure the water is extracted and injected at the depth selected, low permeability layers (eg, clay and loam) that were perforated have to be sealed during backfilling.

9.3.3.3 Well development

During well drilling, drilling mud (or fines from overlying ground layers) can penetrate the part of the formation adjacent to the borehole. A variety of well development techniques is used to remove these mud remnants and the fines present in the aquifer. An overview of well development techniques is available in the literature (eg, Sterrett, 2007; Houben and Treskatis, 2007). Since water has to be infiltrated and extracted for many years without significant clogging, there are a number of strict requirements that the drilling company has to meet: a specific well capacity of at least 80% of the theoretical value, an MFI value below 1–2 s/L^2 and less than 0.01–0.1 mg/L of suspended solids (fines) in the produced water. The effort required for the development of wells for open-loop geothermal systems (and the associated cost), is considerably higher than for wells that are only used for production. A couple of weeks of development time is not exceptional. Even after this effort, there are many examples of wells (extraction or dual purpose) that show improvement of the specific capacity after some time of use. Apparently, parts of the drilling mud remnants were not removed during well development and were mobilized during the operational phase. During preventive maintenance in the operational phase, these mobilized fines can be removed by extracting groundwater at maximum capacity from the injection well for a short period of time and disposal of the extracted water and the suspended fines.

In coarse grained ($D_{10} > 0.25$ mm), poorly sorted aquifers ($D_{60}/D_{10} > 3$) the development of a natural gravel pack may be the best option (Misstear et al., 2006). A natural gravel pack is created by removing the fine fraction from the formation close to the well during well development. The remaining course fraction of the formation is left behind, creating a natural gravel pack around the well screen. In this type of well, it may be needed to repeat well development actions several times during the operational phase when the amount of fines in the extracted water increases too much.

9.3.4 Groundwater loop design

When well development is completed, a number of components have to be installed in the well (Fig. 9.9) and the wells have to be connected to the plant room in the building. In the plant room the heat and cold is transferred from the groundwater loop to the building loop. A number of requirements for the groundwater loop are given below.

9.3.4.1 Air tightness

When the open-loop system pumps reduced, iron- and/or manganese-containing water, it is very important to prevent entry of oxygen. When oxygen enters the system, it will react with the iron and/or manganese in the groundwater. The hydroxides that are formed attach to the inside of the pipes and the heat exchanger and clog the heat exchanger(s) and the injection well(s). It is therefore essential to keep the open-loop system perfectly air tight and inject the water below the water table in the injection well.

9.3.4.2 Flow control

If groundwater contains high concentrations of dissolved gas, the pressure reduction that occurs during upward transport of the water can result in the formation of gas bubbles that can rapidly clog the infiltration well. Maintaining sufficient overpressure will prevent degassing and the associated risk of this type of well clogging. This is achieved with a pressure-controlled automatic valve located in the infiltration well. Variable speed pumps are useful to match load conditions and to ensure water is sufficiently chilled or heated. In this way, the pumping energy is minimized and the amount of energy produced from a given volume of water pumped is maximized.

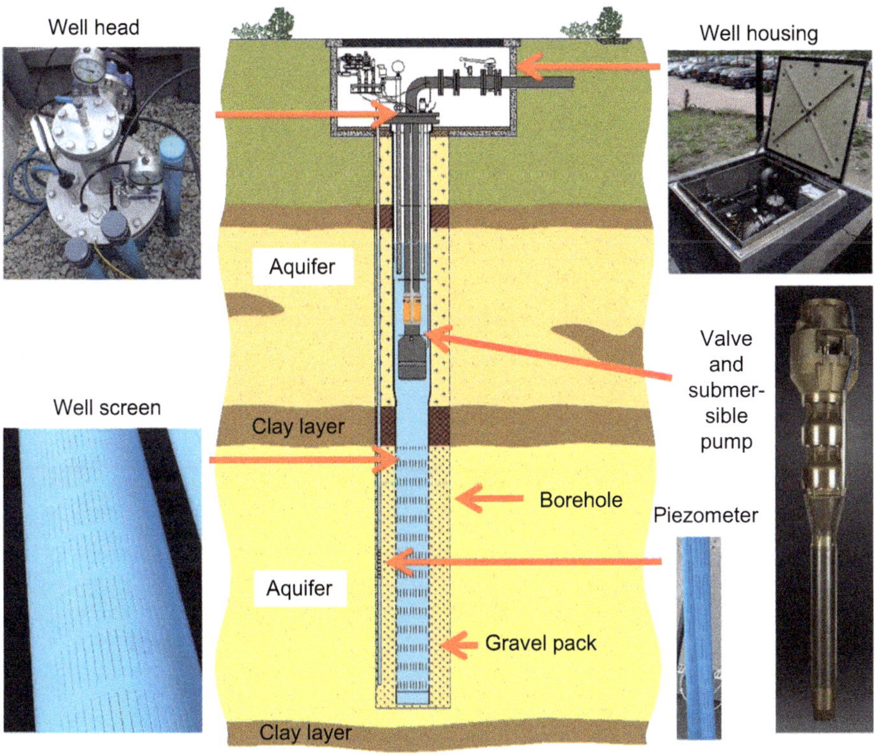

Figure 9.9 Schematic of a well and the associated components.

However, maintaining sufficient overpressure in the groundwater loop will be a greater challenge.

9.3.4.3 Choice of materials

To prevent corrosion, the material selection has to take into account the groundwater composition. The main aspects that influence the required corrosion resistance are salinity and the presence of oxygen. Many open-loop and ATES systems that apply low injection temperatures (below 30°C) use PVC for the well screens and well risers and plastic (mainly polyethylene) for the connecting piping from wells to plant room and vice versa. In general, stainless steel is used for the wellheads and for the piping in the well housing and around the heat exchanger. In fresh water without oxygen, generally grade SS304 is applied, also for the heat exchanger. In case the groundwater is salty or brackish (and contains no oxygen), grade SS316 is used. In case of salty/brackish groundwater also containing oxygen, plastic is the preferred choice in combination with a titanium alloy for the heat exchanger. Use of other metals like carbon steel has led to corrosion problems (Pyne, 2005).

9.4 System operation

9.4.1 System efficiency

The well pump is by far the major electricity consumer in the groundwater loop of an open-loop GSHP or ATES heat pump (ATES-HP) system. The electric power required by the well pump is given by

$$P_e = \frac{\Phi \cdot \Delta p}{\eta}$$

where: P_e = Electrical power pump (W); Φ = Flow rate groundwater in groundwater loop (m^3/s); Δp = Pressure difference provided by the pump (Pa); η = Combined pump and motor efficiency (-).

Due to construction limitations, the efficiency of an in-line deep well pump is relatively low (in the range 0.60–0.70) as compared to a high efficiency circulation pump. To improve the overall energy efficiency of the groundwater loop, it is accordingly necessary to minimize the pressure difference provided by the pump. However, the possibilities to realize this are rather limited because part of the pump pressure required is more or less fixed, especially the static head from groundwater level to surface and the drawdown in the production well.

The other important factor influencing the pump energy demand is the loop flow rate. The groundwater flow rate can be reduced significantly by:

- Increasing the temperature difference between groundwater extracted and groundwater injected. This is not only a groundwater loop issue, but also a function of the temperatures and flow rates in the building loop. So the groundwater loop efficiency cannot be fully optimized without taking into account the interaction with the building system.
- Maintaining the temperature difference in the groundwater loop for building partial load conditions. This can be achieved by applying variable speed well pumps (in combination with pressure maintaining valves in the infiltration wells; see Section 9.3.4).

To have only little effect on the overall system energy efficiency, the ratio between thermal energy provided by (or absorbed by) the groundwater loop and the electricity required by the well pump(s) should be in the range of 30–40. This implies that providing 30–40 kWh of thermal energy requires one kWh of electricity.

To optimize the energy performance of an ATES-HP system, the interaction between groundwater loop and building loop is even more important. This is illustrated in Fig. 9.10, showing the relevant temperatures for an ATES-HP system in cooling mode. For this example it is assumed that:

- The natural ambient groundwater temperature in the aquifer is 14.0°C.
- The supply temperature of the building cooling system is 8.0°C and constant.
- The return temperature of the building cooling system is 16.0°C and constant.
- Charging the cold well(s) in winter is at a constant temperature of 7.0°C.
- The temperature difference over the heat exchanger between groundwater loop and building loop is 1.0°C.

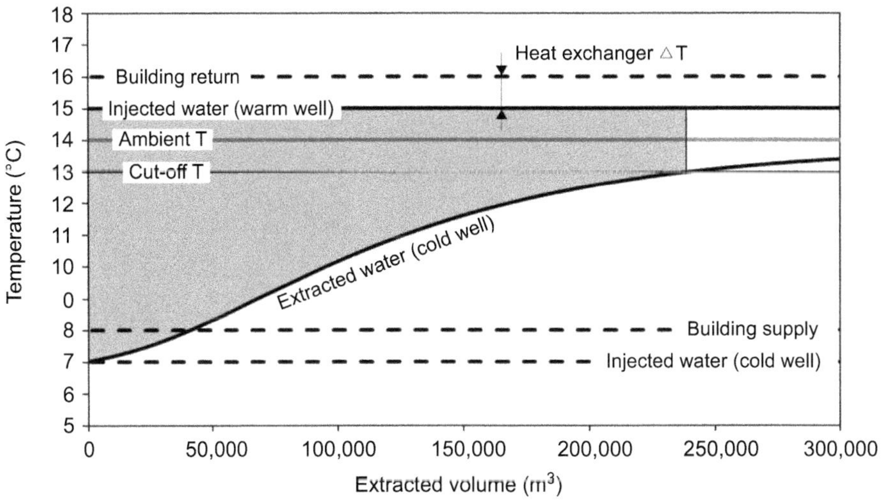

Figure 9.10 Example of the evolution of the discharge temperature for an aquifer thermal energy storage heat pump (ATES-HP) system in cooling mode.

The extraction temperature (discharging temperature) from the cold well(s) will increase slowly during the summer season as a result of thermal losses in the underground and eventually go to the ambient groundwater temperature. As long as this temperature is well below the building return temperature, part of the cooling can be provided directly from the groundwater loop (*dashed area* in Fig. 9.10), which is very energy efficient due to the low electricity consumption. The remainder of the cooling should be provided by the heat pump in cooling mode.

Fig. 9.10 shows that direct cooling is stopped when the abstraction temperature exceeds 13.0°C (the so-called cut-off temperature). This implies that the return water in the building system still might be cooled down 2.0°C, taking into account the 1.0°C temperature drop over the heat exchanger.

As can be seen from this example the optimization of the integral ATES-HP system is important for the performance of the ATES loop. In particular, the charging temperature of the cold well(s) in winter and the building loop return temperature in summer are strongly influencing the ATES thermal efficiency in cooling mode by influencing the ratio between direct ATES cooling and heat pump cooling. After the low temperature groundwater has been charged into the cold well(s), the return temperature from the cold well(s) is determined by the aquifer characteristics, the storage period and the amount of water that has been extracted.

9.4.1.1 Monitoring

Monitoring an open-loop GSHP system may have various objectives. First, monitoring may be required as part of the groundwater abstraction and discharge licence or consent. In general, this monitoring is restricted to drawdowns in the wells and the water and energy flows in the groundwater loop (eg, m^3/month and MWh/month) and focused on the impacts of the open-loop system on the aquifer (and via the aquifer on nearby interests).

Second, monitoring can be focused on preventing/scheduling maintenance. For the groundwater loop, the major issues are detecting a leakage in the loop and deterioration of a well at an early stage. A leakage is detected from pressure loss in the loop during standstill periods. A good way to monitor the well quality is by trending the specific well capacity, ie, yield (m^3/h) divided by drawdown (m), as derived from monitoring results of the groundwater flow rate and the water level in the well. An early warning system of potential well clogging is important, since the clogging process accelerates over time. When a clogged well is redeveloped in time, it can be used for at least several decades. However, when a well has been severely clogged once, it will keep clogging rapidly after each cleaning action.

Finally, monitoring data can be used to determine and optimize the energy performance of the open-loop system. Because of the interaction with the building part of the system, this will require monitoring data from the building system also. For this purpose, the water and energy flows, the supply and return temperatures, as well as the electricity consumption of the heat pump and circulation pumps need to be monitored with a five or ten minute interval for the groundwater loop and the building heating and cooling loops.

In practice, the actual load and/or energy demand of the building often turns out to be (significantly) different from the values available in the design stage. This may negatively influence the thermal efficiency of the open-loop GSHP system:

- When the actual thermal load is lower than designed for, the GSHP system will have more part load operation, resulting in a higher electricity consumption and increased thermal losses from transport piping per MWh delivered.
- When the thermal balance in the underground is significantly changed due to the different heating and/or cooling demand of the building, provisions might be required to restore the balance in order to avoid interference between extraction and injection wells.

Furthermore, the return temperature from the building loop might turn out to be higher (heating mode) or lower (cooling mode) than designed for. This will result in additional electricity consumption for transportation pumps (including well pumps) and, in case of an ATES-HP system, in a reduction of the direct cooling contribution.

Performance monitoring and optimization of larger open-loop GSHP systems, at least during the warranty period, is consequently important to be able to deliver the energy efficiency that the system was designed for.

9.4.1.2 Maintenance

In order to guarantee trouble-free operation of the groundwater loop of an open-loop GSHP system for many years, preventive maintenance will be required. This preventive maintenance will require one to two site visits a year, depending on the level of remote monitoring. The focus of the preventive maintenance is on avoiding well deterioration and well pump failure.

The preventive maintenance of the groundwater loop includes:

- Visual inspection of well heads, valves, transmitters and heat exchangers on leakage and corrosion;
- Removing dirt from well housings;
- Checking the state (deterioration) of the well pumps by measuring the electrical resistance between the phases;
- Twice a year back-flushing of each of the injection wells at maximal capacity for about 1 h per well to remove collected fines;
- Assessment of the proper functioning of the well pump frequency drives, the valves, the sensors and indicators (temperature, pressure and flow), the safeguards, as well as the control unit.

Back-flushing is scheduled preferably near the end of the winter season and the end of the summer season. To remove the fines from the system, the groundwater extracted during back-flushing is not returned to the aquifer, but discharged to, for example, the sewer. During back-flushing the specific capacity of the well is also assessed. This action can be performed automatically or from remote using the GSHP control system.

9.5 Evaluation

This chapter has described the application of open-loop wells to GSHP systems and ATES systems for heating and cooling. Where suitable aquifers can be exploited, open-loop GSHP systems and ATES systems can have significant advantages over other forms of heating and cooling and may be more economical than closed-loop approaches. ATES technology has been demonstrated to be a technically and economically viable form of seasonal energy storage in many large-scale projects.

The requirements for proper aquifer characterization and design, implementation and operation of the well field have been presented in this chapter along with examples of projects.

Open-loop heat sources/sinks have formed the basis of a wide variety of hybrid system concepts. These may include other forms of ground heat exchange or alternative heating and cooling sources and novel system configurations are continuing to be developed. As long-term sustainability and economic viability of these forms of systems has been demonstrated in recent decades, application of the technology continues to grow. Developments in the marketplace that support the deployment of the open-loop technology include:

- Government policies to mitigate CO_2 emissions.
- Growing recognition of the role thermal storage can play in sustainable energy networks and, as an alternative to electrical storage, in smart grids.
- Utility companies and Energy service companies owning and operating GSHP and ATES systems, reducing the barriers for application of these systems at larger developments with more than one building owner.
- Private companies standardizing the most common GSHP and ATES systems by developing heat pump skids, standardized controls, etc.

References

Andersson, O., Bjelm, L., 2013. Geothermal energy use, country update for Sweden. In: European Geothermal Congress 2013, Pisa, Italy.

Bakr, M., van Oostrom, N., Sommer, W.T., 2013. Efficiency of and interference among multiple aquifer thermal energy storage systems; a Dutch case study. Renewable Energy 60, 53–62.

Banks, D., 2009. Thermogeological assessment of open-loop well-doublet schemes: a review and synthesis of analytical approaches. Hydrogeology Journal 17, 1149–1155.

Buik, N., Willemsen, A., 2002. Clogging rate of recharge wells in porous media. In: Proceedings ISAR-4, Adelaide, South-Australia.

CBS, 2014. Hernieuwbare Energie in Nederland 2013 (Renewable Energy in the Netherlands 2013). Centraal Bureau voor de Statistiek, Den Haag.

De Zwart, A., 2007. Investigation of Clogging Processes in Unconsolidated Aquifers Near Water Supply Wells (Ph.D. thesis). TU-Delft.

Diersch, H.-J.G., 1998. FEFLOW Interactive, Graphics Based Finite Element Simulation System for Modelling Groundwater Flow, Contaminant Mass and Heat Transport Processes. WASY Institute for Water Resource Planning and Systems Research Ltd., Berlin.

Hoes, P., Hensen, J.L.M., Loomans, M.G.L.C., de Vries, B., Bourgeois, D., 2009. User behaviour in whole building simulation. Energy and Buildings 41 (3), 295–302.

Houben, G., Treskatis, C., 2007. Water Well Rehabilitation and Reconstruction. McGraw-Hill.

Kipp, K.L., 1997. Guide to the Revised Heat and Solute Transport Simulator: HST3D – Version 2. US Geological Survey Water-Resources Investigations Report 97-4157, 149 pp.

Konikow, L.F., Sanford, W.E., Campbell, P.J., 1997. Constant-concentration boundary condition: lessons from the HYDROCOIN variable-density groundwater benchmark problem. Water Resources Research 33 (10), 2253–2261.

Kruseman, G.P., de Ridder, N.A., 1994. Analysis and Evaluation of Pumping Test Data, second ed. Publication 47, International Institute for Land Reclamation and Improvement, Wageningen, The Netherlands. 370 pp.

Lund, J.W., Freeston, D.H., Boyd, T.L., 2010. Direct utilization of geothermal energy. Worldwide review. In: Proceedings World Geothermal Congress 2010, Bali, Indonesia.

Misstear, B.D.R., Banks, D., Clark, L., 2006. Water Wells and Boreholes. Wiley, Chichester, UK.

NVOE (Nederlandse Vereniging voor Ondergrondse Energieopslagsystemen (Dutch Association for Underground Energy Storage Systems)), 2006. NVOE Guidelines for Underground Energy Storage (in Dutch).

Pyne, R.D. (Ed.), 2005. Aquifer Storage Recovery: A Guide to Groundwater Recharge through Wells. ASR Press, Gainesville, FL.

Sommer, W.T., 2015. Modelling and Monitoring of Aquifer Thermal Energy Storage - Impacts of Heterogeneity on Thermal Interference and Bioremediation (Ph.D. thesis). Wageningen University.

Sommer, W.T., Valstar, J., van Gaans, P.F.M., Grotenhuis, J.T.C., Rijnaarts, H.H.M., 2013. The impact of aquifer heterogeneity on the performance of aquifer thermal energy storage. Water Resources Research 49 (12), 8128–8138.

Sommer, W.T., Valstar, J., Leusbrock, I., Grotenhuis, J.T.C., Rijnaarts, H.H.M., 2015. Optimization and spatial pattern of large-scale aquifer thermal energy storage. Applied Energy 137, 322–337.

Sterrett, R.J., 2007. Groundwater and Wells, third ed. Johnson Screens, St. Paul, Minnesota. 812 pp.

Van Beek, C.G.E.M., 2010. Cause and Prevention of Clogging of Wells Abstracting Groundwater from Unconsolidated Aquifers (Ph.D. thesis). VU Amsterdam.

VDI 4640 Part 2, 2001. Thermal use of the Underground. Ground Source Heat Pump Systems. Beuth verlag GmbH, Berlin.

Standing column wells

10

P. Pasquier, A. Nguyen, F. Eppner, D. Marcotte, P. Baudron
Polytechnique Montréal, Montréal, QC, Canada

Abstract

Standing column well (SCW) systems present a strong potential for energy savings, especially in dense urban areas with suitable geological conditions where lack of space constitutes an impediment to the use of closed-loop systems. This chapter aims at presenting some design considerations and recent advancements relative to the thermal, hydraulic, and chemical simulation of SCWs and identifies specific research needs to foster their use. It is shown how a thermal resistance and capacity model can assess the performance of a hybrid system operated under different bleed ratios in a fractured aquifer. Additionally, this chapter shows how to perform coupled thermo-hydrogeochemical simulation to predict the dissolution and precipitation of calcite that occurs along the well as a function of operating conditions. The approach helps to foresee possible operation problems with the SCW and to select suitable mitigation measures to sustain long-term performance of SCW.

10.1 Introduction

10.1.1 Overview

Closed-loop ground heat exchanger (GHE) technology is currently the preferred approach to exchange heat between the geological environment and a ground-source heat pump (GSHP) system. The relative popularity of the closed-loop GHE stems mainly from its long history of successes in terms of energy savings and reliability as well as the fact that designing a closed-loop GHE requires only few underground thermal parameters readily provided by a thermal response test (see chapters: Vertical borehole ground heat exchanger design methods and In situ estimation of ground thermal properties).

In contrast, the design of an open-loop system made of pumping and injection wells may require a heavy field investigation program involving many exploratory boreholes, stratigraphic analyses, pumping, and tracer tests. If suitable hydrogeological conditions are observed and indicate that the local aquifer is productive enough to sustain the operation of GSHPs, an open-loop system may prove to be a more cost-effective solution than closed-loop system (see chapter: Open-loop heat pump and thermal energy storage systems). However, completion of hydrogeological investigations requires time and resources, and many designers will prefer the more linear design process associated with closed-loop GHE approaches.

Advances in Ground-Source Heat Pump Systems. http://dx.doi.org/10.1016/B978-0-08-100311-4.00010-8
Copyright © 2016 Elsevier Ltd. All rights reserved.

An alternative approach to the well-known closed- and open-loop GHEs is the standing column well (SCW), which consists of a long open borehole (75−450 m long) mostly drilled in bedrock and usually having a diameter of 150 mm (Orio et al., 2005). To prevent the surface unconsolidated soils falling down into the open hole, the overburden is cased from the surface to the rock basement with a steel or PVC pipe as shown in Fig. 10.1. The most common design for commercial systems pumps groundwater at the base of the well with a submersible pump installed within a riser pipe screened or slotted at its base. The pumped groundwater is then directed to the building heating, ventilation, and air-conditioning (HVAC) system (often via an intermediate heat exchanger). Unlike open-loop systems, the pumped groundwater is reintroduced in the same well below the dynamic level in the annular space of the SCW.

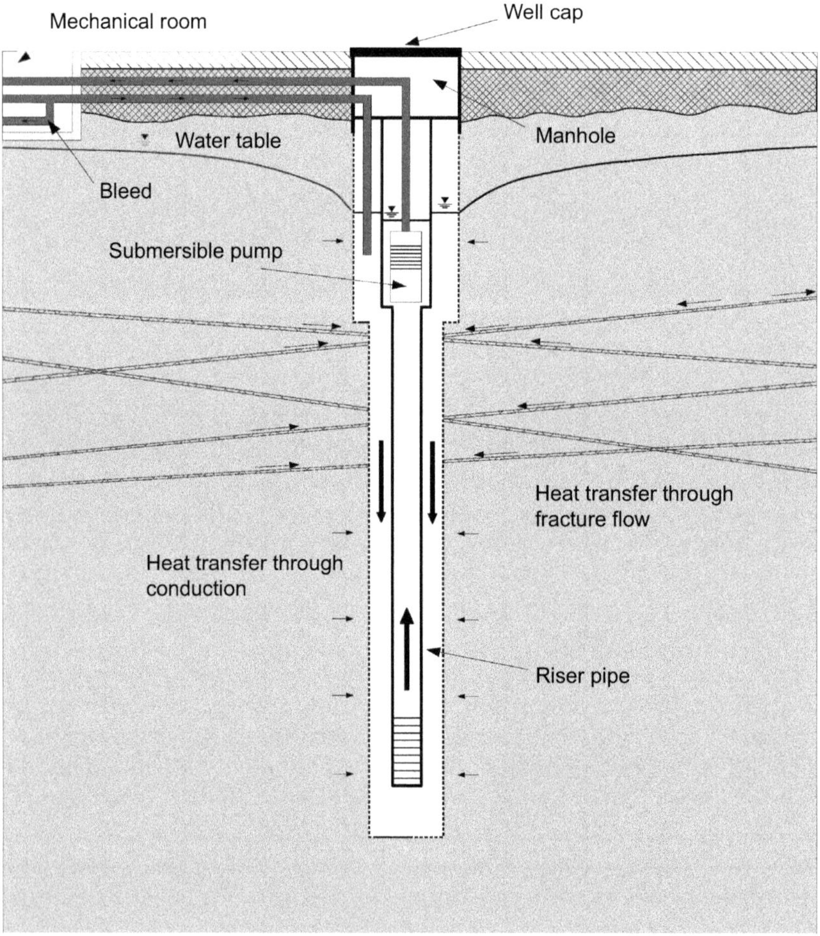

Figure 10.1 Illustration of a standing column well system.

The large volume of water contained in the well and its long residence time induce a significant thermal inertia that can damp the impact of peak building demands on fluid temperature and ensure longer operation periods for the GSHPs. The key component of SCWs is their capacity to renew their water content through an operation commonly known as bleed (Banks, 2008; Rees et al., 2004; O'Neill et al., 2006). Indeed, during peak periods, part of the pumped water is not recirculated into the SCW but discharged outside the well. This creates a net pumping rate and a cone of depression around the well that stimulates flow in the surrounding fracture network and promotes advective heat transfer. Controlling the bleed process gives an ability to manage long-term temperature swings.

10.1.2 Design considerations

10.1.2.1 Installation practices

Different designs and configurations are possible for completion of an SCW in terms of well diameter, pump location, use of riser pipe, well screen, and so forth. The design retained is usually a function of local practices, regulations, and geological conditions. However, geological environments having a thin overburden are often preferred for installation of SCWs since less casing is required to reach the rock basement, which may lead to a significant reduction of the drilling costs. Similarly, competent rocks are less prone to a collapse of the borehole wall and may help reduce the risk associated with drilling operations.

The use of a polyvinyl chloride (PVC) riser pipe (also called inner pipe or porter shroud) allows placement of the submersible pump in a pumping chamber located near the surface. This configuration facilitates installation and maintenance operations while reducing the length of the pump discharge pipe and head losses due to friction. Installing the pump near the surface in the riser pipe also allows reducing the diameter of the borehole and riser pipe below the pumping chamber, which also helps to lower construction costs. However, such design could be difficult to implement in practice, and innovative solutions (Koenig, 2009) have recently started to emerge to ease construction of SCWs.

10.1.2.2 Bleed control strategies

The bleed of an SCW temporarily boosts the performances of an SCW by allowing warmer (colder) water getting into the well in heating (cooling) mode. The bleed flow rate is typically between 5% and 25% of the total pumping rate (Spitler et al., 2002) and can be modulated to reduce the volume of discharged water (Nguyen et al., 2015a). Bleed control is usually performed according to the temperature difference within the GSHPs or dead-band temperature. For dead-band control, bleed is triggered when entering water temperature (EWT) reaches approximately 5°C in heating mode and 30°C in cooling mode. Although it has not yet been reported for SCWs, use of predictive controls based on weather forecasts could also prove useful to activate the bleed and bring the well to a more suitable temperature before peak events.

Submersible variable speed pumps can also be used to modulate the total pumping rate and reduce pumping energy when GSHPs are operating at part load. Additionally, if the system is composed of several wells operating at part load, sequential operation of the SWCs can be envisioned to reduce pumping energy (Koenig, 2010).

10.1.2.3 Discharge point

Increasing the bleed flow rate helps stabilize the temperature at the GSHPs inlet but increases the volume of water discharged outside the well. This water can be directed to a sewer, an infiltration ditch or toward the surface water if permitted by local regulations, or simply reinjected in the origin aquifer with a relatively small injection well. Increasing the bleed flow rate increases the well drawdown and care must be taken to keep the pump intake below the lowest dynamic level reached during dry years.

10.1.2.4 Water quality

The operation of an SCW modifies the temperature and pressure of the groundwater circulating in the well and in the surrounding aquifer, and promotes its oxygenation through gaseous diffusion along the dynamic level. Consequently, the chemical signature of the water changes over time, which can foster undesirable effects such as scaling and growth of bacteria or algae in the system's components. Scaling problems are well known for open-loop systems (Criaud and Fouillac, 1989; Thomas and Gudmundsson, 1989; Vuataz et al., 1989; Kristmannsdottir, 1989; Lipping et al., 1997; Stanasel et al., 2001, 2006; Gunnlaugsson, 2012; Ma et al., 2012) and can also jeopardize operation of SCWs. If concerns exist regarding water quality, groundwater can be transferred to a primary heat exchanger to prevent scaling and clogging within the heat pumps and ease maintenance. Several private users report using their water well both as an SCW and a drinking water source. As the operation of an SCW could also promote development of pathogen microorganisms in groundwater, such dual use of SCWs should be permitted only after proving the local drinking water standards are met all year round.

Iron bacteria is another possible nuisance of SCW systems (Citernesi et al., 1985; Cahalan-Roach et al., 2010; Douglas et al., 2015) that can clog a well and affect its productivity. If groundwater has naturally low iron concentration, designers should be aware that using steel casing and steel pipes introduces a source of iron that can foster growth of iron bacteria and accumulation of ocherous precipitates. If needed, one can avoid the use of steel casing or cover the casing with an appropriate paint to reduce bioaccessibility of iron.

10.1.3 Potential

Several operational successes observed with SCW systems in the United States (Spitler et al., 2002; Orio et al., 2005) have generated a growing interest for SCWs among the scientific and commercial communities in the last decade. In particular, the analysis performed by Orio et al. (2006) on a 200-ton installation relying on six SCWs (33 tons/SCW) in a New England public school has probably been an eye-opener for many. Based on a nine-year metering of the installation, Orio et al. (2006) monitored stable energy savings of 686,820 kWh/year for a GHE made of only six SCWs. Similar observations were done by Orio and Patnaude (2014) on a 615-ton installation made of 326 ground-coupled heat pumps and 16 SCWs (38 tons/SCW).

Several numerical and parametric analyses have been conducted to improve our understanding of SCWs. For instance, O'Neill et al. (2006) have shown that, by comparison to a closed-loop GHE of similar capacity, reductions of 49−78% of the size of the GHE were possible with the use of SCW systems, thereby translating into significant reductions in construction costs. Comparable results have been reported by Yavuzturk and Chiasson (2002) and Yuill and Mikler (1995). Financial simulations carried out over a 20-year life cycle showed that despite higher maintenance costs, the use of SCW systems may result in overall costs between 27% and 51% lower than for closed-loop systems (O'Neill et al., 2006).

Beyond the financial gains associated to SCWs, their real potential lies in their ability of being integrated to existing buildings located in dense urban areas where the land required for a wide closed-loop GHE is not available. Indeed, a single SCW may prove enough to cover a load of 30−40 tons and can be easily installed near an existing building. Examples of astute installations of an SCW in a building's front sidewalk (where a small service box is the only visible trace of the SCW) prove that SCWs may offer an interesting cost-effective alternative to closed- and open-loop GHEs.

10.1.4 Outline

Despite their lower construction costs and the significant energy savings SCWs can generate, their adoption by building developers, owners, and professionals around the world is no easy matter. According to many practitioners, this is mostly due to a difficulty to size these systems when no local expertise is available, a misunderstanding of the geochemical risks associated to the use of groundwater and a difficulty to assess the maintenance costs of these systems. This chapter will present some recent advancements and findings relative to the thermal, hydraulic, and chemical simulation of SCWs and identify specific research needs to foster their use worldwide.

10.2 Thermal and hydraulic simulation

Designers of GSHP systems need to forecast the performances of the system they envision to estimate the energy savings generated by their design. To estimate accurately energy savings, modeling of the system's thermal response is required. Several analytical approaches were developed over the years for this purpose. The first published model proposed by Oliver and Braud (1981) was limited to steady-state heat conduction. Orio (1995, 1999) used the infinite line source model to predict the transient temperature of an SCW. In their model, they represented the bleed by an enhanced ground thermal conductivity.

The analytical models are relatively easy to implement. However, they cannot account for many important SCW features such as the vertical flow and the thermal capacity of groundwater circulating within the well. More important, the bleed control used to increase advective heat transfer in the ground makes the problem nonlinear, hence preventing the use of the superposition principle (Marcotte and Pasquier, 2008; Pasquier and Marcotte, 2013). As a consequence analytical models cannot represent faithfully the operation of an SCW with bleed control.

This section aims at presenting available SCW models published in the literature, but also to point out important design and geological parameters that can affect system performance.

10.2.1 Modeling of SCW

Different numerical approximations can be used to solve the governing equations describing heat transfer and groundwater flow within the domain. For instance, finite element or finite volume methods can be used to solve the general heat transfer partial differential equation, given by:

$$\rho C_{\mathrm{P}} \frac{\partial T}{\partial t} = \nabla \cdot (k \nabla T) - \rho C_{\mathrm{P}} u \cdot \nabla T \qquad [10.1]$$

where T is the temperature, k is the equivalent thermal conductivity, ρC_{P} is the volumetric heat capacity, and u is the Darcy's velocity field.

To include the groundwater flow and obtain the velocity field u, conservation equation coupled with Darcy's law has to be solved as well. The governing equation for flow in equivalent porous media is expressed by:

$$S_{\mathrm{s}} \frac{\partial H}{\partial t} = \nabla \cdot (K \nabla H) \qquad [10.2]$$

where H, S_{s}, and K are the hydraulic head, specific storage, and hydraulic conductivity, respectively.

Numerical approaches can account for coupled transient heat transfer and groundwater flow within complex geometries, something not available with the analytical models. The first numerical model was proposed by Yuill and Mikler (1995). It was a simple quasi-two-dimensional representation of an SCW that did not account for bleed. More recently, numerical and analytical methods were used jointly to emulate arrays of nonbleeding SCWs (Woods and Ortega, 2011). Other complex models either fully solved Eq. [10.1] by finite element or finite volume methods (Abu-Nada et al., 2008; Croteau, 2011; Nguyen et al., 2012) or coupled an ordinary differential equation (ODE) model with a finite volume model (Rees et al., 2004; Deng, 2004; Ng et al., 2011). These models fully incorporate groundwater flow caused by bleed and peak damping caused by the thermal capacity and residence time of the water circulating within the well. Nonetheless, although these methods are robust and reliable, they are too computationally intensive for a routine use by practitioners.

Thermal resistance and capacity models (TRCMs) are now widely used to represent closed-loop GHEs (De Carli et al., 2010; Bauer et al., 2011a,b; Zarrella et al., 2011; Eslami-Nejad and Bernier, 2012; Pasquier and Marcotte, 2012, 2014). Recent contributions suggested various TRCM schemes to model SCWs (Nguyen et al., 2013, 2015a,b) or use of the Haar wavelet method to solve more rapidly the resulting ODE system (Nguyen and Pasquier., 2015).

Basically, the TRCM approach seeks to approximate Eq. [10.1] by means of an ODE system of the form:

$$C_j \frac{dT_j}{dt} = \sum_{k=1}^{n_j} \frac{T_k - T_j}{R_{T,k}} \quad \forall j = 1, \ldots, n \qquad [10.3]$$

where R_T is the thermal resistance, j, k, n, and n_j are, respectively, the node index, the index of the neighboring node, the total number of nodes in the network, and the number of neighboring nodes to node j. An example of a TRCM is provided in Fig. 10.2 for a case where the SCW system is discretized into seven annular regions. The main advantage of this approach lies in its computation efficiency as it can rapidly emulate dynamic systems with an accuracy often comparable to the one obtained by finite element methods (for instance, Nguyen et al., 2015a).

10.2.2 Dynamic simulations

The use of hourly simulations is a simple way for estimating a system's performance. For hybrid systems, results from an energy simulation can easily yield crucial information such as the heat pump's EWT over time and the total system's energy consumption (heat pumps and auxiliary systems), assuming that a typical load profile is available for the building.

For instance, Fig. 10.3 illustrates the evolution of the EWT for a system composed of four commercially available four-ton GSHPs used in conjunction with a backup heating system for a typical multiresidential building located in a northern climate (see Table 10.1 for the simulation parameters). As shown in Fig. 10.4, in this case a single 150-m SCW provides most of the heating and all the cooling to the building through GSHPs having a total nominal capacity of 56 kW. Here, the installed GSHP capacity cannot cover all heating loads and a significant backup system must be used to cover the peak heating loads. After obtaining the system's EWT, one can thus derive the load provided by the GSHPs and auxiliary system and compute the energy consumption of each subsystem (Fig. 10.5).

Notice that the system is highly dynamic as it includes a three-level bleed control and a heat pump on−off control sequence. To illustrate clearly the dynamic behavior of the temperature and drawdown, Fig. 10.6 shows a close-up view of the EWT and leaving water temperature (LWT) and the corresponding well drawdown for the first

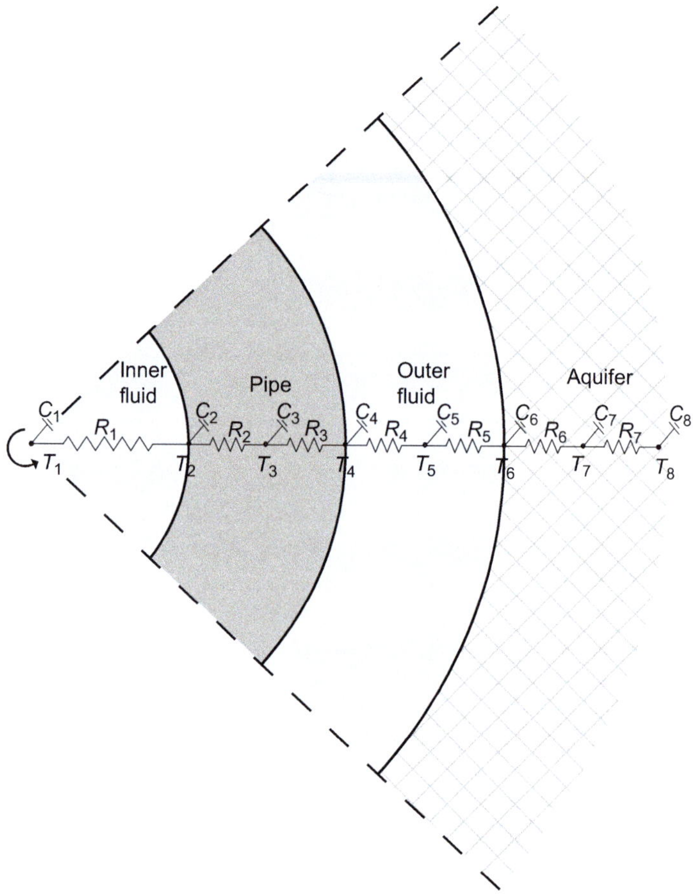

Figure 10.2 Thermal resistance and capacity network in the radial direction in a standing column well.

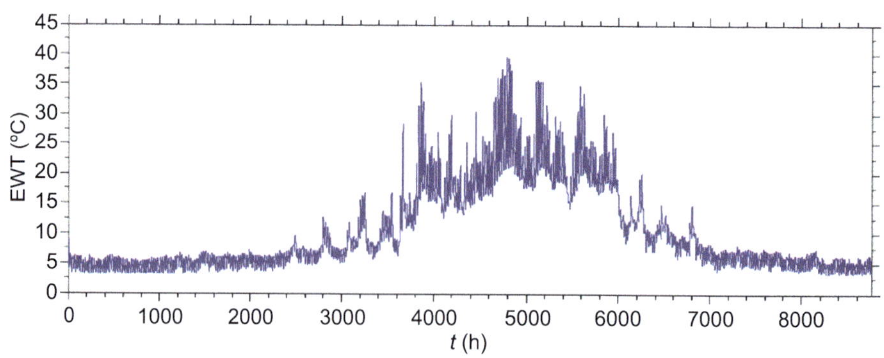

Figure 10.3 Temporal evolution of the heat pump's entering water temperature for a 56-kW (16 ton) standing column well system.

Table 10.1 Thermal, hydrogeological, and geochemical properties and geometry

Properties	Description	Symbol	Unit	Value
Thermal	Pipe thermal conductivity	k_p	W/(mK)	0.1
	Aquifer thermal conductivity	k_a	W/(mK)	2.5
	Fluid volumetric heat capacity	$C_{s,f}$	kJ/(m^{3}K)	4200
	Pipe volumetric heat capacity	$C_{s,p}$	kJ/(m^{3}K)	1560
	Aquifer volumetric heat capacity	$C_{s,a}$	kJ/(m^{3}K)	2500
Hydrogeological	Pipe hydraulic conductivity	K_p	m/s	1e-9
	Aquifer hydraulic conductivity	K_a	m/s	1e-6
	Fluid specific storage	$S_{s,f}$	1/m	4e-6
	Pipe specific storage	$S_{s,p}$	1/m	1.3e-7
	Aquifer specific storage	$S_{s,a}$	1/m	5e-5
Geochemical	Molecular diffusion - fluid	$D_{f,f}$	m^2/s	1e-2
	Molecular diffusion - soil	$D_{f,s}$	m^2/s	1e-9
	Ratio A/V, in the well	$(A/V)_f$	1/m	37.1
	Ratio A/V, in the aquifer	$(A/V)_a$	1/m	$1.35\cdot 10^4$
Geometry	Pipe inner radius	r_{pi}	m	0.038
	Pipe outer radius	r_{po}	m	0.045
	SCW radius	r_b	m	0.076
	Aquifer radius	r_a	m	40
	Aquifer thickness	b	m	150
	Pumping flow rate	$\dot{V}$	L/min	151

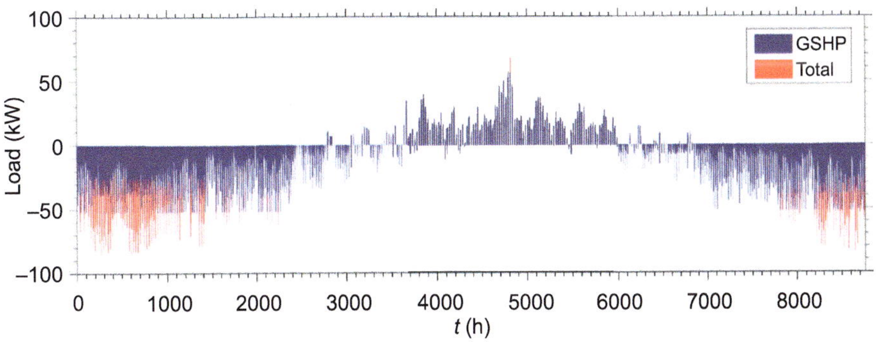

Figure 10.4 Load provided by the ground-source heat pumps and total building loads.

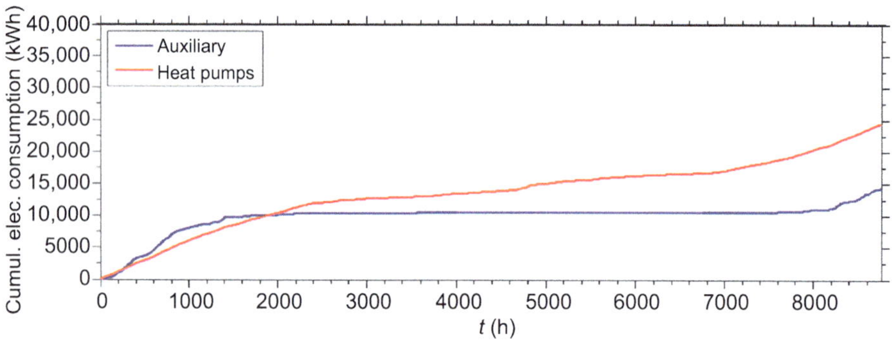

Figure 10.5 Evolution of the cumulative energy consumption of the ground-source heat pumps and auxiliary heating system.

24 h of the simulated period. The increase of well drawdown corresponding to EWT at 7, 6, and 5°C shows the implementation of the bleed control. Notice also that the on–off sequences were triggered at 4 °C, which explains why the auxiliary heating system goes on and off during the first 24 h of operation. Of course, use of a deeper SCW would lessen the use of auxiliary system when the building loads are smaller than the GSHP's capacity.

10.2.3 Effect of design parameters

Results from dynamic simulations are also useful for determining the effect of key design parameters on the performance of SCW systems. Spitler et al. (2002) compared the maximum and minimum EWT for various parameters but one can also use other information such as the system's electrical consumption as comparison criteria. For instance, Fig. 10.7 presents the total electrical consumption for various borehole lengths, well diameters, and maximum bleed ratios (fraction of the bleed flow rate to the total pumping rate) for the building loads presented in Fig. 10.4. Note that an increase of the total electrical consumption can be due to a degradation of the heat pumps coefficient of performance (COP) which is here temperature dependent, a closure of a GSHP if the EWT reaches the temperature of 4°C and a higher use of auxiliary system whose COP is here assumed to be equal to one.

10.2.3.1 Well depth

In absence of a bleed, conduction is the main heat transfer mechanism for SCWs. In that sense, Fig. 10.7 confirms that increasing the borehole depth can have a significant, almost linear, effect on the performance of the system. Recall that when bleed is activated, however, heat advection becomes the primary driving force and the effect of well depth becomes significantly less important. Indeed, as shown in Fig. 10.7, a 100-m SCW operating with a maximum bleed ratio of 30% can present a performance

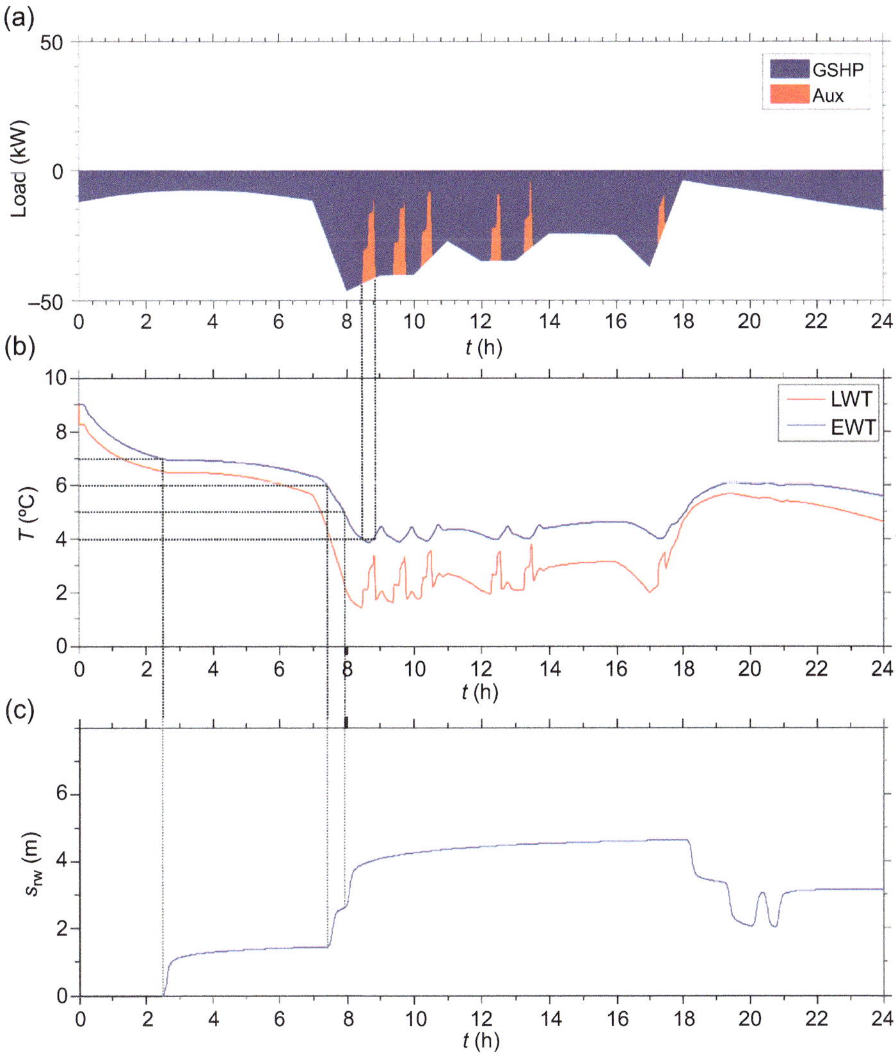

Figure 10.6 Close-up view of the first 24 h of the evolution of (a) the load provided by the ground-source heat pumps and backup heating system, (b) the corresponding entering water temperature and leaving water temperature, and (c) the well drawdown.

similar to a 200-m SCW operating with a maximum bleed of 22.5%. Nonetheless, an increase of the well depth can significantly increase the volume of water contained in the SCW, which can be beneficial to damp the peaks in fluid temperature. Note that if the pumping cost is significant an increase of the well depth may reduce the well's drawdown due to bleed if additional fractures are found below.

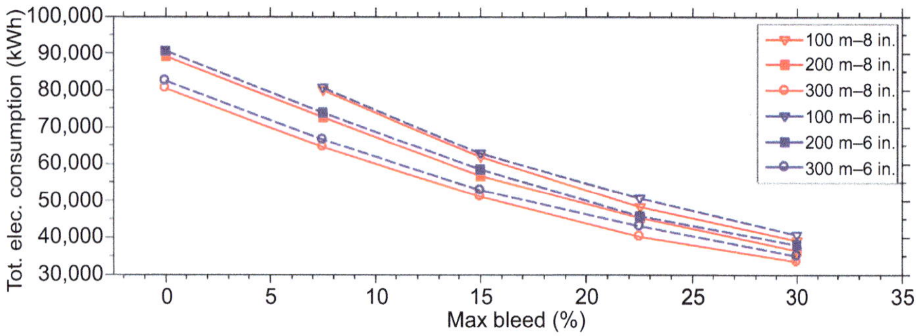

Figure 10.7 Annual total electrical consumption curves for various well depths, well diameters, and maximum bleed ratios.

10.2.3.2 Well diameters

Similarly to the well depth, increasing the well diameter also increases the overall conductive heat transfer and borehole thermal capacity, resulting in better performances. However, as shown in Fig. 10.7, increasing the drilling diameter from 150 mm (6 in.) to 200 mm (8 in.) has a limited influence on the energy consumption, especially for high bleed ratios. Note also that, depending on local drilling contractors, drilling costs can significantly increase with well diameter.

10.2.3.3 Maximum bleed ratio

It has been well demonstrated that a bleed control strategy is a key feature for maintaining the heat pump's EWT within its operational limits (Spitler et al., 2002; Deng, 2004; Choi et al., 2012; Nguyen et al., 2015a), especially during peak load periods. When bleed is activated, the induced groundwater flow instantly boosts the thermal efficiency of the SCW as advection becomes the dominant heat transport mechanism. As shown in Fig. 10.7, the effect of bleed on the energy consumption is major, almost linear, and dominates the effect of the well depth and diameter. The implementation of a multilevel bleed control can also be an efficient way to reduce the volume of discharged water or to reduce the drawdown for unproductive wells.

10.2.4 Presence of fractures

For simplicity, most contributions assumed that the surrounding aquifer is homogeneous and isotropic. Under this assumption, the groundwater flow induced by the bleed is almost uniform along the borehole wall, which is not realistic for deep boreholes. In fact, when drilling in the rock basement, it is common to intercept fracture zones which are likely to affect the performances of the SCW system. The effect of fracturation on the performance of the SCW was first addressed by Deng (2004) and Deng et al. (2005) with a simplified one-dimensional model and later by Ramesh and Spitler (2012) with a quasi-two-dimensional model. Nguyen et al. (2015b) developed a coupled TRCM and studied the influence of groundwater flow in the presence of a single horizontal fractured zone at the center of the aquifer. They observed that a fracture can have a significant beneficial effect on the performance of an SCW. In fact, for a fixed hydraulic conductivity, the benefit is proportional to the contrast ratio ϕ between the hydraulic conductivity of the fracture and porous matrix (Fig. 10.8) as more groundwater coming from the far field will rapidly reach the SCW at high ϕ ratios. To illustrate this, one can observe the effect of a single fracture in Fig. 10.9.

From a practical point of view, conducting hydrogeological investigations and fracture characterization might be too heavy for a routine design of SCW systems. Nonetheless, the presence of fractured zones will always benefit the system and therefore represent a safety factor for designers.

10.3 Coupled geochemical simulation

Historically, most successful installations of SCWs have been located along the Appalachian Mountains in the Northeastern United States (Spitler et al., 2002; Orio et al., 2005) in geological environments composed mainly of igneous and metamorphic sequences. The relatively inert nature of these rocks usually leads to a good groundwater quality that does not promote scaling or corrosion of the system's components. However, some sedimentary rocks, such as carbonates, are more soluble and may chemically react with ions in solution in groundwater through the so-called geochemical reactions (see Brantley et al. (2008) or Appelo and Postma (2005)).

These reactions are often reversible and may favor mineral dissolution or precipitation. The direction of the reaction depends on the local concentration of the products and reactants, but also on the temperature and pressure of the solution. This means that a mineral can locally undergo dissolution and release ions in solution that may precipitate elsewhere in the system if the local temperature or pressure is modified. In an SCW, the frequent temperature changes may affect the solubility of a variety of minerals. One of the most common minerals composing carbonates and frequently observed in the scales found in open-loop GSHP systems is calcite ($CaCO_3$) (Lipping et al., 1997; Izgec et al., 2005; Gunnlaugsson, 2012).

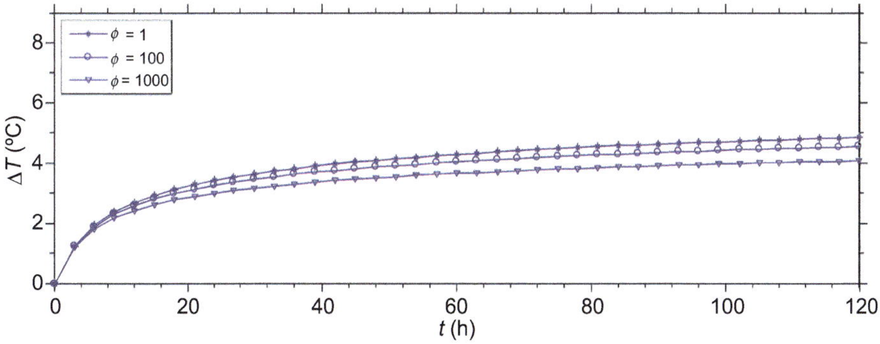

Figure 10.8 Comparison of the fluid's thermal response for various contrast ratios for a case with a single horizontal fractured zone at the center of the model.

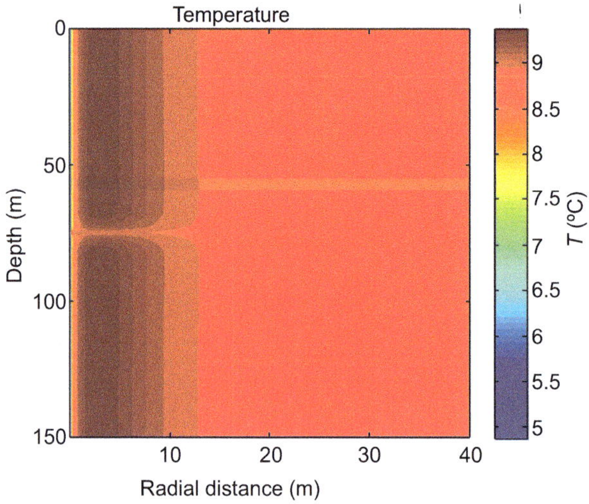

Figure 10.9 Effect of various contrast ratio ϕ on the fluid's thermal response.

The chemical reactions involving calcite and its associated chemical species can be regrouped in two classes. The first one involves the relatively slow surface reactions occurring at the interface between the mineral and the water and can be described by reaction kinetics. The second category involves ions in solution which can be described by equilibrium reactions since they occur almost instantaneously. Both classes of reactions are temperature dependent and any serious attempt to predict their reaction rates must integrate the temperature.

This section aims at presenting the reactions involving calcite, how to perform coupled geochemical simulations for SCW applications, and which useful information such simulations can bring to a designer. The work presented hereinafter is based on recent developments and additional details can be found in Eppner et al. (2015) and Pandey et al. (2014).

10.3.1 Transport and geochemical reactions

Diffusion and advection of dissolved chemical species subject to kinetics and equilibrium reactions in groundwater can be modeled, following the notation of Holzbecher (2005, 2012), by

$$\varphi \frac{\partial \mathbf{c}}{\partial t} = \varphi(\nabla \cdot (D\nabla \mathbf{c}) - \nabla \cdot (u\mathbf{c})) + \mathbf{S}_{\mathbf{k}}'\mathbf{r}_{\mathbf{k}} + \mathbf{S}_{\mathbf{e}}'\mathbf{r}_{\mathbf{e}} \qquad [10.4]$$

where $\mathbf{c}$ is a vector containing species concentration, φ is the porosity, D is the diffusion coefficient, $\mathbf{S_k}$ is the stoichiometric matrix for reaction kinetics, $\mathbf{r_k}$ is the vector of reaction rates for reaction kinetics, $\mathbf{S_e}$ is the stoichiometric matrix for equilibrium reactions, $\mathbf{r_e}$ is the vector containing the rates required to maintain the equilibrium and $'$ is the transpose operator. Notice that in Eq. [10.4], u is the vector of groundwater velocity computed by solution of Eq. [10.2] while the reaction rates defined by $\mathbf{r_k}$ and $\mathbf{r_e}$ are temperature dependent.

Solution of Eq. [10.4] requires evaluation of the term $\mathbf{S'_e r_e}$, which is not possible since vector $\mathbf{r_e}$ can't be evaluated directly. Indeed, the direct and reverse rate constants of the equilibrium reactions are unknown and can't be provided by explicit expressions. To work around this problem, Saaltink et al. (1998) suggested to solve the transport problem for the total concentrations with the reaction kinetics modeled by a source term and the equilibrium reactions expressed by a system of nonlinear equations designed to ensure local mass balance. The transformed equation is obtained by multiplying Eq. [10.4] by a transformation matrix $\mathbf{U}$ to obtain:

$$\varphi \frac{\partial \mathbf{\Gamma}}{\partial t} = \varphi(\nabla \cdot (D \nabla \mathbf{\Gamma}) - \nabla \cdot (u \mathbf{\Gamma})) + \mathbf{U S'_k r_k} \quad [10.5]$$

where $\mathbf{\Gamma} = \mathbf{Uc}$ is a vector containing the total concentrations. For the carbonate system considered in this work, use of Eq. [10.5] instead of Eq. [10.4] greatly simplifies the simulation burden by allowing solution of only three advection–diffusion equations instead of nine.

The transformation matrix $\mathbf{U}$ is nonunique and many techniques can be used to find it (see Saaltink et al., 1998; Holzbecher, 2005; or Holzbecher, 2012). In this work, matrix $\mathbf{U}$ is constructed according to the Tableaux method (Morel and Hering, 1993) and incorporates the following nine chemical species $\mathbf{c} = \left[\mathbf{c}_{H^+}\ \mathbf{c}_{HCO_3^-}\ \mathbf{c}_{Ca^{2+}}\ \mathbf{c}_{OH^-}\ \mathbf{c}_{H_2CO_3}\ \mathbf{c}_{CO_3^{2-}}\ \mathbf{c}_{CaHCO_3^+}\ \mathbf{c}_{CaCO_{3(aq)}}\ \mathbf{c}_{CaOH^+}\right]'$. Now, using H^+, HCO_3^- Ca^{2+} as basis components in the Tableaux method leads to the following transformation matrix (Eppner et al., 2015):

$$\mathbf{U} = \begin{array}{c} \\ H \\ HCO \\ Ca \end{array} \begin{array}{ccccccccc} H^+ & HCO_3^- & Ca^{2+} & OH^- & H_2CO_3 & CO_3^{2-} & CaHCO_3^+ & CaCO_{3(aq)} & CaOH^+ \\ \end{array}$$

$$U = \begin{bmatrix} 1 & 0 & 0 & -1 & 1 & -1 & 0 & -1 & -1 \\ 0 & 1 & 0 & 0 & 1 & 1 & 1 & 1 & 0 \\ 0 & 0 & 1 & 0 & 0 & 0 & 1 & 1 & 1 \end{bmatrix} \quad [10.6]$$

	H^+	HCO_3^-	Ca^{2+}	OH^-	H_2CO_3	CO_3^{2-}	$CaHCO_3^+$	$CaCO_{3(aq)}$	$CaOH^+$
H	1	0	0	−1	1	−1	0	−1	−1
HCO	0	1	0	0	1	1	1	1	0
Ca	0	0	1	0	0	0	1	1	1

Here, the vector of total concentrations $\mathbf{\Gamma} = \mathbf{Uc} = [\mathbf{\Gamma}_H\ \mathbf{\Gamma}_{HCO}\ \mathbf{\Gamma}_{Ca}]'$ is then defined by:

$$\mathbf{\Gamma}_H = \mathbf{c}_{H^+} - \mathbf{c}_{OH^-} + \mathbf{c}_{H_2CO_3} - \mathbf{c}_{CO_3^{2-}} - \mathbf{c}_{CaCO_{3(aq)}} - \mathbf{c}_{CaOH^+} \quad [10.7]$$

$$\mathbf{\Gamma}_{HCO} = \mathbf{c}_{HCO_3^-} + \mathbf{c}_{H_2CO_3} + \mathbf{c}_{CO_3^{2-}} + \mathbf{c}_{CaHCO_3^+} + \mathbf{c}_{CaCO_{3(aq)}} \quad [10.8]$$

$$\mathbf{\Gamma}_{Ca} = \mathbf{c}_{Ca^{2+}} + \mathbf{c}_{CaHCO_3^+} + \mathbf{c}_{CaCO_{3(aq)}} + \mathbf{c}_{CaOH^+} \quad [10.9]$$

10.3.2 Reaction kinetics for calcite

The reaction rates occurring at the calcite surface are often described by the widely accepted Plummer–Wigley–Parkhurst (PWP) geochemical model (Plummer et al., 1978). This reaction model was developed after an impressive experimental program and provides the reaction rates at various temperatures for three elementary chemical reactions. These reversible reactions, which can be simultaneous, are:

$$
\begin{aligned}
&CaCO_{3(s)} + H^+ \underset{k_{-1}}{\overset{k_1}{\rightleftharpoons}} Ca^{2+} + HCO_3^- \\
&CaCO_{3(s)} + H_2CO_3^* \underset{k_{-2}}{\overset{k_2}{\rightleftharpoons}} Ca^{2+} + 2HCO_3^- \\
&CaCO_{3(s)} + H_2O \underset{k_{-3}}{\overset{k_3}{\rightleftharpoons}} Ca^{2+} + HCO_3^- + OH^-
\end{aligned}
\qquad [10.10]
$$

where k_1, k_2, and k_3 are the direct rate constants of each reaction and k_{-1}, k_{-2}, and k_{-3} are the corresponding reverse rate constants. Note that in Eq. [10.10], $H_2CO_3^*$ corresponds to $CO_{2(aq)} + H_2CO_3^0$ since at 25°C, the concentration of $CO_{2(aq)}$ is usually 600 times higher than the concentration of $H_2CO_3^0$ (Appelo and Postma, 2005).

In the PWP model, the direct rate constants of Eq. [10.10] were found empirically and can be linked to temperature through the following expressions:

$$
\begin{aligned}
&\log k_1 = 0.198 - (444/T) \\
&\log k_2 = 2.84 - (2177/T) \\
&\log k_3 = -5.86 - (317/T) \text{ for } T < 298.15\text{K} \\
&\log k_3 = -1.10 - (1737/T) \text{ for } T > 298.15\text{K}
\end{aligned}
\qquad [10.11]
$$

The equilibrium constant K_{eq} of each reaction (Eq. [10.10]) is also temperature dependent and can be expressed as a function of T (Brantley et al., 2008; Appelo and Postma, 2005). Now, since the direct (k_{+j}) and reverse (k_{-j}) rate constants are linked through $K_{eq,j} = k_{+j}/k_{-j}$, k_{-j} can also be expressed as a function of T, allowing linking dissolution and precipitation processes to temperature.

Using Eq. [10.10] to construct matrix $\mathbf{S_k}$ and Eq. [10.11] to derive vector $\mathbf{r_k}$ leads to the following source term:

$$
\mathbf{U S'_k r_k} = \begin{bmatrix} -R_1 - R_2 - R_3 \\ R_1 + R_2 + R_3 \\ R_1 + R_2 + R_3 \end{bmatrix} \qquad [10.12]
$$

with R_j given by:

$$
\begin{aligned}
R_1 &= k_1 \cdot \mathbf{c}_{H^+} - k_{-1} \cdot \mathbf{c}_{ca^{2+}} \cdot \mathbf{c}_{HCO_3^-} \\
R_2 &= k_2 \cdot \mathbf{c}_{H_2CO_3} - k_{-2} \cdot \mathbf{c}_{ca^{2+}} \cdot \mathbf{c}^2_{HCO_3^-} \\
R_3 &= k_3 \cdot \mathbf{c}_{H_2O} - k_{-3} \cdot \mathbf{c}_{ca^{2+}} \cdot \mathbf{c}_{HCO_3^-} \cdot \mathbf{c}_{OH^-}
\end{aligned}
\qquad [10.13]
$$

10.3.3 Equilibrium reactions for calcite

The equilibrium reactions for the carbonate system involve six elementary reversible reactions coupled together through nine chemical species by:

$$
\begin{aligned}
& H^+ + OH^- \rightleftharpoons H_2O \\
& H^+ + CO_3^{2-} \rightleftharpoons HCO_3^- \\
& Ca^{2+} + HCO_3^- \rightleftharpoons CaHCO_3^+ \\
& H^+ + HCO_3^- \rightleftharpoons H_2CO_3 \\
& Ca^{2+} + CO_3^{2-} \rightleftharpoons CaCO_{3(aq)} \\
& H^+ + CaOH^+ \rightleftharpoons Ca^{2+} + H_2O
\end{aligned}
\qquad [10.14]
$$

Assuming unity for the concentration of H_2O and using a positive coefficient for the reactants and a negative one for the reaction products, the equilibrium reactions (Eq. [10.14]) can be expressed in a compact form by the following stoichiometric matrix:

$$
\mathbf{S_e} =
\begin{array}{c}
\\ R_1 \\ R_2 \\ R_3 \\ R_4 \\ R_5 \\ R_6
\end{array}
\begin{array}{ccccccccc}
H^+ & HCO_3^- & Ca^{2+} & OH^- & H_2CO_3 & CO_3^{2-} & CaHCO_3^+ & CaCO_{3(aq)} & CaOH^+ \\
\end{array}
\left[
\begin{array}{ccccccccc}
1 & 0 & 0 & 1 & 0 & 0 & 0 & 0 & 0 \\
1 & -1 & 0 & 0 & 0 & 1 & 0 & 0 & 0 \\
0 & 1 & 1 & 0 & 0 & 0 & -1 & 0 & 0 \\
1 & 1 & 0 & 0 & -1 & 0 & 0 & 0 & 0 \\
0 & 0 & 1 & 0 & 0 & 1 & 0 & -1 & 0 \\
1 & 0 & -1 & 0 & 0 & 0 & 0 & 0 & 1
\end{array}
\right]
\qquad [10.15]
$$

These equilibrium reactions are all reversible and have their own equilibrium constant to establish ions speciation. Noting $\mathbf{K_{eq}}$, a vector containing the equilibrium constant of each reaction, the equilibrium reactions described by Eq. [10.14] can then be expressed by:

$$\mathbf{S_e} \log\mathbf{c} = \log\mathbf{K_{eq}} \quad [10.16]$$

To efficiently solve Eq. [10.5] under the constraint of the equilibrium reactions described by Eq. [10.16], Saaltink et al. (1998) suggested solving locally a system of equations simultaneously to the three advection–diffusion equations of Eq. [10.5]. The resulting system of equation is designed to ensure mass balance ($\mathbf{\Gamma} = \mathbf{Uc}$) and at equilibrium is given by:

$$\mathbf{Uc} - \mathbf{\Gamma} = 0 \quad [10.17]$$

and

$$\mathbf{S_e} \log\mathbf{c} - \log\mathbf{K_{eq}} = 0 \quad [10.18]$$

The system defined by Eqs. [10.17] and [10.18] is made of nine independent equations and solved locally for **c**. The whole solution procedure involves the Newton method to solve Eqs. [10.17] and [10.18] and the finite element method to solve the state equations of this coupled thermo-hydrogeochemical problem (Eqs. [10.1], [10.2] and [10.5]). The solution contains the concentration of each chemical species over the whole simulation domain.

10.3.4 Coupled simulation of an SCW

In order to illustrate how a change in temperature can modify the chemistry of groundwater in a calcareous aquifer, the methodology presented previously was used with the thermal, hydraulic, and geochemical parameters depicted in Table 10.1. However, to highlight the effect of temperature, the hourly loads illustrated in Fig. 10.4 were aggregated over a one-week period. The resulting heat pump's EWTs and LWTs are illustrated in Fig. 10.10(a) along with the concentration of calcium ion (Ca^{2+}) in Fig. 10.10(b)).

First, one can observe how temperature and calcium concentration are well negatively correlated together. As the temperature of the water within the SCW increases, calcite's solubility decreases which leads to its precipitation and to the consumption of Ca^{2+}. Inversely, during heating periods, the calcite present in the rock surrounding the SCW is dissolved and releases Ca^{2+} ions susceptible to precipitate later in the system's components.

A useful tool to identify if calcite is locally undergoing dissolution or precipitation is the evaluation of the overall reaction rate given by (Plummer et al., 1978):

$$\widetilde{R} = R_1 + R_2 + R_3 \quad [10.19]$$

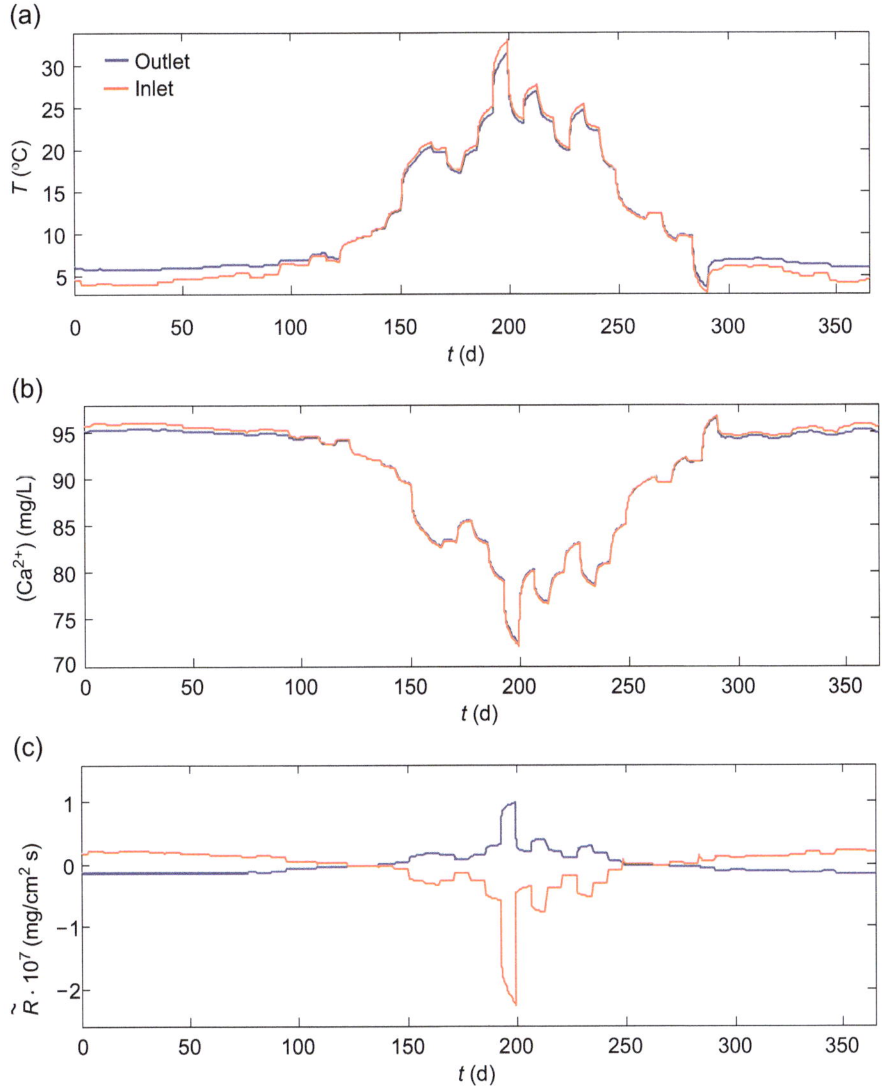

Figure 10.10 Evolution of (a) the temperature, (b) the concentration of Ca^{2+}, and (c) the reaction rate of calcite $(\widetilde{R})$ at the outlet and at the inlet of the well.

A positive $\widetilde{R}$ value indicates that dissolution of calcite is a dominant process while a negative value indicates that calcite precipitation is likely to occur. Fig. 10.10(c) shows the evolution of the overall reaction rate at the well inlet (LWT) and outlet (EWT). One can see that dissolution and precipitation occur simultaneously at different locations in the well all year round.

Basically, this behavior is explained by the fact that as the water descends in the SCW in cooling mode (see Fig. 10.11(a,c,e)), calcite precipitates $\left(\widetilde{R} < 0\right)$ in the first 50 m or so, which consumes Ca^{2+} until an equilibrium is reached $\left(\widetilde{R} = 0\right)$. Then, the subsequent cooling will bring the water to an undersaturated state with respect to calcite $\left(\widetilde{R} > 0\right)$, which will promote calcite dissolution for depths greater than 50 m and explain the release of small but continuous amounts of Ca^{2+} in the water. The inverse behavior is observed during the heating season (see Fig. 10.11(b,d,f)).

From a practical point of view, the shift from a dissolution to a precipitation state explains why in cooling mode scale is likely to form within the GSHPs, heat exchangers, or horizontal pipes downstream of the GSHPs.

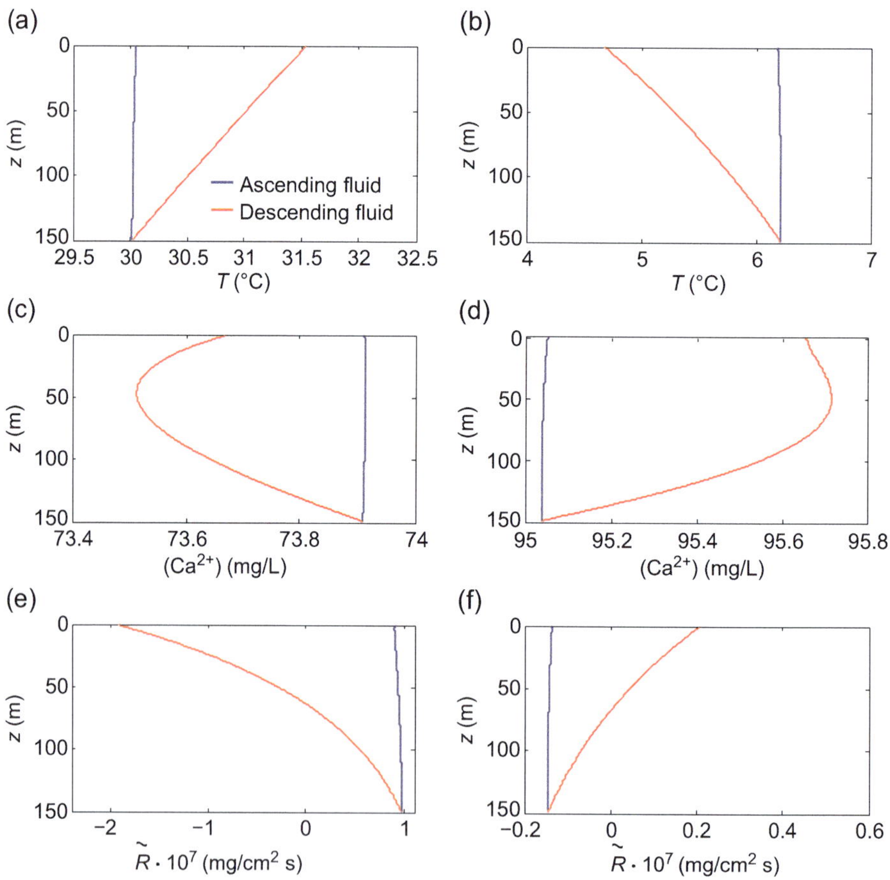

Figure 10.11 Temperature, Ca^{2+} concentration, and overall reaction rate of calcite profile in the ascending and descending fluid after 195 days (left column: cooling) and 365 days of simulation (right column: heating).

10.4 Conclusions

GSHP systems using SCWs have been widely used in the Northeastern United States where several operational successes have been reported. These successes can be largely explained by the significant thermal inertia of SCWs that can damp peak loads and by the bleed that allows groundwater renewal during extended operation of the heat pumps. In the end, this combination helps maintain the GSHPs within their operational limits, especially for heating, and ensures a longer operation of the GSHPs.

Despite the research that demonstrates the financial benefits of SCWs, these GHEs are still barely used outside the areas where they historically emerged. Some practitioners find it risky to propose an SCW system. An argument frequently invoked is the geochemical risk associated with the direct use of groundwater. Although water quality problems are not widely reported in the Northeastern United States, these problems are likely to become more frequent as SCWs are used in different geological environments prone to dissolution and precipitation. Indeed, the operation of an SCW modifies the groundwater temperature, which contributes to change the concentration of the various ions in solution. These ions are likely to precipitate elsewhere in the system's components and cause operational problems.

This chapter showed how to perform coupled thermo-hydrogeochemical modeling to predict the changes in the geochemistry of the water circulating within an SCW. The methodology presented herein is among the first attempts to predict the evolution of the geochemistry as a function of the bleed and building thermal loads. Such approach could be used to develop mitigation measures or identify water treatment systems adapted to SCWs. Despite their potential usefulness, coupled geochemical models will have to be extended to include more than one mineral and will require a thorough experimental validation before they can be used with confidence by practitioners.

Simulation of SCW is complex as it includes both advection and conduction as heat transfer mechanisms. Several noteworthy attempts have been accomplished to simulate SCWs. However, reducing computing time while maintaining an acceptable accuracy level remains a challenge especially if one wants to use optimization-based approaches requiring SCW models. Use of simplified models is a good way to reduce the simulation burden especially if a field made of several SCWs must be simulated. A recent contribution using a wavelet-based method was highly successful at improving simulation speed for TRCMs. However, development of analytical approaches incorporating the various beneficial features of SCWs (bleed, thermal capacity, and vertical advection) might be the best way to improve simulation speed of a field made of several SCWs. In the end, every model should be confronted to real temperature measurements in order to validate their assumptions and implementation.

Several technical challenges have yet to be overcome to promote use of SCWs toward building developers and owners. However, the biggest challenges to overcome are probably the need to improve the reciprocal understanding between specialists coming from different technical areas and demystify the potential of SCWs.

Nomenclature

b	Aquifer thickness (m)
$\mathbf{c}$	Vector containing the species concentration (kg/m^3)
C	Thermal capacity (J/K)
C_s	Volumetric heat capacity (J/(m^3 K))
D	Dispersivity coefficient (m)
H	Hydraulic head (m)
K	Hydraulic conductivity (m/s)
k	Thermal conductivity (W/(m K))
k_{+j}, k_{-j}	Direct and reverse reaction rate constants (variable units)
k_1 to k_3	Direct reaction rate constants of the three kinetic reactions (m/s)
k_{-1} to k_{-3}	Reverse reaction rate constants of the three kinetic reactions (m^4/(kg s)), (m^7/(kg^2 s)), and (m^7/(kg^2 s))
K_{eq}	Equilibrium constant (variable units)
$\mathbf{K_{eq}}$	Vector of equilibrium constants (variable units)
n	Number of nodes (–)
n_j	Number of neighboring nodes to node j (–)
r	Radius (m)
$\mathbf{r_e}$, $\mathbf{r_k}$	Vector of reaction rates for equilibrium and kinetic reactions (–)
R_j	Rate of reaction (variable units)
R_1 to R_3	Reaction rates of three kinetic reactions (kg/(m^2 s))
$\widetilde{R}$	Overall reaction rate of calcite (kg/(m^2 s))
$\mathbf{S_e}$, $\mathbf{S_k}$	Stoichiometric matrix for equilibrium and kinetic reactions (–)
S_s	Specific storage (1/m)
t	Time (s)
T	Temperature (°C)
u	Darcy's velocity field (m/s)
$\mathbf{U}$	Transformation matrix
$\dot{V}$	Pumping flow rate (m^3/s)
z	Depth (m)
$'$	Transpose operator (–)

Greek symbols	
Γ	Vector containing the total concentrations (kg/m^3)
ϕ	Hydraulic contrast ratio (–)
Subscripts	
a	Aquifer
b	Borehole
f	Fluid
p	Pipe
pi	Pipe inner
po	Pipe outer
Acronyms	
COP	Coefficient of performance
EWT	Entering water temperature
GHE	Ground heat exchanger
GSHP	Ground-source heat pump
LWT	Leaving water temperature
ODE	Ordinary differential equation
PWP	Plummer–Wigley–Parkhurst model
SCW	Standing column well
TRCM	Thermal resistance and capacity model

Acknowledgment

The authors would like to acknowledge the Natural Sciences and Engineering Research Council of Canada (NSERC) for its financial support.

References

Abu-Nada, E., Akash, B., Al-Hinti, I., Al-Sarkhi, A., Nijmeh, S., Ibrahim, A., Shishan, A., 2008. Modeling of a geothermal standing column well. International Journal of Energy Research 32, 306–317.

Appelo, C.A.J., Postma, D., 2005. Geochemistry, Groundwater and Pollution, second ed. A.A. Balkema, Leiden, The Netherlands.

Banks, D., 2008. An Introduction to Thermogeology: Ground Source Heating and Cooling. Blackwell Publishing Ltd.

Bauer, D., Heidemann, W., Mller-Steinhagen, H., Diersch, H.-J.G., 2011a. Thermal resistance and capacity models for borehole heat exchangers. International Journal of Energy Research 35 (4), 312–320.

Bauer, D., Heidemann, W., Diersch, H.-J., 2011b. Transient 3D analysis of borehole heat exchanger modeling. Geothermics 40 (4), 250–260.

Brantley, S.L., Kubicki, J.D., White, A.F., 2008. Kinetics of Water-rock Interaction. Springer-Verlag, New York.

Cahalan-Roach, S., Campbell, K., Cho, Y., 2010. Hydrogeologic, geologic, and sustainable challenges and interpretations for the installation of a geothermal well. World Environmental and Water Resources Congress 868–879.

Choi, H.K., Yoo, G.J., Lim, K.B., Lee, S.H., Lee, C.H., 2012. Characteristic analysis of bleeding effect on standing column well (SCW) type geothermal heat exchanger. Journal of Central South University 19, 3202–3207.

Citernesi, U., Benvenuti, G., Ferrara, G.C., 1985. Microbiological aspects of concrete and iron deterioration in geothermal power-plants. Geothermics 14 (2–3), 315–326.

Criaud, A., Fouillac, C., 1989. Sulfide scaling in low enthalpy geothermal environments: A survey. Geothermics 18 (1–2), 73–81.

Croteau, J., 2011. Évaluation des paramètres influençant les températures d'opération des puits à colonne permanente (Master's thesis). École Polytechnique de Montréal (In French).

De Carli, M., Tonon, M., Zarrella, A., Zecchin, R., 2010. A computational capacity resistance model (CaRM) for vertical ground-coupled heat exchangers. Renewable Energy 35 (7), 1537–1550.

Deng, Z., 2004. Modeling of Standing Column Wells in Ground Source Heat Pump Systems (Ph.D. thesis). Oklahoma State University.

Deng, Z., Rees, S., Spitler, J., 2005. A model for annual simulation of standing column well ground heat exchangers. HVAC&R Research 11 (4), 637–655.

Douglas, G.B., Trefry, M.G., Wyglie, J.T., Wiles, P.G., Puzon, G.J., Kaksonen, A.H., 2015. Potential biogeochemical impacts of heat reinjection in the Mullaloo aquifer, Western Australia. Geothermics 53, 429–445.

Eppner, F., Pasquier, P., Baudron, P., 2015. Development of a thermo-hydro-geochemical model for low temperature geoexchange applications. In: Boston, USA. Comsol Conference.

Eslami-nejad, P., Bernier, M., 2012. Freezing of geothermal borehole surroundings: a numerical and experimental assessment with applications. Applied Energy 98, 333–345.

Gunnlaugsson, E., 2012. Scaling in Geothermal Installation in Iceland. Short Course on Geothermal Development and Geothermal Wells. Santa Tecla, El Salvador.

Holzbecher, E., 2005. Reactive transport in porous media – concepts and numerical approaches. In: Ingham, D.B., Pop, I. (Eds.), Transport Phenomena in Porous Media III. Pergamon, Oxford, pp. 305–340.

Holzbecher, E., 2012. Modeling Using Matlab, second ed. Springer, Berlin Heidelberg, Germany.

Izgec, O., Demiral, B., Bertin, H., Akin, S., 2005. Calcite precipitation in low temperature geothermal systems: an experimental approach. In: 30th Workshop on Geothermal Reservoir Engineering. Stanford University, Stanford, California. TR-176.

Koenig, A.A., 2009. Ground Water Heat Transfer Systems and Deployment Thereof. US Patent 7,597,138.

Koenig, A.A., 2010. Ground Source Heat Pump Well Field Design and Control Strategy for Large Tonnage. US Patent 7,647,773.
Kristmannsdottir, H., 1989. Types of scaling occurring by geothermal utilization in Iceland. Geothermics 18, 1983–1990.
Liping, B., Xianji, M., Thorhallsson, S., 1997. The study of an allowable calcite supersaturation in low temperature geothermal systems. Transactions of Tianjin University 3 (1), 60–65.
Ma, Z., Yan, H., Zho, X., Hou, C., 2012. Impact of carbonate scaling on the efficiency of used geothermal water reinjection from low-middle temperature geothermal fluid in Xianyang porous geothermal field, NW China. Advanced Materials Research 307, 614–615.
Marcotte, D., Pasquier, P., 2008. Fast fluid and ground temperature computation for geothermal ground-loop heat exchanger systems. Geothermics 37, 651–665.
Morel, F.M.M., Hering, J.G., 1993. Principles and Applications of Aquatic Chemistry. Wiley, Hoboken, USA.
Ng, B.M., Underwood, C.P., Walker, S.L., 2011. Standing column wells – modeling the potential for applications in geothermal heating and cooling. HVAC&R Research 17 (6), 1089–1100.
Nguyen, A., Pasquier, P., Marcotte, D., 2012. Multiphysics modelling of standing column well and implementation of heat pumps off-loading sequence. In: Boston, USA. Comsol Conference.
Nguyen, A., Pasquier, P., Marcotte, D., 2013. Development of an ODE model featuring a three level bleed control and an off-loading sequence for standing column wells. In: Proceedings of BS2013, Chambéry, France, pp. 26–28.
Nguyen, A., Pasquier, P., Marcotte, D., 2015a. Thermal resistance and capacity model for standing column wells operating under a bleed control. Renewable Energy 76, 743–756.
Nguyen, A., Pasquier, P., Marcotte, D., 2015b. Influence of groundwater flow in fractured aquifers on standing column wells performance. Geothermics 58, 39–48.
Nguyen, A., Pasquier, P., 2015. An adaptive segmentation Haar wavelet method for solving thermal resistance and capacity models of ground heat exchangers. Applied Thermal Engineering 89, 70–79.
O'Neill, Z.D., Spitler, J., Rees, S., 2006. Performance analysis of standing column well ground heat exchanger systems. ASHRAE Transactions 112, 633–643.
Oliver, J., Braud, H., 1981. Thermal exchange to earth with concentric well pipes. Transactions of ASAE 24 (4), 906–910.
Orio, C.D., 1995. Design, use & example of standing column wells. In: IGSPHA Technical Meeting.
Orio, C.D., 1999. Geothermal heat pump applications industrial/commercial. Energy Engineering 96 (3), 58–66.
Orio, C., Johnson, C., Rees, S., Chiasson, A., Deng, Z., Spitler, J., 2005. A survey of standing column well installations in North America. ASHRAE Transactions 111, 109–121.
Orio, C.D., Johnson, C.N., Poor, K.D., 2006. Geothermal standing column wells: ten years in a New England school. ASHRAE Transactions 112 (2), 57–64.
Orio, C.D., Patnaude, Z.J., 2014. Eight years of operation of 615 ton geothermal nursing home in Northern tier. ASHRAE Transactions 120, 8 pages.
Pandey, S.N., Chaudhuri, A., Kelkar, S., Sandeep, V.R., Rajaram, H., 2014. Investigation of permeability alteration of fractured limestone reservoir due to geothermal heat extraction using three-dimensional thermo-hydro-chemical (THC) model. Geothermics 51, 46–62.
Pasquier, P., Marcotte, D., 2012. Short-term simulation of ground heat exchanger with an improved TRCM. Renewable Energy 46, 92–99.

Pasquier, P., Marcotte, D., 2013. Efficient computation of heat flux signals to ensure the reproduction of prescribed temperatures at several interacting heat sources. Applied Thermal Engineering 59, 515–526.

Pasquier, P., Marcotte, D., 2014. Joint use of quasi-3D response model and spectral method to simulate borehole heat exchanger. Geothermics 51, 281–299.

Plummer, L.N., Wigley, T.M.L., Parkhurst, D.L., 1978. The kinetics of calcite dissolution in CO_2-water systems at 5 to 60°C and 0 to 1 atm CO_2. American Journal of Science 278, 179–216.

Ramesh, A., Spitler, J., 2012. A Quasi-two-Dimensional standing column well model for ground source heat pump systems. In: Innostock (Lleida, Spain).

Rees, S.J., Spitler, J.D., Deng, Z., Orio, C.D., Johnson, C.N., 2004. A study of geothermal heat pump and standing column well performance. ASHRAE Transactions 110, 3–13.

Saaltink, M.W., Ayora, C., Carrera, J., 1998. A mathematical formulation for reactive transport that eliminates mineral concentrations. Water Resources Research 34, 1649–1656.

Spitler, J.D., Rees, S.J., Deng, Z., Chiasson, A., Orio, C.D., Johnson, C., 2002. R&D Studies Applied to Standing Column Well Design (No. ASHRAE 1119-RP).

Stanasel, O., Gilau, L., Stanasel, I., 2001. Types of scaling occurring by geothermal utilization in a low temperature geothermal field. In: Twenty-sixth Workshop on Geothermal Reservoir Engineering. Stanford University, Stanford, USA.

Stanasel, O., Aurel, I., Kristmannsdottric, H., Stanasel, I., 2006. Physical-chemistry studies on geothermal waters and scale formations. Revue Roumaine de Chimie 51 (3), 179–185.

Thomas, D.M., Gudmundsson, J.S., 1989. Advances in the study of solids deposition in geothermal systems. Geothermics 18 (1–2), 5–15.

Vuataz, F.D., Czernichowski-Lauriol, I., Fouillac, C., Detoc, S., 1989. Chemical study of a low temperature geothermal fluid in a triassic sandstone aquifer: Scaling potential and fluid treatment (Melleray, France). Geothermics 18 (1–2), 313–320.

Woods, K., Ortega, A., 2011. The thermal response of an infinite line of open loop wells for ground coupled heat pump systems. International Journal of Heat and Mass Transfer 54, 5574–5587.

Yavuzturk, C., Chiasson, A., 2002. Performance analysis of U-tube, concentric tube, and standing column well ground heat exchangers using a system simulation approach. ASHRAE Transactions 108, 925–938.

Yuill, G.K., Mikler, V., 1995. Analysis of the effect of induced groundwater flow on heat transfer from a vertical open-hole concentric-tube thermal well. ASHRAE Transactions 101 (1), 173–185.

Zarrella, A., Scarpa, M., De Carli, M., 2011. Short time step analysis of vertical ground-coupled heat exchangers: the approach of CaRM. Renewable Energy 36 (9), 2357–2367.

Borehole thermal energy storage 11

S. Gehlin
Swedish Centre for Shallow Geothermal Energy, Lund, Sweden

Abstract

Borehole thermal energy storage (BTES) systems utilize boreholes in rock, soil, or clay to transfer heat and cold to the surrounding ground material, so that the thermal energy may be seasonally stored. BTES systems have been used for more than 35 years in diverse applications. This chapter reviews characteristics of BTES systems and their applications.

11.1 Introduction

Borehole thermal energy storage (BTES) systems store sensible heat (or cold) in the ground surrounding individual boreholes. In a sense, all systems that use boreholes for heat or cold extraction could be considered BTES systems, even single borehole residential systems. However, this chapter will focus on systems with multiple vertical boreholes used for some sort of active and intentional storage of heat and/or cold in the ground.

11.1.1 Definition of borehole thermal energy storage

There is no uniform definition of a BTES system, and there are a number of other terms used in literature for systems that could be regarded as BTES systems. The abbreviation BTES first shows up in literature in a report by the International Energy Agency Energy Conservation through Energy Storage working group (IEA ECES, 1997). It is largely through the various IEA ECES workgroups that the term has been most widely used and spread. In proceedings from the international conferences on thermal energy storage ("Stock" conferences) arranged by IEA ECES since 1981, the abbreviation BTES did not appear until the Terrastock conference in Stuttgart in August and September 2000. In that conference, it suddenly appeared as a widely used and established term (Benner and Hahne, 2000).

Prior to that, BTES applications were included in the more general terms seasonal thermal energy storage (STES) or underground thermal energy storage (UTES). The abbreviation DTES (duct[1] thermal energy storage) was also used to distinguish storage

[1] The term "Duct Thermal Energy Storage" results from an unfortunate translation of the Swedish word *kanal*. Although the word duct in English can mean a channel through which liquid flows, it is used in the North American heating and cooling industry to mean a channel through which air flows. Nevertheless, as used in this chapter in conjunction with "thermal energy storage" it always means a channel through which liquid flows.

Advances in Ground-Source Heat Pump Systems. http://dx.doi.org/10.1016/B978-0-08-100311-4.00011-X
Copyright © 2016 S. Gehlin. Published by Elsevier Ltd. All rights reserved.

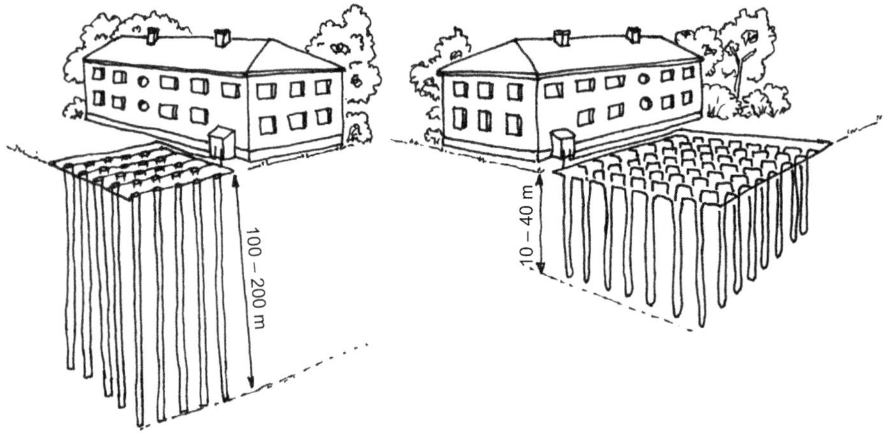

Figure 11.1 Borehole thermal energy storage drilled in rock (left) and with pipes inserted in clay (right).
Illustration by S. Gehlin (2015).

in rock or clay from aquifer thermal energy storage (ATES) systems. Even after the introduction of BTES as a term, DTES is sometimes still used to distinguish systems where vertical heat exchangers are inserted in soft formations such as clay or soil without drilling (ie, there are no boreholes, only a heat exchanger that has been driven into the soft formation). Fig. 11.1 shows BTES systems in rock and soft formations.

BTES systems are referred to by many names in the literature including "ground heat store," "borehole store," "borehole heat store," "seasonal thermal energy storage in rock ducts or ground," "energy storage with borehole heat exchangers (BHEs)," "ground-coupled thermal energy storage (GCTES)," "storage system in rock," "multiple-well storage system," "vertical heat exchanger store," and "vertical loops." BTES systems do not necessarily use heat pumps, but when they do, they are sometimes referred to as "geothermal heat pump systems," "ground-coupled heat pump (GCHP) systems," "ground-source heat pump (GSHP) systems," or "geo-exchange systems."

There are diverse definitions of BTES in the literature. The IEA ECES report from 1997 (p. 33) defined BTES as systems where rock or soil is the energy storage medium accessed by closed-loop heat exchangers placed in boreholes. The energy recovered or stored in the subsurface environment is used for heating and cooling. Nordell (2000) gave a wide but detailed definition of BTES. He divided the technology into small-scale, large-scale, and seasonal storage. His definition of BTES comprised systems that provide cooling and/or heating, with or without heat pumps and with or without recharge, at all temperature levels, and any number of boreholes or ducts in rock, soil, or clay. In Paksoy (2007), BTES is defined differently by the various chapter authors, but the given examples are all multiple-borehole systems for seasonal storage of heat and cold. Banks (2012, pp. 395–396) defined BTES as "deliberate thermal energy storage" in multiple-borehole systems with closed loops. Lee (2013, p. 98) makes a distinction between BTES and "true BTES," which in the latter case implies systems where the cold is actively stored and not passively recharged by heat transfer from the surroundings. He defined BTES as multiple-borehole systems in rock, soil, or clay. In Cabeza (2015, p. 11),

BTES is said to consist of several closely spaced boreholes, but also that many countries have thousands of BTES systems with one to a few 100 boreholes, usually for heating and cooling of buildings. The distinction between vertical GSHP systems and BTES is vague. A distinction suggested by Sanner and Stiles (1997) is that if less than 25% of the annual thermal turnover from the borehole system is being exchanged with the surroundings, the BHE system would qualify as a BTES system.

This chapter on BTES will define BTES in the wide context as given by IEA ECES (1997) and Nordell (2000), but will focus on larger applications with multiple boreholes and active storage, with or without the assistance of heat pumps.

11.1.2 Some borehole thermal energy storage history

One of the first descriptions of vertical BHEs in the ground is found in Kemler (1946). In Kemler (1947), he suggested nine different heat exchanger designs for extraction of heat from the ground to serve as heat sources for a domestic heat pump. Four of these were horizontal coils, of which one is placed in a trench surrounding the building foundation; four were vertical open or closed (U-tube or coaxial) loops, and one was a vertical helical coil heat exchanger. Kemler concluded that all nine methods are potentially feasible. IEA ECES (1997) and Nordell (1994) give credit to Brun (1965) for conceptualizing borehole or duct ground heat storage. Brun, however, had a rather different approach to how the heat be transferred in the ground. He suggested injection of solar-heated high temperature (450°C) steam into boreholes drilled in rock, and recovering the stored heat by injecting water into the boreholes. His impressive proposed field of 800,000 boreholes to a depth of 200 m was never realized.

It was not until the oil crisis in the 1970s that the interest in large-scale storage of heat in the ground took off. The objective was seasonal storage of high temperature solar heat or waste heat, initially in storage tanks, rock caverns, or aquifers, and later also in boreholes or ducts in clay and rock. These systems were intended to work without heat pump assistance. By the end of the 1980s, interest increased in thermal energy storage at lower temperatures for heating and cooling. These systems typically involve heat pumps for extraction of heat, while cooling may be provided without heat pumps. This type of application has proven to be very efficient and feasible, as the demand for comfort cooling increases with improved building insulation.

The first multiple-borehole BTES system was possibly a 12-borehole system built for seasonal storage of solar energy in the Jura mountains of France in 1976 (Guimbal, 1976; Hellström, 1991). Large-scale BTES systems were built in Sweden around 1980 and reported at the first International Conference on Seasonal Thermal Energy Storage and Compressed Air Energy Storage in Seattle, Washington, in October 1981. A full-scale experimental BTES system was built for a single-family house in Sigtuna, Sweden, in the late 1970s (Platell et al., 1981; Platell and Wikström, 1983). The research project was called Sunstore, and the system consisted of 42 boreholes drilled in rock to a depth of 23 m, and connected to 162 m^2 of solar collectors. It was constructed in sections with drilling starting in November 1977 and was completed in 1981. During this period the ground was preheated with electrical heating. In May 1981 the solar collector circuit was connected and solar heat was stored in the ground. The first solar heat extraction started in the fall of 1982 and the store provided temperatures of 10–40°C.

About the same time, a small-scale pilot plant for high-temperature BTES was set up close to the Luleå University of Technology campus in the north of Sweden, in 1980–1981 (Andersson et al., 1983; Nordell, 1994). Nineteen boreholes with a diameter of 52 mm and 1.3 m spacing were drilled to a depth of 21 m. The soil overburden was 6 m. The boreholes were arranged in a hexagonal configuration and fitted with open loops. The plant was used to operate and evaluate the thermal behavior of the store during five cycles, and was in operation from July to November 1981. Positive results from this pilot project led to construction of the experimental full-scale high-temperature demonstration BTES system Lulevärme Heat Store in 1982–1983 (Nordell, 1994). The store was in operation between 1983 and 1989. It consisted of 120 boreholes in a 10 × 12 rectangular configuration in granitic rock to a depth of 65 m. The 152-mm diameter boreholes were spaced 4 m apart and fitted with open loops. Waste heat at a temperature of 70–82°C from a steel industry plant was transferred to the storage via the district heating network during the summers. The heat was recovered at 35–55°C in the winters for heating of one of the university buildings.

Early BTES systems in clay and soil were constructed in Sweden, Switzerland and the Netherlands in the 1980s. In 1979 a single-family building in Utby, Sweden, was connected to a heat storage consisting of 37 vertical tubes inserted in clay to 10-m depth. The heat store was charged with low temperature heat from the outdoor air with an air-to-fluid heat exchanger in the summers (Rosenblad, 1983). The Sunclay heat store in Kungsbacka, Sweden, consisted of 612 vertical plastic single U-tubes inserted in clay to a depth of 35 m and at 2-m spacing (Hultmark, 1981). The storage was heated by 1500 m^2 of unglazed roof-integrated solar collectors in the summers and operated in the temperature interval 15–30°C. It was built and tested in 1980–81. In Mont de Pitié, Cortaillod-Neuchâtel, Switzerland (Matthey and Pillonel, 1985), a system was being constructed in 1978–81 where 12 family houses were heated with solar heat from 320 m^2 solar panels, stored in a BTES system with 400 coaxial tubes in sandy loams, to a depth of only 6–8 m. The system was assisted by gas-driven heat pumps. Other similar systems with a large number of shallow vertical loops inserted in clay, sand, or soil were built in Groeningen (Wijsman, 1985), Genève (Matthey, 1988), and Cormontreuil (Baudoin, 1988).

When interest in storing low temperatures for cooling increased by the end of the 1980s, BTES systems that combined heat extraction with heat pumps and extraction of cold from boreholes without heat pump (free-cooling) developed. In the 1990s, this type of application, which could be constructed at a smaller scale, spread rapidly in Europe and North America. Hellström (1991, p. 17) lists early BTES systems, and examples of German BTES systems are found in Sanner (2005). Bakema et al. (1995, pp. 14–17) provides an extensive table of UTES systems, of which 22 are BTES systems in Sweden and Germany. Early BTES systems reported in the literature are compiled in Table 11.1.

11.2 Typical features of borehole thermal energy storage

BTES systems are said to be the most general type of UTES system and are most efficient for large energy loads with slow changes over time (Nordell, 2000). As discussed earlier in this chapter, the term BTES is subject to wide interpretation, and though the

Table 11.1 **Early full-scale borehole thermal energy storage systems reported in the literature**

First year of operation	Plant name/ location	Ground	Configuration	Collector type	Temperature	References
1979	Utby Sweden	Clay	37 boreholes of 10 m	Two-channel tube	2–12°C Ambient air	Rosenblad (1983)
1981	Sunstore Sigtuna Sweden	Rock	42 boreholes of 23 m	Coaxial	10–40°C Solar	Platell et al. (1981) and Platell and Wikström (1983)
1981	Sunclay Kungsbacka Sweden	Clay	612 ducts of 35 m	U-tube	15–30°C Solar	Hultmark (1981)
1981	Mont de Pitié Cortaillod, Neuchâtel Switzerland	Sandy loams	400 boreholes of 6–8 m	Coaxial	7–25°C Solar	Matthey and Pillonel (1985)
1982	Treviglio Italy	Sand and gravel	220 + 194 boreholes of 11 m	U-tubes	Solar	Dalenbäck (1990)
1983	Kerava Solar Village Finland	Rock	54 boreholes of × 25 m surrounding a 1500 m^3 water pit storage	n.a	Solar	Peltola et al. (1985) and Dalenbäck (1990)
1983	Lulevärme Luleå Sweden	Rock	120 boreholes of 65 m	Open	35–82°C Waste heat	Andersson et al. (1983) and Nordell (1994)

Continued

Table 11.1 Continued

First year of operation	Plant name/ location	Ground	Configuration	Collector type	Temperature	References
1983	Kullavik Kungsbacka Sweden	Clay	130 + 156 ducts of 12 m	U-tube	10–50°C Solar	Hultmark (1983) and Olsson (1984)
1984	CSHPSS Groeningen Netherlands	Sand and clay	Unreported number of boreholes within 38 m diameter, 20 m depth	U-tube	30–50°C Solar	Wijsman (1985) and Wijsman and Havinga (1988)
1984	Viberga Finspång Sweden	Rock	24 boreholes of 110 m	U-tube	Heat/cold	Edstedt and Nordell (1994) and Energiverk (1986b)
1985	Grosvad Finspång Sweden	Rock	126 boreholes of 110 m	U-tube	10–35°C Low temp heat from hockey rink	Edstedt and Nordell (1994) and Energiverk (1986a)
1986	Cormontreuil France	Rock	24 boreholes of 25 m	Coaxial	30–60°C Solar	Baudoin (1988)
1986	Höstvetet Suncourt Stockholm Sweden	Rock	25 boreholes of 80 m	U-tube	6–15°C Solar	Kellner et al. (1986) and Werner (1988)

1987	Ramunderskolan Sports Hall Söderköping Sweden	Clay	382 boreholes of 18 m	Double U-tube	10–31°C Solar	Magnusson et al. (1992)
1988	Meyrin Genève Switzerland	Soil	258 boreholes of 15 m	Double U-tube	4–32°C Solar	Matthey (1988)
1988	GLG-center Upplands Väsby Sweden	Rock	64 boreholes of 110 m	U-tube	Heat/cold	Edstedt and Nordell (1994)
1988	Capella Kristinehamn Sweden	Rock	17 boreholes of 110 m	U-tube	Heat/cold	Edstedt and Nordell (1994)
1988	Infra city Upplands Väsby Sweden	Rock	64 boreholes of 110 m	U-tube	9–20°C Heat/cold	Nordell (1994)
1990	Technorama Düsseldorf Germany	Sand and gravel	77 boreholes of 35 m	Steel tube coaxial	0–25°C Heat/cold	Sanner and Knoblich (1991)
1990	Onoff Järfälla Sweden	Rock	20 boreholes of 110 m	U-tube	Heat/cold	Edstedt and Nordell (1994)

focus in this chapter lies on larger systems for active thermal energy storage, most features listed in this section apply for large as well as small borehole systems, with or without active storage. BTES systems use hard rock and soft formations such as clay, sand, and soil as a storage medium. The boreholes perforating the ground volume are fitted with pipes[2] circulating a heat carrier fluid, and serve as heat exchangers between heat carrier and ground. Heat transport between the heat exchanger and the ground takes place primarily by conduction. Thermal energy is stored in the ground between the BHEs. The ground is an inexpensive storage medium and enables large amounts of heat and cold to be stored over short- or long-term periods at relatively low cost.

11.2.1 Ground properties and storage

The function of BTES is based on the ground material's ability to store and conduct heat. A high volumetric heat capacity is desirable for sensible heat storage, as done with BTES systems. Volumetric heat capacity of the ground is in the order of 1.3–2.8 MJ/m^{3}K for unconsolidated ground material and 1.8–3 MJ/m^{3}K for solid rock. In comparison, volumetric heat capacity of water is 4.2 MJ/m^{3}K. A rock volume with a typical volumetric heat capacity of 2.2 MJ/m^{3}K will hold about 0.6 kWh/m^3 if heated 1°C. This means that a storage volume measuring 100 m × 100 m × 100 m has the capacity to hold 600 MWh/K. Fig. 11.2 shows the storage capacity as a function of

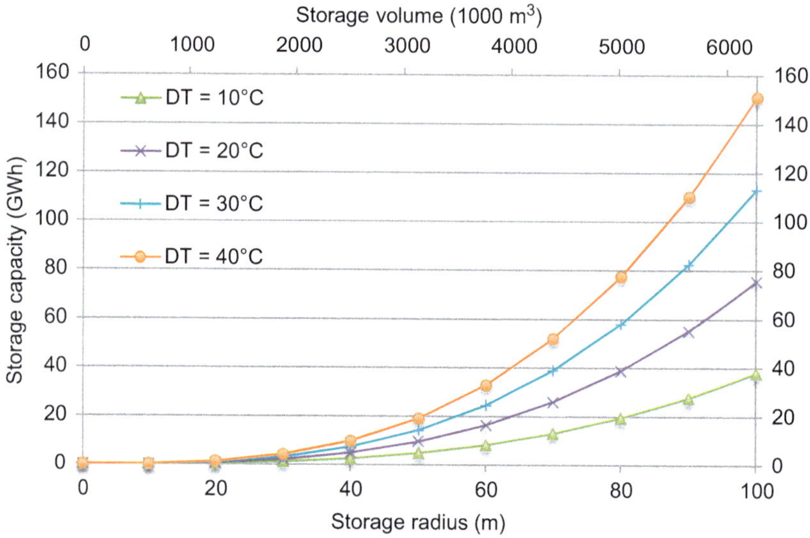

Figure 11.2 Storage capacity as a function of radius (R) and storage temperature range for a cylindrical heat storage in rock with depth $H = 2R$ and volumetric heat capacity of 2.2 MJ/m^{3}K. DT denotes the difference between highest and lowest temperature in the store.

[2] The pipes are a type of heat exchanger sometimes referred to as "collectors" but are more commonly called "ground heat exchangers" or "borehole heat exchangers."

storage radius and the difference between highest and lowest storage temperature, $\Delta T = T_{max} - T_{min}$, for a cylindrical heat storage in rock with depth $H = 2R$ and a volumetric heat capacity of 2.2 MJ/m^3K.

To achieve efficient heat transport in the ground and to/from the ground heat exchangers (GHEs), high thermal conductivity of the ground is desirable. However, high thermal conductivity will also cause higher heat losses from the store. Thermal conductivity of the ground is typically on the order of 1–5 W/m·K. Higher conductivity values are typically found in formations with high contents of quartz minerals, high bulk density and water saturation.

11.2.2 Ground heat exchangers

GHEs for BTES systems are drilled to a certain depth in hard or soft formations, and then fitted with the collector pipes. In soft formations the collector pipes may also be pressed or vibrated into the shallow formation. A typical depth for shallow installations where the collector pipes are pressed or vibrated in is 10–40 m. Common borehole depths are 100–200 m. In most countries, boreholes are backfilled or grouted. The grout serves to stabilize and seal the boreholes and also to achieve good thermal contact between the ground material and the collector pipes. In the Scandinavian countries where geology and hydrology are characterized by hard rock and high groundwater levels, boreholes are typically left ungrouted. Natural groundwater will then fill the borehole to the groundwater table level. Groundwater offers excellent thermal contact between borehole wall and collector pipe and enhances heat transfer due to natural convection (Kjellsson and Hellström, 1997; Gustafsson and Gehlin, 2008).

A wide range of GHE types are used for BTES: single and double U-tubes are most commonly used due to their reliability, simple installation, and low cost. Various types of coaxial collectors, with hard or soft outer shell are used occasionally, due to their low thermal resistance and pressure drop but are more expensive, more complicated to install, and involve higher risk for leakage. Low thermal resistance in the BHEs is particularly desirable in BTES systems with active storage of heat and cold as poor heat transfer in the collectors will affect the storage efficiency both at injection and rejection.

In high-temperature BTES and low-temperature BTES systems permanently operated at temperatures above +4°C, pure water may be used as heat carrier fluid in the borehole collectors (Reuss, 2015). Water mixed with an antifreeze solution—typically ethanol or propylene glycol, and sometimes ethylene glycol—is used for systems where the heat carrier fluid is allowed to fall below 0°C, such as some applications of combined heat and cold storage, or if there is a risk for freezing of the horizontal pipes interconnecting the BHEs and connecting the boreholes with the building. Ethanol cannot be used for high temperature applications.

11.2.3 Storage geometry

The geometry of the BTES system is important for the heat loss. While storage capacity is proportional to storage volume, heat losses are proportional to surface area. The

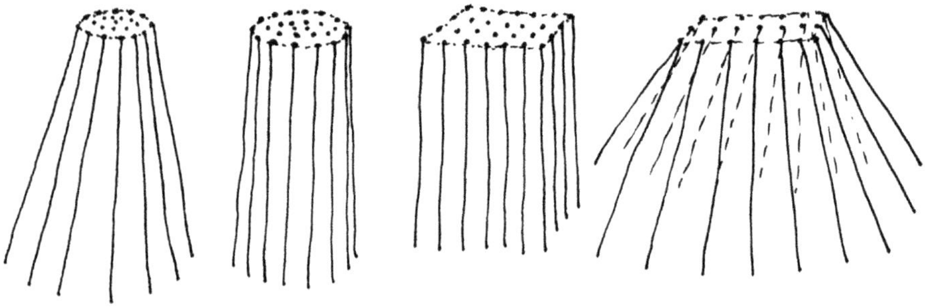

Figure 11.3 Circular configurations (left) and rectangular configurations (right) with or without diverging boreholes.
Illustration from S. Gehlin (2015).

relative heat loss decreases with increasing storage volume. Accordingly, a design with small surface-to-volume ratio is desirable. Annual heat loss for a given annual mean temperature of the store (during steady—periodic seasonal operation) is a function of the volume, shape, and ground thermal conductivity. Thermal conditions at, and distance to, the ground surface are also important factors for the operational heat loss. Hellström (1991) gives a comprehensive analysis of heat loss from BTES systems of various shapes. Storage efficiency is defined by Nordell (1994, p. 12) as given by the ratio between stored and extracted thermal energy. This efficiency is favored by large storage volume and compact geometry, ie, a small surface-to-volume ratio.

The ideal storage shape in an infinite medium would, from a heat loss perspective, be a sphere. However, this is obviously a complicated geometry from a construction perspective. Therefore the most commonly used storage geometries are the cylinder or parallelepiped (box) shapes. If available ground surface is limited, the storage can be given a larger volume by letting the boreholes diverge from the vertical line, so that the boreholes form a broom-like shape (Fig. 11.3).

Boreholes are placed in a symmetrical arrangement, usually in a circular, rectangular, or hexagonal pattern (Fig. 11.4), with borehole distance typically in the range 4–6 m. Borehole depth, distance, and configuration must be carefully optimized for each project, using advanced design software such as Earth Energy Designer (Hellström et al., 1997) and GLHEPRO (Spitler, 2000). Hellström and Sanner (2001) describe and compare a number of available design software of this type.

BTES systems are not insulated on the sides or at the bottom, but top insulation is sometimes used to limit heat loss to the atmosphere (Reuss, 2015).

Figure 11.4 Boreholes may be arranged in rectangular, hexagonal, or circular drilling patterns.
Illustration from S. Gehlin (2015).

11.2.4 Groundwater flow

Heat loss calculations typically only take into account the conductive heat transport in the ground. Convective heat transport with groundwater flow may also affect the heat storage performance (van Meurs, 1986; Claesson and Hellström, 2000). Åberg and Johansson (1988) studied the effects of groundwater flow on BTES systems, both in porous formations and in fractured hard rock. They conclude that for natural groundwater flow in an evenly fractured ground volume, and with common hydraulic conductivity of the ground, the convective heat transport from the storage will be insignificant. For highly permeable ground and large hydraulic gradients, the stored body of heat would move on the order of a few meters downstream over one year. In the case of highly fractured rock or large fractures intersecting the rock volume, the groundwater transported in the fracture may cause significant heat loss from the store. The authors recommend cement injection to eliminate this problem in such rare cases. Chiasson et al. (2000) made numerical simulations of the ten-year performance of a cooling-dominated borehole field in groundwater flow. They found that even moderate groundwater flow affects the year-by-year ground temperature increase compared to pure conductive conditions. Effects of groundwater flow on long-term performance of a BTES system with unbalanced winter and summer loads are discussed by Zanchinia et al. (2012). They conclude from their study that groundwater flow has little effect on short-term peak loads but positively influences the long-term performance by enhanced thermal exchange with the surroundings.

For high temperature BTES systems, convective heat transport induced by the thermal gradients becomes important and must be considered (van Meurs, 1986; Hadorn, 1990).

11.2.5 Storage temperature

An important factor in the design and operation of BTES systems is the average storage temperature in relation to the average temperature of the surrounding ground. If the annual average temperature of the store is lower than that of the ground surrounding the store, heat will be gained to the store (Hadorn, 1990). The amplitude of the temperature variation over the year ($T_{max} - T_{min}$) does not affect heat loss from the store, as the net heat flow through the storage boundaries becomes zero for an annual cycle. Steady-state heat loss depends on the temperature difference between the annual average storage temperature and the undisturbed ground temperature ($T_{avg} - T_o$).

Hadorn (1990, p 57–58) classified UTES systems according to storage temperature level, and identified technical difficulties related to storage temperature. For storage temperatures below 0°C phase change problems related to freezing occur. Within the interval 0–30°C few temperature-related problems occur. Heat loss from the store is low and the store may be left uninsulated, but a heat pump must typically be used if the storage is used for space heating. Above 30°C the store may be used without a heat pump to provide heat to low and medium temperature distribution circuits. However, heat loss to the surrounding ground becomes an issue, and above 60°C problems related to materials and the environment may appear. Storage temperatures above 100°C may be desirable for existing urban district heating circuits, but there will be technical issues related to boiling groundwater, material properties, and the environment (Hadorn, 1990).

Examples of suitable heat sources for low-temperature BTES systems are excess heat from cooling machines in office buildings, commercial buildings, and other facilities. Exhaust air is an excellent heat source as it may also allow for decreased installed heat pump capacity for the winter period. Industrial waste heat and solar heat are suitable for high-temperature BTES applications.

Various schemes of operation are used for BTES. The most common scheme involves parallel-coupled boreholes, with simultaneous charging and rejection of all boreholes in the ground volume. Another strategy is to charge the storage beginning at the center and continuing outwards. This is done by either connecting the boreholes serially in the radial direction, or by dividing the store into circular sections. Heat extraction from a heat storage system would take place in the opposite direction. That means that the heat storage volume would always be warmest in the middle.

11.3 Environmental aspects

Protection of the environment is important at a global level as well as locally. Work has been carried out within IEA ECES (Bakema et al., 1995; IEA ECES, 1997) to evaluate environmental benefits as well as risks related to UTES. BTES systems help to reduce emissions of CO_2, NO_x, and SO_2 to the atmosphere by replacing fossil fuels and increasing the efficiency of energy utilization. By replacing conventional chillers with direct cooling from BTES, the risk for release of refrigerants into the atmosphere is decreased. BTES systems increase the potential for storage of solar heat and waste heat in both the short and long term. Compared to conventional systems, BTES systems are visually unobtrusive (essentially invisible), take up little space, and are quiet.

Like all types of construction work, BTES systems have potential risks for the local environment. Problems may occur during construction as well as during operation. Reuss (2015), Paksoy (2007), and Banks (2012) give advice on environmental concerns related to BTES and how to avoid them.

11.3.1 Risks related to temperature

In some countries (eg, Germany) concerns regarding disturbance to the ground temperature have been raised, and restrictions to how much the ground temperature may be disturbed have been formulated. The temperatures in the underground are, however, already disturbed by urbanization. We like to keep our buildings at a constant temperature around 20°C, and temperature logs in urban areas show clearly the contribution of heat leakage from buildings to the ground temperature profile. This thermal front may reach more than 100 m below ground surface, depending on how long the building has been in place. Similarly, paved streets, sidewalks, and parking lots also raise the underground temperature. The temperature change in the ground caused by low-temperature BTES systems would be of the same order of magnitude.

Frost heaving has caused problems in rare occasions where horizontal connection piping has been placed near the surface in saturated fine sands, silt, or clay and the BTES system has been allowed to operate at temperatures below 0°C for long periods of time. Geotechnical problems related to freezing of vertical pipes in clay have been investigated by Gabrielsson et al. (1997). Severe settling around the boreholes were observed due to structural changes in the clay. The clay collapsed during the thawing process. Freezing around permanent steel casing of boreholes penetrating overburden clay layers may also cause problems with structural changes and collapse of the clay. Proper design and construction should minimize this risk.

In closed systems such as BTES the risk of geochemical problems due to altered ground temperatures around the boreholes is limited, especially when operating at temperatures close to or below undisturbed ground temperature. BTES systems operating at temperatures higher than 40°C are more likely to experience geochemical effects, and a more detailed consideration of geochemistry is recommended (IEA ECES, 1997).

Changes in ground and groundwater temperature may result in alterations in the balance of microorganism species; however, temperature is not the only controlling factor for microorganisms. Light, oxygen, nitrogen, phosphorous, and sulfur are examples of other factors that affect microorganisms. Microbiological risks with BTES systems are unlikely to occur or cause problems, especially at low operating temperatures. Studies on possible microbiological effects may be considered for BTES operating at temperatures above 40°C (IEA ECES, 1997).

11.3.2 Risks related to geology and geohydrology

During drilling, in situ testing, construction, and operation, care must be taken to prevent surface contaminants from the surface entering the borehole and mixing with groundwater. It is essential to avoid interconnecting two aquifers with different pressure conditions and water quality. This is prevented by proper backfilling of boreholes in such cases.

BTES systems are unsuitable for areas with artesian aquifers. Consequences of large uncontrolled artesian flow are severe and costly and must be avoided. In areas with geological formations containing layers of evaporites, such as halite, sylvite, gypsum, and anhydrite, BTES systems should be avoided. Consequences of penetrating such layers and exposing them to water may cause severe settling over large areas (Reuss, 2015).

11.3.3 Risks related to construction

Although the high-density polyethylene pipe used for BTES construction is extremely robust, there is the possibility of leakage of the heat carrier fluid from the heat exchanger pipes into the natural environment. Antifreeze solutions used as a heat transfer (carrier) fluid should therefore be nontoxic and not adversely affect the physical, metallurgical, or chemical integrity of the piping system. Today ethylene glycol, propylene glycol, or ethanol are used as heat transfer fluids. Of these three, ethylene glycol is the most toxic, but biodegrades quickly. Propylene glycol and ethanol are relatively nontoxic and rapidly biodegradable. Denaturation additives degrade more slowly. If leakage would occur, the quantities of carrier fluid reaching the environment would typically be moderate, and the system pressure drop would soon trigger a shutdown of the circulation pump.

Drilling through contaminated layers of soil or groundwater should be avoided. If contaminated layers have been penetrated, boreholes must be thoroughly grouted to seal the borehole and eliminate risks for migration of pollutants.

Before drilling starts, precautions must be taken to eliminate risks of hitting other subsurface constructions and infrastructure such as gas pipes, water, and sewage pipes, electric or telecommunication cables, tunnels, natural or man-made underground cavities and constructions, and possible archaeological remains.

It is important to make appropriate arrangements for the disposal of drill cuttings so that the material does not affect the local environment.

Other risks may be evaluated (and often eliminated) by investigation of the existence of other water wells or energy wells in the neighborhood, overhead cables or constructions in reach of the drill rig, protection zones for water or nature, and adjacent buildings.

11.3.4 Regulation to protect the environment

Guidelines, standards, and codes can all help avoid environmental concerns related to BTES. Furthermore, in many locales, approval from local authorities for specific projects must be obtained. The existence and comprehensiveness of legislation concerning BTES and protection of groundwater and the underground environment varies between countries and within countries. Banks (2012, p. 433) provides a list of legislation and codes that may or may not have impact on closed-loop BTES systems. This includes water resources legislation, energy efficiency legislation, and codes of good practice by industry and environmental bodies. Den Braven (1998, 2000) reported on state-to-state regulations on antifreeze and grouting (backfilling) of boreholes in the United States.

Considering the large number of BTES systems constructed over the years, few accidents with environmental impact have occurred and been reported. When BTES systems are being considered, the very low risks of environmental hazards from a properly designed, constructed, and maintained BTES system should be weighed against the considerable environmental benefits from reduction of emissions and energy conservation.

11.4 Worldwide borehole thermal energy storage applications

BTES systems can be applied in a large variety of ways, with different storage temperature levels, sizes, and system configurations. This section gives examples of various

BTES applications around the world over the last few decades. Early BTES applications are listed in Table 11.1 and examples of BTES systems around the world are summarized in Table 11.2.

BTES systems can be categorized in terms of storage temperature:

1. High temperature storage: solar or waste heat, typically large scale and often connected to district heating network or industrial processes.
2. Low temperature storage: temperature levels close to undisturbed ground temperature for heating and/or cooling—typically for residential buildings, office buildings, commercial, or institutional buildings.

Applications can also be categorized according to what type of user the BTES system serves:

1. residential heating and/or cooling
2. heating and/or cooling of commercial and institutional buildings
3. cooling and/or heating of industrial processes
4. heating of roads and other paved surfaces

The most abundant type of BTES system is low-temperature BTES for combined heating and cooling of office buildings, commercial, or institutional buildings. Design and construction techniques are well understood and the system size may be readily scaled as needed. With growing awareness of the need for energy storage as a crucial component in sustainable energy systems based on renewable energy resources, there is now a budding renewed interest in solar energy usage and high-temperature storage in district heating networks. High-temperature BTES applications are appealing as they do not require heat pumps to make the stored heat useful. To keep relative heat losses acceptably low, high temperature BTES systems need to be large scale with many boreholes and are therefore usually appropriate for district heating systems.

11.4.1 High-temperature solar heat storage

Storage of high-temperature solar heat was the intention that first led to development of BTES systems, and many examples exist. Gao et al. (2015) discuss this type of BTES application and provide lists of full scale and pilot plant BTES systems used with solar heating. More descriptions of early solar heating BTES systems are found in Section 11.1.2, and Table 11.1.

The solar district heating BTES system in Neckarsulm-Amorbach in southwest Germany was designed to provide space heating and domestic hot water for a new-built housing area comprising some 700 residential buildings and a school (Reuss, 2015). The total annual heating load is 3 GWh. Five thousand square meter flat-plate solar collectors are connected to two short-term storage water tanks, each 100 m^3, and the seasonal storage BTES system. The Neckarsulm-Amorbach plant was the first large-scale BTES system in Germany, and has been built in two steps. In a first stage, 36 boreholes were drilled in 1997, and later extended to a total of 528 boreholes in 2001. The borehole field is laid out in a rectangular configuration to allow for future extension. Geological formations at the site

Table 11.2 **Examples of borehole thermal energy storage systems worldwide**

Country	City	System type	Year	Number of boreholes and borehole depth	Heat/cold	References
Canada	Oshawa, Ontario	University campus	2004	384 × 213 m Limestone	Heat/cold	Dincer and Rosen (2007) and Wong et al. (2006)
Canada	Okotoks Drake Landing Solar Community	Residential	2007	144 × 35 m	Solar heat High temp	Sibbitt et al. (2012) and Wong et al. (2006)
China	Tianjin	Business center	2011	3789 × 120 m	Heat/cold + ice storage	Yin et al. (2015)
China	China Academy of Building Research Beijing	Office building Net zero energy	2014	20 × 100 m + 50 × 60 m	Solar + heat/cold	Yu et al. (2015) and Li et al. (2015)
China	Zhungguancon International Center Beijing	Office building	2008	1060 × 123 m	Heat/cold	Zang and Xu (2014)
Denmark	Brædstrup	District heating	2012	48 × 45 m	Solar heat and district heating	Miedaner et al. (2015)

Finland	Sibbo	Logistics center	2012	150 × 300 m +159 in Phase 2 Rock	Heat/cold	Huusko and Valpola (2014)
Germany	Neckarsulm Solar district heating	Residential	1997 + 2001	528 × 30 m Clay	Solar district heating	Reuss (2015, pp. 138–140)
Germany	Attenkirchen	Residential	2002	90 × 30 m	Hybrid solar district heating with central tank	Reuss et al. (2006)
Germany	Crailsheim	Residential + school	2007	80 × 55 m	Solar	Bauer et al. (2007) and Mangold (2007)
Norway	Akershus	Hospital	2007	228 × 200 m Rock	Heat/cold	Midttomme et al. (2010) and Bäcklund (2009)
Poland	Atrium 1	Office building	2014	50 × 200 m	Heat/cold without heat pump	Skanska (2014a)
Romania	Bucharest	VW Bucharest Auto Showroom	2009	112 × 72 m	Heat/cold	Polizu and Hanganu-Cucu (2011)
Romania	Bucharest-Marguele	ELI-NP research center	2015	1080 × 125 m	Heat/cold	Bendea et al. (2015)
South Korea	Lotte World Tower Seoul	Skyscraper	2015	720 × 200 m	Heat/cold	Viessmann (2012)

Continued

Table 11.2 **Continued**

Country	City	System type	Year	Number of boreholes and borehole depth	Heat/cold	References
Sweden	Karlstad	Karlstad University campus	2015	204 × 240–250 m	Heat/cold	Olsson (2014) and Gehlin et al. (2015)
Sweden	Entré Lindhagen	Office building	2014	144 × 220 m	Heat/cold without heat pump	Skanska (2014b)
Sweden	Näsbypark	Historical building	2004	48 × 180 m	Heat/cold recharge with lake water heat	Lund et al. (2004)
Sweden	Luleå	University building	1981–1989	120 × 60 m	Industrial high temp	Nordell (1994)
Sweden	Anneberg	Residential	2002	99 × 65 m	Solar high temp	Dalenbäck et al. (2000), Lundh and Dalenbäck (2008), and Heier et al. (2011)
Sweden	Lund University	Astronomy House	2001	20 × 200 m Clayey soil and shale	Heat/cold	Andersson (2007)
Sweden	Emmaboda Xylem	Industrial waste heat	2011	140 × 150 m Rock	Industrial high temp	Nordell et al. (2015)
Switzerland	Därlingen SERSO	Road heating	1994	91 × 65 m Rock	Heat from road	Eugster (2002) Eugster (2007)

United Kingdom	Croydon	Office and warehouse	2000	30 × 100 m Chalk	Heat/cold	Witte and van Gelder (2007)
United Kingdom	DMU Leicester	Hugh Aston University building	2009	56 × 100 m	Heat/cold	Naicker and Rees (2011) and Cullin et al. (2015)
United States	Richard Stockton College Pomona, New Jersey	Campus buildings	1994	400 × 135 m Sand/clay	Heat/cold	Stiles (1998)
United States	Oakland University Rochester Michigan	Human Health building	2013	256 × 100 m	Heat/cold	Kistler and Karidis (2015)
United States	Ball State University Muncie Indiana	University Campus	2013	1800 × 140–150 m + 1800 in Phase 2	Heat/cold	BSU (2015)

consist of a 30−35-m thick low-permeable clay layer on top of a highly permeable dolomite with large groundwater flow. Therefore the boreholes were drilled to 30-m depth to prevent high heat loss at the bottom. The boreholes are fitted with double U-tube heat exchangers, and the borehole spacing is 2 m. The top of the BTES field is thermally insulated. The system is designed for 50% solar fraction, and a gas boiler provides supplementary heating.

The Kerava Solar Village in Finland is an early large-scale, high-temperature solar-heating system built as a pilot project in 1983 (Peltola et al., 1985; Dalenbäck, 1990). The village containing 44 apartments with large south-facing windows and flat-plate solar panels has a total annual heating demand of 495 MWh, and taking into account heat loss, the solar heating system (including heat pumps and back-up heating) was designed to deliver 550 MWh annually. The BTES system is a hybrid utilizing 54 tilted boreholes of 25-m depth surrounding a 1500 m^3 cavity in the rock, filled with water. The water-filled cavity reached 20-m depth and provided stratified heat storage, with 55−65°C water at the top and as low as 8°C in the winter at the bottom. The water at the top is used for short-term storage and domestic hot water. The boreholes are arranged in two circles around the central water cavity, with 18 boreholes in the inner circle and 36 in the outer circle (Fig. 11.5). A 240 kW heat pump supports the system in the winter, and two electrical boilers of 200 kW each provide back-up heat. A solar fraction of 40−50% was obtained by the system.

The Attenkirchen hybrid solar district heating system was constructed in 2002 to serve 20 residential buildings with a total annual heating demand of 490 MWh (Reuss et al., 2006). It was designed for a solar fraction of 50% and collects solar heat from 765 m^2 of flat-plate solar collectors. The hybrid storage consists of a cylindrical

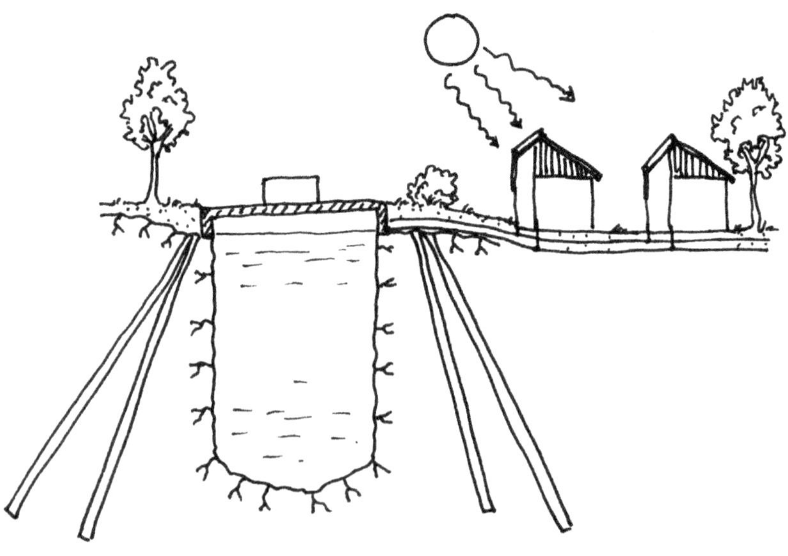

Figure 11.5 The Kerava Solar Village BTES and water-filled rock cavity. Illustration from S. Gehlin (2015).

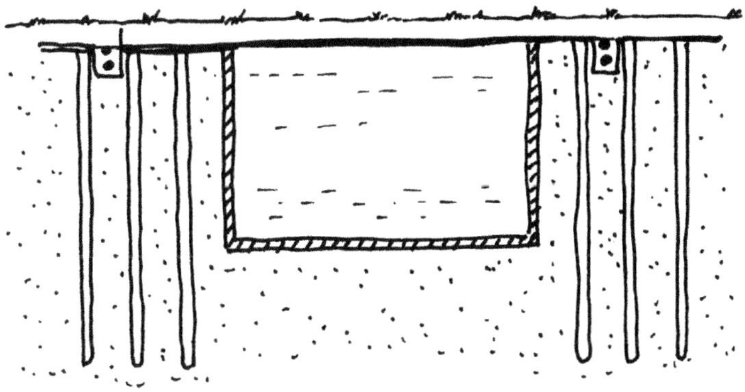

Figure 11.6 Three circles of boreholes surround the top-insulated concrete water tank in Attenkirchen.
Illustration from S. Gehlin (2015).

concrete water-storage tank measuring 9 m in diameter and 8.5 m depth (volume 500 m^3), surrounded by three circles of 30-m deep boreholes (Fig. 11.6). In total, there are 90 boreholes fitted with double U-tube heat exchangers. The top of the storage area is thermally insulated with 20-cm polystyrene.

The Anneberg high-temperature solar-heating BTES (Fig. 11.7) without heat pumps in Stockholm, Sweden (Dalenbäck et al., 2000; Lundh and Dalenbäck, 2008; Heier et al., 2011), was completed in late 2002 and serves a residential area with 50 houses. The buildings have an annual heating demand of 565 MWh, including domestic hot water. The BTES system consists of 99 boreholes drilled to 65 m into hard rock, and fitted with double U-tube heat exchangers. The low-temperature space heating system has three ways in which heating can be supplied: directly from the 2400 m^2 of flat-plate solar collectors, stored heat from the BTES, or from individual electrical backup-heating units when the first two options are

Figure 11.7 The Anneberg high-temperature BTES.
Illustration from S. Gehlin (2015).

insufficient. The system was the first system in Europe with seasonal solar storage in rock and not utilizing a heat pump during discharge. The borehole field is rather small; hence heat loss is significant (40%).

Drake Landing Solar Community in Okotoks, Alberta, Canada (Fig. 11.8), has been in operation since 2007 and supplies space heating to 52 detached houses through a local district heating network, without heat pumps (Sibbitt et al., 2012; Wong et al., 2006). Solar heat from 2293 m^2 of roof-mounted flat-plate solar collectors on the detached garages is stored in the 144-borehole BTES system in soil. The boreholes are 35-m deep and fitted with single U-tube heat exchangers. The top of the BTES system is thermally insulated. The cylinder-shaped borehole field is configured to maintain the center of the field at the highest temperature to maximize heating capacity and the outer edges at the lowest temperature to minimize heat loss. A 240 m^3 short-term thermal storage water tank is used to interconnect the solar collectors, distribution network, and BTES subsystems. The Drake Landing Solar Community BTES system has undergone detailed monitoring since it was brought into service in July 2007. The system met a solar fraction of 97% in its fifth year of operation and should inspire more projects of its kind, in Canada and beyond.

In Crailsheim, Germany, a high-temperature solar-heated BTES system serves 260 apartments, a school, and a gymnasium (Bauer et al., 2007; Mangold, 2007; Miedaner et al., 2015). Solar collectors are mounted on the roofs and along a noise protection wall. The solar collectors are connected to a diurnal storage tank of 100 m^3 and the BTES for seasonal heat storage consisting of 80 boreholes of 55 m depth. The borehole field is connected to a 480 m^3 buffer tank, which serves to even out the solar heat gain peaks in the summer. Heat is distributed to the buildings either directly from the BTES, or via heat pumps if needed. Total heating demand for the area is 4100 MWh annually, and 485 kW of electricity is used for heat pump operation. The measured solar fraction during 2012–2013 was 51%.

Figure 11.8 The Drake Landing Solar Community local district heating with BTES. Illustration from S. Gehlin (2015).

The Brædstrup high-temperature solar BTES district heating plant in Denmark (Miedaner et al., 2015) began operation in May 2012, and is part of Brædstrup Total Energy Plant, which is the production plant for the Brædstrup district heating system. The production plant, of capacity 40,000 MWh, was originally a natural gas–fired CHP unit with a 2000 m^3 steel buffer tank, but is now converting to 100% renewable heat production in several steps. In addition to the waste heat from the gas-fired CHP, heat is collected from 18,600 m^2 of solar collectors, connected to two steel tanks of 5500 m^3 and 2000 m^3 volume. Solar heat is stored in the BTES consisting of 48 boreholes drilled to 45 m depth spaced 3 m apart in a triangular configuration. Double U-tube heat exchangers are used. The top of the storage is insulated to reduce heat loss. The system also includes a 1.2 MW heat pump and a 10-MW electrical boiler. Heat loss from the BTES is estimated at 24%.

11.4.2 High-temperature industrial heat storage

There are few BTES applications that store industrial heat at high temperatures, even though the potential for such systems is significant.

The Emmaboda Xylem high-temperature BTES system began operation in 2010 (Nordell et al., 2015). As of March 2015 approximately10 GWh has been stored and a storage temperature of 40–45°C has been reached. Only a fraction of the stored heat has been extracted so far, and the storage is expected to reach full capacity in 2015. The purpose of the system is to make use of the waste heat from the industrial processes in the molding factory for space heating of the factory and office buildings. When the demand for space heating is lower than the waste heat production, surplus heat is stored in the BTES. The system works without a heat pump, but heat pumps may be added later to improve the system and reduce heat loss. The system consists of 140 boreholes in a rectangular configuration drilled to a depth of 150 m, and with a borehole spacing of 4 m. The storage is divided into seven sections with 20 boreholes in each section. The sections are individually operated for injection or extraction of heat depending on storage temperature. Coaxial BHEs are used, where the circulated water is in direct contact with the borehole wall. The BTES has reduced the amount of bought district heating for space heating by approximately 4 GWh/year.

The Lulevärme Heat Store in Luleå, northern Sweden (Nordell, 1994), was in operation from 1982 to 1989. The 120 boreholes drilled in granitic rock to a depth of 65 m in a 10 × 12 rectangular configuration were used to store high-temperature (70–82°C) waste heat from a steel plant. The heat was transferred to the storage via the district heating network during the summers. The heat was recovered at 35–55°C in the winters for heating a university building.

11.4.3 Low-temperature solar heat storage

Low-temperature solar heat storage is attractive for small-scale solar heating systems, as heat loss is low and potential problems related to high temperatures are eliminated. In most cases, however, heat pumps are needed. There are several interesting examples of applications of this type, including deicing of road surfaces.

The Höstvetet Suncourt project in Stockholm, Sweden, in 1986 was an early low-temperature BTES solar heating project (Kellner et al., 1986; Werner, 1988). The 71-apartment residential building contains a large glazed courtyard that works like a solar collector in the summers. When the air temperature inside the courtyard exceeds 20°C, the air is used as a heat source for heat pumps providing domestic hot water. In this process the courtyard is cooled down. When heat produced from the courtyard air exceeds the hot water demand, excess heat is stored in a BTES system beneath the building, consisting of 25 boreholes of 80-m depth in hard rock. Storage temperature varies between 6 and 15°C.

The Grosvad Finspång BTES system built in 1985 in Sweden (Energiverk, 1986a) offers an interesting system combination with a 7000 m^2 bandy[3] field used as a heat source in combination with BTES to provide heat for 550 apartments and a school. The 126 boreholes, placed below a parking lot, are drilled to 110 m depth. In the summers, 3 GWh solar heat from the bandy pitch is stored in the borehole field, and in the winters the heat is recovered and supplies the building with the aid of three heat pumps with total heating capacity of 2.5 MW. Storage temperature varies between 10 and 35°C. In wintertime when the bandy field is artificially frozen, phase change heat from the freezing process is used for space heating of the residential buildings, through the same heat pump unit.

The SERSO Därlingen Solar heat for road deicing (Eugster, 2002; Eugster, 2007) is a well-known and well-documented BTES installation in central Switzerland (Fig. 11.9). It began operation in 1994 and uses horizontal piping beneath the road surface to collect solar heat from the road in the summer. The heat is stored in the 91 boreholes drilled in hard rock to a depth of 65 m. In the winter, the stored heat is used to stabilize the road temperature just above 0°C, preventing ice formation and freezing of compacted snow. The annual runtime is less than 1000 h in the winter and another 1000 h in the summer. The supply temperature is regulated based on ambient air temperature, and is generally below 10°C. Typical average heat output of the system is around 100 W/m^2 of road surface.

11.4.4 Low-temperature storage for heating and cooling

BTES systems used for combined heating and cooling are the most widely spread applications and exist in a range of sizes. Many buildings such as offices and public buildings have an energy load profile with fairly balanced heating and cooling loads, making these systems very energy efficient. Almost every GSHP system with both significant heating and cooling demands can be considered BTES systems of this kind. Here, just a few examples are presented.

The Croydon building in Sussex is one of the larger BTES systems in the United Kingdom (Witte and van Gelder, 2007). The building is a three-story office building also hosting warehouse facilities. Annual cooling and heating loads are

[3] Bandy is a team winter sport played on ice in which skaters use sticks to direct a ball into the opposing team's goal. It is played on a bandy field, the size of a football pitch.

Figure 11.9 The SERSO Därlingen borehole thermal energy storage system for road deicing. Illustration from S. Gehlin (2015).

100–125 MWh and 90–100 MWh, respectively, with summer peak cooling loads reaching up to 130 kW. A BTES system comprising 30 boreholes of 100 m depth, drilled in chalk and fitted with U-tubes, is combined with a dry cooler that stores cold in the ground in early spring. An under-floor heating system and the 85 distributed GSHPs in the building allow for simultaneous heating and cooling in different parts of the building if needed.

In 2004, University of Ontario Institute of Technology in Oshawa, Canada, installed a BTES system for heating and cooling of the entire campus (Dincer and Rosen, 2007; Wong et al., 2006). Three hundred eighty-four boreholes were drilled to 213 m in limestone and hard rock and are water-filled, as in the Scandinavian practice, instead of the typical North American practice of grouted BHEs. The BTES system is charged in the summer by chillers and heat pumps of total nominal capacity of approximately 7 MW cooling. In the winter, the heat pumps provide warm water for campus use at a temperature of 52.5°C. Supplemental heating is provided by condensing boilers.

Ball State University in Muncie, Indiana, is going through a large-scale BTES conversion project in two phases (BSU, 2015), which will result in its four aging coal-fired boilers being shut down. In the first phase, completed in 2013, 1800 boreholes were drilled to 140–150-m depth in the North District Energy Station. The second phase will add another 1800 boreholes to the South District Energy Station. Two 8.8 MW heat pumps are served by the BTES borehole field, and provide cooling and heating via two separate district loops that run throughout the campus: a 5°C cold water loop and a 65°C hot water loop.

The Akershus University Hospital in Ahus, Norway, was constructed in 2007 (Midttomme et al., 2010; Bäcklund, 2009). The borehole field, comprising 228 groundwater-filled boreholes of 200-m depth, drilled in hard rock, provides heat and cold to the hospital buildings supported by a combined ammonia chiller and heat pump system.

11.4.5 Low-temperature combined systems

There are many examples of BTES systems that are combined with other sources of heat and cold, such as lake or river water, ice storage, water tanks, biofuel, and so forth. Below are some examples.

For a long time the BTES at Richard Stockton College in Pomona, New Jersey (Stiles, 1998), built in 1994, was the largest BTES system in the world. The 400 boreholes, with a depth of 135 m, are fitted with U-tubes and penetrate three aquifers in saturated sands and clays. The borehole field is located under a 16,000 m^2 parking lot. The ventilation and air-conditioning system design, including heat pumps with a cooling capacity of over 5-MW, did not balance the thermal load, so the field was slowly heating up. A supplemental ATES was therefore added in 2008 to provide additional cooling and reduce the cooling demand on the BTES system.

Tianjing Cultural Center in Beijing was completed in 2011 and is, at present, the largest BTES system in the world. Placed below a lake in front of the building are 3789 boreholes of 120-m depth fitted with double U-tube heat exchangers (Yin et al., 2015). The BTES system is combined with ice storage cooling, a cooling tower, and district heating.

Lotte World Tower in Seoul, South Korea, is the second-tallest skyscraper building in the world and was scheduled for completion in 2015 (Viessmann, 2012). The tower is 555-m high and has 123 floors above ground and six floors below ground, hosting private apartments, offices, retail stores, and a hotel. The hybrid BTES system consists of 720 boreholes, 200-m deep, serving six heat pumps providing 1.7-MW heating and 1.9-MW cooling. Another six heat pumps provide 2-MW heating and 1.7-MW cooling from river water outside the building.

The Näsby Park hybrid BTES system was built in 2004, for the Näsby Castle buildings close to Stockholm, Sweden. The 48 boreholes of 200-m depth are recharged with 15–20°C surface water from a nearby lake during the summer. A 400-kW heat pump is used for heat load operation (Lund et al., 2004).

China Academy of Building Research is a net-zero energy building (NZEB) in Beijing, China, built as a demonstration building for NZEB in 2014 (Yu et al., 2015; Li et al., 2015). It has several storage systems for heat and cold; BTES, solar heat, thermal storage in the building material, and heat and cold storage in water tanks. The BTES system consists of two borehole fields on each side of the building. The smaller field has 20 boreholes of 100-m depth, arranged in two rows, fitted with double U-tube heat exchangers. The larger field consists of 50 boreholes of 60 m depth, arranged in five rows, and fitted with single U-tube heat exchangers. The BTES system is charged with solar heat from solar panels on the roof.

The Polish office building Atrium 1 in Warsaw (Skanska, 2014a) and the Swedish Skanska headquarters office building Entré Lindhagen in Stockholm (Skanska, 2014b) are two buildings with high-temperature cooling and low-temperature heating provided by BTES systems. The construction company Skanska has patented a special BTES concept called "Deep Green Cooling." It provides office cooling in the summer and preheating of incoming air in the winter without the use of heat pumps, and operates at temperatures close to undisturbed ground temperature. Free-cooling with outdoor air is used in combination with the BTES system, and additional winter heating demand is provided from the district heating network. Atrium 1 was completed in January 2014 and consists of 50 boreholes, 200-m deep, providing 232-MWh cooling and preheating annually. Entré Lindhagen was built in 2014 and has 144 boreholes of 220-m depth.

SOK Sibbo Bastukärr Logistics Center in Finland (Huusko and Valpola, 2014) is a hybrid BTES system built in 2012 and is the largest BTES system in Finland and the third largest BTES system in Scandinavia. The borehole field is part of a so-called GeoBio hybrid system and consists of 150 closed-loop boreholes drilled to 300-m depth. It covers 100% of the cooling demand and 50% of the heating demand for the logistics center building. The other 50% heating demand is covered by wood-pellet boilers. Temperatures in the storage vary between 18 and 32°C. In the summer the borehole field is charged with solar heat from horizontal pipes under a 2.5 ha parking lot. The BTES system provides free-cooling in the summer and, if needed, reversible heat pumps are used for peak load cooling. Two 1-MW heat pumps provide maximum 50°C heat to a low temperature heating system, and two 2-MW wood-pellet burners provide maximum 120°C heat to the local district heating net.

11.5 Conclusions

Low-temperature BTES systems provide efficient heating and cooling of large buildings, such as office buildings, commercial, and institutional buildings. The technology has a great potential to contribute significantly to energy efficient and sustainable energy systems in most regions in the world. In cooler climates, it offers significant reduction of costs for cooling and low costs for heating. With increasing demand for energy efficiency, buildings tend to be built with improved insulation. These well-insulated buildings, combined with increasing internal heat gains, create an increased demand for comfort cooling in cooler climates. BTES systems can take advantage of the increased demand for cooling and provide both heating and cooling in a very elegant and cost-effective way. Since the 1990s the number and types of buildings using BTES for combined heating and cooling has increased steadily as experience and proof of cost effectiveness has grown.

After the first decades of enthusiastic development of high-temperature seasonal storage, the interest in such systems has faded in favor of less complex low-temperature BTES. However, in recent years a renewed interest in high-temperature BTES applications with seasonal storage of solar heat or waste heat from industries and cogeneration plants has emerged. The number of such applications worldwide is still small, but the success of the moderate-sized district heating project Drake Landing Solar Community in Canada, and the industrial waste heat storage system in Emmaboda, Sweden, show definite potential for such applications.

The heat buffering ability of BTES systems has potential for improving the efficiency and robustness of district heating systems on city scales as district heating providers adjust to a building stock refurbished with low-temperature heating systems and thermally active building systems.

Several very large BTES systems with more than a 1000 boreholes have been constructed in recent years (see Table 11.2), indicating a trend toward larger BTES systems. Top five largest BTES systems in the world to date, counted in number of meter boreholes drilled in total, are:

- Tianjin Cultural Centre, Beijing, China: 3789 boreholes of 120 m (total 454,680 m)
- Ball State University, Indiana, United States: 1806 boreholes of 135 m (total 243,810 m)
- Lotte World Tower, Seoul, South Korea: 720 boreholes of 200 m (total 144,000 m)
- ELI-NP Marguele, Bucharest, Romania: 1080 boreholes of 125 m (total 135,000 m)
- Zhungguancon International Center, Beijing, China: 1060 boreholes of 123 m (total 130,380 m)

While the first BTES systems constructed in the 1980s were drilled to a modest depth, less than 100 m, today's boreholes are typically drilled to a depth of 120−200 m, or even deeper. Especially in the Scandinavian countries, borehole depths down to 300 m are not uncommon for new BTES projects.

It is yet to be seen if the potential for use of BTES for infrastructure applications can be economically feasible in practice. Keeping roads, bridge decks, platforms, and parking areas free from ice by storing solar heat in a BTES system deserves more attention in the future.

As long as BTES systems compete economically in a favorable way with systems that are less efficient, and have lower CO_2 emissions, there is a bright future for its application and development.

References

Åberg, B., Johansson, S., 1988. Vattenströmning till och från borrhålsvärmelager (Water flow to and from borehole heat stores). In: Swedish. Report R6:1988. Swedish Council for Builidng Research, Stockholm.

Andersson, O., 2007. BTES for heating and cooling of the Astronomy House in Lund. In: Paksoy, H.Ö. (Ed.), Thermal Energy Storage for Sustainable Energy Consumption. Fundamentals, Case Studies and Design. In: NATO Science Series. Springer, Dordrecht, The Netherlands, pp. 229−234 (Chapter 13).

Andersson, S., Johansson, A., Nordell, B., Åbyhammar, T., 1983. Värmelagring i berg med borrhålssystem. Pilotförsök i Luleå och förprojektering av fullskaleanläggning (Heat Storage in Rock with a Borehole System. Pilot Plant in Luleå and Pre-study of a Full-scale Plant). Report R15:1983. BFR. Swedish Council for Building Research, Stockholm, Sweden (In Swedish).

Bäcklund, A.-S., 2009. Analys av en grundvärmebaserad värmepumpsanläggning för uppvärmning och kylning av ett sjukhus (Analysis of a Ground Source Heat Based Heat Pump Facility for Heating and Cooling of a Hospital) (Master's thesis). NTNU, Norway (In Swedish).

Bakema, G., Snijders, A.L., Nordell, B., 1995. Underground Thermal Energy Storage – State of the Art 1994. Report IEA ECES.
Banks, D., 2012. Thinking Big: Large-Scale Heat Storage and Transfer. An Introduction to Thermogeology. Ground Source Heating and Cooling, second ed. John Wiley & Sons, Ltd, Chichester, West Sussex, UK, pp. 383–409.
Baudoin, A., 1988. Stockage intersaisonnier de chaleur dans le sol par batterie d'èchanger baionette verticaux; modèle de predimensionnement (thesis). L'Universite de Reims Champagne-Ardenne, France.
Bauer, D., Heidemann, W., Müller-Steinhagen, H., 2007. Central solar heating plants with seasonal heat storage. In: CIBSAT, Renewables in a Changing Climate – Innovation in the Built Environment. Lausanne, September 4–5, 2007.
Bendea, C., Antal, C., Rosca, M., 2015. Geothermal energy in Romania: country update 2010–2014. In: Proc. World Geothermal Congress 2015 Melbourne, Australia, 19–25 April, 2015.
Benner, M., Hahne, E.W.P. (Eds.), 2000. Proceedings of Terrastock 2000, the 8th International Conference on Thermal Energy Storage. Stuttgart, Germany. August 28–September 1, 2000.
Brun, G., 1965. La régularisation de lènergie solaire par le stockage thermique dans la sol. Revue Générale de Thermique 44 (France).
BSU, 2015. Going Geothermal. http://cms.bsu.edu/About/Geothermal/FAQ.aspx#whatis, 09.09.15.
Cabeza, L.F., 2015. Advances in Thermal Energy Storage Systems. Methods and Applications. Elsevier, Woodhead Publishing Series in Energy, p. 66.
Chiasson, A.C., Rees, S.J., Spitler, J.D., 2000. A preliminary assessment of the effects of ground-water flow on closed-loop ground-source heat pump systems. ASHRAE Transactions 106 (1), 380–393.
Claesson, J., Hellström, G., 2000. Analytical studies of the influence of regional groundwater flow on the performance of borehole heat exchangers. In: Proc. of 8th Int. Conf. on Thermal Energy Storage, Terrastock 2000, August 28–September 1, 2000, Stuttgart, Germany, pp. 195–200.
Cullin, J.R., Spitler, J.D., Montagud, C., Ruiz-Calvo, F., Rees, S.J., Naicker, S.S., 2015. Validation of vertical ground heat exchanger design methodologies. Science and Technology for the Built Environment 21 (2), 137–149.
Dalenbäck, J.-O., June 1990. Central Solar Heating Plants with Seasonal Storage – Status Report. IEA Solar Heating and Cooling Programme. Report D14:1990. Swedish Council for Building Reseach, Stockholm, Sweden.
Dalenbäck, J.-O., Hellström, G., Lundin, S.-E., Nordell, B., Dahm, J., 2000. Borehole heat storage for the anneberg solar heated residential district in Danderyd, Sweden. In: Proc. Terrastock 2000, Stuttgart, Germany, August 28–September 1, 2000, pp. 201–206.
Den Braven, K.R., 1998. Antifreeze acceptability for ground-coupled heat pump ground loops in the United States. ASHRAE Transactions 104 (1B), 938–943.
Den Braven, K.R., 2000. Regulations on grouting for closed-loop ground-coupled heat pumps in the United States. ASHRAE Transactions 106 (2), 447–452.
Dincer, I., Rosen, M.A., 2007. A unique borehole thermal storage system at University of Ontario Institute of Technology. In: Paksoy, H.Ö. (Ed.), Thermal Energy Storage for Sustainable Energy Consumption. Fundamentals, Case Studies and Design. In: NATO Science Series. Springer, Dordrecht, The Netherlands, pp. 221–228 (Chapter 12).
Edstedt, U., Nordell, B., 1994. Borrhålslager. Förutsättningar, Erfarenheter, Möjligheter. (Borehole Storage. Conditions, Experience, possibilities.). Swedish. Byggforskningsrådet (The Swedish Building Council), Stockholm, Sweden.
Eugster, W.J., 2002. SERSO PLUS – Neue Wege in der Belagsbeheizung. GtE 38/39/2002. Mitteilungsblatt der Geothermischen Vereinigung e.V. 10–14.

Eugster, W.J., 2007. Road and Bridge Heating Using Geothermal Energy. Overview and Examples. In: Proceedings European Geothermal Congress 2007, Unterhaching, Germany, 30 May–1 June, 2007.

Gabrielsson, A., Lehmets, M., Moritz, L., Bergdahl, U., 1997. Heat Storage in Soft Clay. Field Tests with Heating (70°C) and Freezing of the Soil. Report 53. Swedish Geotechnical Institute, Linköping.

Gao, L., Zhao, J., Tang, Z., 2015. A review on borehole seasonal solar thermal energy storage. Energy Procedia 70, 209–218.

Gehlin, S., Andersson, O., Bjelm, L., Alm, P.-G., Rosberg, J.-E., 2015. Country update for Sweden. In: Proceedings World Geothermal Congress 2015, Melbourne, Australia, 19–25 April, 2015.

Guimbal, 1976. Lotissement de 16 villas a chauffage solaire integral. Société Générale d'Etudes et construction, Saint-Etienne, France.

Gustafsson, A.-M., Gehlin, S., 2008. Influence of natural convection in water-filled boreholes for GCHP. ASHRAE Transaction 114 (1), 416–423.

Hadorn, J.C., 1990. Guide to Seasonal Heat Storage. SIA, Swiss Association of Engineers and Architects, Federal Energy Office. Documentation D 028 in the series "Planning, Energy and Buildings".

Heier, J., Bales, C., Sotnikov, A., Ponomarova, G., 2011. Evaluation of a high temperature solar Thermal seasonal borehole storage. In: ISES Solar World Congress 2011, Kassel, Germany.

Hellström, G., 1991. Ground Heat Storage. Thermal Analyses of Duct Storage Systems. I. Theory (Ph.D. thesis). Department of Mathematical Physics. University of Lund, Sweden.

Hellström, Sanner, G.B., 2001. PC-Programs and modelling for borehole heat exchanger design. In: International Geothermal Days Germany 2001. Bad Urach.

Hellström, G., Sanner, B., Klugescheid, M., Gonka, T., Mårtensson, S., 1997. Experiences with the borehole heat exchanger software EED. In: Proc. The 7th International Conference on Thermal Energy Storage Megastock'97. June 18–21, 1997. Sapporo, Japan, pp. 247–252.

Hultmark, G., 1981. Sunclay project – first year operation with a seasonal storage of 80,000 m^3 in clay. In: Proceedings of the International Conference of Seasonal Thermal Energy Storage and Compressed Air. Seattle, Washington. October 19–21, 1981, vol. 1, pp. 171–176.

Hultmark, G., 1983. Earth storages with vertical pipes. In: Proceedings, International Conference on Subsurface Heat Storage in Theory and Practice. Stockholm, June 6–8, 1983, pp. 176–183.

Huusko, A., Valpola, S.E., November 2014. Insight into shallow geothermal energy research in Finland. The Newsletter of the ENeRG Network (30), 3.

IEA ECES, May 1997. Underground Thermal Energy Storage. Procedures for Environmental Impact Assessment. Working report.

Kistler, D., Karidis, G., Winter 2015. Flagship for Health. High Performing Buildings. ASHRAE.

Kellner, et al., 1986. Stockholmsprojektet kv Höstvetet. Överglasad gård, värmepumpar och borrhålslager i flerbostadshus – system Suncourt. Report R81:1986. Swedish Building Council, Stockholm (In Swedish).

Kemler, N.E., October 1946. Heat-pump heat sources. Edison Electric Institute (EEI) Bulletin 14, 339–346.

Kemler, E.N., 1947. Methods of Earth Heat Recovery for the Heat Pump. Heating and Ventilating, September 1947, 69–72.

Kjellsson, E., Hellström, G., 1997. Laboratory study of the heat transfer in a water-filled borehole with a single U-pipe. In: Proc. of Megastock '97, 7th International Conference on Thermal Energy Storage, Sapporo, Japan, June 18–21, 1997, pp. 509–514.

Lee, K.S., 2013. Borehole Thermal Energy Storage. Underground Thermal Energy Storage. Springer, London, pp. 95–124 (Chapter 5).

Li, H., Yu, Z., Wu, J., Xu, W., Zhang, S., 2015. Application of ground source heat pump system in a nearly zero energy building of China. In: Proc. Greenstock 2015, Beijing, China, 19–21 May, 2015.

Lund, J., Sanner, B., Rybach, L., Curtis, R., Hellström, G., September 2004. Geothermal (ground-source) heat pumps – a world overview. GHC Bulletin 25 (3), 1–10.

Lundh, M., Dalenbäck, J.O., 2008. Swedish solar heated residential area with seasonal storage in rock: initial evaluation. Renewable Energy 33 (4), 703–711.

Magnusson, C., et al., 1992. Värmelager i lera. Utvärdering Söderköping. (Heat store in clay. Evaluation Söderköping). Report R21:1992. BFR, Swedish Council for Building Research, Stockholm. Sweden (In Swedish).

Mangold, D., 2007. Seasonal storage – a german Success Story. Sun & Wind Energy 1 (2007), 48–58.

Matthey, B., Pillonel, B., 1985. Solar assisted gas heat pump with ground coupled storage for twelve family houses. In: van Hattem, D. (Ed.), Proc. 2nd Workshop on Solar Assisted Heat Pumps with Ground Coupled Systems. May 1985. Vienna, Austria, pp. 203–211.

Matthey, B., 1988. Accumulateur saisonnier de chaleur solaire de 20 000 m^3 par sondes verticales sous un bâtiment industriel à Meyrin – Genève. In: Proc. 4th Int. Conf. Energy Storage Jigastock 88, Versailles, France, October 18–21, 1988, pp. 561–565.

van Meurs, G.A.M., 1986. Seasonal Heat Storage in Soil (thesis). Dept. of Applied Physics, University of Delft, Delft, Netherlands.

Miedaner, O., Mangold, D., Sørensen, P.A., 2015. Borehole Thermal energy storage systems in Germany and Denmark – construction and operation experiences. In: Proc. Greenstock 2015, Beijing, China, 19–21 May, 2015.

Midttømme, K., Berre, I., Hauge, A., Musæus, T.E., Kristjansson, B.R., 2010. Geothermal energy – country update for Norway. In: Proceedings World Geothermal Congress 2010 Bali, Indonesia, 25–29 April, 2010.

Naicker, S.S., Rees, S.J., 2011. Monitoring and performance analysis of large non-domestic ground source heat pump installation. In: Proceedings of 2011 CIBSE Technical Symposium, Leicester, UK, September 6–7.

Nordell, B., 1994. Borehole Heat Store Design Optimization (Ph.D. thesis, 1994:137D). Division of Water Resources Engineering. Luleå University of Technology, Sweden.

Nordell, B., 2000. Large-scale Thermal energy storage. In: Proceedings WinterCities'2000, Energy and Environment, 14 February, 2000, Luleå Community, Sweden. Available at: http://pure.ltu.se/portal/files/1172884/Large-scale_TES__WC2000.pdf.

Nordell, B., Andersson, O., Rydell, L., Liuzzo-Scorpo, A., 2015. Long-term performance of the HT-BTES in Emmaboda, Sweden. In: Proc. Greenstock 2015, Beijing, China, 19–21 May, 2015.

Olsson, J., 2014. Svensk Geoenergi 2014, vol. 1, p. 12. www.geoenergicentrum.se.

Olsson, S., 1984. The sunclay and Kullavik projects — Heat storage in Clay at low and high temperature. In: Proceedings of the First E.C. Conference on Solar Heating, pp. 894–898.

Paksoy, H.Ö. (Ed.), 2007. Thermal Energy Storage for Sustainable Energy Consumption. Fundamentals, Case Studies and Design. In: NATO Science Series. Springer, Dordrecht, The Netherlands.

Peltola, S.S., Lund, P.D., Routti, J.T., 1985. First year operating experience from Kerava solar village. International Journal of Ambient Energy 6 (3), 117–122.

Platell, O., Nilsson, T., Margen, P., 1981. Sunstore – the complete solar energy system for heating. In: Proceedings of the International Conference of Seasonal Thermal Energy Storage and Compressed Air. Seattle, Washington. October 19–21, 1981, vol. 1, pp. 177–185.

Platell, O., Wikström, H., 1983. Sunstore experimenthus i Sigtuna. Mätning och utvärdering 1981–1982 (Sunstore Experimental Building in Sigtuna. Measurement and Evaluation 1981–1982). Report R32:1983. BFR, Swedish Council for Building Research, Stockholm. Sweden (In Swedish).

Polizu, R., Hanganu-Cucu, R., 2011. A case study of good practice in ground source heating/cooling: auto showroom, offices and workshop, VW Bucharest – Romania. In: McCorry, M., Jones, G. Ll. (Eds.), Geotrainet Training Manual for Designers of Shallow Geothermal Systems. Geotrainet, EFG, Brussels, Belgium, pp. 128–132 (Chapter 13).

Reuss, M., 2015. The use of borehole thermal energy storage (BTES) systems. In: Cabeza, L. (Ed.), Advances in Thermal Energy Storage Systems. Methods and Applications, Woodhead Publishing Series in Energy, vol. 66. Elsevier, pp. 117–148 (Chapter 6).

Reuss, M., Beuth, W., Schmidt, M., Schoelkopf, W., 2006. Solar district heating with seasonal storage in Attenkirchen. In: Proc. 10th International Conference on Thermal Energy Storage Ecostock 2006. May 31–June 2, 2006. The Richard Stockton College of New Jersey, USA, pp. 193–200.

Rosenblad, G., 1983. Seasonal heat storing 1979–1983 in utby ground heat project. In: Proceedings, International Conference on Subsurface Heat Storage in Theory and Practice. Stockholm, June 6–8, 1983, pp. 670–678.

Sanner, B., 2005. Examples of GSHP and UTES systems in Germany. In: Proceedings World Geothermal Congress 2005. Antalya, Turkey, 24–29 April, 2005.

Sanner, B., Knoblich, K., 1991. Subsurface cold storage in ground coupled heat pump systems. In: Proc. 5th International Conference on Thermal Energy Storage Thermastock'91. 13–16 May 1991. Scheveningen, The Netherlands, pp. 2.4-1–2.4-7.

Sanner, B., Stiles, L., 1997. Status of seasonal cold storage in ground source heat pumps. In: Proc. The 7th International Conference on Thermal Energy Storage Megastock'97. June 18–21, 1997. Sapporo, Japan, pp. 13–18.

Sibbitt, B., McClenahan, D., Djebbar, R., Thornton, J., Wong, B., Carriere, J., Kokko, J., 2012. The performance of a high solar fraction seasonal storage district heating system – five years of operation. Energy Procedia 30, 856–865.

Skanska, March 2014a. Atrium 1, Poland. Case Study 120. Skanska AB.

Skanska, April 2014b. Entré Lindhagen, Sweden. Case Study 122. Skanska AB.

Spitler, J.D., 2000. GLHEPRO – A design tool for commercial building ground loop heat exchangers. In: Proceedings of the 4th International Heat Pumps in Cold Climates Conference, Aylmer, Quebec. August 17–18.

Energiverk, S., 1986a. Värme från borrhålsmagasin försörjer 550 lägenheter. Teknisk rapport. (Heat from borehole storage serves 550 apartments. Technical report). In Swedish. Energi i utveckling 2, 8–11.

Energiverk, S., 1986b. Butiksvärme säsongslagras. Teknisk rapport. (Warehouse heat is seasonally stored. Technical report). In Swedish. Energi i utveckling 2, 12–14.

Stiles, L., 1998. Underground thermal energy storage in the US. IEA Heat Pump Centre Newsletter 16 (2), 22–23.

Viessmann, 2012. Environmentally friendly supply for the world's second-tallest building. Twelve heat pumps for supertall skyscraper in South Korea. Aktuell Viessmann - Magazine for Heating Technology 44 (1), 22–23.

Werner, G., 1988. Seasonal heat storage in borehole heat store in the rock beneath an apartment building in the Stockholm Project. In: Proc. 4th Int. Conf. Energy Storage Jigastock 88, Versailles, France, October 18–21, 1988, pp. 571–576.

Wijsman, A.J.T.M., 1985. Groningen project: 96 solar houses with seasonal heat storage in the soil. In: Proc. 3rd Int. Conf. Energy Storage for Building Heating and Cooling Enerstock 85, Toronto, Canada, September 22–26, 1985, pp. 518–523.

Wijsman, A.J.T.M., Havinga, J., 1988. Monitoring results of the Groningen CSHPSS. In: Proc. 4th Int. Conf. Energy Storage Jigastock 88, Versailles, France, October 18–21, 1988, pp. 543–547.

Witte, H.J.L., van Gelder, A.J., 2007. Three years monitoring of a borehole thermal energy store of a UK office building. In: Paksoy, H.Ö. (Ed.), Thermal Energy Storage for Sustainable Energy Consumption. Fundamentals, Case Studies and Design. In: NATO Science Series. Springer, Dordrecht, The Netherlands, pp. 205–219 (Chapter 11).

Wong, B., Snijders, A., McClung, L., 2006. Recent Inter-seasonal underground Thermal energy storage applications in Canada. In: 2006 IEEE EIC Climate Change Conference, May 2006, pp. 1–7.

Yin, B., Wu, X., Wang, Y., Song, C., 2015. Performance analysis on a large scale borehole ground source heat pump combined with ice storage in Tianjin Cultural Center. In: Proc. Greenstock 2015, Beijing, China, 19–21 May, 2015.

Yu, Z., Li, H., Wu, J., Xu, W., 2015. Application of thermal energy storage in a nearly zero energy office building. In: Proc. Greenstock 2015, Beijing, China, 19–21 May, 2015.

Zanchinia, E., Lazzaria, S., Priaroneb, A., February 2012. Long-term performance of large borehole heat exchanger fields with unbalanced seasonal loads and groundwater flow. Energy 38 (1), 66–77.

Zhang, S., Xu, W., September 2014. Research and development of large scale ground source heat pump system. Asian Heat Pump and Thermal Storage Technologies Network Newsletter 3, 4–14.

Part Three

Building integration and hybrid systems

Hybrid ground-source heat pump systems

12

Y. Kim, J.S. Lee, S.W. Jeon
Korea University, Seoul, Republic of Korea

Abstract

Hybrid ground-source heat pump (HGSHP) systems have been proposed in a number of configurations to achieve more economic and efficient solutions than ground-source heat pump (GSHP) systems using borehole heat exchangers alone, especially in situations where a building's heating and cooling demands are not well balanced. HGSHP can adopt various supplemental heat sources and sinks according to the climatic conditions and the balance of energy demands. For buildings located in hot climate regions, the HGSHP can use a cooling tower as a supplemental heat rejecter to improve the ground thermal environment and cooling performance. The design and control of the HGSHP have to be optimized properly to balance the ground thermal environment and minimize the life cycle cost. Important issues in the optimization are the control strategy, the loop configuration between the ground flow loop and supplemental flow loop, and the sizing of the ground heat exchanger and supplemental equipment. Well designed and optimized HGSHPs can provide substantial economic benefits and improved ground energy balance. Borehole heat exchanger sizing procedures for hybrid systems are also discussed in chapter "Vertical borehole ground heat exchanger design methods," Section 2.7. Dual-source GSHP equipment is discussed in chapter "New trends and developments in ground-source heat pumps."

12.1 The hybrid ground-source heat pump concept

In commercial and residential buildings a large portion of energy is consumed for space cooling, heating, and ventilation. Ground-source heat pump (GSHP) technology has been widely used for cooling and heating of buildings due to its excellent energy saving potential, but in certain situations, GSHP can have unacceptable capital costs. The performance of a GSHP depends highly on local conditions such as ground thermal properties or borehole thermal resistance, but particularly on the imposed heat rejection and extraction profile—as has been commented on in chapter "Vertical borehole ground heat exchanger design methods." The most economic borehole array designs (shortest borehole lengths) tend to be found where the heating and cooling demands are well balanced. Maintaining sustainable performance (accounting for reduced performance after a number of years of operation) can require expensive ground heat exchanger (GHE) designs where the demands are not well balanced. This can be the case for buildings in hot

Advances in Ground-Source Heat Pump Systems. http://dx.doi.org/10.1016/B978-0-08-100311-4.00012-1
Copyright © 2016 Elsevier Ltd. All rights reserved.

climate regions and/or where internal heat gains are high. The hybrid ground-source heat pump (HGSHP) approach is to adopt a supplemental heat rejecter and/or extractor and only use the ground heat exchanger (GHE) to reject or extract part of the building's thermal energy demands. This can be a more economic solution when the whole life cycle is considered (Yavuzturk and Spitler, 2000).

Fig. 12.1 shows the hybrid ground-source heat pump that includes a cooling tower as an auxiliary heat rejecter (Midwest Cooling Towers). In hot climate regions, the cooling tower can reduce the excessive heat rejection through the GHE, mitigating either unacceptable changes in the ground temperature or excessive borehole depths (Man et al., 2008, 2010). As a result, the HGSHP can both operate efficiently and have reduced initial cost because the cooling tower and GHE combination is more cost-effective than a GHE designed to deal with the whole of the demands. In contrast, in cold climate regions and where heating demands are dominant, the ground temperature gradually decreases during the operation of the GSHP. The use of the supplemental heat source can mitigate the ground temperature decrease, resulting in more effective operation of the HGSHP (Dai et al., 2015). The possible supplemental equipment used in the HGSHP includes a cooling tower, an air source, a waste heat source, a boiler, or a solar collector, according to the building type and climate condition and consequential balance of thermal demands (Section 12.2).

In HGSHP system design and operation a key question is the integration with the control system and the strategy adopted to switch between (or mix) the heat exchanger devices. Optimization of the design and control parameters is very important. The key objectives of the optimization are minimizing the life cycle cost and balancing the ground thermal environment. Some important design factors include the control strategy, the size of the GHE and supplemental equipment, and the configuration of the

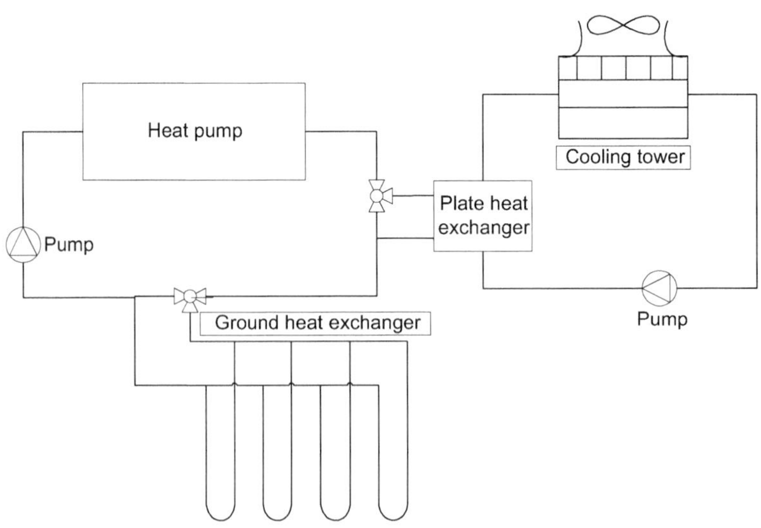

Figure 12.1 A schematic diagram of hybrid ground–ground source heat pump with a cooling tower as a supplemental heat rejecter.

GHE and supplemental heat rejecter circuits. The optimization of the design and control of the HGSHP is discussed in Section 12.3 of this chapter. The evaluation of the life cycle costs of HGSHP systems is discussed in Section 12.4.

12.2 Hybrid ground-source heat pump system types

The supplemental equipment in a HGSHP system is used for both rejecting and extracting heat. Cooling tower (evaporative) and dry-air cooler technologies are commonly used as the supplemental heat rejecters in cooling dominated applications. The solar collector, waste heat source, and boiler technologies are widely used as the supplemental heat source in heating dominated buildings. HGSHP systems can be classified into various types based on the supplemental equipment employed, and these are discussed in the following subsections.

12.2.1 Cooling towers

The cooling tower is the most typical supplemental heat sink in HGSHP systems. The cooling tower is connected with the GHE and isolated from the building and ground loops with a plate heat exchanger, as shown in Fig. 12.1. In the cooling tower, water is evaporated to remove heat and cool the working fluid so that the outlet temperature approaches the ambient wet-bulb temperature (Midwest Cooling Towers). Cooling towers can be divided into open and closed circuits according to whether the fluid in the circuit is evaporated directly (open-loop design) or whether the circulating fluid passes through a heat exchanger loop and a secondary fluid is circulated to achieve evaporation (closed-loop design). In addition, according to air flow generation methods, they can be classified into natural and mechanical draft cooling towers. The open circuit type with mechanical draft is the most commonly used cooling tower in HGSHP systems.

Fig. 12.2 shows cross-flow and counter-flow cooling towers of the open circuit type with mechanical draft. In the cross-flow cooling tower, Fig. 12.2(a), air flow is directed perpendicular to the water flow. Dry air enters one or more vertical faces of the cooling tower and moves horizontally through fill material. Hot water enters from the top distribution basin and flows through the fill by gravity. The air passes through the fill and past the water flow into an open plenum. Finally, a fan forces the air out into the atmosphere. Fig. 12.2(b) shows the counter-flow cooling tower. The air flow is directly opposite to the water flow in the counter-flow design. The air flow enters an open area beneath the fill media and is drawn up vertically. The water is sprayed through water distribution nozzles and flows downward through the fill material.

There are a lot of simulation studies to examine the performance of a HGSHP using a cooling tower as the supplemental heat rejecter. Yavuzturk and Spitler (2000) used a cooling tower in cooling-dominated commercial building applications and investigated the advantages and disadvantages of various operating and control strategies under different climate conditions. Man et al. (2008, 2010) reported a

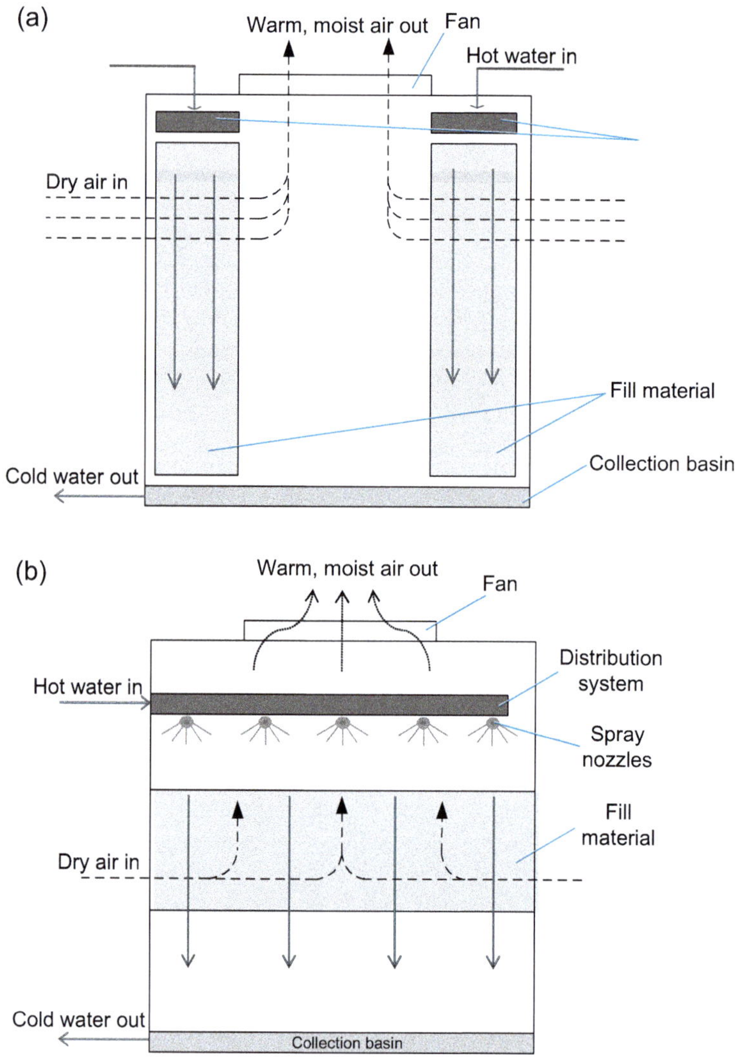

Figure 12.2 (a) Cross-flow and (b) counter-flow cooling towers.

practical simulation model of a HGSHP by modeling the heat transfer process of its main components on an hour-by-hour simulation of a system located in China. A cooling tower was integrated with the GSHP as a supplemental heat rejecter, and economic comparisons were conducted between the conventional GSHP and HGSHP with four different control strategies. They found that the HGSHP can effectively mitigate excessive ground temperatures and reduce both the initial and operating costs. Lubis et al. (2011) conducted a thermodynamic analysis of a HGSHP. The system had a cooling tower as a supplemental heat rejecter and the system performance was evaluated in terms of coefficient of performance (COP) and exergy efficiency. The heating COP for the overall system was found to be 5.34, while the corresponding exergy efficiency was 63.4%. The results indicate that the performance of HGSHPs is superior to air-source heat pumps.

Some approaches attempted to predict and simulate the performance of a HGSHP system installed in actual buildings. Sagia et al. (2012) computed the energy demands of a HGSHP in a five-floor office building with a total area of 1000 m^2. The building was located in Greece and the system was cooling dominated. The simulation was conducted using the TRNSYS simulation tool (Klein et al., 2010) considering two different scenarios based on different glazing properties. A ground loop consisting of a rectangular field of 15 borehole heat exchangers and a cooling tower was utilized to cover building loads. Assuming different cooling tower capacities, the desired flow rate was estimated for a cooling range of 5.7°C. The cooling range means the difference between water inlet and outlet temperature which flows through the cooling tower. A MATLAB code was developed to calculate the required pressure drop per packing height of the cooling tower for four packing in different operating conditions. Yang et al. (2014) simulated the annual dynamic load of an eight-floor hotel in Wuhan, China with a total floor area of approximately 11,200 m^2. They proposed a HGSHP system with two cooling towers. Apart from a cooling tower being used as the supplemental heat rejecter to reject the excess heat, another cooling tower was added which was only used a few days per week to replace the GHE so that the ground could be recharged naturally. The HGSHP with these double-cooling towers not only restored the ground temperature but also improved the efficiency of heat pump units and reduced the operating cost.

There are limited experimental studies on the HGSHP system using a cooling tower as a supplemental heat rejecter. Hern (2004) and Gentry et al. (2006) measured the performance of a HGSHP to validate the simulation results. The experimental facility consisted of a GHE, a pond loop heat exchanger, and an evaporative cooling tower. The validation was performed both for each component and the entire system. The measured data were matched with the simulation results in the acceptable range. Park et al. (2012) measured the performance of a HGSHP with a parallel configuration. The performance of the HGSHP was optimized with the variation of the refrigerant charge and secondary fluid flow rates in the ground fluid loop and supplemental fluid loop. The supplemental flow loop containing a cooling tower was connected separately with the primary refrigerant flow loop. The COP of the HGSHP with the parallel configuration was found to be 21% higher than that of the conventional GSHP at the entering fluid temperature of 40°C in the GHE and the entering fluid temperature of 28°C in the supplemental heat exchanger. In addition, Park et al. (2013) also conducted a performance comparison between the GSHP and the HGSHP with parallel and serial configurations by varying the leaving fluid temperature of the GHE, the fluid flow rate of the supplemental fluid loop, the mean outdoor bin temperature, and the switching temperature of the hybrid operation. At the leaving fluid temperature in the GHE of 40 °C, the COP of the HGSHP with the parallel and serial configurations were 18% and 6% higher than that of the GSHP respectively. The cooling seasonal performance factors of the HGSHPs with the parallel and the serial configurations were 6.5% and 2.0% higher than that of the GSHP located in Daegu, South Korea, respectively.

Lee et al. (2014) investigated the performance characteristics of the HGSHP in a parallel configuration under transient conditions. They applied a thermal storage bath to simulate ground thermal environment by the heat rejection through the GHE. They reported that the average COP of the HGSHP with the parallel configuration increased 7.2% compared to the GSHP. In addition, Hackel and Pertzborn (2011) monitored and analyzed two actual HGSHP systems using a cooling tower as a supplemental heat rejecter. The buildings were monitored for a year and the measured data was used to validate models of each system. The model was used to analyze further improvements on the hybrid approach, showing that the HGSHP had positive impacts economically and environmentally as well.

12.2.2 Air source

In a HGSHP, an air source can be used as both a supplemental heat sink and source. In order to achieve optimal operation of the HGSHP, it is necessary to develop a system utilizing both ground and air sources according to the building load conditions. During intermediate seasons (such as spring and autumn) with reduced heating and cooling loads, an air-source heat pump can be more efficient than the conventional GSHP. Nam et al. (2010) developed a HGSHP using dual groundwater and air heat sources. The annual performance of the developed system was calculated, and several case studies were conducted on the effects of location, refrigerant, and pumping rates. As a result, the annual performance of the developed HGSHP was improved by 2–7% compared to the water cooling system, and 4–18% compared to the air cooling system.

Pardo et al. (2010) presented a heating, ventilation, and air-conditioning (HVAC) system which combines a GSHP, an air-to-water heat pump, and a thermal storage device. The HVAC system was linked to an office building in a cooling dominated area. The results showed that the energy consumption of the combined HVAC system was 60% lower than the air-to-water heat pump and 82% lower than the conventional GSHP. In addition, Dikici and Akbulut (2008) presented an exergy evaluation of an air-source heat pump, a GSHP, a solar-assisted heat pump, and a combination of these systems, such as a ground-source/air-source heat pump (GSASHP), a solar-assisted ground-source heat pump (SAGSHP), and a solar-assisted air-source heat pump (SAASHP). The COPs of the GSASHP and SAGSHP were 2.14 and 3.36, respectively. The exergy losses of the GSASHP and SAGSHP were 3.11 and 2.13 kW, respectively.

12.2.3 Waste heat

Waste water from saunas, public baths, and buildings can be effectively applied to water and space heating. The US Department of Energy (DOE) reported that the equivalent of 350 billion kWh worth of hot water was discarded annually through drains. A large portion of the unused energy is, in fact, recoverable by the use of a gravity film heat exchanger (Tomlinson, 2005). The gravity film heat exchanger is a simple heat exchanger design for heat recovery that was developed under a grant from the DOE

Inventions Program. This straightforward design is a vertical, counter-flow heat exchanger that extracts heat out of drain-water (usually warm) and applies it to preheat the cold water entering the building. The gravity film heat exchanger is a simple and effective method for significantly reducing the energy needed to produce hot water.

Until now, there were very limited studies using waste heat as the supplemental heat source in a HGSHP. Li et al. (2014) proposed a novel CHP (combined heat and power) and GSHP integrated system in which the waste heat of flue gas could be utilized. The novel system can increase total energy efficiency by 0.8–34.7%, and the COP of the GSHP subsystem can be increased by 3.6–6.

12.2.4 Solar

The annual thermal load imposed on the ground can be balanced by using the solar collector in heating dominated buildings. The average solar energy incident on any horizontal surface is around 150–200 W/m^2 in the United States (Rabl, 1985). The solar collector converts the absorbed solar energy into the heat, and transfers the heat to the circulating fluid with minimal heat loss (Fig. 12.3). The solar collectors are connected to the ground loop to provide additional heat during the heating operation of the HGSHP.

Theoretical, experimental, and numerical methods have been used to study the characteristics of a solar-assisted GSHP. Bi et al. (2004) conducted theoretical and experimental studies on a solar assisted GSHP with vertical double-spiral coil GHE.

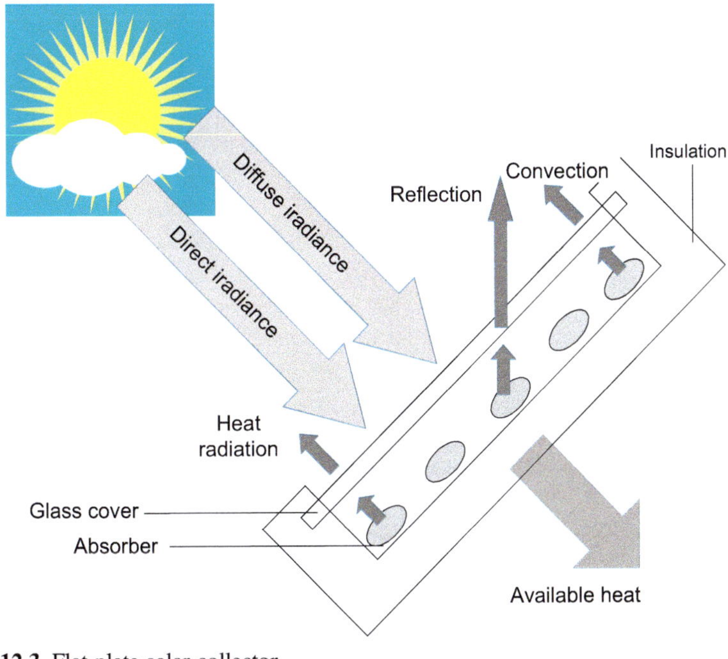

Figure 12.3 Flat plate solar collector.

The average heating load and COP of the GSHP were 2298 W and 2.83, respectively, and those of the solar assisted GSHP were 2316 W and 2.78, respectively. The underground temperature of the GHE was simulated by using the volume-control method, and the results were compared with the experimental data. Wang and Qi (2008) and Wang et al. (2009) simulated the performance of a solar-assisted GSHP system through a long period operation based on the experimental data. They also investigated the effects of the intensity of solar radiation, water tank volume, and area of solar collectors on the system performance. The suggested optimum ratio between the tank volume and the area of solar collectors was in the range of 20−40 L/m^2.

Yang et al. (2015) performed both experiments and dynamic simulations in different heating modes of a solar-assisted GSHP. Four operation modes were investigated during the winter: a single GSHP operation mode, combined operation mode, a day and night alternate operation mode, and solar U-tube feeding heat alternate operation mode. The system efficiency was shown to be improved by the assistance of solar energy, and the excess solar energy collected during the day could be stored in the ground by the GHE to improve the operation performance of the GSHP during the night. A dynamic simulation program was constructed to investigate the seasonal performance of the solar-assisted GSHP operated in different heating modes. The seasonal average COPs were 3.67, 3.64, 3.52, and 3.48 for the single GSHP operation mode, day and night alternate operation, solar U-tube feeding heat, and GSHP mode, respectively.

There are a lot of experimental studies on the performance of the solar-assisted GSHP. Bakirci et al. (2011) conducted experiments on a solar-assisted GSHP system in the Erzurum province which has a cold climate. The experimental apparatus consisted of solar collectors, a GHE, a liquid-to-liquid vapor compression heat pump, water circulating pumps, and other measurement equipment. The experimental results were obtained from October 2008 to May 2009, and the COP of the heat pump and system were found to be in the range of 3.0−3.4 and 2.7−3.0, respectively. Stojanović and Akander (2010) presented the buildup and long-term performance tests of a full-scale solar-assisted GSHP for residential heating in Nordic climatic conditions. Despite unfavorable building conditions, the solar-assisted GSHP was in full operation (for about 2 years) for low energy use while satisfying heating requirements. The seasonal performance factors of the heat pump and the solar-associated GSHP were 2.85 and 2.90, respectively, with the consideration of all supplied energies to the system. In addition, Dai et al. (2015) performed an experimental study on the influence of operation modes on the heating performance of a solar-assisted GSHP. They showed that solar heat will improve the soil temperature recovery rate.

Mathematical modeling is an effective method to predict the performance of a solar-assisted GSHP. Han et al. (2008) developed a mathematical model on the solar-assisted GSHP with latent heat energy storage tank. The operation characteristics of the solar-assisted GSHP was analyzed during the heating period in Harbin (45.75°N, 126.77°E), China. The COP and operation stability of the solar-assisted GSHP were improved by solar energy and soil alternately or together as the heat source of the heat pump. The simulation tool TRNSYS has frequently been used to estimate the performance of HGSHP systems. Wang et al. (2012) developed a simulation model to predict the multiyear performance of a solar-assisted GSHP used for a 4954 m^2 office building for heating and cooling in Tianjin, China. The simulated results showed that the proposed solar-assisted GSHP was reasonably designed to resolve the ground temperature imbalance problem on an annual basis. It is very important to store sufficient solar thermal energy in the GHE by enlarging the spacing between boreholes. Small spacing does not provide enough energy storage between boreholes, so the ground temperature drops quickly. The electrical energy consumption can be saved by 32% when the load circulation pump is turned off when no fan-coil is running. In addition, Kjellsson et al. (2010) conducted a simulation of a solar-assisted GSHP of a family house in Stockholm (59.32°N, 18.05°E), Sweden, using the TRNSYS. As a result, the optimal design was obtained when solar heat produced domestic hot water during summertime and recharged the borehole during wintertime.

12.2.5 Boiler

In a HGSHP system, a boiler can be used as the supplemental heat source with fossil fuels. Ni et al. (2011) carried out the energy saving and economic analysis of a boiler assisted GSHP in which weather data and fuel prices in Xinjiang, China, were applied. They found the optimal design was obtained by allocating a heating load of 60% for the GSHP and 40% for the gas boiler, not considering both the variation of primary energy price and site energy consumption. Michopoulos et al. (2007) presented a HGSHP used for air conditioning in a public building in northern Greece. The HGSHP used a 120 kW oil-fired boiler as a backup for the GHE. In addition, Hackel and Pertzborn (2011) monitored and analyzed a boiler assisted GSHP located in a multifamily house in Madison. The system was cost-effective compared to the conventional GSHP.

12.3 Optimization of hybrid ground-source heat pump

The objectives of the optimization in a HGSHP are balancing the ground thermal condition and reducing the total cost of operation and installation. The objective functions are dependent on the number of design parameters. The main design variables which will be considered in this section are the following: the control strategy, sizing between the GHE and supplemental heat rejecter, and loop configuration.

12.3.1 Optimal control strategies

The strategy used to control the supplemental heat rejecter has a significant effect on the performance of a HGSHP. Probably the simplest strategy is set-point temperature control in which the cooling tower starts to operate when the temperature of the entering or exiting fluid becomes higher than a set-point value (Yavuzturk and Spitler, 2000; Gentry et al., 2006). The optimum set-point temperature can be changed according to climate and operating conditions.

Yavuzturk and Spitler (2000) carried out a comparative analysis with various control strategies of a HGSHP in prototypical office buildings in Houston, Texas, and Tulsa, Oklahoma. They investigated the effects of installation configuration, energy use, and operation cost with eight control strategies. The control strategies can be divided into three main categories: set-point temperature control, scheduled control, and temperature difference control. The length of the GHE in a baseline GSHP was designed without consideration of the supplemental heat rejecter. Cost analysis was carried out to evaluate which control strategy yields the smallest expense in each substrategy. The details are as follows:

1. **Set-point temperature control**: The cooling tower is activated when the temperature of the exiting fluid is higher than 35.8°C.
2. **Schedule control**: The cooling tower is activated between 12:00 am and 06:00 am for all year-round and operates when the temperature of the entering fluid exceeds 35.8°C.
3. **Temperature difference control**: The cooling tower is activated when the difference between the exiting fluid temperature of the heat pump and the wet-bulb temperature of ambient air is greater than 2.0°C, and turned off when the difference is smaller than 1.5°C.

Tables 12.1 and 12.2 show the results of cost analysis for three control strategies. All of the control strategies were beneficial in two buildings considering both the initial installation and operating costs. Especially, the hybrid control scheme based on the temperature difference showed the lowest energy consumption and operating cost in two buildings.

Table 12.1 Summary of the cost analysis for each control strategy in Houston, Texas (Yavuzturk and Spitler, 2000; Oak Ridge National Laboratory, 2001)

	GSHP "no cooling tower"	HGSHP (set-point temp. control)	HGSHP (schedule control)	HGSHP (temp. difference control)
Total length of ground heat exchanger (m)	2743.2	914.4	914.4	914.4
Design capacity of the cooling tower (kW)		79	30	40
Total cost of ground heat exchanger installation ($)[a]	54,000	18,000	18,000	18,000
Total cost of ground heat exchanger and cooling tower ($)[b]	54,000	26,662	21,272	22,427
Present value of 20 years of electricity costs ($)[c]	32,062	23,671	24,874	20,375
Present value of total costs ($)	86,062	50,333	46,146	42,803

Table 12.1 Continued

	GSHP "no cooling tower"	HGSHP (set-point temp. control)	HGSHP (schedule control)	HGSHP (temp. difference control)
Annual energy consumption (20 year average) Total energy consumption (kWh)	40,423	29,573	31,018	25,335

[a]Estimated as $22.92/m of borehole, including horizontal runs and connections.
[b]Estimated as $99.52/kw of cooling, including controls.
[c]$0.07/kWh is assumed for cost of electricity. A 6% annual percentage rate is used for life cycle cost analysis.

Table 12.2 Summary of cost analysis for each control strategy in Tulsa, Oklahoma (Yavuzturk and Spitler, 2000; Oak Ridge National Laboratory, 2001)

	GSHP "no cooling tower"	HGSHP (set-point temp. control)	HGSHP (schedule control)	HGSHP (temp. difference control)
Total length of ground heat exchanger (m)	1,170	658	658	658
Design capacity of the cooling tower (kW)		60	19	39
Total cost of ground heat exchanger installation ($)[a]	23,040	12,960	12,960	12,960
Total cost of ground heat exchanger and cooling tower ($)[b]	23,040	19,505	15,078	17,195
Present value of 20 years of electricity costs ($)[c]	21,587	19,254	20,814	18,248
Present value of total costs ($)	44,627	38,759	35,892	35,443
Annual energy consumption (20 year average) Total energy consumption (kWh)	27,117	24,036	25,959	22,699

[a]Estimated as $22.92/m of borehole, including horizontal runs and connections.
[b]Estimated as $99.52/kw of cooling, including controls.
[c]$0.07/kWh is assumed for cost of electricity. A 6% annual percentage rate is used for life cycle cost analysis.

Fan et al. (2014) conducted a theoretical analysis on the combination of control strategies. They proposed six control methods for a HGSHP. The first method is the same as the set-point temperature control strategy suggested by Yavuzturk and Spitler (2000), and the set-point temperature varies from 22 to 32°C. In the second method, the cooling tower operates when the difference between the entering fluid temperature in the heat pump and the ambient dry-bulb temperature exceeds the set-point temperature. The set-point of the temperature difference is between 3 and 7°C. The third method is the same as the temperature difference control strategy suggested by Yavuzturk and Spitler (2000), and the set-point varies from 2 to 6°C. In the fourth method, the heat pump and the cooling tower operate during the day and night, respectively. The operating time of the cooling tower varies between 6 and 14 h. The fifth method is similar to the fourth method except that the cooling tower operates during the transitioning seasons. The operation time of the cooling tower varies from 8 to 20 h for all days during the transitioning seasons. In the last method, the cooling tower only operates during the period of the peak cooling load. The operation time of the cooling tower is as follows: 12:00−15:00, 11:00−16:00, 10:00−17:00, and 09:00−18:00. In addition, as shown in Table 12.3, Fan et al. (2014) investigated the effects of control strategies on the energy consumption and the increase in the ground temperature. The energy consumption was the lowest in the combination of first and third control strategies, but the ground temperature increase was the highest in this combination. In the other three combinations, the fifth control strategy is combined with the first, third, and sixth. In the other combinations, the ground temperature increase was smaller than 3°C, but the energy consumption was higher than the combination of the first and third due to operation of the cooling tower during the transitioning seasons.

Moreover, Yang et al. (2014) proposed intermittent operation strategies of a HGSHP by using double-cooling towers. In the intermittent control strategies, another

Table 12.3 Effect of the control strategies on the energy consumption and ground temperature (Fan et al., 2014)

Strategy	Set-point	Soil temp. rise in the 10th year (°C)	Total system energy consumption in 10 years (MWh)
1st + 3rd (temp. + temp. diff., WB)	30°C/2°C	3.65	934.9
	30°C/3°C	3.82	932.8
1st + 5th (temp. + run in transition)	22°C + 20 h	1.54	1066.2
	28°C + 20 h	2.38	1051.7
3rd + 5th (temp. diff. (WB) + run in transition)	3°C + 20 h	1.92	1055.2
	4°C + 20 h	2.16	1052.9
5th + 6th (timely run + run in transition)	09:00−18:00 + 20 h	1.58	1068.2

cooling tower was installed apart from the primary cooling tower and operated only a few days in a week to replace the GHE so that the ground could be recharged naturally. The intermittent operation strategies showed enhanced performance of the heat pump and ground thermal conditions compared to the continuous operation.

This review shows that a promising strategy is one based on the difference between the exiting temperature of the heat pump and the wet-bulb temperature of ambient air. However, the most important issue is determining the optimum set-point temperature for the minimum total cost and the maximum performance of a HGSHP for particular operating conditions. Several conditions including climate, internal load, system design, and type of the supplemental equipment affect the choice of optimum set-point temperature. The determination of an optimum set-point temperature still remains as an important issue in the HGSHP design—not only for improving the system performance but also for reducing the total cost.

12.3.2 Optimization of ground heat exchanger and supplemental equipment capacities

In an HGSHP, size design of the GHE and supplemental heat rejecter (cooling tower) is very important in the cooling dominated buildings. The size of the GHE and the capacity of the supplemental heat rejecter have significant effects on the initial and operating costs, system performance, and ground thermal conditions especially for the long-term operation. Commercial programs, such as Ground Loop Design (Geothermal Bore Technologies Inc., 2006) and GLHEPRO (Spitler, 2006), can be used to predict the size of the GHE. However, the size of the GHE should also be determined by considering the capacity of the supplemental heat rejecter. A method for the combined sizing of the GHE and supplemental heat rejecter in an HGSHP was first introduced by Kavanaugh and Rafferty (1997) and Kavanaugh (1998). Kavanaugh and Rafferty (1997) proposed a detailed procedure for determining the size of the GHE in cooling and heating modes. They calculated the required length of the GHE in terms of the borehole thermal resistance, soil properties, ground thermal conductivity heat transfer rate, power consumption, and fluid temperature.

The required length of a GHE in the cooling mode can be calculated by

$$L_c = \frac{q_a R_{ga} + \left(q_{lc} - \overline{W_c}\right)\left(R_b + PLR_m R_{gm} + R_{gd} F_{sc}\right)}{t_g - \frac{t_{wi} + t_{wo}}{2} - t_p} \qquad [12.1]$$

In addition, the required length of a GHE in the heating mode can be calculated by Eq. [12.2]. After determining the required length in the cooling and heating modes, the larger value should be selected as the required length of the GHE.

$$L_h = \frac{q_a R_{ga} + \left(q_{lh} - \overline{W_h}\right)\left(R_b + PLR_m R_{gm} + R_{gd} F_{sc}\right)}{t_g - \frac{t_{wi} + t_{wo}}{2} - t_p} \qquad [12.2]$$

In Eqs. [12.1] and [12.2], F_{sc} is the short-circuit heat loss factor, and L_c and L_h are the required length (m) of the GHE for cooling and heating, respectively. PLR_m denotes the part-load factor which is calculated by dividing the peak load by the total load during design month. q_a is the net annual average heat transfer to the ground (W). q_{lc} and q_{hc} are the building design cooling and heating block load (W), respectively. R_b is the thermal resistance of the borehole, and R_{ga}, R_{gm}, and R_{gd} are the effective thermal resistance of the ground based on the annual pulse (m-K/W), monthly pulse (m-K/W), and daily pulse (m-K/W), respectively. t_g and t_p are the undisturbed ground temperature (°C) and temperature penalty for interference of adjacent borehole (°C), respectively, and t_{wi} and t_{wo} denote the inlet and outlet fluid temperatures (°C) in the heat pump, respectively. W_c and W_h are the power inputs at the design cooling load and heating load, respectively.

In addition, the net annual average heat transfer to the ground is expressed by

$$q_a = \frac{C_{fc} \times q_{lc} \times EFLH_c + C_{fh} \times q_{lh} \times EFLH_h}{8760} \quad [12.3]$$

Where C_{fc} and C_{fh} are the correction factors that account for the amount of the heat rejection and absorption by the heat pump. The values are dependent on the energy efficiency rate (EER) and the COP of the heat pump. Table 12.4 shows C_{fc} and C_{fh}. $EFLH_c$ and $EFLH_H$ are the annual equivalent full-load cooling and heating hours, respectively.

Kavanaugh and Rafferty (1997) also proposed the required nominal capacity of the cooling tower in cooling dominated operation. Their calculation was based on the assumption that the capacity of the cooling tower is related to the length difference of the boreholes between the cooling and heating modes. In cooling dominated buildings, the required borehole length in the cooling mode is larger than that in the heating mode. Therefore, the length of borehole was designed based on the heating load, and the additional cooling load was handled by the supplemental heat rejecter. Conversely, for heating dominated buildings, the length of the GHE was determined based on the cooling load, and the additional heating load was assigned to the supplemental heat source.

Table 12.4 Heat pump correction factors for Eq. [12.3] (Kavanaugh, 1998)

Cooling EER (COP)	C_{fc}	Heating COP	C_{fh}
11 (3.2)	1.31	3.0	0.75
13 (3.8)	1.26	3.5	0.77
15 (4.4)	1.23	4.0	0.80
17 (5.0)	1.20	4.5	0.82

Kavanaugh (1998) also proposed the design procedure of the cooling tower in the HGSHP. He tried to balance the heat flow of ground for an annual basis to limit the heat accumulation. The details are as follows:

1. The required length of the borehole in the GHE is calculated in the cooling and heating modes, and then the required flow rate is obtained.
2. The calculation is repeated with the entering fluid temperature of the heat pump in the heating mode. This process can reduce the temperature of the entering fluid as low as possible. Typically, the system capacity is sufficient to cover the thermal load with the low entering temperature because the cooling load of commercial buildings tends to be higher than the heating load. In warm climate regions, the entering fluid temperature is recommended as at least 7°C to avoid freezing in the heat pump. This procedure can reduce the required length of the GHE under the acceptable limit.
3. The first iteration for determining the size of the cooling tower is carried out based on the assumption that all of the annual condenser output is rejected to the ground, since the size of the cooling tower is unknown at this point. The size of the cooling tower can be obtained by

$$LPM_{\text{cooler}} = LPM_{\text{system}} \times \frac{L_c - L_h}{L_c} \quad [12.4]$$

The rated flow rate in the cooling tower is expressed by liters per minute (LPM) based on a 5.5°C range.

4. The amount of thermal load that is rejected into the ground can be obtained based on the current capacity of the cooling tower. For typical buildings, Eqs. [12.3]–[12.5] are proposed to correct the equivalent full load hours (*EFLH*) in the cooling mode ($EFLH_{\text{cooler}}$) by considering the contribution of the cooling tower.

$$EFLH_{\text{cooler}} = EFLH_c \times \left(1 - \frac{LPM_{\text{cooler}}}{2 \times LPM_{\text{system}}}\right) \quad [12.5]$$

By using $EFLH_{\text{cooler}}$ instead of $EFLH_c$, steps 2 and 3 are repeated until the calculation for the size of the cooling tower converges. In this calculation, the cooling tower is activated when the loop temperature is higher than the set-point temperature, typically 27–32°C.

5. This procedure has to be revised in order to balance the annual thermal load of the ground. The required lengths of the GHE in the cooling and heating modes are calculated without the consideration of annual heat buildup in the ground.
6. The required operating time of the cooling tower ($\text{Hours}_{\text{Cooler}}$) is calculated by Eqs. [12.3]–[12.6]. $\text{Hours}_{\text{Cooler}}$ represents the total operating time of the cooling tower to balance the heat flow of ground for an annual basis.

$$\text{Hours}_{\text{Cooler}} = \frac{C_{fc} \times q_{lc}(W) \times EFLH_c - C_{fh} \times q_{lh}(W) \times EFLH_h}{\text{Specific heat} \times \text{Mass flow rate} \times \text{Range }(^\circ\text{C})} \quad [12.6]$$

Kavanaugh (1998) reported that in an HGSHP with the cooling tower, the size of the GHE can be designed as small as possible when the required length of the GHE for cooling is much larger than that of heating. Hackel et al. (2009) also suggested an optimal design method for a combined GHE and cooling tower in an HGSHP. They found that the optimal length of the GHE increased almost linearly with the peak heating load, which was consistent with the results of Kavanaugh and Rafferty (1997). In their analysis, the length of GHE was inversely proportional to the ground temperature (Kavanaugh and Rafferty, 1997).

The optimal size of the GHE can be obtained by regression analysis, as given in Eq. [12.7], as a function of normalized peak heating load.

$$L_{tot} = C_1 \frac{\dot{q}_{peak,\ heat}}{\Delta T_{ground}} \qquad [12.7]$$

Where, L_{tot} is the total length of the GHE in m, and C_1 is 147 m-K/kW. $\dot{q}_{peak,\ heat}$ denotes the peak heating load in kW. ΔT_{ground} is the difference between the ground temperature and the minimum allowable fluid temperature, which was assumed to be 1.7°C. The total length of the GHE is 46.3 m per ton of heating at a ground temperature of 12.8°C and 26.5 m per ton of heating at a ground temperature of 21.1°C. In these estimations, the thermal conductivity of the ground was assumed to be 2.4 W/m K.

As mentioned earlier, the size of the GHE was determined based on the peak heating load, and the size of the cooling tower was designed based on the additional cooling load. The unmet cooling load of the GHE is calculated by

$$\dot{q}_{unmet,\ cool} = \dot{q}_{peak,\ cool} - \left(\frac{L_{tot}}{C_1 (T_{ground} + 18°C)} \right) \qquad [12.8]$$

Where, q_{unmet} is the unmet cooling load (tons) by the GHE. C_1 is 0.476 m/kW °C, and T_{ground} denotes the initial ground temperature in °C. The second term in Eqs. [12.3]–[12.8] represents the cooling capacity of the GHE. It should be noted that the cooling capacity of the GHE is proportional to the length of the GHE, and inversely proportional to the initial ground temperature.

The optimal capacity of the cooling tower can be expressed as

$$C_{CCCT} = 2.1 \dot{q}_{unmet,\ cool} \qquad [12.9]$$

Where, C_{CCCT} is the capacity of the cooling tower (tons), and 2.1 is a factor applied to oversize the cooling tower. Half-speed operation with the oversized cooling tower was found to be more effective than full-speed operation (Hackel et al., 2009). Hackel et al. (2009) also suggested that the optimum size of a single-speed cooling tower can be calculated by

$$C_{CCCT} = 1.3 \dot{q}_{unmet,\ cool} = 1.3 \left(\dot{q}_{peak,\ cool} - \frac{L_{tot}}{C_1 (T_{ground} + 18°C)} \right) \qquad [12.10]$$

Where, C_1 is 0.736 m/kW °C. The optimal size of a single-speed cooling tower is 1.3 times larger than the unmet cooling load of the GHE.

12.3.3 Optimal loop configuration

The loop configuration is another important factor effecting the performance of an HGSHP. The HGSHP is composed of three flow loops: a refrigerant flow loop, a ground flow loop, and a supplemental flow loop. The ground thermal load of the HGSHP is also strongly dependent on the choice of loop configuration.

The flow loop configuration can be classified on the basis of the combination of flow loop, flow direction, and types of supplemental heat rejecter. Fig. 12.4 shows

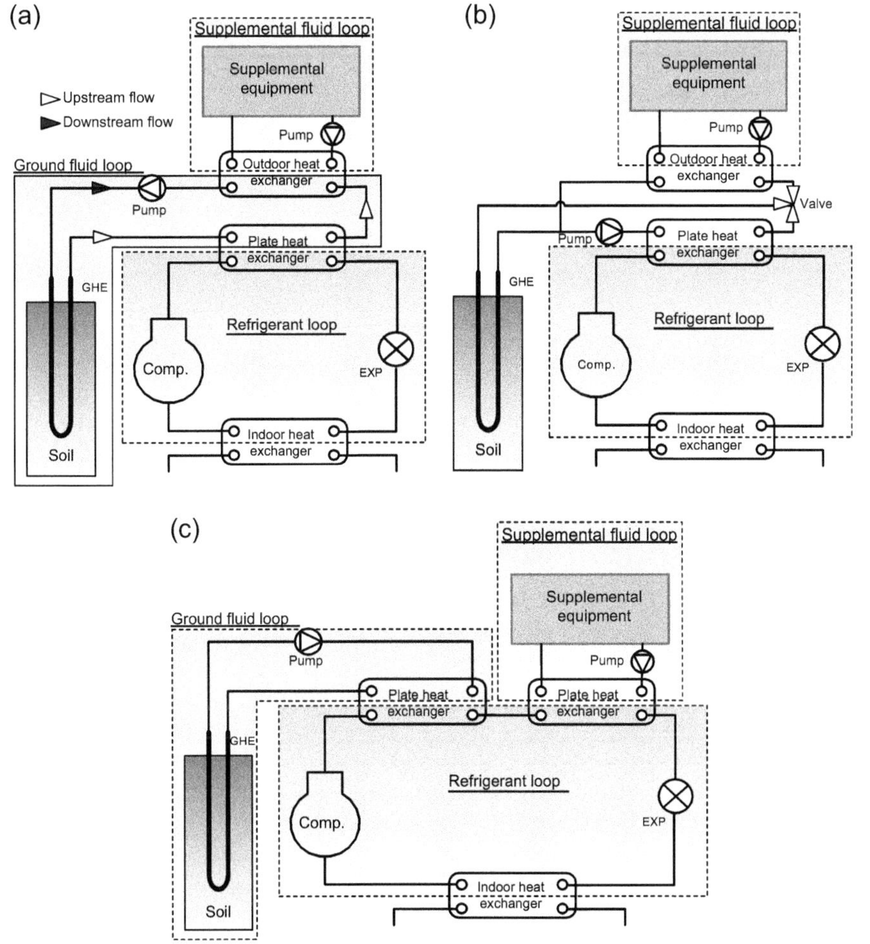

Figure 12.4 Schematics of hybrid ground-source heat pump with various loop configurations. (a) Hybrid-GSHP with serial configuration, (b) hybrid-GSHP with parallel configuration (A-type), (c) hybrid-GSHP with parallel configuration (B-type).

schematics of prototypical HGSHP systems with three loop configurations: (a) serial configuration, (b) parallel configuration of A-type, and (c) parallel configuration of B-type. The loop configuration is classified based on the arrangement between the GHE and supplemental heat rejecter. In the serial configuration, a GHE was connected serially with the outdoor heat exchanger in the supplemental side. The HGSHP of the serial configuration can also be classified into upstream and downstream flow configurations according to the flow direction of the secondary fluid in the main loop. In the upstream flow configuration, the secondary fluid flows in the order of the GHE, plate heat exchanger in the heat pump side, and outdoor heat exchanger in supplemental side, while in the downstream flow configuration, the secondary fluid flows in the reverse direction. Consequently, for the upstream flow the supplemental heat rejecter is activated as the primary heat rejecter, while for the downstream flow the GHE operates as the primary heat rejecter.

The GHE and supplemental heat rejecter are connected in parallel in the parallel configuration ($HGSHP_P$) of A-type. In this loop configuration, the main flow is split into the outdoor heat exchanger and GHE, and then the mixing valve is applied to control the flow rate of the GHE and the supplemental heat rejecter. Therefore, in the $HGSHP_P$ of A-type, the heat rejection is distributed to the GHE and the supplemental heat rejecter. In the $HGSHP_P$ of B-type (Park et al., 2012), the ground fluid loop including the GHE and the supplemental fluid loop containing a supplemental heat rejecter are connected with the refrigerant loop with separate liquid-to-refrigerant heat exchangers. The refrigerant leaving the compressor is initially rejected to the GHE, and then subsequently rejected to the supplemental heat rejecter.

Park et al. (2013) compared the performance of an HGSHP between the serial configuration of the downstream flow and the parallel configuration of B-type. The performance was investigated by varying the secondary fluid flow rate of the supplemental flow loop and the entering fluid temperature of the heat pump under steady-state conditions. Fig. 12.5 shows the COP and total power consumption of

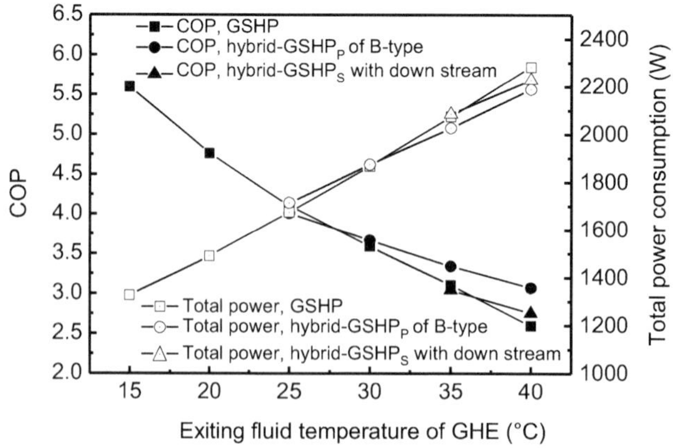

Figure 12.5 Variations of COP and total power with the leaving fluid temperature of the ground heat exchanger in the ground-source heat pump and hybrid ground-source heat pump with various configurations (Park et al., 2013).

the GSHP, $HGSHP_S$ of the downstream flow, and $HGSHP_P$ of B-type according to the leaving fluid temperature of the GHE. When the leaving fluid temperature of the GHE was smaller than 25°C (the background ground thermal condition), the COP of the $HGSHP_P$ of B-type was lower than that of the GSHP. However, in the case of degraded ground thermal condition (leaving fluid temperature of the GHE higher than 30°C), the COP of the $HGSHP_P$ of B-type was higher than that of the GSHP in spite of increased total power consumption due to additional pump power in the supplemental fluid loop. As the ground thermal condition gets worse (leaving fluid temperature of the GHE higher than 40°C), the COP of the $HGSHP_S$ with the downstream flow and $HGSHP_P$ of B-type were 6% and 18% higher than that of the GSHP, respectively.

Furthermore, Lee et al. (2015) conducted the optimization of the $HGSHP_S$ by varying the flow direction of the secondary fluid in the main loop: upstream and downstream flow (Fig. 12.4(a)). Lee et al. (2015) reported that the degradation of the ground thermal condition was more severe in the $HGSHP_S$ with the downstream flow compared with the upstream flow. In addition, the COP of the $HGSHP_S$ with the upstream flow was higher than that of the $HGSHP_S$ with the downstream flow.

Lee et al. (2015) also compared the performance of the $HGSHP_S$ with the upstream flow and $HGSHP_P$ of B-type under the transient state. Fig. 12.6 shows the performance improvement ratios in the $HGSHP_S$ with the upstream flow and the $HGSHP_P$ of B-type over the GSHP. The performance data for each operation mode was averaged after the steady-state which was defined when the heat accumulation in the thermal storage bath reached thermal equilibrium under the degraded ground thermal condition. The $HGSHP_S$ with the upstream flow showed lower heat accumulation than the $HGSHP_P$ of B-type, because the heat rejection rate of the GHE in the $HGSHP_S$ with the upstream flow was relatively lower than that in the $HGSHP_P$ of B-type. The $HGSHP_P$ of B-type showed relatively higher cooling capacity than the $HGSHP_S$ with the up-stream flow, because the enthalpy difference in the indoor heat exchanger of the $HGSHP_P$ of B-type increased by the higher subcooling level at the expansion device inlet. However, the total power consumption in the $HGSHP_S$ with the upstream flow was significantly

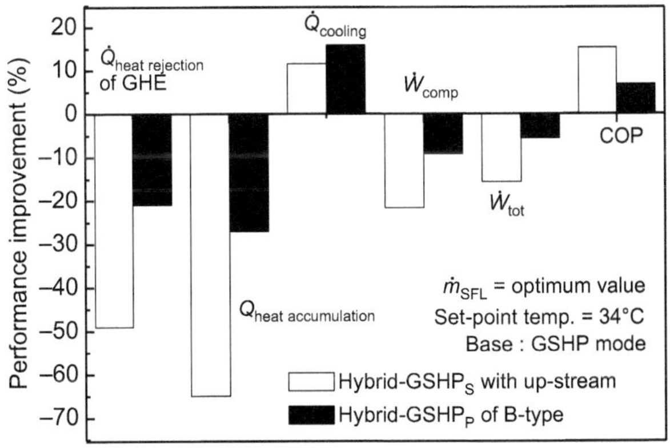

Figure 12.6 Comparison of the performance improvement of the hybrid ground-source heat pump with serial and parallel configurations over the ground-source heat pump (Lee et al., 2015).

lower than that in the $HGSHP_P$ of B-type because of the lower compressor power consumption according to an improved ground loop temperature. Accordingly, the COPs in the $HGSHP_S$ and $HGSHP_P$ increased by 15% and 7%, respectively, compared to that in the GSHP. Lee et al. (2015) concluded that the $HGSHP_S$ (upstream flow) can be recommended as the most stable and effective arrangement in terms of the COP and ground thermal condition under transient state.

Thornton (2000) investigated the economic feasibility of the HGSHP with various loop configurations ($HGSHP_S$ with upstream flow and $HGSHP_P$ of A-type) and control strategies using the cooling tower as the supplemental heat rejecter. In the first case, the cooling tower is connected in series with the GHE, which is not isolated from the ground flow loop. The set-point temperature control method is adopted in this case: the cooling tower is activated when the fluid temperature of the main flow loop exceeds 26.7°C. The length of the GHE was calculated to be 2286 m in this case while the length required for the GHE alone, was calculated to be 3475 m. In the second case, the GHE and cooling tower are connected in parallel as shown in Fig. 12.4(b). The cooling tower is activated when the fluid temperature of supplemental flow loop exceeds 21°C. The main fluid loop is split between the outdoor heat exchanger in the supplemental side and the GHE. Approximately 90% of the flow rate is distributed to the outdoor heat exchanger in the supplemental side to achieve ground thermal balance. The length of the GHE was calculated to be 1417 m in the second case. In the third case the loop configuration, control strategy, and size of GHE are the same with the second case. However, the flow rate is split between the outdoor heat exchanger in the supplemental side and the GHE to maintain maximum temperature of the entering fluid in the heat pump below 35°C.

Table 12.5 summarizes the results of the cost analysis reported by Thornton (2000). The overall energy use of the HGSHP was higher than that of the GSHP without the supplemental heat rejecter. In the first case, the system initial cost

Table 12.5 The summary of cost analysis for HGSHP (Thornton, 2000; Oak Ridge National Laboratory, 2001)

	Base case "no cooling tower"	Hybrid case 1	Hybrid case 2	Hybrid case 3
Total length of ground heat exchanger (m)	3,475	2,286	1,417	1,417
Design capacity of the cooling tower (kW)		334	334	334
Total cost of ground heat exchanger installation ($)[a]	92,454	60,825	37,712	37,712
Total cost of ground heat exchanger, plate heat exchanger, and cooling tower pump ($)	92,454	60,825	41,512	41,512

Table 12.5 Continued

	Base case "no cooling tower"	Hybrid case 1	Hybrid case 2	Hybrid case 3
Present value of 10 years of electricity costs ($)[b]	29,069	29,492	32,898	31,906
First-year electricity costs ($)	3,950	4,007	4,470	4,335
Present value of initial costs plus electricity costs ($)	121,523	90,317	74,410	73,418
Annual energy consumption Total energy consumption (kWh)	70,027	71,047	79,251	76,861

Heating degree-days = 3639; cooling degree-days = 1485; annual heating load = 57.96 MWh; annual cooling load = 127.5 MWh.
[a]Actual cost of test bores drilled at Oceana site for ground thermal conductivity testing was $26.61/m of borehole, including horizontal runs and connections (personal communication with B. Koshka, Trane Co., December 2000).
[b]Electricity cost $0.0564/kWh (rate charged to Oceana buildings by US Navy Public Works Center, Norfolk); no price escalation rate assumed. A 6% discount rate is used for present value computation.

decreased by 34%, and the annual total energy consumption increased by 1.4% compared to the GSHP. In the second and third cases, the system initial costs decreased about 55%, and the annual total energy consumption increased by 13.2% and 9.7%, respectively.

12.4 Efficiency and cost

In order to evaluate the effect of optimization in a HGSHP, it is necessary to define an evaluation indicator. Life cycle cost is the most suitable method because it includes both the capital and operating costs of the energy system. The P_1/P_2 method was proposed by Duffie and Beckman (2006) to determine the life cycle cost associated with the energy system. In their method, the life cycle cost can be calculated by

$$LCC = P_1 C_{F1} + P_2 C_E \quad [12.11]$$

Where, C_{F1} means the cost of fuel in the first year, and the C_E is the initial cost of the energy system equipment. P_1 and P_2 denote the ratio of the fuel cost and system cost in the life cycle to the first year, respectively.

In addition, Hackel (2008) proposed the life cycle savings between the optimized system and the baseline system, which is expressed by

$$LCS_{\text{hybrid}} = LCC_{\text{case2}} - LCC_{\text{case1}} \quad [12.12]$$

Where LCS_{hybrid} is the life cycle savings in an HGSHP. LCC_{case1} and LCC_{case2} mean the life cycle cost of a conventional heat pump (GSHP) and HGSHP, respectively.

Hackel et al. (2009) also compared the relative life cycle cost among the HGSHP, GSHP, and conventional heat pump with boiler and/or cooling tower, for each building and climate scenarios. The relative life cycle costs were used to determine the most cost-effective system, and the parametric study considered the optimized HGSHP, GSHP, and the conventional heat pump boiler and/or cooling tower. Table 12.6 shows the life cycle cost of the HGSHP and GSHP with various climate conditions (Hackel et al., 2009). The life cycle cost of the GSHP was found to increase with the decrease in the ratio of the annual heating load to the cooling load, while the life cycle cost of the HGSHP decreased with the decrease in the ratio of the annual heating load to the cooling load. The life cycle savings increased with decreasing the ratio of the annual heating load to the cooling load. The life cycle savings arose from the reduction in the life cycle cost of the HGSHP by decreasing the size of the GHE and enhancing the performance by enhanced ground thermal condition. It should be noted that the life cycle savings became negligibly small when the ratio of the annual total heating load to the cooling load was higher than 0.9 (Seattle).

Moreover, Hackel et al. (2009) reported that the life cycle savings associated with the reduction of the GHE was $75/m^2$ for a building in a climate such as that in Atlanta, Georgia, and $32/m^2$ for a building in a climate such as that in St. Louis, Missouri. Hackel et al. (2009) also investigated the life cycle savings of the HGSHP compared to the conventional water-to-water heat pump system using the cooling tower and the boiler as heat sink and heat source, respectively. The life cycle savings of the HGSHP were proportional to the peak heating load, which were $11–$22/m^2$ in the buildings as shown in Table 12.6. The life cycle savings of the HGSHP was more significant

Table 12.6 Life cycle cost of hybrid ground-source heat pumps, ground-source heat pumps, and conventional heat pump system for 20 years (Hackel et al., 2009)

Climate (ratio of annual heating to cooling load)	Life cycle cost ($), HGSHP	Life cycle cost ($), GSHP	Life cycle cost ($), conventional heat pump	Life cycle savings ($), HGSHP versus GSHP	Life cycle savings ($), HGSHP versus conventional heat pump
Atlanta (0.37)	609,892	873,000	683,069	263,108	73,177
Salt Lake City (0.82)	755,061	759,051	780,900	3,990	25,839
Phoenix (0.1)	677,239	2,349,957	721,877	1,672,718	44,638
St. Louis (0.71)	738,561	747,399	815,356	8,838	76,795
Seattle (0.93)	487,884	493,286	594,352	5,402	106,468

when compared to the GSHP than when compared to the conventional heat pump with the boiler and/or cooling tower, because the installation cost of the conventional heat pump system was cheaper than that of the GSHP system.

Yavuzturk and Spitler (2000) compared the life cycle cost of HGSHP and GSHP systems and investigated the effects of various design parameters and control strategies. Table 12.1 shows the size of the GHE and capacity of the cooling tower in each case of their study. The respective control strategies of the set-point temperature, night time schedule, and temperature difference control adopted in each HGSHP case were discussed in 12.1.

Figs. 12.7–12.9 show the system initial cost, electricity cost, and life cycle cost of the HGSHP and GSHP in Houston, Texas, and Tulsa, Oklahoma, calculated by Yavuzturk and Spitler. The initial cost of the GSHP in Houston was higher than that in Tulsa because the required length of the GHE was higher in Houston than that in Tulsa (the annual heating load of Houston was smaller than that of Tulsa). As noted earlier in Section 12.2, in the case of the HGSHP the required length of GHE is dependent on the peak heating load. Therefore, as shown in Fig. 12.7, the initial cost savings of the HGSHP was higher in Houston than in Tulsa because the length of the GHE can be significantly decreased in Houston by the supplemental heat rejecter. The HGSHP with the schedule control strategy showed the smallest initial cost due to the lowest capacity of the supplemental heat rejecter, while it had the largest annual energy use. The HGSHP with the temperature difference control yielded the lowest annual energy use and life cycle cost. The life cycle cost of the HGSHP with the temperature difference control strategy was about 47% and 17% lower than that of the GSHP in Houston and Tulsa, respectively. In addition, the temperature difference control improved the ground thermal condition by decreasing the exiting fluid temperature of GHE. Compared to the GSHP, the HGSHP with any of the control strategy yielded significant economic benefits, as can be seen from the life cycle cost in Fig. 12.9.

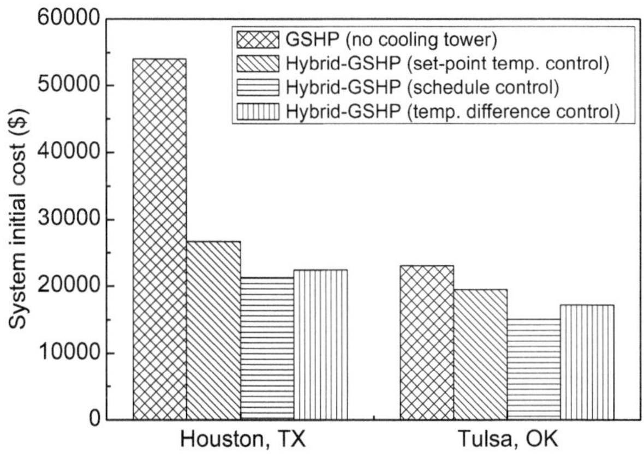

Figure 12.7 System initial cost of hybrid ground-source heat pump and ground-source heat pump in Houston, TX and Tulsa, OK (Yavuzturk and Spitler, 2000; Collins et al., 2001).

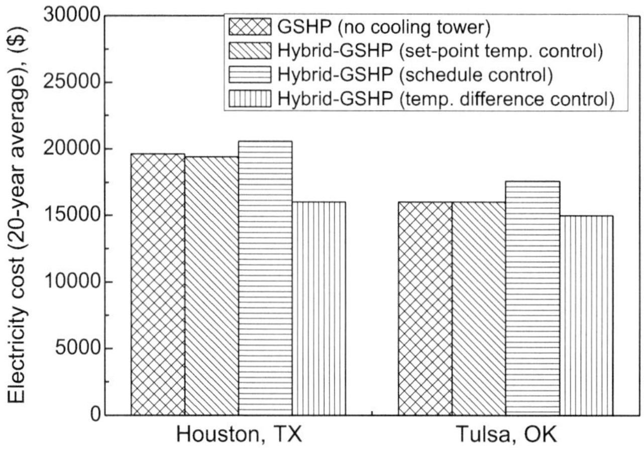

Figure 12.8 Electricity cost of hybrid ground-source heat pump and ground-source heat pump in Houston, TX and Tulsa, OK (Yavuzturk and Spitler, 2000; Collins et al., 2001).

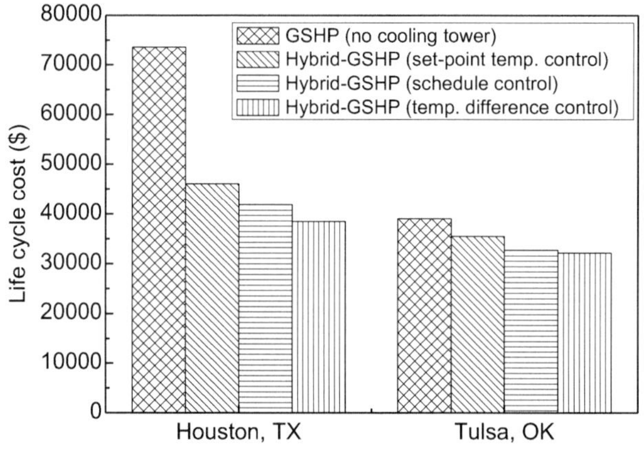

Figure 12.9 Life cycle savings of hybrid ground-source heat pump and ground-source heat pump in Houston, TX and Tulsa, OK (Yavuzturk and Spitler, 2000; Collins et al., 2001).

12.5 Conclusion

HGSHP systems have been suggested for their economic and efficient performance especially in situations where the building heating and cooling demands are not well balanced. The HGSHP can adopt various supplemental heat sources and sinks according to the climatic conditions and the balance of energy demands. In hot climate regions, the cooling tower can reduce the excessive heat rejection through the GHE, mitigating either unacceptable changes in the ground temperature or excessive borehole depths. In cold climate regions, the use of the supplemental heat extractor can mitigate the ground temperature decrease, resulting in more effective operation of the HGSHP. In the HGSHP system, the optimization of the design and control parameters is very important. The key objectives of the optimization are minimizing the life cycle cost and balancing the ground thermal environment. The hybrid control scheme based on the temperature difference showed the lowest energy consumption and operating cost. In cooling dominated buildings, the length of the GHE should be determined based on the cooling load, and the additional heating load should be assigned to the supplemental heat source. In addition, the $HGSHP_S$ (upstream flow) can be recommended as the most stable and effective arrangement in terms of the COP and sustainable ground thermal conditions. Well designed and optimized HGSHP can provide substantial economic benefits and improved ground energy balance.

Acknowledgment

This work was supported by the Human Resources Development Program (No. 20144010200770) of the Korea Institute of Energy Technology Evaluation and Planning (KETEP) grant funded by the Korea Government Ministry of Trade, Industry and Energy.

References

Bakirci, K., Ozyurt, O., Comakli, K., Comakli, O., 2011. Energy analysis of a solar-ground source heat pump system with vertical closed-loop for heating applications. Energy 36, 3224–3232.

Bi, Y., Guo, T., Zhang, L., Chen, L., 2004. Solar and ground source heat-pump system. Appl Energy 78, 231–245.

Collins, T., Parker, S.A., Baxter, V., 2001. Assessment of hybrid geothermal heat pump systems, technology installation review. The U.S. Department of Energy.

Dai, L., Li, S., DuanMu, L., Li, X., Shang, Y., Dong, M., 2015. Experimental performance analysis of a solar assisted ground source heat pump system under different heating operation modes. Appl Therm Eng 75, 325–333.

Dikici, A., Akbulut, A., 2008. Exergetic performance evaluation of heat pump systems having various heat sources. Int J Energy Res 32, 1279–1296.

Duffie, J.A., Beckman, W.A., 2006. Solar engineering of thermal processes, 3rd ed. John Wiley and Sons, Inc.

Fan, R., Gao, Y., Hua, L., Deng, X., Shi, J., 2014. Thermal performance and operation strategy optimization for a practical hybrid ground-source heat-pump system. Energy Build 78, 238–247.

Gentry JE, Spitler JD, Fisher DE, Xu X. Simulation of hybrid ground source heat pump systems and experimental validation. In: Proceedings of 7th international conference on system simulation in buildings, Liège, December 11–13, 2006.

Geothermal Bore Technologies Inc., 2006. Ground loop design (GLD).

Hackel, S., Pertzborn, A., 2011. Effective design and operation of hybrid ground-source heat pumps: three case studies. Energy Build 43, 3497–3504.

Hackel, S., Nellis, G., Klein, S., 2009. Optimization of cooling-dominated hybrid ground coupled heat pump systems. ASHRAE Trans 115, 565–580.

Hackel, S., 2008. Development of design guidelines for hybrid ground-coupled heat pump systems. Department of Mechanical Engineering, University of Wisconsin, Madison (WI).

Han, Z., Zheng, M., Kong, F., Wang, F., Li, Z., Bai, T., 2008. Numerical simulation of solar assisted ground-source heat pump heating system with latent heat energy storage in severely cold area. Appl Therm Eng 28, 1427–1436.

Hern, S.A., 2004. Design of an experimental facility for hybrid ground source heat pump systems [M.S. thesis]. School of Mechanical and Aerospace Engineering, Oklahoma State University.

Kavanaugh, S.P., Rafferty, K., 1997. Ground-source heat pumps: design of geothermal systems for commercial and institutional buildings.

Kavanaugh, S.P., 1998. A design method for hybrid ground-source heat pump. ASHRAE Trans 104 (2).

Kjellsson, E., Hellström, G., Perers, B., 2010. Optimization of systems with the combination of ground-source heat pump and solar collectors in dwellings. Energy 35, 2667–2673.

Klein, S., Beckman, W., Mitchell, J., 2010. TRNSYS – Reference manual. Technical report. Solar Energy Laboratory, University of Wisconsin-Madison.

Lee, J.S., Park, H., Kim, Y., 2014. Transient performance characteristics of a hybrid ground-source heat pump in the cooling mode. Appl Energy 123, 121–128.

Lee, J.S., Song, K.S., Ahn, J.H., Kim, Y., 2015. Comparison on the transient cooling performances of hybrid ground-source heat pumps with various flow loop configurations. Energy 82, 678–685.

Li, H.Q., Kang, S.S., Yu, Z., Cai, B., Zhang, G.Q., 2014. A feasible system integrating combined heating and power system with ground-source heat pump. Energy 74, 240–247.

Lubis, L.I., Kanoglu, M., Dincer, I., Rosen, M.A., 2011. Thermodynamic analysis of a hybrid geothermal heat pump system. Geothermics 40, 233–238.

Man, Y., Yang, H.X., Fang, Z.H., 2008. Study on hybrid ground-coupled heat pump systems. Energy Build 40, 2028–2036.

Man, Y., Yang, H.X., Wang, J.G., 2010. Study on hybrid ground-coupled heat pump system for air-conditioning in hot-weather areas like Hong Kong. Appl Energy 87, 2826–2833.

Michopoulos, A., Bozis, D., Kikidis, P., Papakostas, K., Kyriakis, N.A., 2007. Three-years operation experience of a ground source heat pump system in Northern Greece. Energy Build 39, 328–334.

Midwest Cooling Towers. http://midwesttowers.com/cooling-tower-fundamentals-cooling-tower-types/.

Nam, Y., Ooka, R., Shiba, Y., 2010. Development of dual-source hybrid heat pump system using groundwater and air. Energy Build 42, 909–916.

Ni L, Song W, Zeng F, Yao Y. Energy saving and economic analyses of design heating load ratio of ground source heat pumps with gas boiler as auxiliary heat source. In: Conference on Electric Technology and Civil Engineering (ICETCE) 2011, Lushan, China, p. 1197–1200.

Oak Ridge National Laboratory, 2001. Assessment of hybrid geothermal heat pump systems. DOE/EE-0258, S.l.

Pardo, N., Montero, Á., Martos, J., Urchueguía, J.F., 2010. Optimization of hybrid – ground coupled and air source – heat pump systems in combination with thermal storage. Appl Therm Eng 30, 1073–1077.

Park, H., Lee, J.S., Kim, W., Kim, Y., 2012. Performance optimization of a hybrid ground source heat pump with the parallel configuration of ground heat exchanger and a supplemental heat rejecter in the cooling mode. Int J Refrig 35, 1537–1546.

Park, H., Lee, J.S., Kim, W., Kim, Y., 2013. The cooling seasonal performance factor of a hybrid ground-source heat pump with parallel and serial configurations. Appl Energy 102, 877–884.

Rabl, A., 1985. Active solar collectors and their applications. Oxford University Press.

Sagia, Z., Rakopoulos, C., Kakaras, E., 2012. Cooling dominated hybrid ground source heat pump system application. Appl Energy 94, 41–47.

Spitler, J.D., 2006. GLHEPRO 3. International Ground Source Heat Pump Association.

Stojanović, B., Akander, J., 2010. Build-up and long-term performance test of a full-scale solar-assisted heat pump system for residential heating in Nordic climatic conditions. Appl Therm Eng 30, 188–195.

Thornton, J., 2000. A detailed TRNSYS analysis of building 137. Oak Ridge National Laboratory, Oceana Naval Air Station, Virginia Beach (Virginia).

Tomlinson, J.J., 2005. Heat recovery from wastewater using a gravity-film heat exchanger. Federal Energy Management Program. The U.S. Department of Energy.

Wang, H., Qi, C., 2008. Performance study of underground thermal storage in a solar-ground coupled heat pump system for residential buildings. Energy Build 40, 1278–1286.

Wang, H., Qi, C., Wang, E., Zhao, J., 2009. A case study of underground thermal storage in a solar-ground coupled heat pump system for residential buildings. Renew Energy 34, 307–314.

Wang, E., Fung, A.S., Qi, C., Leong, W.H., 2012. Performance prediction of a hybrid solar ground-source heat pump system. Energy Build 47, 600–611.

Yang, J., Xu, L., Hu, P., Zhu, N., Chen, X., 2014. Study on intermittent operation strategies of a hybrid ground-source heat pump system with double-cooling towers for hotel buildings. Energy Build 76, 506–512.

Yang, W., Sun, L., Chen, Y., 2015. Experimental investigations of the performance of a solar-ground source heat pump system operated in heating modes. Energy Build 89, 97–111.

Yavuzturk, C., Spitler, J.D., 2000. Comparative study of operating and control strategies for hybrid ground-source heat pump systems using a short time step simulation model. ASHRAE Trans 106, 192–209.

New trends and developments in ground-source heat pumps

13

J.M. Corberan
Polytechnic University of Valencia, Valencia, Spain

Abstract

This chapter summarizes the main differences between ground-source heat pumps (GSHP) and other kinds of packaged vapor-compression heat pumps (HPs) and air-conditioning equipment, and reviews new trends and developments currently observed in the field. The chapter firstly comments upon the typical performance of GSHPs both at full load and at partial load. Secondly, current issues regarding compressor technology and its adaptation to GSHP-specific characteristics are reviewed. Thirdly, the current situation regarding preferred refrigerants for this kind of HP is discussed with an analysis of the most probable short-term and long-term solutions. This is followed by a discussion on the characteristics of each type of heat exchanger employed in GSHPs, presented with a comment on recent developments. Finally, with the current trend toward hybrid ground source systems, the potential advantages of dual source (ground and air) HP equipment is outlined.

13.1 Introduction

Ground-source heat pump (GSHP) systems have seen significant market expansion in recent years and a considerable number of unitary air-conditioning and heat pump (HP) equipment manufacturers have developed HPs for ground-coupled applications for both residential and commercial building sectors.

Packaged GSHPs for heating and cooling are very similar to other HPs for comparable building applications. Given that their heat source is the ground, the external heat exchange can be either directly with the ground (direct-expansion evaporator) or with a heat transfer fluid in a refrigerant-to-liquid heat exchanger. If the geothermal system is based on an open loop the heat transfer fluid is the ground water. If the geothermal system is based on a closed loop the heat transfer fluid is water or an antifreeze mixture if there is any possibility of freezing. Consequently, the main difference between a GSHP and an air-source heat pump (ASHP) is the source temperature and the external heat exchanger design (the evaporator in heating operation).

In North America, heating and cooling is delivered on the load side by ducted air systems in most residential and small commercial applications using packaged (unitary) equipment. Accordingly, the load-side heat exchanger (ie, evaporator in cooling operation) is designed like other forms of air-conditioning equipment—with due consideration for dehumidification. In Europe, residential equipment is most often

Advances in Ground-Source Heat Pump Systems. http://dx.doi.org/10.1016/B978-0-08-100311-4.00013-3
Copyright © 2016 Elsevier Ltd. All rights reserved.

designed for heating applications using water distribution systems such as underfloor heating or low-temperature convectors/radiators.

A consequence of the smaller temperature lifts (difference between source and load sides) found in GSHP applications, in comparison with ASHPs, is that the compressor works under lower pressure ratios than found in ASHPs. Given that scroll compressors are normally used in this application for their superior efficiency, and that these compressors have an optimum pressure ratio at which they provide their maximum efficiency, the smaller working pressure ratios of a GSHP indicates that the compressors for these HPs should be designed with a lower optimum pressure ratio than the one required for ASHPs or air-conditioning. This is discussed in detail in Section 13.3 devoted to compressors.

13.2 Ground-source heat pump performance

In this section, a review of the current performance, the main characteristics of its components, and the new trends and developments of this technology is presented.

13.2.1 Full load performance

Staffell et al. (2012) compiled performance data for a large number of domestic air-to-water and water-to-water HPs. The author compiled the information from different data sources and combined it with catalog data from different manufacturers. He then presented the results of the data survey in the form of a graph of declared coefficient of performance (COP) versus the temperature difference between the supply temperature and the inlet water temperature at the source side. The author found a COP ranging from 5 to 2, decreasing with the increase of the temperature difference and reported that the COP falls by 0.67−1.07 for every 10K temperature difference rise.

In BINE (2013), the authors present an analysis of HP performance tested in situ under real life conditions is presented in the form of monthly average COP and average seasonal performance factors (SPF). The authors also present the COP of a water-to-water HP relative to the source inlet temperature and the heating circuit output temperature.

Fig. 13.1 presents the results from the two above mentioned studies inside an envelope of best and worse values for current GSHPs. The COP values only include the compressor and parasitic losses of the HPs. The figure gives an estimation of the expected COP for current GSHPs. The COP of any unit will obviously depend on the temperature difference between the sink and source temperatures as it is clearly seen in the figure. The decrease does not only depend on the temperature difference but also the value of the source or sink temperatures. For example, the COP results reported in BINE (2013) are plotted as two lines: one corresponds to the case of supply water temperature of 35°C while the other corresponds to 55°C. As it can be observed,

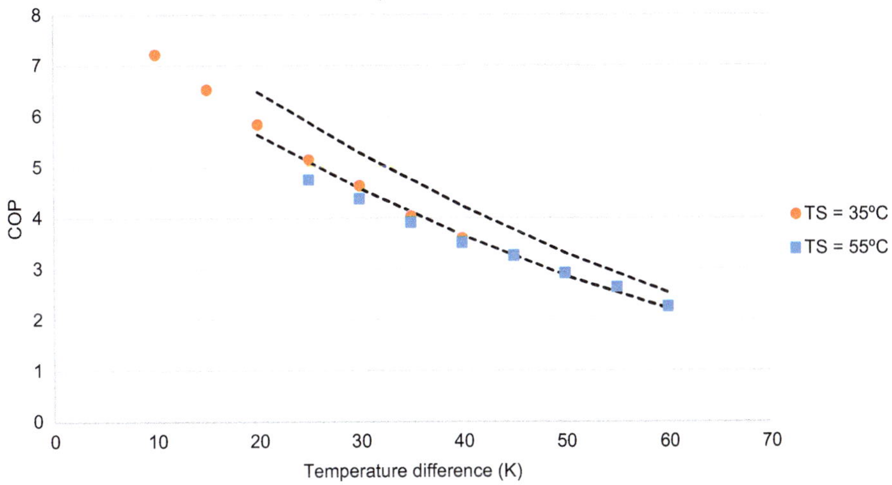

Figure 13.1 Coefficient of performance (COP) of ground-source heat pumps as a function of the temperature difference between the heat sink and source. Nonsolid lines indicate the band of maximum and minimum COP values compiled by Staffell et al. (2012) while the other values correspond to the results in the BINE report (2013) for two different building supply water temperatures as indicated.

the values reported in the BINE report agree well with the bottom line of the values compiled by Staffell. The bottom line of the figure represents the minimum value of the COP that should be attained by a reasonably good GSHP, and can be simply evaluated by the following expression:

$$COP_{min} = 0.00068 \times \Delta T^2 - 0.140 \times \Delta T + 8.17 \quad [13.1]$$

The rate of variation of the COP with the increase of the temperature difference ΔT ranges from −0.12, at 20K difference, to −0.10, at 60K difference, so that −0.11 is a good value to estimate the decrease in COP when the temperature difference is increased by 1K. This decrease in percentage is more important at operating conditions where the temperature difference is higher, ranging from −2.2% to −4.5% in the range of the figure.

In general the performance of a water-to-water HP depends on the conditions of temperature and flow rate at both source and sink types. A meta-model of HP performance in terms of heating capacity $\dot{Q}_{cond}(W)$ and the electric consumption $W_{HP}(W)$ can be fitted by polynomials depending on those variables, ie,

$$\dot{Q}_{cond} = \dot{Q}(\dot{m}_b, T_{b,ret}, \dot{m}_w, T_{w,ret}) \quad [13.2]$$

$$\dot{W}_{HP} = \dot{W}_{HP}(\dot{m}_b, T_{b,ret}, \dot{m}_w, T_{w,ret}) \quad [13.3]$$

where $T_{b,ret}$ is the brine temperature at the inlet of the evaporator (return from the borefield) and $T_{w,ret}$ is the temperature of the water at the inlet of the condenser (return from the building).

For instance, in Corberan et al. (2011) the authors presented a meta-model for the HP performance for a water-to-water HP of 16 kW based on experimental results. Fig. 13.2 shows the COP of that GSHP as a function of the two mass flow rates for a supply temperature of 45°C and a return temperature from the ground of 10°C, while Fig. 13.3 shows the COP at flow rates providing 5K temperature difference at both evaporator and condenser at the nominal point (10,50), for different temperatures on the source and load sides. Again it should be noted that the COP in Figs. 13.2 and 13.3 do not include the circulation pumps consumption. As can be seen in Fig. 13.2, the COP of the GSHP depends on the mass flow rate at the source and load water loops. The greater the mass flow rate the lower the temperature difference between the source/load temperatures and the corresponding evaporation/condensation temperatures. Consequently, the greater the mass flow rate the higher the COP. However, the greater the mass flow rates the greater the electricity consumption of the circulation pumps. This leads to an optimum flow rate which not only depends on the HP but also on the hydraulic pressure losses. The variation of the circulation flow rates also influences the heat transfer at the ground heat exchanger (GHE) and at the room terminals or heat emitters in the building. It can be seen that the optimization of overall system performance and the control of the circulation pumps flow rate is a complex problem, which involves a lot of coupled effects. An in-situ optimization method for this problem is presented in Montagud et al. (2014) and Cervera-Vázquez et al. (2015).

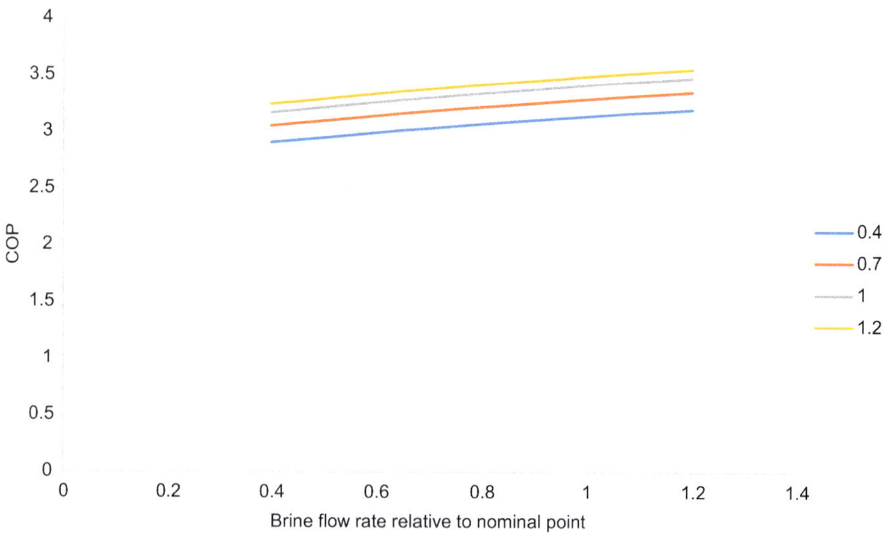

Figure 13.2 Coefficient of performance at different load and source loop flow rates. Mass flow rates are shown relative to the nominal point $\left(\frac{\dot{m}_w}{\dot{m}_{w,nom}}\right)$. Variation with load-side flow rate is shown on the horizontal axis and source-side flow rates vary for each line.

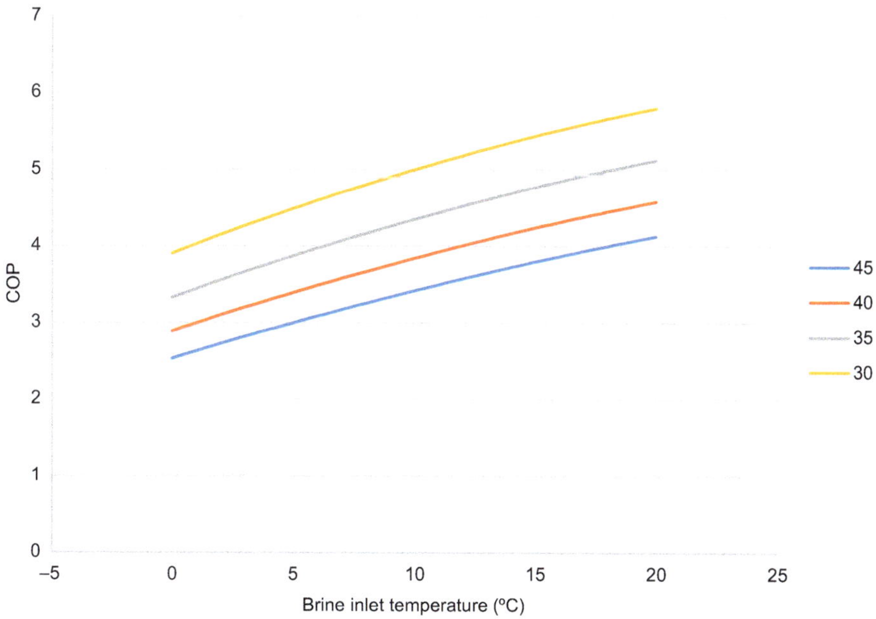

Figure 13.3 Coefficient of performance at different source (brine inlet) and load (supply water) temperatures.

Another interesting way to characterize the performance of the equipment is to analyze the energy certification or qualification programs. For instance, the Energy Star qualifying program in the United States (EPA, 2009a) currently requires the following COP values as the minimum to get the qualification for closed-loop ground source (CLGS) HPs (Table 13.1).

The units must be qualified under rating conditions in accordance with ISO 13256-1 for water-to-air models (Inlet water at 0°C and 20°C dry bulb [DB] entering air temperature) or ISO 13256-2 for water-to-water models (Inlet water at 0°C and entering load temperature at 40°C). Assuming a 5K temperature difference for the water though the condenser, this rating point for water-to-water models corresponds to a temperature difference of 45K (given that supply water is then the return temperature + 5K) which,

Table 13.1 Efficiency requirements for qualifying closed loop HP products as Energy Star (single stage units, single phase current)

Heat pump type	COP_{min}
Water-to-air	3.6
Water-to-water	3.1
Direct geoexchange	3.6

as can be seen in Fig. 13.1, lies below the bottom part of the performance band. Notice that the COP following those standards includes a certain penalization for the consumption of the circulation pump of the GHE loop.

In Europe, the Eurovent certification program does not include a COP labeling yet for GSHPs. It is only available for the category Heat Pump/Water Package, requiring a COP greater than 4.45 for Class A for water cooled case (10/7°C) providing heating water at the condenser at 30/35°C for floor heating.

13.2.2 Part load performance

The degradation of the performance of refrigeration equipment at partial loads has become an important topic mainly due to its implication in the seasonal performance evaluation of the complete building heating, ventilation, and air-conditioning (HVAC) system but also because of the interest in comparing different system concepts. For example, in comparison of ground source versus air source systems, or on–off versus inverter-driven capacity control. A lot of work has been done for the characterization of the seasonal performance of the air-conditioning and heating equipment, including the estimation of the performance at part load, for example, that reported by Henderson et al. (2000), Schibuola (2000), Anglesio et al. (2001), Bettanini et al. (2003), Riviere et al. (2005), and Cecchinato et al. (2010). The PhD work of Riviere (2004) contains a very detailed analysis of the operation of different systems, of their part load performance, and finally of the evaluation of the corresponding seasonal performance.

Most of these studies were concentrated on air-to-air and air-to-water systems. The dynamic performance and part load degradation of water-to-water and water-to-air systems, which are the HPs employed for ground source systems, have been studied very little. Corberan et al. (2013) investigated the cycling losses of water-to-water units working under the classic thermostat on–off control applied at the return temperature, and reached the following conclusions:

- The dynamic performance of the HP is almost quasi-steady, ie, the HP can be considered as working at each instant under steady conditions at the instantaneous values of the inlet water temperatures to the evaporator and condenser. In fact, the instantaneous quasi-steady COP agrees very well with the instantaneous dynamic COP.
- The cycling losses are small and are mainly due to the stand-by losses. For the unit tested they only represented a 0.9% of the average consumption of the compressor, so they are negligible when the compressor is switched on, but even with that low value they degrade the performance significantly for load ratios below 20%.
- The degradation of the performance can be evaluated by the following expression, which is identical to that employed in EN (2010): EN-14825: Rating at part load conditions of refrigeration and heat pump equipment.

$$\mathrm{PLF} = \frac{1}{1 + \frac{\dot{W}_{\mathrm{sb}}}{\dot{W}_{\mathrm{HP}}} \frac{(1-\propto)}{\propto}} \qquad [13.4]$$

where PLF is the part load factor which is defined as the COP at a given load ratio $\propto$ with respect the COP at full load $\left(\mathrm{PLF} = \frac{\mathrm{COP}(\propto)}{\mathrm{COP}(\propto=1)}\right)$, $\dot{W}_{sb}$ is the stand-by electrical consumption of the unit, $\dot{W}_{HP}$ is the electrical consumption of the HP when the compressor is ON, and $\propto$ is the Load ratio (null load $\propto = 0$; full load $\propto = 1$).

- The standby losses $\dot{W}_{sb}$ are mainly caused by the consumption of the electronic control system components, the consumption of the valves (which remain energized during the "off" period), and the consumption of the crankcase heater. These last two consumptions should be critically reviewed and solutions with null or low energy consumption adopted; otherwise, standby losses can be very high and significantly reduce the performance at low partial loads.

The formula [13.4] is also employed in EN 14825 for air-to-water systems so it could also be of application for GSHP water-to-air systems in the case that the fan at the condenser is switched off when the compressor is off. Fig. 13.4 shows the variation of PLF, as defined by Eq. [13.4], with the load ratio ($\propto$) for several values of the relative standby losses. As it can be observed, at load ratios below 0.2 the degradation of COP could be very fast.

The study was focused on the classic thermostat on–off control applied at the load-side return temperature but conclusions are applicable to all on–off control techniques, such as floating hysteresis or degree minute, as studied by Madani (2012).

Capacity regulation with variable speed compressors usually lead to PLFs greater than one at intermediate part loads since the regulation is made through the adaptation of the refrigerant flow rate and therefore the temperature difference between the evaporation/condensation and the source/sink temperatures decreases for lower refrigerant flows, and correspondingly the COP tends to improve. The efficiency of the compressor is also affected by the speed so the PLF curve must

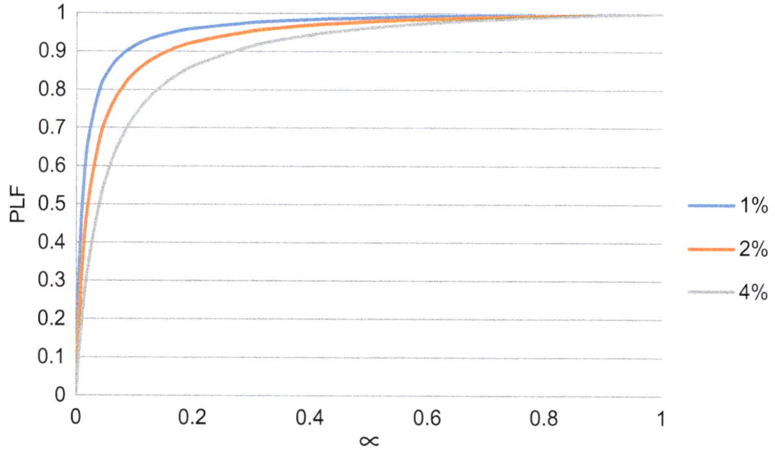

Figure 13.4 Part load factor (PLF) versus the load ratio ($\propto$) for different values of the relative parasitic losses $\frac{\dot{W}_{sb}}{\dot{W}_{HP}}$ (%).

be determined experimentally. For very low loads, below the one provided by the compressor at its minimum speed the regulation becomes on−off again and the PLF will follow Eq. [13.4].

13.3 Compressors for ground-source heat pumps

13.3.1 Compressor technology: scroll, rotary, piston

Compressor technology for refrigeration, air-conditioning, and heat pump equipment (RAC&HP) has significantly improved in the past years, and the most suitable compressor technology for each application has been changing depending on the evolution of the different technologies, the availability of adequate sizes, and the cost. Two decades ago the majority of RAC&HP equipment employed piston compressors. Piston compressors were a mature and reliable technology able to adapt to any operating condition, and cost was relatively low since they were manufactured in large quantities. Screw and centrifugal compressors were employed for medium and large capacity installations.

More recently, scroll compressors have improved and offer superior performance, and are also available in a very wide range of capacities. Nowadays scroll compressors are available in capacities suitable for the full range of building applications—from 6 to almost 200 kW—and combined in duo or trio they can reach capacities around 0.5 MW. Some small size scroll compressors have also been developed with capacities around 3 kW.

Current scroll compressors have a number of advantages versus piston types: efficiency is higher; they have fewer moving parts; they have quieter operation and higher reliability, and they are able to start under any system load without start components. On the other hand, they are more expensive.

Compressor efficiency depends on the size of the compressor, with a general trend to slightly increasing with size. Scroll compressors are very efficient around the design pressure ratio. Fig. 13.5 shows the compressor efficiency for one piston and one scroll compressor of the same size. As it can be observed in Fig. 13.5, the efficiency of the scroll compressor is higher than that of the piston compressor. The pressure ratio at which the scroll compressor shows its maximum efficiency is related with the designed volume compression ratio. Essentially, at higher pressure ratios a reverse flow from discharge to the scroll volume occurs when the discharge orifice opens, producing a backflow which, in the end, must be pumped out again. At lower pressure ratios, the pressure at the scroll volume is higher than the discharge pressure when the discharge orifice is opened and a reexpansion of the overcompressed gas occurs. Both phenomena bring irreversibilities that decrease the compression efficiency. The loss of efficiency produced by those phenomena makes the compressor efficiency of the scroll compressor drop relatively quickly for pressure ratios different from the optimum.

Piston compressors also have a maximum of efficiency around a certain pressure ratio but the efficiency characteristic is very flat. On the other hand, piston compressors have a much lower degradation of efficiency at high pressure ratios. Therefore, as it

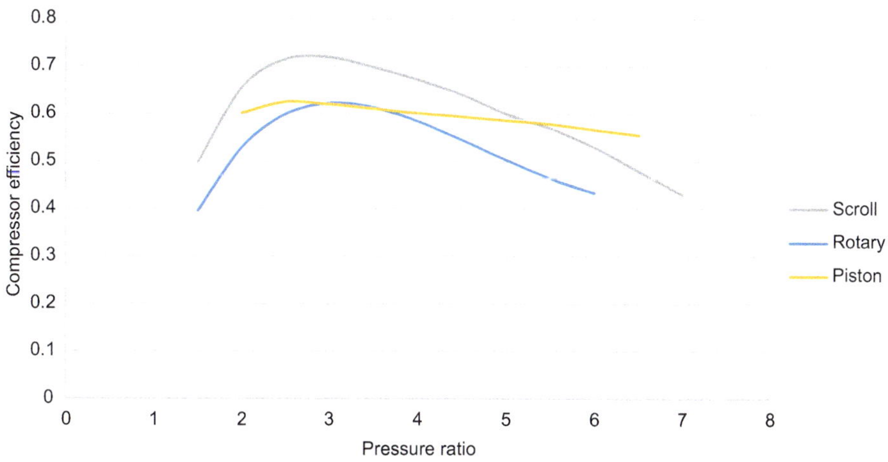

Figure 13.5 Comparison of compressor efficiencies of different compressor technologies.

can be observed in Fig. 13.5, at high pressure ratios the efficiency of the piston compressor can be higher than the efficiency of the scroll compressor. The most advanced scroll compressors nowadays incorporate a discharge valve and some other special features that reduce these losses and make the efficiency characteristic flatter. This helps to improve the efficiency of the scroll compressors when working at pressure ratios far from the optimum.

Rotary compressors have also improved a lot in recent years. Their efficiency cannot compete with scroll compressors but, for small sizes, they are reaching efficiencies close to those of piston compressors. Containing fewer parts than piston compressors they have become a reliable and cheap technology which is spreading from small capacities to the low range of commercial applications.

13.3.2 Optimal pressure ratio

The high efficiency attained by current scroll compressors suggests they are the best option for GSHPs. However, as shown earlier, the position of pressure ratio corresponding to the maximum of efficiency of a scroll compressor can be adjusted to the application by adjusting the compression volume ratio. The manufacturing of the scroll vanes is one of the most delicate and expensive parts of the scroll compressors. Consequently, manufacturers tend to produce a reduced set of sizes and designs optimized for the most common capacities, which normally are unitary air-conditioners or ASHPs. GSHPs normally work under pressure ratios that are much lower than those found in these types of equipment.

This is illustrated in the Groundmed Project (GROUNDMED, 2009) where the pressure ratio variation of a GSHP operating along the winter and summer season for Mediterranean climates was evaluated. Fig. 13.6 shows the compressor efficiency versus the pressure ratio for a R410A scroll compressor of commercial size of around 20-kW heating capacity. Heating water supply temperature in winter is 45°C and cooling water supply temperature in summer is 10°C.

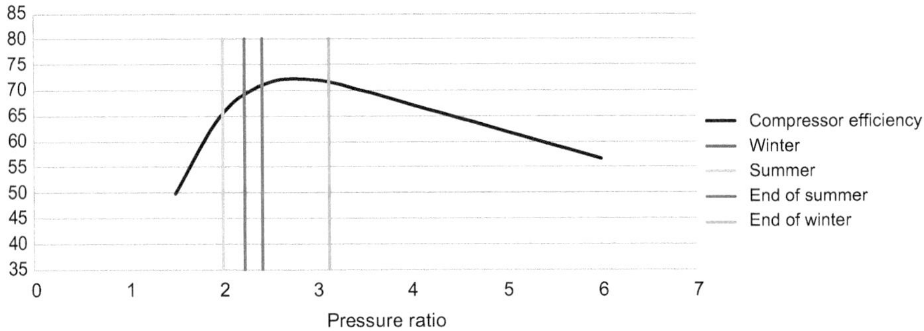

Figure 13.6 Compressor efficiency and operation pressure ratios.

The pressure ratio of a GSHP slightly changes along one day since the ground loop return temperature is generally quite stable and the supply temperature is kept constant by the control system. In this way, the pressure ratio evolves from low values at the beginning of the season to higher values at the end of the season. Only at the end of the winter are pressure ratios above 3 reached. As can be observed in Fig. 13.6, the optimum pressure ratio for this compressor is around 2.8. This value is a bit higher than the optimum for heating operation since the number of running hours at intermediate pressure ratios (winter in Fig. 13.6) is much higher than at the end of the season (end of winter in Fig. 13.6). Heat supply temperature compensation also contributes to having mild pressure ratios most of the season. In addition, notice that the drop in efficiency from the maximum is more pronounced from the optimum pressure ratio to the left than to the right (higher pressure ratios). On the other hand, it is clear that the compressor optimum pressure ratio is too high for the cooling season if the HP is reversible.

It should also be pointed out that this evaluation was done for heating water supplied at 45°C, ie, for using low temperature radiators or fan-coil units. If the heating water is used for a radiant floor, then pressure ratios are lower, and clearly the compressor has an optimum pressure ratio much greater than the one optimal for the application. It should furthermore be pointed out that if capacity modulation via tandem or variable speed is employed at partial loads, condensation and evaporation temperatures become closer to the source and sink temperatures so that working pressure ratios become even smaller.

Therefore, taking into account that the current design trend is to employ supply temperatures as close to the conditioned space temperature as possible, and that the use of capacity modulation is becoming a usual practice, it is clear that a series of scroll compressors with a smaller optimum pressure ratio is required for future high efficiency GSHP technology.

13.3.3 Fixed speed, tandem, variable speed

It was noted in Section 13.2.2 that HPs in GSHP systems work most of the time at relatively low load factors. If the HP is built with a single compressor and the control is on—off, then the cycles can become quite short, depending on the thermal capacity of the system in relation to the HP output. Provided that a buffer tank is included in the load-side loop and has been adequately sized, the minimum on time will be sufficient to insure the return of the oil to the compressor and no major reliability trouble arises from "short cycling." As pointed out in Section 13.2.2 there is almost no deterioration of performance of the ground HP due to partial load operation when the compressor is on.

The use of two or three compressors in tandem or a variable speed compressor will allow the HP to adapt the refrigerant mass flowrate, and hence the capacity to closely match the actual thermal load. In heating mode, this allows the reduction of the temperature difference between the evaporation and source, and between the condensation and load: the lower the load, the higher the reduction of the temperature differences. This will bring, on the one hand, a significant improvement on the cycle COP. On the other hand, this will also bring a reduction of the compressor pressure ratio, which, if the operation pressure ratio is above the optimum pressure ratio, will also lead to an increase of compressor efficiency. However, this could also lead to a decrease if the operation pressure is below the optimum. Usually, the overall effect on the COP and SPF is quite positive and the HP unit COP with capacity regulation is considerably better than one with on—off regulation of a single compressor. Fig. 13.7 illustrates part load performance for different capacity regulation methods.

In Fig. 13.7, it has been assumed that the efficiency of the two half-size compressors working in tandem is 98% of the single large compressor at full load, and that the

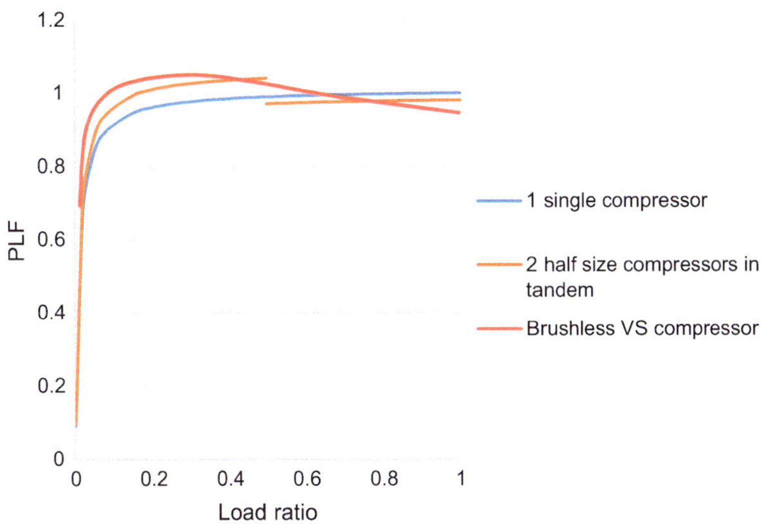

Figure 13.7 Part load factor (PLF) versus the load ratio ($\propto$) for different compressor configurations. (Relative parasitic losses: $\frac{\dot{W}_{sb}}{\dot{W}_{HP}} = 1\%$.)

brushless variable speed compressor has a lower efficiency at maximum speed due to the inverter losses and possibly higher mechanical losses.

The increase of the HP COP at low loads, ΔCOP, depends on the reduction of the temperature difference between the evaporation and condensation temperatures in reference to the sink and source temperatures respectively (see Section 13.2.2). These temperature differences mainly depend on the size and effectiveness of evaporator and condenser. If their effectiveness is high then the COP at full load will be high and the potential for improvement at partial load ΔCOP will be low because the temperature differences with sink and source will be already relatively small. Therefore, the lower the effectiveness of the evaporator and condenser the larger the potential for COP recovery (ΔCOP) at partial loads with variable speed compressor technology. This means that HPs employing variable speed compressors reduce cost by slightly undersizing the heat exchangers and still reaching high COPs most of the operating hours.

The actual influence of the capacity regulation method is difficult to estimate because in practice the efficiencies are not the same. A single fixed speed compressor normally has a slightly better efficiency than a half-size compressor used in tandem since usually efficiency increases with size. The efficiency of the variable speed compressor is also usually lower than the fixed speed one at maximum speed. First, this is mainly due to the inverter losses intrinsically associated to the variation of the speed. The loss in efficiency due to the inverter is estimated to be 2–5% (Cuevas and Lebrun, 2009). Second, at higher speeds the mechanical losses normally increase. The same is true for lower speeds: the mechanical losses decrease so efficiency should increase although in scroll compressors, in practice, this is not the case because the relative losses, due to the leakage of refrigerant through the seals of the vanes, increase with the decrease of the refrigerant flow rate. These losses can be large at low speeds and could reduce the HP COP at low loads. Therefore, in practice the benefit of variable speed operation suggested in Fig. 13.7 at low load factors can be very little and at low speeds can be even worse than fixed speed operation.

The use of fixed speed scroll compressors with external (nonintegrated) inverters for speed modulation is limited by lubrication problems. The lubrication at low speed can become very poor and the sealing by the lubricant is lost, leading to poor performance and excessive wear. Additionally, at high speeds friction losses become very high and also lubrication may fail due to the high temperatures produced. Normally, fixed speed compressors rated at 50 Hz could start to have problems below 30 Hz and above 75 Hz. Given the penalty on efficiency due to the inverter, the decrease of efficiency at high speeds due to higher mechanical losses, and the reduced margin for variation, the use of this combination for load modulation in ground HPs has seen little application.

New variable speed scroll compressors, eg, EMERSON (2015a,b) bring a series of improvements which make them much more suitable for capacity control. The lubrication problem has been solved with a new lubrication system, which allows them to cover a large range of speeds with reliable operation. Some have incorporated brushless electronically controlled motors that have higher efficiency than the conventional motor technology and have an efficient speed regulation. This new technology could significantly contribute to obtain better seasonal performance than that indicated in Fig. 13.7.

With a relatively small displacement, hence small size, the new variable speed compressors can provide very high efficiency in the most frequent operation conditions at medium and low loads, and still be able, through high speeds, to cope with high loads when required. The efficiency at high speeds may be reduced, but such conditions have a low influence on overall seasonal performance. Therefore, this new technology may allow a decrease in size of both compressor and heat exchangers with still high seasonal COP.

Additionally, the use of variable speed compressors can lead to a further improvement of the COP of the system at partial loads if it is combined with variable speed circulation pumping of the heat transfer fluid through the ground loop. The COP of the combination of the HP with the ground loop circulation pump can be written in the following way for both fixed speed and variable speed compressor options (Montagud et al., 2014):

$$\mathrm{PLF} = \frac{1}{1 + \frac{\dot{W}_{\mathrm{P}}}{\dot{W}_{\mathrm{HP}}} + \frac{\dot{W}_{\mathrm{sb}}(1 - \propto)}{\dot{W}_{\mathrm{HP}} \propto}} \qquad [13.5]$$

$$\mathrm{PLF}_{\mathrm{VSC}} = \frac{1}{1 + \frac{\dot{W}_{\mathrm{P}}}{\propto \dot{W}_{\mathrm{HP@MS}}} + \frac{\dot{W}_{\mathrm{sb}}(1 - \propto)}{\propto \dot{W}_{\mathrm{HP@MS}}}} \qquad [13.6]$$

where $\dot{W}_{\mathrm{P}}$ is the electric consumption of the circulation pump and $\dot{W}_{\mathrm{HP@MS}}$ is the electric consumption of the HP at the maximum velocity of the compressor at which it delivers the maximum capacity ($\propto = 1$).

If one considers now that the ground loop circulation pump is also modulated proportionally to the load keeping the temperature variation of the fluid through the evaporator constant, and that the circulation pump consumption is proportional to the cube of the fluid flow rate, then it is proportional to $\propto^3$ and the following expression is obtained:

$$\mathrm{PLF}_{\mathrm{VSC}} = \frac{1}{1 + \frac{\dot{W}_{\mathrm{P@MS}}}{\dot{W}_{\mathrm{HP@MS}}} \propto^2 + \frac{\dot{W}_{\mathrm{sb}}}{\dot{W}_{\mathrm{HP@MS}}} \frac{(1 - \propto)}{\propto}} \qquad [13.7]$$

where $\dot{W}_{\mathrm{P@MS}}$ is the consumption of the circulation pump at maximum speed that we assume is synchronized with the maximum speed of the compressor.

The standby losses on the PLF were shown in Fig. 13.4. Fig. 13.8 furthermore shows the effect of additionally considering the consumption of the circulation pump in PLF, for both cases: fixed speed and variable speed compressor. It has been assumed that the fraction of the circulation pump consumption referred to the HP electric consumption (at maximum compressor speed in the case of a variable speed compressor) is the same for both cases and equal to 5%, which is a value that

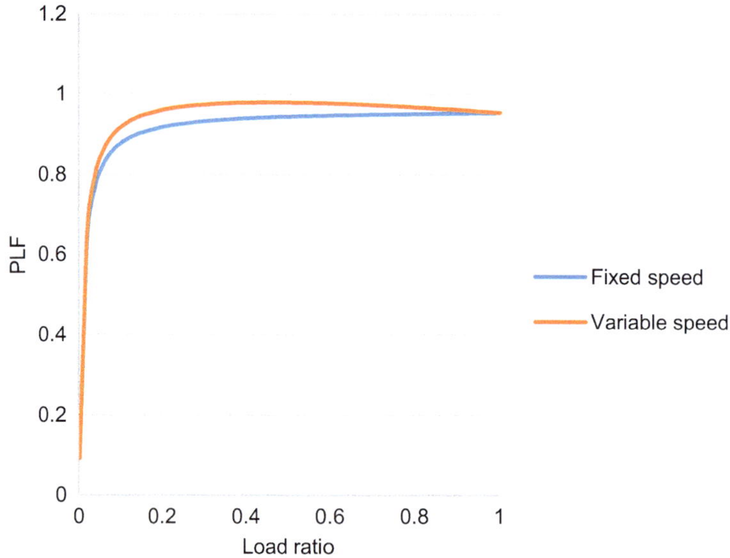

Figure 13.8 Part load factor (PLF) taking also into account the consumption of brine circulation pump for the single speed compressor and for a variable speed compressor.

is feasible in well-designed systems. The relative parasitic losses $\frac{\dot{W}_{sb}}{\dot{W}_{HP}}$ have been assumed again as 1%. The comparison in Fig. 13.8 does not take into account the benefit on the PLF of the reduction of the temperature lift due to variable speed compressor operation shown in Fig. 13.7.

This data also shows the possibility of adapting the mass flow rate of the brine to the capacity of the HP in case the variable speed compressor leads to a significant improvement of the PLF and therefore of the seasonal COP. The greater the relative consumption of the circulation pump (assumed 5% in the figure) the higher the advantage.

13.4 New refrigerants

13.4.1 Current use of refrigerants for heat pumps

Chlorofluorocarbon (CFC) refrigerants started to be phased out after the Montreal Protocol in 1987. Hydrochlorofluorocarbon (HCFC) refrigerants have been in use as substitutes to CFCs in the following years but they were also included in the phaseout of chlorine refrigerants as ozone depletion substances to be completely ruled out by 2020 in industrialized countries and by 2030 in developing ones. Hydrofluorocarbons (HFCs) then came into play as substitutes of HCFCs and their use has been spreading very fast. However, HFCs have a very high global warming potential (GWP) and in 1997 they were included in the Kyoto Protocol as greenhouse gases (GHGs). Further restrictions are currently being considered (IIR, 2015).

Early in 2006, the European Union (EU) published the F-gas regulation (842/2006) which started to place some restrictions to the use and control of GHGs in a number of applications. Recently, the EU has updated this regulation and published a new version: F-gas regulation (517/2014) (F-gas regulation). This new regulation (EU, 2014) imposes a large reduction in the use of HFCs in the EU, and will lead to a progressive drop on the availability of HFC refrigerants and probably to a considerable rise of their cost. Some European countries have implemented even more restrictive measures in order to penalize emissions of GHGs: adding special taxes to the use of HFCs, or advancing the phaseout of HFCs. This "phasing down," which mainly affects the refrigeration sector, will certainly influence the air-conditioning and HP sector, and will push the industry to move into low GWP solutions.

Most of the HPs in the European market currently employ R410A, and this seems too the trend observed in the United States and Asia. After the phase down of CFC and HCFCs in Europe, the industry employed for a while R407C, but in the end most of the manufacturers changed to R410A. In the study done by Clodic et al. (2013) about the future refrigerants for the air-conditioning and refrigeration sectors in the French market, they report the distribution of refrigerants in the HPs sector in 2011, shown in Table 13.2. In the same study, they argue that the HP sector will probably follow the same trends, which now are observed in the air-conditioning sector based on low GWP refrigerants, ie, R32 and HFC and hydrofluoroolefin (HFO) blends based on R32. For industrial HPs, R1234ze and R1233zd, which have relatively high critical temperatures, offer interesting possibilities for high temperature applications.

For substitution of R407C in HPs, some mixtures of R32 and HFOs have been proposed. Bentivegni and Clodic (2011) suggest mixtures of R1234yf together with R32 and R152a. In Domanski et al. (2014) the authors performed a comparison of the cycle COP and volumetric cooling capacity of a wide set of refrigerants, including HFCs, HFOs, hydrocarbons (HCs), and four different vapor-compression cycles, and showed R32 to be the most promising alternative—performing better than R410A.

R32 has a GWP around 700, which is approx. 1/3 that of R410A. This represents a significant reduction of direct CO_2 emissions but still is a high GWP. Mixtures of R32 with HFOs such as R1234yf and R1234ze can lead to GWPs around 500 with characteristics similar to R410A but not less.

Flammability is a further factor that has to be considered when seeking alternative refrigerants. R32 is classed as an A2 flammable fluid. HFOs have been inserted in the new ASHRAE category of mildly flammable refrigerants A2L. These refrigerants are flammable but have a low combustion energy and low flame speed. Fig. 13.9, extracted from Pham and Rajendran (2012), shows the position of different refrigerants in relation to heat of combustion and burning velocity.

Table 13.2 Distribution of heat pump refrigerant in France in 2011

R410A	R407C	R134a
90%	5%	5%

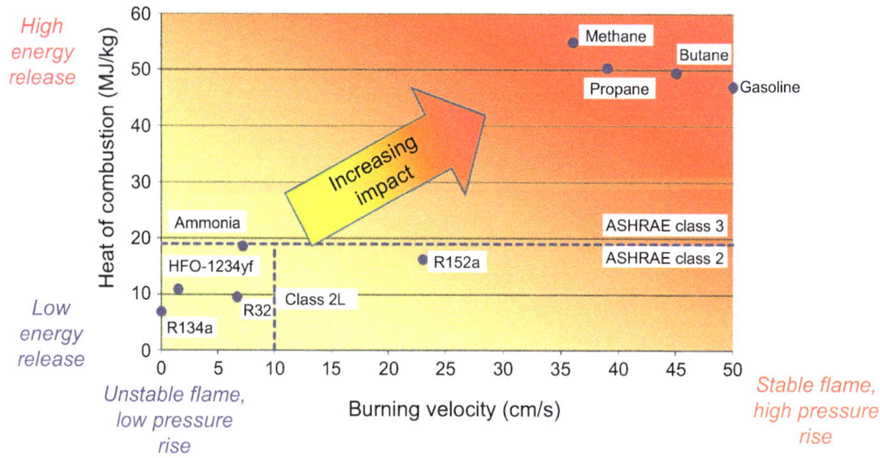

Figure 13.9 Flammability characteristics of refrigerants.
Extracted from Pham, H.M., Rajendran, R., 2012. R32 and HFOs as low-GWP refrigerants for air conditioning. In: International Refrigeration and Air Conditioning Conference at Purdue, Paper 1235.

Longo et al. (2015) measured the heat transfer characteristics of R32 in a brazed plate heat exchanger (BPHE) and compared it with R410A. They concluded that R32 heat transfer coefficient is much higher for the same mass flux while the pressure drop is only slightly higher. This could be attributed to the fact that the thermal conductivity of R32 is around 30% higher than that of R32. Finally, the authors conclude that R32 seems to be a very promising low GWP substitute for R410A in medium size chillers and HPs.

R32 is already being employed by several manufacturers in Japan and China for air conditioners. In Pham and Rajendran (2012) the authors conclude that R32 offers an attractive Low-GWP/Low-LCCP (life cycle climate performance) solution for mainstream air-conditioning and HP applications with performance comparable to R410A and can serve as the initial candidate for new equipment to meet any potential HFC phasedown proposal for at least until 2020+. R32 therefore appears to be a good, moderately low GWP transition refrigerant for GSHPs. In the long-term—if the push for the lowest GWP continues—HFOs and natural refrigerants are the only alternatives.

13.4.2 Natural fluids for heat pump applications

After two decades of considerable research on natural refrigerants, only small capacity refrigeration and HP equipment are commercially available using them. There is almost no commercially available HP equipment of commercial size employing natural refrigerants. This has prevented in practice the implementation of natural fluids for heating and cooling in the building sector.

Although ammonia is an excellent fluid for refrigeration, and could be employed for heat pumping too (although with limitations due its high discharge temperature) its utilization is complex and in practice it is only cost-effective for large capacity units in the industrial refrigeration and district heating sector, where it has been in use with excellent performance since the early years of refrigeration. Some effort has been done to employ ammonia as a fluid for small HP equipment (Palm, 2008). However, the lack of compressors of small size and the practical difficulties of the refrigerant makes it difficult to adopt as a solution for domestic or commercial HPs.

A lot of effort has recently been made—especially by the European automotive industry—to develop an air-conditioner utilizing CO_2, since this solution is known to be optimal from the point of view of reducing direct emissions of refrigerant to the atmosphere, the total of which are very large in the automotive sector. One of the other sectors in which the CO_2 technology has found growing implementation is refrigeration, mainly on low temperature industrial refrigeration, alone or combined with ammonia, and in supermarkets where the distribution of cold using CO_2 as the secondary fluid and also the generation of cold in a transcritical cycle is becoming a very well accepted technology. The other important sector in which CO_2 technology has become accepted is in HPs for sanitary water production, where the transcritical CO_2 cycle has proved to be a very good alternative for production of hot water at high temperature; it is being successfully commercialized in Japan as an efficient electric domestic water heater. This is precisely due to the excellent capability of the transcritical cycle of CO_2 to adapt with good efficiency to the high temperature variation occurring from the low temperature of the city water up to the high temperature required for domestic hot water (Nekså et al., 2010). However, the transcritical CO_2 cycle is not able to offer good efficiencies for heat pumping at moderate water temperatures with a small difference between return and supply temperatures, so it is not an alternative for GSHPs for heating of buildings.

HCs, on the other hand, are excellent refrigerants and could be an efficient alternative for GSHPs. Quite a lot of research has been devoted to fully understand the features of HCs as refrigerants and prototypes; many different applications have been developed and successfully demonstrated. However, their flammability has always been argued as a disqualifying criterion for its use. Currently, HCs are being extensively used in domestic refrigerators, freezers, display cabinets, bottle coolers, and so forth. The 2010 Report of the United Nations Environment Program (UNEP, 2010) estimates that 100 million household refrigerators and freezers are manufactured annually worldwide. One-third of these now use either isobutane or an isobutane/propane blend, and this proportion is expected to increase to 75 percent by 2020. The US Environmental Protection Agency (EPA) already approved a list of HCs as substitutes for CFCs and HCFCs for refrigerators and freezers (EPA, 2009b).

The utilization of HC refrigerants in the commercial air-conditioning and HP sector has been very limited. One very important reason for this has been the unavailability of commercial size compressors certified for HCs until very recently. Emerson Technologies presented at the ATMOsphere Europe Conference in 2015 a new series of HC compressors for commercial refrigeration (2−9 kW at MT) both variable and fixed speed compressors suitable for HP applications (Kandu, 2015). Additionally, Danfoss Commercial Compressors is participating in the NxtHPG project (NxtHPG 2013) providing the compressors for propane for the developed HPs. This project includes the development of a water-to-water GSHP working with propane and a capacity of 50 kW.

HC refrigerants, and in particular propane, offer very good performance for HP applications: good efficiency and low discharge temperatures. Factory sealed HPs which are installed outdoors could easily satisfy the current standards (Corberan, 2008), retrofit with HC refrigerant with very low extra cost, if any, and become a very suitable and safe solution. The new European F-gas regulation seems to offer opportunities for solutions of this kind in Europe, at least for air-to-water HPs. GSHPs with distribution by water could also be easily designed for propane provided that the unit is installed in a machinery room or outdoors. Indoor installations are possible but the restrictions imposed in the standards are very important and in practice make this not cost-effective. GSHPs, which are water-to-air HPs, cannot employ HCs since it is forbidden in the standards for direct air-conditioning and heating.

13.4.3 Refrigerant charge minimization

The minimization of charge in equipment designs allows for the reduction of direct emissions of refrigerant gases along the whole life of the unit, as well as the reduction of emissions generated along the whole chain of the refrigerant production, transport, and service. In addition, the reduction of the refrigerant charge implies a lowered unit cost, and fewer associated risks in the case of refrigerants with harmful properties such as flammable refrigerants, and, in fact, the standards impose severe restrictions to the maximum charge of those refrigerants. Nevertheless, minimization of charge cannot be achieved at the expense of unit efficiency because this would imply higher electricity consumption and therefore higher CO_2 indirect emissions. An overall optimization of the equipment design and charge is therefore necessary in order to achieve minimal CO_2 equivalent total emissions. Accordingly, minimization of charge becomes one of the most important requirements for all future refrigeration equipment, and it will be a key factor in the potential deployment of a natural refrigerant technology (IIR, 2014).

13.5 GSHP heat exchangers

13.5.1 Liquid to refrigerant heat exchangers

Water-to-water GSHPs that distribute heating and cooling to the building via water include two liquid-to-refrigerant heat exchangers. In heating mode, the heat exchanger coupled to the ground loop acts as the evaporator and the heat exchanger coupled to the building heating and cooling system acts as the condenser. This kind of equipment is typical of that found in Europe where the heating and cooling distribution is done by water to the room terminal units. In Europe, both heat exchangers are usually BPHEs. In the United States, the liquid-to-refrigerant heat exchanger could be a BPHE but usually the heat exchanger coupled to the ground loop is a double-pipe coaxial type. This solution is preferred in the United States basically because double tube devices are much more resistant to fracture in the case of freezing conditions than BPHEs. BPHEs are, in contrast, much more compact than the double tube heat exchangers.

BPHEs are a compact and effective solution for refrigeration and heat pumping applications and show high effectiveness with a small size. Sizing of a BPHE for evaporator or condenser duty is based on first choosing the right size of plate. Each manufacturer offers a range of different sizes depending on the application and the required heating and cooling capacity. The number of plates is selected in order to fit the required capacity and other specifications (as for instance, superheat or subcooling required inside the BPHE). There has not been very much development of this technology in recent years. The manufacturers usually offer three different corrugation angles, which provide a different combination of heat transfer and pressure drop. Often a higher pressure drop configuration is accepted in the interests of good heat transfer performance.

Maldistribution of refrigerant in BPHE evaporators is an important drawback for a good performance in these heat exchangers, since some of the refrigerant channels can become flooded with refrigerant while others starve of liquid and produce a considerable superheating. This implies an inefficient use of the available plate area and produces a deterioration of performance, which normally it is distinguished by an evaporation temperature lower than expected.

Fig. 13.10 shows the infrared thermography of an evaporator with around 60 plates. As can be observed, at moderate superheats (case *a* and *b*) the channels at the left side, close to the inlet, are flooded with refrigerant, while in the channels at the right the

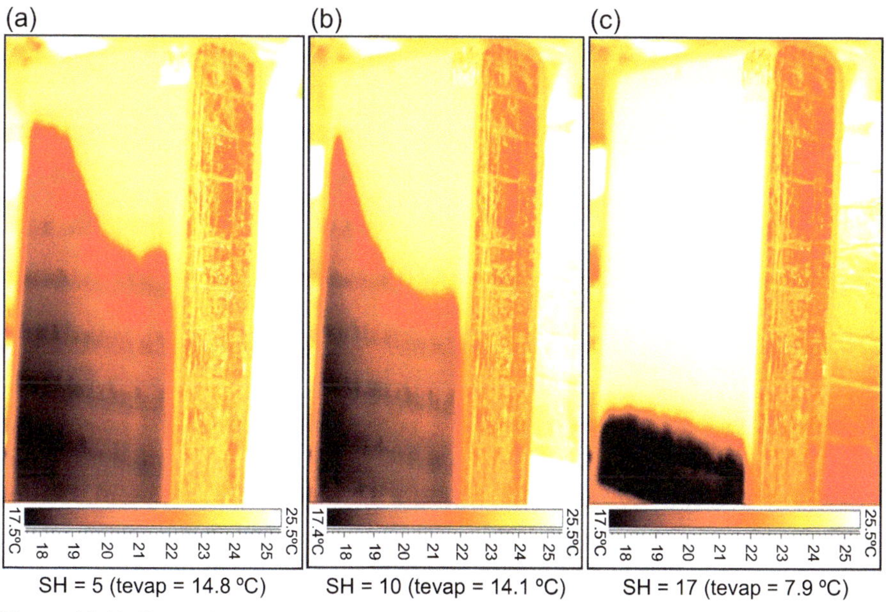

Figure 13.10 External temperature distribution at a BPHE evaporator at different superheats (SH) for the same water inlet conditions.

refrigerant dry out earlier producing a high superheat. This represents a nonefficient use of the available area. If the superheat is increased to high values (case *c*) the distribution improves but, as can be seen, most of the area of the evaporator is employed for the production of the superheat and not for the evaporation and this leads to a significant drop in the evaporation temperature.

Refrigerant maldistribution prevents the adequate minimization of the superheat since—except for very high superheats able to have a dry vapor outlet from all the channels—they normally are prone to have liquid at the outlet of the evaporator. This makes it difficult to sense the actual superheat of the refrigerant flow and can also induce control instabilities because of the fluctuations in the sensed temperature due to slugs of liquid refrigerant.

Manufacturers normally incorporate a distribution system (usually in the form of calibrated orifices at the entrance of each channel) that improves the distribution of refrigerant in evaporators. This is the most important difference between the design of a BPHE evaporator and a BPHE condenser. The problem of maldistribution of refrigerant becomes worse with the increase of the number of plates. BPHEs with a small number of plates do not need to employ a distributor.

The design of distributors is focused on typical operation conditions. Refrigeration and HP equipment employing variable speed compressors suffer from a great variation of the conditions at the evaporator, and both mass flow rate and inlet quality vary considerably. This is a challenge for the design of the distributors, and manufacturers are developing new solutions trying to improve the refrigerant distribution along the wide range of possible operating conditions: for example, the AlfaDist system from Alfa Laval (Christensen, 2015).

The basic rule to minimize the refrigerant mass in a unit, and therefore of the required charge, is to minimize the internal volume. BPHEs offer important advantages from this point of view and contribute to a minimum charge technology. This is very important in small compact units where most of the charge is in the heat exchangers (HEs). Minimizing the charge in this way is especially important in equipment employing flammable refrigerants since the standards limit the maximum refrigerant charge of the unit.

In order to maximize this advantage, some manufacturers have started to develop BPHEs with a lower corrugation depth in the refrigerant side compared with the one at the liquid side. SWEP has developed a new plate design called SWEP AsyMatrix, which can lead to a significant reduction of the refrigerant internal volume and enhanced heat transfer but with a higher pressure drop (Gullapalli, 2012).

Some manufacturers are also exploring new plate designs offering a better compromise between heat transfer and pressure drop while reducing the refrigerant charge. For instance, Danfoss has recently presented new plate designs based on dimpled surfaces that offer interesting advantages (Wei et al., 2015).

13.5.2 Refrigerant-to-air heat exchangers

GSHP units in the United States are typically are water-to-air units whereby heated or cooled air is distributed via duct systems. These units consequently employ a refrigerant-to-air heat exchanger as a condenser/evaporator coupled to the duct system.

The typical design of these condenser/evaporators is a find and tube coil. Refrigerant-to-air heat exchangers are typically quite large in comparison with BPHEs so this type of GSHP employs a refrigerant charge much larger than water-to-water GSHP of similar capacity. GSHPs of this kind that are reversible on the refrigerant side (in order to provide heating in winter and cooling in summer) require a considerable extra amount of charge since they need to incorporate a liquid receiver in order to allocate the excess of refrigerant when the coil passes from condenser to evaporator duty.

The trend toward minimization of charge is expected to force water-to-air GSHPs to evolve and employ lower and lower tube diameters. Some years ago, the typical diameter of the coil tubes was 12 mm but there has been a gradual move to 9 mm and then to 7-mm diameters. Some manufacturers are already offering solutions with 5-mm tube diameters providing very good performance and reduction of refrigeration charge (Filippini and Merlo, 2011). More compact solutions and higher reductions in refrigerant charge are expected from the microchannel technology developed in the automotive sector this is now starting to spread throughout the HVAC sector.

Units which are reversible in the refrigerant side need the microchannel heat exchanger to work as an evaporator. However, achieving good refrigerant distribution in this kind of heat exchanger is very difficult and the degradation of performance due maldistribution can be considerable. Although a lot of researchers and manufacturers have tried to find solutions for this problem, no cost-effective general solution seems to have been found yet. This restricts the use of microchannel technology to heating-only applications for the moment.

13.5.3 Ground-source heat pump with domestic hot water production

The ability to generate domestic hot water is an important requirement in many residential markets. Accordingly, many manufacturers of GSHP equipment have come to offer either integrated (multifunctional) domestic hot water (DHW) solutions or means of generating hot water for delivery to a separate storage tank. The typical demand of DHW is at temperatures in the range 50–65°C. Production of hot water at those temperatures at the condenser requires condensation temperatures slightly above those temperatures and so COP may become much poorer than in normal heating operation. Table 13.3 shows the typical COP of a GSHP with a source loop temperature at 5°C and several inlet water temperatures at the condenser and it can be observed that the COP deteriorates very rapidly with the increased water temperature.

Table 13.3 Coefficient of performance of a ground-source heat pump for production of domestic hot water at the prescribed inlet (T_{wi}) and outlet (T_{wo}) water temperatures

T_{wi}/T_{wo}	30/35°C	40/45°C	50/55°C
COP	4.8	4	3.3

In many residential applications, the capacity required for space heating purposes can be much lower than that required for instantaneous domestic hot production. In these cases DHW is produced in batch mode (timed cycle) and accumulated in a water storage tank. It is common for the water tank to additionally incorporate a backup electrical resistance heater that is able to top-up the temperature to the desired value when required. The resistance heater may also be employed to raise the temperature of the tank above the normal delivery temperature for short periods to ensure *Legionella* is controlled.

The heating of the DHW by the HP can be done mainly in two different ways: heat recovery in a desuperheater and heating in a dedicated condenser. The first option is the most common one and it is basically a heat recovery system that consists of installing a small BPHE at the outlet of the compressor upstream of the main condenser, ie, where the refrigerant gas is at its hottest point in the cycle. This heat exchanger normally acts only as a desuperheater capturing the heat transferred from the refrigerant in its superheated state to heat up the water. Condensation of the refrigerant then takes place at the main condenser of the HP in order to condition the condensation pressure. If the temperature of the water at the superheater is sufficiently low, partial condensation could start at the desuperheater.

The desuperheating of the refrigerant varies between 10% and 15% of the total heat rejected so that the contribution to DHW production is around that same fraction of the total heating capacity. As the equipment size is mostly determined by the space heating (or cooling) capacity, this proportion of the total heat may not be enough to deliver the whole of the DHW demand. This is why this technique is better considered as a heat recovery system rather than a DHW one and supplementary heating is often necessary.

A rather different approach is required if the aim is to produce the whole of the DHW demand using the HP alone. In this case, the whole of the heat rejection (condensation heat) is intended to be transferred to the DHW system. This is done by adding a second condenser in parallel for this purpose and adding the necessary valves to direct the refrigerant to one of the two available parallel condensers serving the heating water loop and the DHW loop. The valves are actuated in a way that the refrigerant is sent to either one or the other condenser, but never to both at the same time (HIREF, 2015). This kind of arrangement allows the production of DHW much faster than the desuperheater but, as the condenser operates at higher temperature than it would with a desuperheater, this is at the cost of a lower COP.

13.6 Dual source heat pumps

A number of approaches have been taken to incorporating supplementary heat injection or heat rejection devices into the GSHP system design in order to further optimize efficiency and economic performance. At larger scales this may amount to adding solar collectors or cooling towers to the source side of a water-to-water HP, for example. These approaches are discussed in Chapter "Integration with nonresidential buildings." At smaller scales an alternative approach to hybrid operation is to incorporate both air and liquid heat exchangers on the source side within the same HP equipment. This is referred to here as a dual source approach.

The possibility of having the dual source of heat exchange available right at the HP has several advantages: first, the cost will be certainly lower than the one required for an external (hydronically coupled) air source heat exchanger alternative (eg, dry cooler or cooling tower); second, the unit will be much more compact since it is integrated in the HP equipment package. An integrated dual source approach also takes advantage of superior heat transfer on the refrigerant side and the elimination of the temperature difference produced by any intermediate BPHE in a secondary hydronic circuit. Consequently, COPs of dual source HPs can be expected to be higher than hybrid designs that are hydronically coupled. These advantages come at the marginal cost of adding an extra heat exchanger and the necessary valves and control to switch from one source to another into the HP design.

Furthermore, as with other hybrid options, the dual source HP will significantly help to reduce the size and corresponding cost of the GHE and at the same time will offer an interesting potential for energy reduction if an adequate operation strategy is developed for making use of the most effective heat source according to operating and climatic conditions.

Nam et al. (2010) developed a dual source HP utilizing groundwater and air sources. They found an improvement of the annual performance factor ranging from 2% to 7% compared with the system only employing the groundwater source, and from 4% to 18% when compared with the system only employing the air source. They also pointed out that higher system performance could be achievable with the use of variable speed for the compressor and the circulation pump.

Recently, one company (MHG, 2015) has brought to the market a series of dual ASHP and GSHP units, ranging from 15 to 40 kW of heating capacity, and with COPs of around 4.5. The air-to-refrigerant heat exchanger has been designed as a split component for flexible installation options. They state that the dual operation can allow the ground loop heat exchanger to be considerably reduced.

In an EU project funded under the Horizon 2020 program (GEOTeCH, 2015) the partners have sought to develop a dual source HP in the frame of the development of an innovative plug-and-play geothermal solution for heating and cooling of houses, and to develop a control system able to make optimal use of both and ground heat sources throughout the year.

13.7 Conclusion

Packaged GSHP equipment is very similar to other HP equipment, with a few distinctive characteristics. In comparison to air-to-water HPs the main difference is the design of the source-side heat exchanger (evaporator in heating operation), which is a liquid-to-refrigerant design instead of an air-to-refrigerant heat exchanger. This also makes the unit much more compact and suitable for indoor installation. In Europe the solution usually adopted for a source-side heat exchanger is a BPHE, while in North America it is more common make use of a double coaxial pipe device.

The maldistribution of refrigerant among the channels of the BPHE evaporator has been found to be a challenging problem, especially in units with variable speed compressors because of the large variation range of the refrigerant mass flow. BPHE manufacturers are developing new distributor designs to cope with this problem.

Another important difference between GSHPs and other HPs and air-conditioning units is that the typical values of the operating pressure ratio of GSHPs are comparatively smaller—especially those employed for floor heating which have a low condensing temperature. Usually, scroll compressors are optimized for air-conditioning or air HP applications that have higher operating pressure ratios. The difference in pressure ratio is considerable so scroll compressors in GSHPs tend to work below the maximum efficiency and this leads to a nonnegligible decrease on the COP compared to what could be achieved. A dedicated scroll compressor design for GSHP application would lead to a significant increase in their seasonal performance.

Nevertheless, packaged GSHP equipment has been shown to have relatively high COPs. The minimum reported values of reasonably good designs range from 2.5 to 6.0, depending on the temperature lift they have to cope with between the ground loop temperature at the inlet of the evaporator and the temperature of the heating water supply (usually in the range 20–60K). The rate of COP decrease over this range has been found to be approximately 0.11 per 1K of increase of the temperature lift.

Cycling losses of on–off controlled GSHPs have been found to be small and are mainly due to the standby losses. They normally represent less than 1% of the average consumption of the compressor and are negligible when the compressor is switched on, but even with that low value they degrade the performance very fast for load ratios below 20%. The standby losses are mainly caused by the consumption of the electronics, the valves (which remain energized during the off period), and the crankcase heater. These last two sources of demand should be critically reviewed and solutions with null or low energy consumption adopted; otherwise, standby losses can be very high and particularly deteriorate the performance at low partial loads.

GSHPs normally employ scroll compressors because of their superior efficiency. Duos or trios of compressors working in sequence, or the new brushless electronically controlled compressors can lead to significant improvements in COP at partial loads. These technologies could lead to a considerable increase in the seasonal performance of GSHPs. Variable speed technology also helps to reduce the impact of the ground loop circulation pump consumption on the total system consumption if the pump speed is synchronized with the compressor speed.

Concerning the refrigerants, the situation at this moment is quite fuzzy. Recent EU regulation imposes a large reduction in the use of HFCs in the EU, and will lead to a progressive drop on the availability of HFC refrigerants and probably to a considerable rise in their cost. Some European countries have implemented even more restrictive measures in order to penalize emissions of GHGs: adding special taxes to the use of HFCs, or advancing the phaseout of HFCs. This phasing down, which mainly affects the refrigeration sector, will certainly influence the air-conditioning and HP sector, and will push the industry to move into low GWP solutions.

Most of the HPs in the European market employ R410A, and this seems to be the trend observed in the United States and Asia. Refrigerant R32 has a GWP that is approximately one-third that of R410A. This represents a significant reduction of direct CO_2 emissions but still is a relatively high GWP. Mixtures of R32 with

HFOs such as R1234yf and R1234ze can lead to GWPs around 500 with characteristics similar to R410A but not better. R32 therefore appears to be a good moderately low GWP transition refrigerant for GSHPs. In the long term, if the push for the lowest GWP continues HFOs and natural refrigerants, among them especially propane for open air installations, are the only alternatives.

Given the situation with the refrigerants, minimization of refrigerant charge becomes one of the most important requirements for all future refrigeration equipment, and it will be a key factor in the potential deployment of a natural refrigerant technology. The use of microchannel technology in GSHPs with an indoor refrigerant-to-air condenser could be a good solution to decrease the refrigerant charge and at the same time lead to a more compact and lower weight technology. However, the problem of maldistribution of the refrigerant when the microchannel heat exchanger works as an evaporator has not yet been solved, so this solution is not yet suitable for reversible units providing heating and cooling.

Finally, dual air HPs and GSHPs seem to be a promising solution for residential and small nonresidential applications. Given intelligent control strategies, they could provide the seasonal performance advantages of hybrid GSHP systems but with easier installation and potentially lower cost.

References

Anglesio, P., Caon, S., Caruso, S., 2001, febbraio. Determinazione delle prestazioni energetiche di condizionatori elettrici a due unità in aria invertible: determinazione delle prestazioni energetiche. CDA.

Bentivegni, M., Clodic, D., 2011. R-1234yf mixtures for replacing R-407C in residential heat pumps. In: 10th International Energy Agency Heat Pump Conference - HPC 2011, Tokyo, Japan, 10 pp.

Bettanini, E., Gastaldello, A., Schibuola, L., 2003. Simplified models to simulate part load performances or air-conditioning equipments. In: 18th IBPSA Conference, Eindhoven, Netherlands.

BINE, 2013. BINE Information (A Service from FIZ Karlsruhe). 'Electrically Driven Heat Pumps'. Available from: http://www.bine.info/fileadmin/content/Publikationen/Themen-Infos/I_2013/themen_0113_engl_Internetx.pdf.

Cecchinato, L., Chiarello, M., Corradi, M., 2010. A simplified method to evaluate the seasonal energy performance of water chillers. International Journal of Thermal Sciences 49, 1–11.

Cervera-Vázquez, J., Montagud, C., Corberan, J.M., 2015. In situ optimization methodology for the water circulation pumps frequency of ground source heat pump systems: analysis for multistage heat pump units. Energy and Buildings 88, 238–247.

Clodic, D., et al., 2013. Alternatives to High GWP in Refrigeration and Air-conditioning Applications: Final Report. Study Funded by ADEME, AFCE and Uniclima.

Corberan, J.M., 2008. Review of standards for the use of hydrocarbon refrigerants in A/C, heat pump and refrigeration equipment. International Journal of Refrigeration 31, 748–756 (Special issue on Refrigeration with Ammonia and Hydrocarbons).

Corberan, J.M., et al., 2011. 'A quasi-steady state mathematical model of an integrated ground source heat pump for building space control'. Energy and Buildings 43, 82–92.

Corberan, J.M., et al., 2013. Partialization losses of ON/OFF operation of water-to-water refrigeration/heat pump units. International Journal of Refrigeration 36 (8), 2251–2261.

Christensen, R., 2015. Brazed plate heat exchanger solutions for HC and CO_2. In: European Heat Pump Association 1st NxtHPG Technical Workshop, Brussels, Belgium 29th May.

Cuevas, C., Lebrun, J., 2009. Testing and modelling of a variable speed scroll compressor. Applied Thermal Engineering 29, 469–478.

Domanski, P., et al., 2014. A thermodynamic analysis of refrigerants: performance limits of the vapor compression cycle. International Journal of Refrigeration 38, 71–79.

EMERSON, 2015a. Emerson Technologies, "Copeland Scroll™ Digital Compressor". Available from: http://www.emersonclimate.com/EN-US/PRODUCTS/COMPRESSORS/SCROLL_COMPRESSORS/COPELAND_SCROLL_COMMERCIAL/Pages/copeland_scroll_digital.aspx.

EMERSON, 2015b. Emerson Technologies. "Variable Speed Scroll Compressor". Available from: http://www.emersonclimate.com/en-us/Products/Compressors/Scroll_Compressors/copeland_scroll_residential/Pages/copeland_scroll_variable_speed_compressors.aspx.

EN, 2010. EN-14825: Air Conditioners, Liquid Chilling Packages and Heat Pumps, with Electrically Driven Compressors, for Space Heating and Cooling — Testing and Rating at Part Load Conditions and Calculation of Seasonal Performance.

EPA, 2009a. ENERGY STAR Program Requirements for Geothermal Heat Pumps: Version 3.0. Available from: http://www.energystar.gov/sites/default/files/specs//private/ES_V3.0_Geothermal_Heat_Pump_Specification.pdf.

EPA, 2009b. EPA–HQ–OAR–2009–0286, FRL–9507–7: Protection of Stratospheric Ozone: Listing of Substitutes for Ozone-depleting Substances - Hydrocarbon Refrigerants.

EU, 2014. Regulation (EU) N° 517/2014 of the European Parliament and of the Council of 16 April 2014 on Fluorinated Greenhouse Gases and Repealing Regulation (EC) N° 842/2006.

Filippini, S., Merlo, U., 2011. New finned heat exchanger development with low refrigerant charge. In: ICR 2011, August 21–26-Prague, Czech Republic. Paper No. 296.

GEOTeCH, 2015. Geothermal Technology for €conomic Cooling and Heating'. EU H2020. N. 656889. Project Website. http://www.geotech-project.eu/.

Gullapalli, V.S., 2012. On new plate designs for brazed plate heat exchangers. In: 3rd IIR Workshop on Refrigerant Charge Reduction in Refrigerating Systems, October 25–26, Valencia, Spain.

GROUNDMED, 2009. Ground-med Project. Advanced Ground Source Heat Pump Systems for Heating and Cooling in Mediterranean Climate. Project No. 217985. EU FP7 Programme. http://www.groundmed.eu/.

Henderson, H.I., Parker, D., Huang, Y.J., August 2000. Improving DOE-2's RESYS Routine: User Defined Functions to Provide More Accurate Part Load Energy Use and Humidity Predictions. 2000 ACEEE Summer Study on Energy Efficiency in Buildings (Pacific Grove, CA).

HIREF, 2015. Multifunction Geothermal Heat Pumps for High Temperatures. Available from: http://www.hiref.it/en/high-performance-heat-pumps/gsh.

IIR, 2014. Refrigerant Charge Reduction in Refrigerating Systems. International Institute of Refrigeration, 25th Informatory Note on Refrigeration Technologies.

IIR, January 2015. Overview of Regulations Restricting HFC Use. Focus on the EU F-Gas Regulation. International Institute of Refrigeration, 26th Informatory Note on Refrigeration Technologies.

Kandu, V., 2015. Hydrocarbons: A Viable Natural Refrigerant Solution for Medium Temperature Commercial Refrigeration Applications. Atmosphere Europe 2015, Brussels.

Longo, G.A., et al., February 2015. HFC32, a low GWP substitute for HFC410A in medium size chillers and heat pumps. International Journal of Refrigeration 53.

Madani, H., 2012. Capacity-Controlled Ground Source Heat Pump Systems for Swedish Single-family Dwellings (Doctoral thesis in Energy Technology). KTH, Stockholm, Sweden.

MHG, 2015. MHG Heating Ltd. "ThermSelect Dual Air and Ground Heat Pumps". Available from: www.mhgheating.co.uk.

Montagud, C., Corberan, J.M., Montero, A., 2014. In situ optimization methodology for the water circulation pumps frequency of ground source heat pump systems. Energy and Buildings 68 (Part A), 42–53.

Nam, Y., Ooka, R., Shiba, Y., 2010. Development of dual-source hybrid heat pump system using groundwater and air. Energy and Buildings 42, 909–916.

Nekså, P., Walnum, H.T., Hafner, A., 2010. CO_2-A refrigerant from the past with prospects of being one of the main refrigerants in the future. In: 9th IIR Gustav Lorentzen Conference 2010, Sydney, April 12–14.

Palm, B., 2008. Ammonia in low capacity refrigeration and heat pump systems. International Journal of Refrigeration 31 (4), 709–715.

Pham, H.M., Rajendran, R., 2012. R32 and HFOs as low-GWP refrigerants for air conditioning. In: International Refrigeration and Air Conditioning Conference at Purdue. Paper 1235.

Riviere, P., May 2004. Seasonal Performances of Chillers (Ph.D. thesis). Ecole des Mines de Paris.

Riviere, P., et al., 2005. A method to reduce European chiller hourly load curves to a few points. In: Climamed 2005–2nd Mediterranean Congress of Climatization, Madrid, Spain.

Schibuola, L., 2000. Heat pump seasonal performance evaluation: a proposal for a European standard. Applied Thermal Engineering 20, 387–398.

Staffell, I., et al., 2012. A review of domestic heat pumps. Energy and Environmental Science 5, 9291–9306.

UNEP, 2010. Report of the United Nations Environment Programme (UNEP)'s Refrigeration, Air Conditioning and Heat Pumps Technical Options Committee (RTOC).

Wei, W., Tucker, J., Xu, Y., 2015. Experimental Investigations on Performance of Evaporator and Condenser of R410A Inside a Novel Brazed Plate Heat Exchanger. ICR, China.

Heat pump modelling

14

C.P. Underwood
University of Northumbria, Newcastle upon Tyne, United Kingdom

Abstract

In this chapter, detailed consideration is given to some of the established approaches for constructing mathematical models of heat pump systems, either for use in standalone modelling applications (eg, for implementation in a spreadsheet) or for embedding into a parent simulation code. The treatment is limited to ground-source heat pumps (GSHPs) and, as such, the focus is on liquid-source and water-sink systems (ie, hydronic heating). Section 14.2 considers the steady-state modelling of conventional vapour-compression heat pumps by considering each of the cycle components in turn followed by a practical application example in which the component models are drawn together into a complete plant simulation. Special consideration is given to the steady-state modelling of single-effect absorption cycle heat pumps in Section 14.3. The focus is on the aqueous lithium bromide and water cycle because of its superior performance when using intermediate grade sources such as ground and ground water. In Section 14.4, consideration is given to the use of heat pump models using regression fitting to experimental, field and manufacturers' performance data. Finally, Section 14.5 gives detailed consideration to fully dynamic approaches to modelling heat pumps in which the primary interest is control system design and stability. In each section, a practical interpretation of the modelling methods described is provided by means of a practical illustrative example.

14.1 Introduction

The motivations for heat pump system modelling are many and varied. Some of the everyday routine investigations which rely on a heat pump system model with either simple or moderate complexity include:

- Simulation of seasonal energy use and carbon emissions.
- Simulation of energy costs and tariff selection.
- Design analyses including heat pump sizing.
- Ground heat exchanger (GHE) sizing and performance with the use of various fluids.
- GHE and heat pump performance over extended time horizons.

Some of the more sophisticated areas of design and analysis enquiry which will tend to require a more rigorous modelling treatment include:

- Control system design, including scheduling and controller tuning.
- Buffer store design and sizing and the reduction of cycling.

Advances in Ground-Source Heat Pump Systems. http://dx.doi.org/10.1016/B978-0-08-100311-4.00014-5
Copyright © 2016 Elsevier Ltd. All rights reserved.

- Advanced feature performance evaluation, including variable speed drives, smart control, alternative working fluids, cascading, hybridisation and integration with other renewable energy systems.
- Evaluation of advanced cycles including absorption, adsorption and transcritical cycles.
- Troubleshooting and fault detection and diagnosis.

The established approaches to modelling heat pump systems can be summarised in Table 14.1 according to the importance attached to modelling rigour and the expected degree of confidence required in the results obtainable.

The simplest heat pump model is one in which the coefficient of performance (COP) is assumed to be constant over a defined operating period, there is no restriction on the availability of source energy and the heating demand placed on the heat pump is always fully met. According to the first law of thermodynamics and the definition of the heat pump COP (Eqs [14.1] and [14.2]):

$$Q_c = W + Q_e \qquad [14.1]$$

$$\text{COP} = \frac{Q_c}{W} \qquad [14.2]$$

The inputs are Q_c (the heat load to be met) and COP (the constant coefficient of performance) from which the required energy input, W (heat or electricity depending on the heat pump cycle used), is obtained from Eq. [14.2] and the required source energy, Q_e, from Eq. [14.1]. This model can provide adequate results for simulating seasonal energy use and economics provided that the values of COP used are suitably adjusted to account for variations in plant boundary conditions (mainly source and sink temperatures). The COP data used can be sourced from the heat pump manufacturer or use can be made of laboratory or field measured values (in this sense the model may have a theoretical or empirical basis).

For more elaborate design, performance and system integration analyses however, the balanced energy model lacks both rigour (accuracy) and utility and more detailed

Table 14.1 Approaches to heat pump modelling

Method	Basis	Confidence	Typical application	Computational need
Balanced energy	Theoretical/ empirical	Low	Seasonal energy use and economics	Negligible
Steady state	Theoretical	Moderate	Design and sizing	Low
Fitted models	Empirical	Conditionally high	Seasonal energy use and economics	Low
Dynamic state	Theoretical	High	System integration and control	High

approaches become necessary. These approaches will be considered in detail in the following sections. Section 14.2 will consider the modelling of vapour compression heat pumps which form the overwhelming majority of systems in use today for ground source applications. A component-level treatment will be presented and this will be followed by a consideration of practical approaches to the synthesis of the components into a whole system model for the vapour compression heat pump. Section 14.3 will take a brief look at the special case of heat-driven vapour absorption cycle heat pumps. These systems are developing commercially mainly through the use of the ammonia–water cycle for air-source heat pumps. However, there is strong potential for the long-established (and better performing) water-lithium bromide cycle due to the higher source temperatures available in some ground source applications. Section 14.4 will consider fitted models using data either from manufacturers (so-called 'catalogue' models) or from laboratory or field measurements. This type of model offers simplicity and, potentially, high accuracy and confidence strictly within the bounds of the data used to inform the model. Models of this kind are likely to accumulate growing interest as increasing amounts of quality data begin to emerge from building energy metering and data logging facilities. Section 14.5 will take a look at the most rigorous approach to heat pump modelling: the use of distributed parameter dynamic-state modelling which offers the advantages of high rigour and model utility at the expense of greater complexity and computational demands.

14.2 Steady-state modelling of the vapour compression cycle

Numerous performance metrics for heat pumps have arisen, and care needs to be taken when interpreting data, in particular from manufacturers. As an illustration, further details of European metrics can be found in BS EN 14511 (2013) and BS EN 14825 (2013). A brief summary follows, but the classical thermodynamic metric – the COP (defined in Eq. [14.2]) will be used throughout this chapter.

- Seasonal coefficient of performance (SCOP): accounts for variations in performance across the entire operational cycle.
- Energy efficiency ratio (EER): the cooling-mode counterpart to COP.
- Seasonal energy efficiency ratio (SEER)[1]: the cooling-mode counterpart to SCOP.
- Seasonal performance factor (SPF): a seasonal performance metric which also accounts for ancillary power consumption linked to the heat pump system but not necessarily part of the heat pump. (The SPF is most commonly used in conjunction with measured field performance data.)

A schematic of the vapour compression heat pump in its simplest form is shown in Fig. 14.1 and the cycle is plotted on a temperature-enthalpy basis in Fig. 14.2. In the

[1] In this context, SEER is defined strictly as a non-dimensional ratio. In some data SEER is a ratio of seasonal thermal output and power demand that has the units BTU per Wh (defined in ARHI Standard 210/240: 2008).

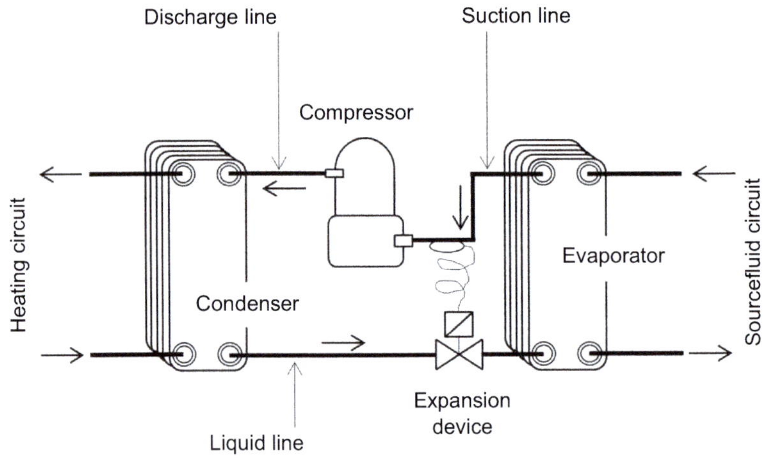

Figure 14.1 Schematic of the vapour compression heat pump.

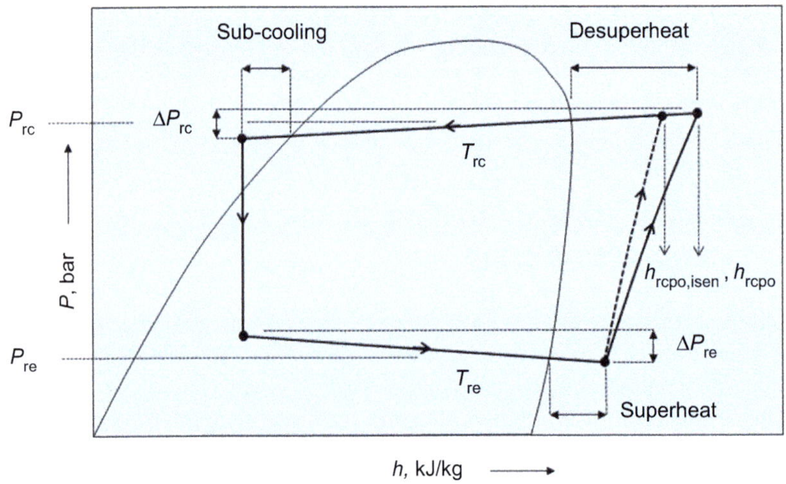

Figure 14.2 Vapour compression heat pump cycle.

following, a steady-state model of the vapour compression heat pump is developed based on a liquid source (water, brine or glycol solution) as might be expected from a GHE and a hot water heating system.

14.2.1 The condenser and evaporator

For a simple steady-state model to get started, the basic cycle of Fig. 14.2 can be simplified as follows:

- Refrigerant pressure losses (ΔP_{re}, ΔP_{rc}) may be neglected – these losses tend to be very small in close-coupled packaged plant.

- The main condensing and evaporating loads usually occupy at least 85% of the heat exchange area (ASHRAE, 2012) and so desuperheating, subcooling and superheating effects can also be neglected in a simple model with the assumption that outlet refrigerant states are saturated.

In the following, we assume that the heat pump will satisfy the load imposed on the condenser, Q_c, which will be known. Similarly, the required heating system supply temperature (which forms the cooling water outlet temperature from the condenser, T_{wco}) will be specified. For the evaporator, the inlet source fluid temperature, T_{fei}, will either be known or will be supplied from the GHE model. These three variables form the boundary conditions of our model.

A particularly convenient approach to modelling these heat exchangers in the steady-state is to use an effectiveness approach as first proposed by Kays and London (1984). The key assumption in this method is the heat exchanger effectiveness, E, which is defined as heat transfer actually achieved divided by the maximum theoretically possible heat transfer. Exact relationships can be derived for the effectiveness of a number of heat exchange geometries through the ratio of heat exchanger thermal transmittance value (AU) to thermal capacitance (mc) − the so-called number of transfer units (NTU) and the capacity ratio of the two fluids involved (Kays and London, 1984). For condensing heat transfer the capacity ratio is zero since the specific heat capacity of a condensing fluid at constant temperature is infinity (ie, $c = dh/dT = \infty$). Thus the number of transfer units and condensing section effectiveness are obtained from Eqs [14.3] and [14.4].

$$\mathrm{NTU_c} = \frac{\mathrm{AU_c}}{m_{wc}c_{wc}} \quad [14.3]$$

$$E_c = 1 - \exp(-\mathrm{NTU_c}) \quad [14.4]$$

The condenser effectiveness leads to the condensing temperature from which the condensing pressure is obtained from refrigerant properties (Eq. [14.5]). (It is convenient to express this in terms of the condenser outlet water temperature which forms the required heating system set point temperature.)

$$T_{rc} = T_{wco} + \frac{Q_c(1 - E_c)}{E_c m_{wc} c_{wc}} \quad \text{thus:} \quad P_{rc} = \mathrm{f}(T_{rc}) \quad [14.5]$$

In a similar manner for the evaporator, $\mathrm{NTU_e}$, E_e, the evaporating temperature, T_{re}, and the corresponding evaporating pressure are obtained (Eqs [14.6], [14.7] and [14.8]). (Here, it is convenient to express the evaporating temperature in terms of the inlet source fluid temperature, T_{fei}, which will be known from the GHE model.)

$$\mathrm{NTU_e} = \frac{\mathrm{AU_e}}{m_{fe}c_{fe}} \quad [14.6]$$

$$E_e = 1 - \exp(-\mathrm{NTU}_e) \qquad [14.7]$$

$$T_{re} = T_{fei} - \frac{Q_e}{E_e m_{fe} c_{fe}} \quad \text{thus:} \quad P_{re} = f(T_{re}) \qquad [14.8]$$

Conventionally, the Antoine equation can be used to obtain vapour pressure from temperature (Eq. [14.9], in which A, B and C are fluid-dependent constants).

$$\log_{10}(P_{rc,re}) = A - \frac{B}{C + T_{re,rc}} \qquad [14.9]$$

The parameters of these simple condenser and evaporator models will be the heat transfer coefficients, AU_c, AU_e and the (usually constant) mass flows rates of condenser cooling (heating system) water and source fluid, m_{wc}, m_{fe}. Optionally, target values of the effectiveness of these heat exchangers can be used, bypassing the need to determine heat transfer coefficients. The issue of heat transfer coefficients will be discussed in more detail in Section 14.4.2.

14.2.2 Expansion devices

Mechanical expansion devices as depicted in Fig. 14.1 continue to be prevalent especially where fixed speed compressors are in use. The bulb in contact with the evaporator outlet (suction) pipe and its connected capillary which links the bulb and the spring-loaded valve contains a small amount of the same refrigerant as in the parent system. Thus the pressure in the bulb and capillary must balance with the evaporating pressure plus the pressure required to act against the spring tension (Eq. [14.10]).

$$P_{bulb} = P_{re} + P_{spring} \qquad [14.10]$$

As an illustration, suppose the refrigerant temperature in the suction line is 8°C then the temperature in the bulb and capillary will also be 8°C and will be saturated. For refrigerant R134a, the saturation vapour pressure at this temperature will be 387.61 kN/m^2. Suppose the spring tension is adjusted to an equivalent pressure of 37.95 kN/m^2 then the balancing evaporating pressure (neglecting any evaporator pressure loss) will be $387.61 - 37.95 = 349.66$ kN/m^2. At this pressure, the evaporating temperature will be 5°C, which means that $8 - 5 = 3$K of superheat will exist in the suction line.

In the absence of any comprehensive treatments of these devices, the orifice equation is used to model the capacity (Eq. [14.11]). This model works well enough at moderate pressure ratios and when the inlet fluid is significantly subcooled, but at higher pressure ratios and when the inlet fluid either contains some gas or is saturated, choked flow conditions arise which introduces uncertainties in behaviour.

$$m_r = C_{ed} A_{ed} \sqrt{2\rho_{rco}(P_{rc} - P_{re})} \qquad [14.11]$$

Further details can be found in Eames et al. (2014). In smaller heat pumps, short tube orifices are sometimes used instead of thermostatic expansion devices due to their simplicity and low cost. Modelling of these devices has been reported for refrigerant R22 by Aaron and Domanski (1990) and for refrigerant R410A by Payne and O'Neal (1999). The modelling of electronic expansion devices of particular interest in inverter-driven compressor systems and networked systems has received attention mainly through experimental studies. Zhang et al. (2006) developed a new correlation for electronic valves based on experimental results for R22 and R407C. Liu et al. (2007) considered the modelling of electronic valves including choked flow conditions and Park et al. (2001) investigated optimum valve opening with compressor operating frequency in an inverter-driven compressor application. However, most of this work has been related specifically to air conditioners.

14.2.3 Compressors

One of the most convenient (and simple) compressor models expresses the compressor power as a function of the suction and discharge pressures, (ie, if interconnecting pressure losses are ignored, P_{re}, P_{rc}) through the use of the polytropic index, n (Eq. [14.12]).

$$W_{cp} = \frac{n}{n-1} P_{re} V_{reo} \left[\left(\frac{P_{rc}}{P_{re}} \right)^{\frac{n-1}{n}} - 1 \right] \qquad [14.12]$$

This model is appealing because it can be applied to reasonably any compressor type and, in the absence of data on the polytropic index, the index of compression of an ideal gas can be used for approximate calculations. The suction volume flow rate (and, hence, mass flow rate for a given suction density or specific volume) is obtained by introducing the compressor volumetric efficiency η_v, which is defined as the volume actually delivered divided by the compressor displacement volume (Eq. [14.13]).

$$\eta_v = \frac{V_{suc}}{N_{cp} v_{disp}} \qquad [14.13]$$

(in which N_{cp} is the compressor speed (revolutions per second) and v_{disp} is the compressor displacement volume in m^3 per revolution).

Where test or manufacturers' data are available for a given compressor, Eq. [14.12] forms a good basis upon which to parameterise a representative model through n. In the model of Li (2013) for instance, the compressor volumetric efficiency is obtained from Eq. [14.14]:

$$\eta_v = b_1 + b_2 \left(\frac{P_{rc}}{P_{re}} \right)^{\frac{1}{\gamma}} \quad \text{in which:} \quad \gamma = f(P_{re}, P_{rc}) \qquad [14.14]$$

in which b_1, b_2 are constants fitted to test data, and γ is the index of compression of a perfect gas which can be obtained as the average of values at suction and discharge conditions.

The compressor power may then be obtained from Eq. [14.15] in which a_1, a_2 and a_3 are parameters fitted to test data and W_{loss} is a fixed component of overall compressor loss (mainly motor windings) which are recovered by the working fluid as heat and, therefore, ultimately contribute to the heat pump heat output (Li, 2013).

$$W_{\text{cp}} = P_{\text{re}} V_{\text{reo}} a_1 \left[\left(\frac{P_{\text{rc}}}{P_{\text{re}}} \right)^{a_2 + \frac{\gamma - 1}{\gamma}} + a_3 / P_{\text{rc}} \right] + W_{\text{loss}} \qquad [14.15]$$

Optionally, the overall electrical load on the compressor is finally obtained by including a loss component to cover auxiliaries. It is common for instance to include as a minimum the power consumed by the source fan or pump. This load is included in the term WAux in Eq. [14.16].

The W_{loss} term in Eq. [14.15] includes casing heat loss to the ambient environment which is not recovered as heat by the working fluid. An estimate for this loss can be made from Eq. [14.17] in which T_{shell} can be approximated as the average of the refrigerant temperatures at the compressor inlet and discharge connections.

$$W_{\text{cp},+\text{aux}} = W_{\text{cp}} + \text{WAux} \qquad [14.16]$$

$$Q_{\text{amb}} = \text{UA}_{\text{shell}} (T_{\text{shell}} - T_{\text{amb}}) \qquad [14.17]$$

In the above, a_1, a_2, a_3, b_1, b_2, W_{loss} and UA_{shell} are parameters that can be fitted to experimental (or other) data applicable to a given compressor. This model proposed by Li (2013) is applicable to reciprocating, scroll and rotary compressors – these are the main types of compressor currently used in vapour compression heat pumps of small to medium capacity. As an example, Underwood (2015) fitted parameters to a scroll compressor operating with refrigerant R410A with a rated displacement volume of 1.48×10^{-3} m^3/s using an optimisation algorithm based on data from a manufacturer. The results are summarised in Table 14.2.

The heat pump performance metrics, COP and η_{isen}, may now be determined (Eqs [14.18] and [14.19]).

$$\text{COP} = \frac{Q_{\text{c}}}{W_{\text{cp},+\text{aux}}} \qquad [14.18]$$

$$\eta_{\text{isen}} = \frac{m_{\text{r}} (h_{\text{cpo,isen}} - h_{\text{cpi}})}{W_{\text{cp}}} \quad \text{in which:} \quad h_{\text{cpo,isen}}, h_{\text{cpi}} = \text{f}(P_{\text{rc}}, P_{\text{re}}) \qquad [14.19]$$

Li (2013) also included a parameterisation method to account for variable speed compressor modelling. Numerous other recommended compressor model formulations (especially for scroll compressors) can be found in the literature (Duprez et al., 2007; Cuevas and Lebrun, 2009; Winandy et al., 2002; Chen et al., 2002a,b).

Table 14.2 Example fitted compressor parameters (Underwood, 2015)

Parameter	Value
a_1	0.3377
a_2	1.1597
a_3	0
W_{loss}	250 W
b_1	1.0661
b_2	−0.0591
UA_{loss}	3.3 W/K

14.2.4 Solution methods in the steady state

The model described in Sections 14.2.1 and 14.2.3 is easily solved to predict the operating performance of a vapour-compression heat pump. There are three inputs and seven main parameters, compressor modelling parameters and a requirement to define the refrigerant properties. The method is illustrated by means of a simple application case study in the following.

14.2.4.1 Application case study: domestic-scale heat pump

Consider a domestic-scale heat pump required to satisfy a heating demand of 10 kW when receiving source water at 8°C. In this illustration, the required effectiveness of the condenser and evaporator are given (which bypasses the need to include Eqs [14.3], [14.4], [14.6] and [14.7] in the solution). The compressor parameters in Table 14.2 are used (conveniently, this compressor happens to be of a capacity suitable for the heating duty). Suppose we want to determine the heat pump performance when operating at alternative heating system supply temperatures of 45, 50 and 55°C.

The model inputs and parameters are listed as follows:

Inputs: $(Q_c, T_{cwo}, T_{efi}) = 10, 45, 8$
Parameters: $(m_{wc}, m_{fe}, E_e, E_c, N_{cp} \times v_{disp}, \text{WAux}, T_{amb}) = 0.5, 0.5, 0.65, 0.65, 1.48 \times 10^{-3}$, 0.1, 18

In this example, refrigerant properties are determined using REFPROP 7.0 (NIST, 2015).

The solution algorithm is as follows:

Step 1: Enter inputs, parameters and refrigerant type (R410A).
Step 2: Calculate T_{rc} from Eq. [14.5] and, hence, P_{rc} from properties.
Step 3: Guess Q_e and solve Eqs [14.8] and [14.12]–[14.17].
Step 4: Update: $Q_e' = Q_c - (W_{cp} - Q_{amb})$.
Step 5: Compare Q_e and Q_e' and if the two values differ by more than an acceptable error margin, refine the estimate of Q_e and repeat from Step 3; otherwise proceed to Step 6.
Step 6: Calculate the performance metrics (Eqs [14.18] and [14.19]) and T_{efo} and exit.

Table 14.3 Results from the application case study

	Result at stated heating water supply temperature (°C)		
Variable	**45**	**50**	**55**
Q_e (kW)	7.4	7.0	6.4
Q_c (kW)	10	10	10
$W_{cp,+aux}$ (kW)	2.7	3.2	3.7
COP	3.64	3.12	2.67
η_{isen}	0.57	0.53	0.49
T_{efo} (°C)	4.44	4.67	4.93
T_{re} (°C)	2.41	2.64	3.21
T_{rc} (°C)	47.6	52.6	57.6

At Step 5 a suitable iterative method is required to refine the estimate of Q_e based on the current estimate. In the present illustration, the bisection method was used; however, other iterative methods could also have been used such as Newton–Raphson's method (see, for example, Venkataraman, 2009).

Results of the simulation for heating water set point temperatures of 45, 50 and 55°C are summarised in Table 14.3. The results confirm that, as might be expected, the power consumption increases as the heating water temperature increases with a consequential fall in COP.

14.3 Vapour absorption cycle

Considerable potential exists for the use of gas absorption and gas adsorption cycle heat pumps in conjunction with streams of medium grade ground source heat such as often found in deep direct geothermal wells. The take-up of this technology has been limited to a few instances of the well-established ammonia–water cycle mainly in air-source applications. The other established cycle – the much better performing water-lithium bromide (H_2O–LiBr) cycle has received far less attention because it is unsuitable for air-source applications on account of its use of water as the refrigerant. However, for ground-sourcing involving higher grade heat such as exists in mine dewatering and deep geothermal well applications the H_2O-LiBr cycle has considerable potential largely because it is able to operate at higher heating temperatures than conventional vapour compression heat pumps. The simplified steady-state modelling of this cycle is considered in this section. The treatment is restricted to the simplest instance of this heat pump – the single-effect cycle typically 'powered' using the combustion of a fuel.

A schematic of the basic single-effect cycle is illustrated in Fig. 14.3. Again, pressure losses due to circulating working fluids are neglected and it is assumed that the refrigerant leaves the condenser and evaporator in a saturated state. The energy consumed and imparted by the solution pump will also be neglected. The model described here broadly follows the method proposed by Bourdouxhe et al. (1998) for absorption cycle chillers.

In the method that follows, it is assumed that the heat source conditions are fully defined. That is, the flow rate of the source fluid, the source fluid plant inlet temperature (ie, GHE or abstraction well outlet temperature) and the required target plant outlet temperature. The latter will be especially important when water is used in order to restrict the minimum temperature at plant outlet in order to protect both the source fluid and the water refrigerant within the heat pump evaporator.

Once again, an effectiveness model is used for all five heat exchangers in this cycle (Fig. 14.3) with the effectiveness determined in a manner similar to that described in Eqs [14.3], [14.4], [14.6] and [14.7] (Section 14.2). With the source fluid details fully specified, the evaporating temperature is obtained from Eq. [14.20]. For a heat pump application, it will be convenient to specify the outlet temperature from the condenser as this will be the target heating system set point temperature. The condensing temperature can therefore be determined from this value and the condenser effectiveness (Eq. [14.21]). The condenser and absorber in these plants are usually cooled in series (Fig. 14.3). Thus the absorber solution temperature is also obtained from the specified

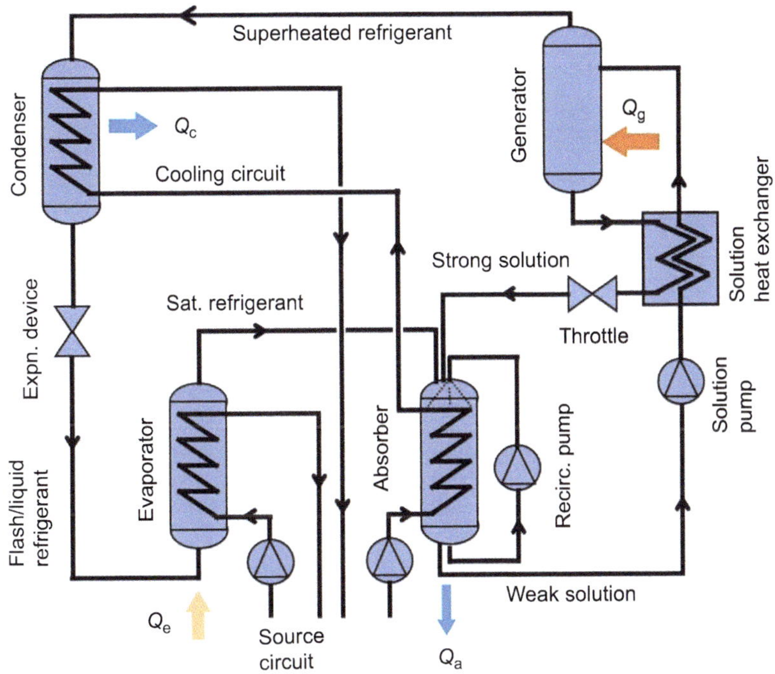

Figure 14.3 Single-effect vapour absorption cycle heat pump schematic.

condenser cooling water outlet temperature by adjusting for the cooling water temperature difference across the condenser and also applying the absorber effectiveness (Eq. [14.22]). The generator is assumed to be heated using hot flue gases resulting from the combustion of a fuel. For convenience, the flue gas temperature at the flue discharge is specified such that this value, the flue gas mass flow rate and the generator effectiveness can be used to obtain the required generator solution temperature (Eq. [14.23]). With the weak absorber solution and strong generator solution temperatures known, the strong solution temperature entering the absorber from the solution heat exchanger can be determined (Eq. [14.24]).

$$T_{\mathrm{re}} = T_{\mathrm{fei}} - \frac{Q_{\mathrm{e}}}{E_{\mathrm{e}} m_{\mathrm{fe}} c_{\mathrm{fe}}} \quad \text{and} \quad P_{\mathrm{re}} = f(T_{\mathrm{re}}) \qquad [14.20]$$

$$T_{\mathrm{rc}} = T_{\mathrm{wco}} + \frac{Q_{\mathrm{c}}(1 - E_{\mathrm{c}})}{E_{\mathrm{c}} m_{\mathrm{wc}} c_{\mathrm{wc}}} \quad \text{and} \quad P_{\mathrm{rc}} = f(T_{\mathrm{rc}}) \qquad [14.21]$$

$$T_{\mathrm{sao}} = T_{\mathrm{cwo}} - \frac{Q_{\mathrm{c}} E_{\mathrm{a}} - Q_{\mathrm{a}}(1 - E_{\mathrm{a}})}{E_{\mathrm{a}} m_{\mathrm{wc}} c_{\mathrm{wc}}} \qquad [14.22]$$

$$T_{\mathrm{sgo}} = T_{\mathrm{ggo}} - \frac{Q_{\mathrm{g}}(1 - E_{\mathrm{g}})}{E_{\mathrm{g}} m_{\mathrm{gg}} c_{\mathrm{gg}}} \qquad [14.23]$$

$$T_{\mathrm{sai}} = T_{\mathrm{sgo}} - E_{\mathrm{shx}}(T_{\mathrm{sgo}} - T_{\mathrm{sao}}) \qquad [14.24]$$

The solvent-strong solution concentration is obtained from the equilibrium properties of lithium bromide solution (ASHRAE, 2013a):

$$X_{\mathrm{s}} = \mathrm{f}(T_{\mathrm{rc}}, T_{\mathrm{sgo}}) \qquad [14.25]$$

The weak solution concentration is obtained similarly (Eq. [14.26]) and the required refrigerant mass flow rate is obtained from the (specified) evaporator load (Eq. [14.27]). The strong solution flow rate is then obtained from a mass balance on either the absorber or generator, which reduces to Eq. [14.28]. This then leads to the required weak solution mass flow rate (Eq. [14.29]).

$$X_{\mathrm{w}} = \mathrm{f}(T_{\mathrm{re}}, T_{\mathrm{sao}}) \qquad [14.26]$$

$$m_{\mathrm{r}} = \frac{Q_{\mathrm{e}}}{(h_{\mathrm{reo}} - h_{\mathrm{rco}})} \quad \text{in which:} \quad h_{\mathrm{reo}}, h_{\mathrm{rco}} = \mathrm{f}(P_{\mathrm{re}}, P_{\mathrm{rc}}) \qquad [14.27]$$

$$m_{\mathrm{ss}} = \frac{X_{\mathrm{w}} m_{\mathrm{r}}}{X_{\mathrm{s}} - X_{\mathrm{w}}} \qquad [14.28]$$

$$m_{\mathrm{sw}} = m_{\mathrm{ss}} + m_{\mathrm{r}} \qquad [14.29]$$

Solution enthalpies can now be determined using fluid properties (ASHRAE, 2013b) (Eqs [14.30]–[14.32]).

$$h_{\mathrm{sgo}} = \mathrm{f}(T_{\mathrm{sgo}}, X_{\mathrm{s}}) \tag{14.30}$$

$$h_{\mathrm{sao}} = \mathrm{f}(T_{\mathrm{sao}}, X_{\mathrm{w}}) \tag{14.31}$$

$$h_{\mathrm{sai}} = \mathrm{f}(T_{\mathrm{sai}}, X_{\mathrm{s}}) \tag{14.32}$$

The temperature of the refrigerant vapour leaving the generator is assumed to be the same as the equilibrium solution temperature at the solution/vapour interface. In turn, this is assumed to be at the lower (entering) weak solution concentration. From this temperature together with the condensing pressure, the enthalpy of this refrigerant vapour is obtained (Eqs [14.33] and [14.34]).

$$T_{\mathrm{rgo}} = \mathrm{f}(T_{\mathrm{rc}}, X_{\mathrm{w}}) \tag{14.33}$$

$$h_{\mathrm{rgo}} = h_{\mathrm{rci}} = \mathrm{f}(P_{\mathrm{rc}}, T_{\mathrm{rgo}}) \tag{14.34}$$

Algorithm:

The solution procedure is broadly in accordance with that developed by Bourdouxhe et al. (1998).

Step 1: Input the source heating load (Q_{e}) and first approximations for the condenser and absorber loads (Q_{c}, Q_{a}). The generator heat load can be determined from an overall heat balance with respect to the entire cycle (ie, $Q_{\mathrm{g}} = Q_{\mathrm{c}} + Q_{\mathrm{a}} - Q_{\mathrm{e}}$). Input the target heating water supply temperature (T_{wco}) and the boundary conditions (T_{fei}, T_{ggo}, m_{wc}, m_{fe}, m_{gg}). Input the heat exchanger effectivenesses (E_{e}, E_{c}, E_{a}, E_{g}, E_{shx}). Solve Eqs [14.21]–[14.34].

Step 2: Calculate zero functions based on energy balances about the absorber and condenser:

$$F_{\mathrm{a}} = Q_{\mathrm{a}} - (m_{\mathrm{r}}h_{\mathrm{reo}} + m_{\mathrm{ss}}h_{\mathrm{sai}} - m_{\mathrm{sw}}h_{\mathrm{sao}}) \tag{14.35}$$

$$F_{\mathrm{c}} = Q_{\mathrm{c}} - m_{\mathrm{r}}(h_{\mathrm{rci}} - h_{\mathrm{rco}}) \tag{14.36}$$

Step 3: Form the 2 × 2 Jacobian matrix as follows:

$$\frac{\partial F_{\mathrm{a}}}{\partial Q_{\mathrm{a}}} = \frac{F_{\mathrm{a},Q_{\mathrm{a}}+\Delta Q} - F_{\mathrm{a},Q_{\mathrm{a}}-\Delta Q}}{2\Delta Q} \tag{14.37}$$

$$\frac{\partial F_{\mathrm{a}}}{\partial Q_{\mathrm{c}}} = \frac{F_{\mathrm{a},Q_{\mathrm{c}}+\Delta Q} - F_{\mathrm{a},Q_{\mathrm{c}}-\Delta Q}}{2\Delta Q} \tag{14.38}$$

$$\frac{\partial F_{\mathrm{c}}}{\partial Q_{\mathrm{a}}} = \frac{F_{\mathrm{c},Q_{\mathrm{a}}+\Delta Q} - F_{\mathrm{c},Q_{\mathrm{a}}-\Delta Q}}{2\Delta Q} \tag{14.39}$$

$$\frac{\partial F_c}{\partial Q_c} = \frac{F_{c,Q_c+\Delta Q} - F_{c,Q_c-\Delta Q}}{2\Delta Q} \qquad [14.40]$$

Thus

$$\mathbf{J} = \begin{bmatrix} \frac{\partial F_a}{\partial Q_a} & \frac{\partial F_a}{\partial Q_c} \\ \frac{\partial F_c}{\partial Q_a} & \frac{\partial F_c}{\partial Q_c} \end{bmatrix} \qquad [14.41]$$

Step 4: Update Q_a, Q_c as follows:

$$\mathbf{Q}^+ = \mathbf{Q} - \mathbf{J}^{-1}\mathbf{F} \qquad [14.42]$$

in which:

$\mathbf{Q}^+ = \begin{bmatrix} Q_a \\ Q_c \end{bmatrix}$ (updated values) and $\mathbf{Q}$ contains the existing values.

$\mathbf{F} = \begin{bmatrix} F_a \\ F_c \end{bmatrix}$ are the zero functions evaluated at the existing Q_a, Q_c.

Step 5: Repeat the procedure from Step 2 until two successive pairs of values of Q_a, Q_c agree to within an acceptable tolerance.

Step 6: Conclude the simulation by calculating the heat pump cycle COP and undefined inlet/outlet heat exchanger fluid temperatures:

$$\text{COP} = \frac{Q_c + Q_a}{Q_g} \qquad [14.43]$$

$$T_{wci} = T_{wco} - \frac{Q_c}{m_{wc}c_{wc}} \qquad [14.44]$$

$$T_{wai} = T_{wci} - \frac{Q_a}{m_{wc}c_{wc}} \qquad [14.45]$$

$$T_{ggi} = T_{ggo} + \frac{Q_g}{m_{gg}c_{gg}} \qquad [14.46]$$

14.3.1 Application case study – higher temperature heat pump

To illustrate the absorption cycle heat pump model, a commercial-scale application is considered using high temperature source water at 30°C which might, for example, come from a deep geothermal well. The source heat is restricted to 200 kW and the heat pump is required to satisfy a conventional heating system at supply temperatures of 70–80°C.

The model inputs and parameters are set as follows. Again, for convenience, heat exchanger effectivenesses were specified which avoided the need to determine AU values:

Inputs: $(Q_e, T_{wco}, T_{fei}, T_{ggo}) = 200, 80, 30, 163$
Parameters: $(m_{fe}, m_{wc}, m_{gg}, E_e, E_c, E_a, E_g, E_{shx}) = 10, 20, 10, 0.65, 0.65, 0.65, 0.65, 0.65$

The strong solution concentration was set at an arbitrary maximum value of 68% to protect against the formation of crystals of lithium bromide. Thus each simulation run was repeated with alternative values of T_{ggo} until the condition for X_s was satisfied. Three alternative simulations were carried out at heating water supply temperatures of 80, 75 and 70°C. The results are given in Table 14.4.

Once again, we see a decline in the COP as the heating supply temperature increases. Also, of particular note, the pumping rate of weak solution increases sharply

Table 14.4 Results from the vapour absorption cycle heat pump application case study

	Result at stated heating water supply temperature (°C)		
Variable	**70**	**75**	**80**
Q_e (kW)	200.0	200.0	200.0
Q_g (kW)	287.0	324.7	459.5
Q_a (kW)	272.1	306.6	460.0
Q_c (kW)	214.9	218.0	199.5
COP	1.697	1.616	1.435
T_{re} (°C)	22.3	22.3	22.3
T_{rc} (°C)	71.4	76.4	81.3
T_{sgo} (°C)	144.6	150.8	157.1
T_{sao} (°C)	69.2	74.4	80.6
X_s (%)	68.0	68.0	68.0
X_w (%)	62.2	64.5	67.2
m_{sw} (kg/s)	1.035	1.741	7.176
m_r (kg/s)	0.089	0.090	0.091
T_{feo} (°C)	25.2	25.2	25.2
T_{wai} (°C)	64.2	68.7	72.1
T_{wci} (°C)	67.4	72.4	77.6

at the higher heating temperature due to the lower concentration gradient. Though the solution pump energy was neglected in the model, this will have implications for performance in practice by further reducing the COP.

14.4 Regression models

In many applications, a much simpler approach to modelling can be adopted through equation or curve fitting if detailed equipment performance data are available. Empirical models of this kind have the advantages of being both easy to generate and use, as well as having high predictive accuracy (though the latter is strictly conditional to within the ranges of available data to which the model was fitted).

Two general sources of data are field or experimental data and manufacturers' catalogue data. It is important to distinguish between the two. Catalogue data arises as a consequence of the equipment test requirements of various international standards such as ANSI/AHRI 551/591 (2012), ANSI 870 (2005) or the European standards BS EN 14511 (2013) and BS EN 14825 (2013). These standards set out detailed testing procedures for heat pump equipment. Manufacturers present extensive catalogue performance data typically consisting of energy balances at a range of boundary conditions (ie, inlet or outlet evaporator and condenser temperatures). In most cases most of these data are themselves generated using simulated models validated using a limited fund of the measured test data described in the standards. These catalogue data are idealised performances unhindered by design and installation factors such as equipment sizing, the design of the balance of system components, control and social (user) factors. Heat pumps that are oversized, or coupled to inappropriately sized or controlled systems and operated by users who have unrealistic expectations of them, will simply not perform as described in the catalogue. Some of these limitations can be addressed by resorting to measured performances of the heat pump in laboratory conditions and all of them can be addressed by measuring the performance in situ in the field. However, in all of these cases, the models that arise are strictly limited to the conditions imposed by the catalogue or the experiment. This latter point is what sets them apart from the deterministic modelling approaches described in the previous two sections.

For heat pumps, we have already seen that performance is dominated by the source and sink conditions – in particular, temperatures. The simplest approach to fitting is to regress one or more of the important heat pump dependent variables onto the two inlet or outlet temperatures using multiple regression. By way of illustration in this section, we will consider COP as the only independent variable of interest. A model of this kind works well if it is assumed that the heating demand is known and always satisfied (as will often be the case in entire system modelling) which means that the COP will lead to the power consumed by the heat pump (ie, Eq. [14.2]) and the source energy required from the ground can be obtained by energy balance (ie, Eq. [14.1]). A more detailed and comprehensive treatment of the subject can be found in Jin and Spitler (2002) in which catalogue data are used to parameterise a detailed deterministic model of a heat pump (similar to that described in Section 14.2) and the result is compared with a directly fitted model to the catalogue data.

A simple multiple regression framework for this application will be the bilinear model of Eq. [14.47].

$$\mathrm{COP} = (a + bT_{\mathrm{feo}}) \times (c + dT_{\mathrm{wco}}) \qquad [14.47]$$

which multiples through to give (Eq. [14.48]):

$$\mathrm{COP} = A + BT_{\mathrm{feo}} + CT_{\mathrm{wco}} + DT_{\mathrm{feo}}T_{\mathrm{wco}} \qquad [14.48]$$

The coefficients ($A \ldots D$) are fitted using multiple regression which can be easily carried out in a spreadsheet based on the available equipment performance data. A goodness-of-fit statistic (eg, Pearson's correlation statistic or the coefficient of determination) can be obtained to see how well the fitted model described the original data and if the result is unsatisfactory, an alternative model may be tested. For example, the biquadratic model of Eq. [14.49], which multiplies to form Eq. [14.50]:

$$\mathrm{COP} = \left(a + bT_{\mathrm{feo}} + cT_{\mathrm{feo}}^2\right) \times \left(d + eT_{\mathrm{wco}} + fT_{\mathrm{wco}}^2\right) \qquad [14.49]$$

$$\begin{aligned}\mathrm{COP} = {} & A + BT_{\mathrm{feo}} + CT_{\mathrm{wco}} + DT_{\mathrm{feo}}T_{\mathrm{wco}} + ET_{\mathrm{feo}}^2 + FT_{\mathrm{wco}}^2 + GT_{\mathrm{feo}}T_{\mathrm{wco}}^2 \\ & + HT_{\mathrm{feo}}^2T_{\mathrm{wco}} + IT_{\mathrm{feo}}^2T_{\mathrm{wco}}^2\end{aligned} \qquad [14.50]$$

14.4.1 Models fitted to catalogue data

An application of model fitting to catalogue data is illustrated through a simple case study example. The data in Table 14.5 gives a small part of the catalogue data for a commercially available liquid-to-water heat pump of medium capacity driven by a screw compressor (Ciat, 2007). The complete data set covers source fluid temperatures

Table 14.5 Example water-to-water heat pump data at $T_{\mathrm{feo}} = 2$°C (Ciat, 2007)

T_{wco} (°C)	W_{cp} (kW)	Q_{c} (kW)	COP
35	105	530	5.05
40	114	515	4.52
45	125	501	4.01
50	138	487	3.53
55	152	473	3.11
60	168	457	2.72

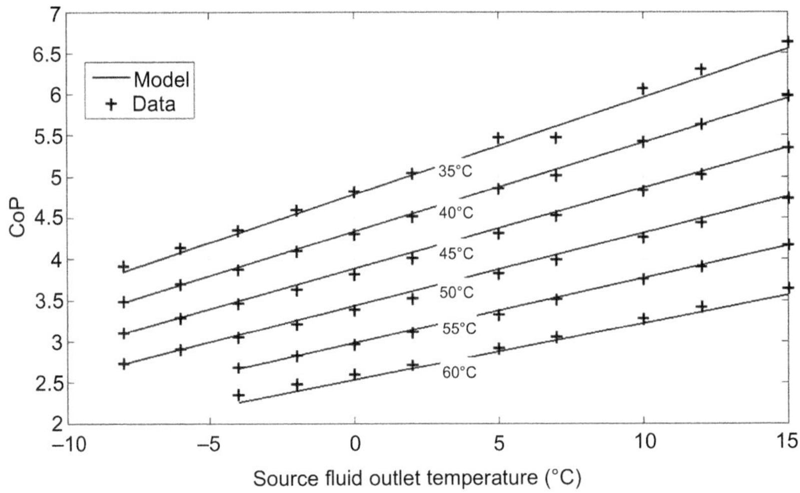

Figure 14.4 Catalogue-fit model example.

from −8°C (ethylene glycol solution) to 15°C (water) and sink water temperatures from 35 to 60°C.

A bilinear regression model was fitted to the data based on Eq. [14.48] and the results are shown in Fig. 14.4. The result with this simple model is good – the R^2 statistic being 0.997 and so there seems little point in attempting a higher-order fit, such as offered by Eq. [14.50]. The fitted coefficients are given in Table 14.6.

14.4.2 Models fitted to laboratory and field data

Little reported literature exists on the regression modelling of heat pumps based on either laboratory or field data. Gupta and Irving (2013) reported on the regression modelling of ground- and air-source heat pumps (GSHPs and ASHPs) using data from test results. Tangwe et al. (2014) developed a regression model for a small ASHP for hot water production. Underwood (2014) reported on the development of a regression model fitted to experimental data from a laboratory GSHP. In the latter, the source and sink temperatures were combined into a single temperature lift variable and a further variable was introduced – the part-load ratio to take account of situations during which the heat pump would be operating at less than rated capacity over time. The resulting bilinear model is given in Eq. [14.51] in which p is the heat pump

Table 14.6 Fitted model coefficients to the Ciat (2007) example and Eq. [14.48]

A	B	C	D
7.9339	0.18640	−0.08996	−0.00196

Table 14.7 **Fitted coefficients for the heat pump model of Eq. [14.51]**

GHE size (m)	*A*	*B*	*C*	*D*
100	2.852525	2.868282	−0.017015	−0.037951
200	7.936727	−1.206375	−0.154520	0.067780
300	3.315421	2.816345	−0.021494	−0.042534
365	3.188746	3.414232	−0.017961	−0.062144

part-load ratio (ie, ratio of actual heating load to rated-capacity heating) and ΔT_{sosi} is the difference between the sink outlet and source inlet temperatures to the heat pump. The fitted model parameters are given in Table 14.7 for a range of vertical closed-loop borehole heat exchanger capacities with a goodness-of-fit statistic of 0.891 (Underwood, 2014).

$$\mathrm{COP} = A + Bp + C\Delta T_{\mathrm{sosi}} + DpT_{\mathrm{sosi}} \qquad [14.51]$$

14.5 Dynamic-state modelling

Dynamic models of thermal systems attempt to describe variations in plant behaviour with time and are therefore of interest when independent variables and boundary conditions are non-stationary. In many cases, satisfactory heat pump simulations can be made using the steady-state methods described in Sections 14.2 and 14.3 when the balance of system components are themselves very damped (ie, react to changes in their input variables very slowly). This will usually be the case when a building model participates in the simulation and performance over seasons or longer time horizons are of interest. Such mixed-dynamic problems are sometimes called 'stiff systems' and it becomes possible to use a steady-state model for the heat pump when its dependent variables stabilise within the time-step of relevance to the parent system. For example, a five minute time step may give accurate simulation results of a building and its plant behaviour over a particular time horizon and if the heat pump that forms part of the system model responds to changes in its input variables within seconds or two or three minutes, then a steady-state model of the heat pump may suffice. Such an approach is sometimes called 'quasi-steady-state' modelling (at least as far as the heat pump is concerned). However, in some modelling applications such as control system design and troubleshooting or fault detection and diagnosis, or in situations where more detailed information about the plant behaviour is required such as pressure transients, superheat behaviour or periodic switching events, then a more detailed dynamic model is likely to be required.

The simplest of these results in linear time-invariant descriptions of the plant within which model parameters can be assumed to be invariant – in which case boundaries can be placed around large part of the heat pump (or indeed to entire plant) and a time-domain model leading to a transfer function becomes possible. This possibility is discussed and illustrated briefly in Section 14.5.1. In many other cases, however, a much more rigorous treatment may be needed and this is addressed in Section 14.5.2.

There are numerous examples of progress in this latter area to be found in the literature, most of them addressing specific issues around control or detailed plant behaviour such as start-up transients. Quite a lot of this work deals with air-conditioning plants but, of course, the principles are the same for heat pumps (eg, Wang et al., 2007; Koury et al., 2001; Browne and Bansal, 2002; Murphy and Goldschmidt, 1986; Hermes and Melo, 2008). However, some historical work has specifically addressed the modelling of transients and control of heat pumps (eg, Sami et al., 1987; Sami and Dahmani, 1996; Vargas and Parise, 1995) and, more recently, advanced applications of heat pump system dynamic modelling (eg, Zhao et al., 2007; Hepbasli et al., 2009; White et al., 2002) though very little emphasis appears to exist regarding the dynamics of GSHPs.

14.5.1 Block diagram modelling with transfer functions

The simplest first-order interpretation for a heat pump arises when a single dependent variable of interest is isolated and all other variables with the exception of one forcing input are assumed to be constant. The most important dependent variable of interest will, in most cases, be the condenser cooling water outlet temperature as this forms the heating system supply temperature and will generally form the main controlled variable for the heat pump. An overall heat pump energy balance can be written as Eq. [14.52].

$$C_{\mathrm{HP}}\frac{dT_{\mathrm{wco}}}{dt} = m_{\mathrm{wc}}c_{\mathrm{wc}}(T_{\mathrm{wci}} - T_{\mathrm{wco}}) + Q_{\mathrm{e}} + W_{\mathrm{cp}} \qquad [14.52]$$

If it is assumed that the energy supplied to the plant (ie, $Q_{\mathrm{e}} + W_{\mathrm{cp}}$) is stepped in when the heat pump is first switched on and then remains constant, then this collective energy input forms a step change the Laplace transform of which is $(Q_{\mathrm{e}} + W_{\mathrm{cp}})/s$.

Exercising the Laplace transform of Eq. [14.52] with reference to variations from an initial steady-state (ie, stationary terms disappear), results in the first-order transfer function for the heat pump of Eq. [14.53].

$$T_{\mathrm{wco}}(s) = \frac{K}{s(\tau s + 1)} \quad \text{in which: } K = \frac{Q_{\mathrm{e}} + W_{\mathrm{cp}}}{m_{\mathrm{wc}}c_{\mathrm{wc}}} \quad \tau = \frac{C_{\mathrm{HP}}}{m_{\mathrm{wc}}c_{\mathrm{wc}}} \qquad [14.53]$$

In Eq. [14.53], K will be the heat pump gain and τ the heat pump time constant. Of course this is a very simple interpretation in that Q_{e} will not behave in such an instantaneous manner but if sufficient allowance is made in the parameter, C_{HP} (the overall plant thermal capacity in J/K), then an adequate simple tool emerges to investigate transient behaviour and controls.

A more accurate approach when transient performance data are available is to 'identify' the plant and fit a transfer function to the response. As an illustration, Fig. 14.5 shows the step response of a small domestic heat pump to which a first-order delay-time model has been fitted. This is a classical process reaction curve of plant response and the type of model fitted includes an early pure time delay component giving details that are often difficult or impossible to model deterministically. This might, for example, describe the response of the heat pump immediately following thermostat closure.

The transfer function which describes the modelled step response in Fig. 14.5 is given in Eq. [14.54]. The exponential time delay function can be approximated by the first-order Padé approximation resulting in Eq. [14.55].

$$G_{\mathrm{p}}(s) = \frac{K_{\mathrm{p}} \exp(-t_{\mathrm{d}}s)}{(\tau s + 1)} \tag{14.54}$$

$$G_{\mathrm{p}}(s) = \frac{K_{\mathrm{p}}(1 - t_{\mathrm{d}}s/2)}{(\tau s + 1) \times (t_{\mathrm{d}}s/2 + 1)} \tag{14.55}$$

The fitted model in this illustration resulted in a delay time value of 6.51 s, a heat pump time constant of 6.71 s and a gain of 2K. A detailed description of the fitting procedure can be found in Underwood (1998).

Transfer functions of this kind can be useful to form part of larger whole system simulations. For example, a block diagram simulation model of a house equipped with a heat pump and a hot water buffer tank is shown in Fig. 14.6. The heat pump model consists of the fitted transfer function described above. Simulation results during the early morning start-up phase of the heating supply temperature on a typical winter day are shown in Fig. 14.7. (Further details of the block diagram modelling method can be found in Underwood, 2014).

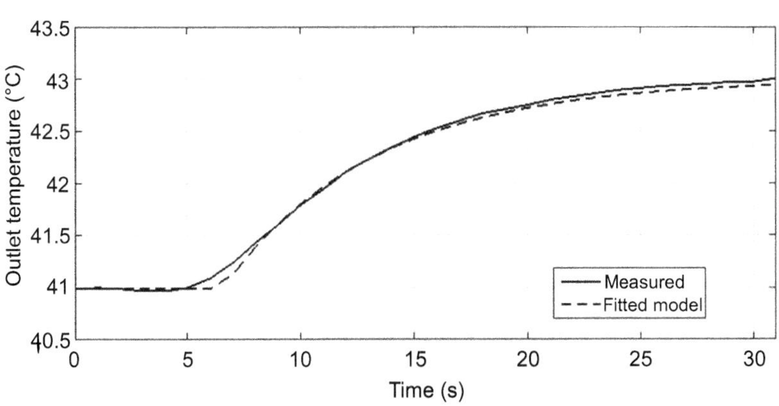

Figure 14.5 Fitted transfer function to heat pump time response.

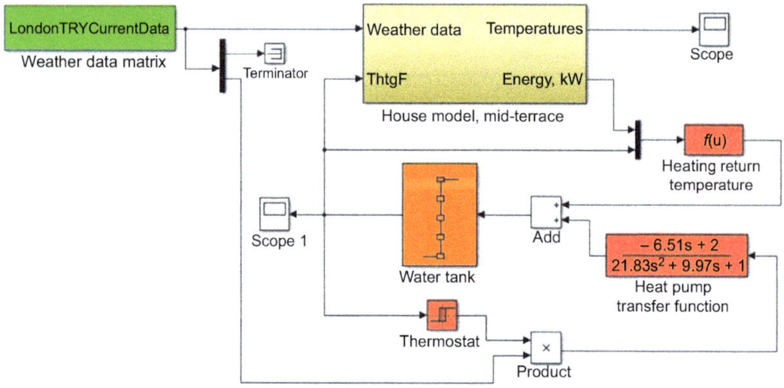

Figure 14.6 Block diagram of a house heating system using a heat pump transfer function.

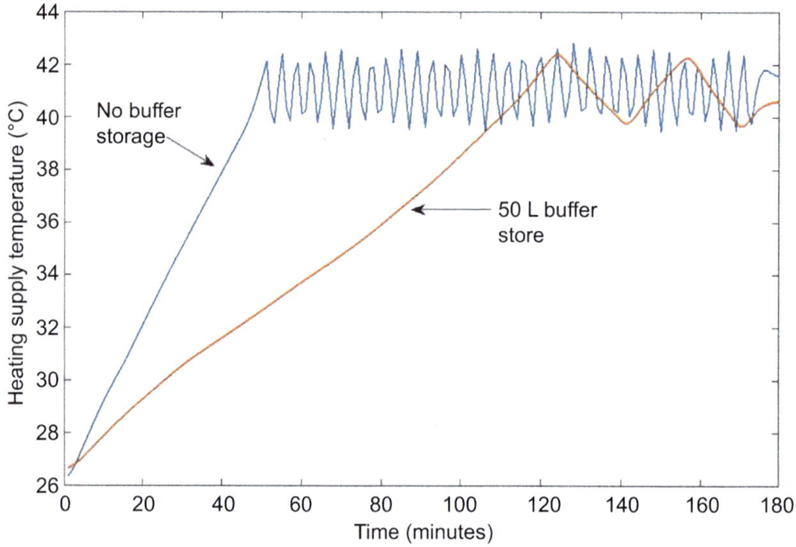

Figure 14.7 Results of house heating simulation using a heat pump transfer function.

The results clearly illustrate the dramatic benefit of using buffer storage. Without the tank, the heat pump thermostat starts the heat pump at a rate of approximately 17 starts per hour which is likely to lead to severe compressor deterioration over time. A 50-L buffer tank reduces the number of starts to two or three per hour.

14.5.2 Distributed parameter modelling

A distributed-parameter modelling method for heat exchange systems considers independent variables in both time and space. This would be particularly relevant to the heat exchangers in a heat pump – the evaporator and condenser. The advantage is that the evolving refrigerant states can be predicted as the refrigerant proceeds along the evaporator (or condenser). In particular, the extent of any superheat and subcooling zones can be predicted. This is a rigorous modelling approach and high computational power will usually be required compared with the other methods discussed earlier in this chapter. Numerous examples of this type of modelling approach can be found in the literature – for examples, see Wang et al. (2007) and MacArthur and Grald (1987). In fact, the latter example is one of the first attempts at distributed parameter modelling of heat pumps and will be considered in detail in the following. It has the advantage of detail and rigour whilst also combining a relatively simple structure.

Generic equations can be developed for both the evaporator and condenser. Neglecting momentum conservation terms and pressure losses, and considering axial variations in the direction of flow only, continuity and energy balances for the evaporator and condenser can be written as Eqs [14.56] and [14.57].

$$\frac{\partial}{\partial t}(\rho_r A_{rcs}) + \frac{\partial m_r}{\partial z} = 0 \quad [14.56]$$

$$\frac{\partial}{\partial t}(\rho_r h_r A_{rcs}) + \frac{\partial}{\partial z}(m_r h_r) + U_r L_{rp}(T_r - T_m) = 0 \quad [14.57]$$

A discretisation scheme is used in which control volumes bounded by interfaces in the axial direction can be used, as shown in Fig. 14.8.

The discretised form of the continuity and energy equations (Eqs [14.56] and [14.57]) for the nth zone can be written as Eqs [14.58] and [14.59] (MacArthur and Grald, 1987). (Note here that the '–' superscript denotes the value at the previous time step and $A_{m,n}$ is the heat exchange surface area in the nth zone.)

$$\frac{(\rho_{r,n}A_{rcs}) - (\rho_{r,n}A_{rcs})^-}{\Delta t} + \frac{m_{r,j} - m_{r,j-1}}{\Delta z} = 0$$
$$\text{ie,} \quad \frac{V_{r,n}}{\Delta t}\left(\rho_{r,n} - \rho_{r,n}^-\right) + m_{r,j} - m_{r,j-1} = 0 \quad [14.58]$$

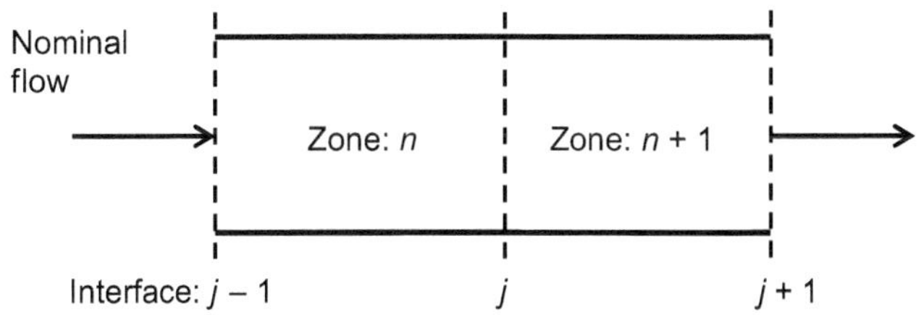

Figure 14.8 Discretisation scheme for the evaporator and condenser.

$$\frac{\left(\rho_{r,n}h_{r,n}A_{rcs}\right)-\left(\rho_{r,n}h_{r,n}A_{rcs}\right)^{-}}{\Delta t}+\frac{m_{r,j}h_{r,j}-m_{r,j-1}h_{r,j-1}}{\Delta z}+U_{r,n}A_{m,n}\left(T_{r,n}-T_{m,n}\right)=0$$

$$\text{ie,}\quad \frac{V_{r,n}}{\Delta t}\left[\left(\rho_{r,n}h_{r,n}\right)-\left(\rho_{r,n}h_{r,n}\right)^{-}\right]+m_{r,j}h_{r,j}-m_{r,j-1}h_{r,j-1}+U_{r,n}A_{m,n}\left(T_{r,n}-T_{m,n}\right)=0 \qquad [14.59]$$

It is convenient to eliminate $\rho_{r,n}$ in the above by multiplying the discretised continuity equation (Eq. [14.58]) by $h_{r,n}$ and then subtracting the result from Eq. [14.59] to give Eq. [14.60] (MacArthur and Grald, 1987).

$$\begin{aligned}&\frac{V_{r,n}\rho_{r,n}^{-}}{\Delta t}\left(h_{r,n}-h_{r,n}^{-}\right)+m_{r,j}\left(h_{r,j}-h_{r,n}\right)+m_{r,j-1}\left(h_{r,n}-h_{r,j-1}\right)\\&+U_{r,n}A_{m,n}\left(T_{r,n}-T_{m,n}\right)=0\end{aligned} \qquad [14.60]$$

The interface variables, $h_{r,j}$, $h_{r,j-1}$, are eliminated using an upwind scheme as follows:

$$\begin{aligned}&\frac{V_{r,n}\rho_{r,n}^{-}}{\Delta t}\left(h_{r,n}-h_{r,n}^{-}\right)+h_{r,n}\max\left(m_{r,j},0\right)-h_{r,n+1}\max\left(-m_{r,j},0\right)-m_{r,j}h_{r,n}\\&+m_{r,j-1}h_{r,n}-h_{r,n-1}\max\left(m_{r,j-1},0\right)+h_{r,n}\max\left(-m_{r,j-1},0\right)\\&+U_{r,n}A_{m,n}\left(T_{r,n}-T_{m,n}\right)=0\end{aligned}$$

ie,

$$\begin{aligned}&h_{r,n}\left[\frac{V_{r,n}\rho_{r,n}^{-}}{\Delta t}+\max\left(m_{r,j},0\right)+\max\left(-m_{r,j-1},0\right)+m_{r,j-1}-m_{r,j}\right]=\\&h_{r,n+1}\max\left(-m_{r,j},0\right)+h_{r,n-1}\max\left(m_{r,j-1},0\right)+\frac{V_{r,n}\rho_{r,n}^{-}h_{r,n}^{-}}{\Delta t}-\\&U_{r,n}A_{m,n}\left(T_{r,n}-T_{m,n}\right)=0\end{aligned}$$

But: $\max\left(m_{r,j},0\right)-m_{r,j}=\max\left(-m_{r,j},0\right)$ and $\max\left(-m_{r,j-1},0\right)+m_{r,j-1}=\max\left(m_{r,j-1},0\right)$, hence the final discretised equations for the refrigerant zones of the evaporator and condenser can be expressed as Eqs [14.61] and [14.62] (MacArthur and Grald, 1987).

$$\text{Continuity:}\quad m_{r,j}=m_{r,j-1}-\frac{V_{r,n}}{\Delta t}\left(\rho_{r,n}-\rho_{r,n}^{-}\right) \qquad [14.61]$$

$$\text{Energy:}\quad h_{r,n}=\frac{Ah_{r,n-1}+Bh_{r,n+1}-U_{r,n}A_{m,n}\left(T_{r,n}-T_{m,n}\right)+V_{r,n}\rho_{r,n}^{-}h_{r,n}^{-}/\Delta t}{C} \qquad [14.62]$$

where:

$$A = \max\left(m_{\mathrm{r},j-1}, 0\right), \ B = \max\left(-m_{\mathrm{r},j}, 0\right) \quad \text{and} \quad C = A + B + \frac{V_{r,n}\rho_{\mathrm{r},n}^{-}}{\Delta t}$$

For the essentially evolving two-phase flow experienced through the evaporator and condenser a void fraction model requires to be defined so that the masses of vapour and liquid and, hence, the overall mass of fluid in each heat exchanger zone can be determined. For a 'no slip' condition both liquid and vapour phases flow through the heat exchanger at the same velocity and the void fraction will be as Eq. [14.63].

$$\alpha = \left(1 + \frac{\rho_{\mathrm{rg}}}{\rho_{\mathrm{rl}}}\left(\frac{1-x}{x}\right)\right)^{-1} \qquad [14.63]$$

Several models that deal with the effects of slip were compared by Rice (1987) from which the simple and convenient model of Eq. [14.64] was suggested by MacArthur and Grald (1987):

$$\alpha = \frac{1}{1 + s\frac{\rho_{\mathrm{rg}}}{\rho_{\mathrm{rl}}}\left(\frac{1-x}{x}\right)}$$

in which the slip ratio, $s = \left(\frac{\rho_{\mathrm{rg}}}{\rho_{\mathrm{rl}}}\right)^{-\frac{1}{3}}$.

$$\text{ie,} \ \alpha = \frac{1}{1 + \left(\frac{1-x}{x}\right)\left(\frac{\rho_{\mathrm{rg}}}{\rho_{\mathrm{rl}}}\right)^{\frac{2}{3}}} \qquad [14.64]$$

Thus the total mass, M_{r}, of refrigerant contained in a heat exchanger of length, L, can be obtained from Eq. [14.65].

$$M_{\mathrm{r}} = \frac{V_{\mathrm{r}}\left(\rho_{\mathrm{rg}}\int_0^L \alpha dL + \rho_{\mathrm{rl}}\int_0^L (1-\alpha) dL\right)}{\int_0^L dL} \qquad [14.65]$$

14.5.2.1 Heat exchanger wall

A heat balance about the heat exchange walls of the evaporator and condenser can be written as Eq. [14.66] in which the subscript 'm' refers to the wall material and 'f' refers to the source or sink fluid (ie, water, brine or glycol solution).

$$\rho_{\mathrm{m}}A_{\mathrm{mcs}}c_{\mathrm{m}}\frac{\partial T_{\mathrm{m}}}{\partial t} + U_{\mathrm{f}}L_{\mathrm{fp}}(T_{\mathrm{m}} - T_{\mathrm{f}}) - U_{\mathrm{r}}L_{\mathrm{rp}}(T_{\mathrm{r}} - T_{\mathrm{m}}) = 0 \qquad [14.66]$$

For a compact heat exchanger such as a brazed plate or shell and tube heat exchanger, it is reasonable to assume that these dimensions are uniform along the entire length of the heat exchanger and that $L_{fp} = L_{rp}$. Thus Eq. [14.66] may be discretised for the nth zone of the heat exchanger and then re-arranged to give Eq. [14.67].

$$T_{m,n} = \frac{(\rho Vc)_{m,n} T^{-}_{m,n} \big/ \Delta t + U_{r,n} A_{m,n} T_{r,n} + U_{f,n} A_{m,n} T_{f,n}}{(\rho Vc)_{m,n} \big/ \Delta t + A_{m,n} \left(U_{r,n} + U_{f,n} \right)} \qquad [14.67]$$

14.5.2.2 Source and sink fluid zones

A heat balance about the source/sink fluid spaces of the evaporator and condenser gives Eq. [14.68], which can be discretised for the nth zone (with the assumption that the heat exchanger is connected in counter-flow configuration) and then rearranged to give Eq. [14.69].

$$\rho_f A_{fc} c_f \frac{\partial T_f}{\partial t} + m_f c_f \frac{\partial T_f}{\partial z} + U_f L_{fp} (T_f - T_m) = 0 \qquad [14.68]$$

$$T_{f,n} = \frac{(\rho Vc)_{f,n} T^{-}_{f,n} \big/ \Delta t + m_f c_f T_{f,n+1} + U_{f,n} A_{m,n} T_{m,n}}{(\rho Vc)_{f,n} \big/ \Delta t + m_f c_f + U_{f,n} A_{m,n}} \qquad [14.69]$$

14.5.2.3 Heat transfer

In the more commonly found smaller heat pump systems, brazed plate evaporators and condensers with corrugated plates are gradually replacing shell and tube heat exchangers (see chapter: Standing column wells). For the distributed heat pump model developed here as an illustration, this type of heat exchanger will be assumed based on refrigerant R410A. Therefore, surface heat transfer coefficients may be determined using the results of Han et al. (2003a) for boiling heat transfer in brazed plate evaporators and Han et al. (2003b) for condensing heat transfer in brazed plate condensers.

In summary, for the evaporator, the flow boiling Nusselt number can be determined from Eq. [14.70] in which the constants, G_{E1}, G_{E2} and the equivalent boiling number, $\mathbf{Bo}_{eq}$ are determined as set out in Han et al. (2003a).

$$\mathbf{Nu}_{E,D_{eq}} = G_{E1} \mathbf{Re}_{eq}^{G_{E2}} \mathbf{Bo}_{eq}^{0.3} \mathbf{Pr}^{0.4} \qquad [14.70]$$

For the condenser, the condensing Nusselt number can be determined from Eq. [14.71] in which the constants, G_{C1} and G_{C2}, can be determined as described in Han et al. (2003b).

$$\mathbf{Nu}_{C,D_{eq}} = G_{C1} \mathbf{Re}_{eq}^{G_{C2}} \mathbf{Pr}^{0.4} \qquad [14.71]$$

Han et al. (2003a) also determined surface convection coefficients applicable to the source/sink fluid channels in the evaporator/condenser. The correlation of Eq. [14.72] was recommended.

$$\mathrm{Nu}_{\mathrm{f},D_{eq}} = 0.295\left(\frac{\pi}{2} - \beta\right)^{0.09} \mathrm{Re}^{0.64}\mathrm{Pr}^{0.32} \qquad [14.72]$$

Of course other heat transfer correlations could be substituted in place of Eqs [14.70]–[14.72] depending on the type of heat exchanger, heat exchanger geometry, refrigerant and other details. However, this should not fundamentally affect the structure of the modelling method described above for the evaporator or condenser. For a detailed review of heat transfer correlations for use in heat pump modelling, see Radermacher and Hwang (2005).

14.5.2.4 Expansion device and compressor

To complete the distributed parameter heat pump model, models of the expansion device and compressor will be needed. For the compressor, a detailed dynamic description is unnecessary because the key variables change extremely rapidly. Hence a steady-state compressor model is quite suitable. The model described in Section 14.2.3, for example, could be used.

For the expansion device, the capacity model given in Section 14.2.2 (ie, Eq. [14.11]) can be used. However, for dynamic modelling, the positioning mechanism of the expansion device needs to be considered and included. It may be assumed that this follows a first-order response the time constant of which will depend on the valve type (protracted for a conventional mechanical valve with a faster response for electro-mechanical devices). The model of Eq. [14.73] may be used (Underwood, 2015) in which the valve positioning time constant is τ (s) and the valve degree of opening is assumed to be dimensioned through an effective port diameter, D_{ed}.

$$\tau\frac{dD_{ed}}{dt} = K_{ed}\varepsilon_{sh} - D_{ed} \qquad [14.73]$$

Eq. [14.73] can be discretised to give Eq. [14.74].

$$D_{ed} = \left(\frac{\tau}{\Delta t + \tau}\right) \times \left(D_{ed}^{-} + K_{ed}\Delta t\varepsilon_{sh}\right) \qquad [14.74]$$

The mass flow rate of refrigerant can then be determined from Eq. [14.11] subject to a finite range of valve opening possibilities, $D_{ed,min}$, $D_{ed,max}$.

14.5.2.5 Application of the distributed heat pump model – an illustration

Algorithm:

A hybrid algorithm may be used in which the two heat exchangers (evaporator and condenser) are solved using a fully implicit algorithm, whereas the other system components are solved using an explicit time-forward-marching procedure. Of course, there are other options – quite possibly leading to economies in computational cost – but this method is practical and workable. After the setting of initial conditions, the following procedure may be adopted:

Step 1: Input parameters, read refrigerant property array* and boundary data.
Step 2: Define output variable arrays and set initial conditions.
Step 3: Open/update an outer simulation time loop.
Step 4: Update boundary conditions for this time row.
Step 5: Call expansion valve and compressor models.
Step 6: Call the evaporator model:
- Update properties and iterate by adjusting pressure until successive zone mass flow vectors converge.
- Iterate using Gauss–Seidel scheme until successive sets of energy vectors converge (refrigerant enthalpies, wall temperatures, water/fluid temperatures).

Step 7: Call the condenser model:
- Similar procedure as for the evaporator.

Step 8: Prepare and store interim reporting variables for this time row.
Step 9: Go to Step 3 (or exit if final time row).

Further details of an interpretation of this modelling method for a small domestic heat pump can be found in Underwood (2015) including a comparison of the model with experimental data.

*In this procedure, refrigerant property data are extracted at each calculation step as a function of the evaporating and condensing pressures. Though thermodynamic property correlations may be used (eg, Eq. [14.9] and similar correlations) an alternative procedure for dynamic modelling is to enter all property data for the given refrigerant as a matrix and interpolate individual property values from within the matrix as a function of prevailing pressure values. This method is computationally demanding due to high memory allocation but it offers a powerful advantage of robustness.

Application:
As an illustration, the distributed parameter model described in this section was applied to a small domestic-scale heat pump using a fixed speed scroll compressor controlled by a thermostat in the heating water connection. A sample of results is shown in Fig. 14.8. The refrigerant properties were evaluated using REFPROP v7.0 (NIST, 2015) for refrigerant R410A (see Fig. 14.9).

The results are for a dynamic simulation from an initial cold heating system state and run through to the first two thermostat switching events. The transients reveal the time taken for the heat pump to raise the system temperature towards the thermostat setting and, during this cold heating system state, the COP is very high but then declines towards its nominal value as the system warms. The degree of superheat and subcooling shows that the model is capable of capturing this detailed information and the GHE return temperature profile may then be used to investigate the impact of short-time dynamics on details such as borehole grout response and so forth. This type of modelling, therefore, provides a high reward in terms of detail and modelling utility but it comes at the cost of complexity and high computational cost. The solution in this illustration used a fixed integration interval of 0.2 s and required 2 h of (quad core) computer time for a 1-day simulation.

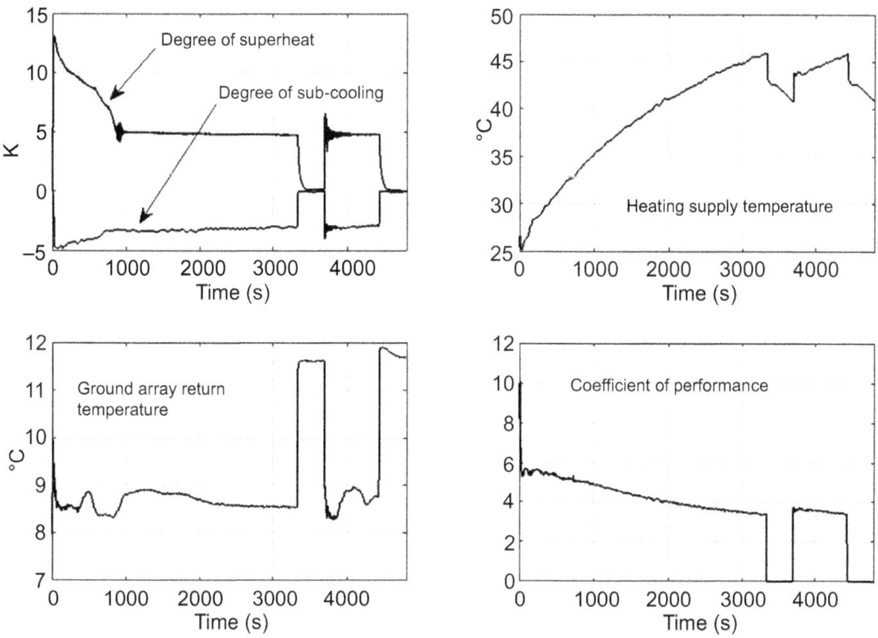

Figure 14.9 Sample of results from a distributed parameter heat pump simulation.

14.6 Conclusions

This chapter has given detailed consideration to the modelling of heat pumps with particular reference to ground source applications in which water or a water-rich solution will form the source fluid. The treatment has focused on three basic approaches to heat pump modelling – steady-state deterministic models, regression-fitted models based on experimental or manufacturers' data and fully dynamic modelling. In the past, considerable emphasis has been focused on GHE design and performance methods with the heat pump model expressed as a very simple model such as a constant effective COP. Increasing sophistication in system control, thermal storage and other balance-of-system components means that a more rigorous approach to heat pump modelling for both design and longer term performance evaluation is essential. Some of the particular themes within which heat pump research and modelling are likely to evolve in the coming years are summarised below:

- Heat pump working fluids – working towards zero global warming potential (GWP) fluids.
- Model validation and model-fitting opportunities arising from the current rise in system field monitoring, data logging and other data mining possibilities.
- A growing interest in absorption cycle and other directly fired heat pump systems.
- Advances in balance-of-system components such as phase change thermal storage.
- An increased use of variable speed heat pump compressor drives using electronic expansion devices.

- Advances in artificial intelligence–based control including fuzzy logic, artificial neural networks and adaptive neuro-fuzzy control.
- An increasing interest in community energy and building stock models particularly in applications dealing with electrical grid stress remediation.

The use of heat pump models in ground source applications forms an essential part of system design as well as performance evaluation over extended time horizons in order to address energy use, carbon emission and financial economics. This chapter has provided an insight into some of the approaches that might be used to address these and many other issues related to the development and application of heat pumps.

Symbols

Symbol	Meaning
A	Area (m^2)
$A\ldots I$; $a\ldots f$	Regression model-fitting constants (various dimensions)
AU	Area-integrated thermal transmittance value (W/K, kW/K)
$a_1\ldots a_3$; b_1, b_2	Compressor model-fitting constants (generally dimensionless)
C	Thermal capacity (J/K; kJ/K)
COP	Coefficient of performance (dimensionless)
D	Diameter (m)
E	Heat exchanger effectiveness (dimensionless)
EER	Energy efficiency ratio (dimensionless)
F	Zero function (W, kW)
f	A function of
G	Transfer function (various dimensions)
h	Enthalpy (J/kg, kJ/kg)
K	Gain (various dimensions)
L	Length (m)
M	Mass (kg)
m	Mass flow rate (kg/s)
N	Speed of rotation (RPS)
NTU	Number of heat exchange transfer units (dimensionless)
n	Polytropic index; modelling zone number (both dimensionless)
P	Pressure (N/m^2, kN/m^2, bar)

Symbol	Meaning
p	Part load ratio (dimensionless)
Q	Heat transfer rate (W, kW)
SCOP	Seasonal coefficient of performance (dimensionless)
SEER	Seasonal energy efficiency ratio (dimensionless)
SPF	Seasonal performance factor (dimensionless)
s	Laplace variable; slip ratio (dimensionless)
T	Temperature (°C, K)
t	Time (s, min)
U	Thermal transmittance value (W/m^2K, kW/m^2K)
V	Volume (m^3)
W	Compressor power (W, kW)
WAux	Auxiliary power consumption (W, kW)
X	Concentration (dimensionless or percent)
x	Vapour quality (dimensionless)
z	Distance (m)
Matrix-vector	
$\mathbf{F}$	Zero function vector (W, kW)
$\mathbf{J}$	Jacobian matrix (dimensionless elements)
$\mathbf{Q}$	Vector of heat transfer rates (W, kW)
$\mathbf{Q}^+$	Updated vector of heat transfer rates (W, kW)
Dimensionless groups	
Bo	Boiling number
Nu	Nusselt number
Pr	Prandtl number
Re	Reynolds number
Greek	
Δ	Difference
α	Void fraction (dimensionless)
β	Corrugation angle (degrees)
γ	Index of compression (dimensionless)
η	Efficiency (dimensionless)
ε	Error (K; various)

Continued

Symbol	Meaning
ρ	Density (kg/m^3)
τ	Time constant (s)
Subscripts	
a	Absorber
amb	Ambient
c	Condenser
cp	Compressor
cp,+aux	Compressor plus auxiliary
cs	Cross section
d	Delay
disp	Displacement
e	Evaporator
ed	Expansion device
f	Fluid (usually relates to evaporator source − water, brine, etc.)
g	Generator, gas
HP	Heat pump
i	Inlet
isen	At isentropic conditions
j	Zone interface index
l	Liquid
m	Material (usually heat exchange wall)
n	Modelling zone index
o	Outlet
p	Plant; perimeter
r	Refrigerant
s	Strong (solution)
ss	Strong solution stream
sw	Weak solution stream
sh	Superheat
shx	Solution heat exchanger
suc	Suction
sosi	Source-to-sink
w	Water; weak (solution)

Symbol	Meaning
Composite subscripts – examples	
fei	Fluid-evaporator-inlet
wco	Water-condenser-outlet
cpo	Compressor-outlet
Other	
–	When used as a superscript, refers to the previous time-row value
+	When used as a superscript, refers to an updated value

References

Aaron, D.A., Domanski, P.A., 1990. Experimental analysis and correlation of refrigerant 22 flow through short-tube restrictors. ASHRAE Transactions 96 (1), 729–742.

ANSI/AHRI, 2005. Standard 870 – Performance Rating of Direct GeoExchange Heat Pumps. Air Conditioning, Heating and Refrigeration Institute, Arlington.

ANSI/AHRI, 2012. Standard 551/591 – Performance Rating of Water-Chilling and Heat Pump Water—Heating Packages Using the Vapour Compression Cycle. Air Conditioning, Heating and Refrigeration Institute, Arlington.

ASHRAE, 2012. Handbook of Systems and Equipment. American Society of Heating, Refrigerating and Air-conditioning Engineers, Atlanta, p. 39.9.

ASHRAE, 2013a. Handbook of Fundamentals. American Society of Heating, Refrigerating and Air Conditioning Engineers, Atlanta, p. 30.71.

ASHRAE, 2013b. Handbook of Fundamentals. American Society of Heating, Refrigerating and Air Conditioning Engineers, Atlanta, p. 30.70.

Bourdouxhe, J.-P., Grodent, M., Lebrun, J., 1998. Reference Guide for Dynamic Models of HVAC Equipment. American Society of Heating, Refrigerating and Air-conditioning Engineers, Atlanta.

Browne, M.W., Bansal, P.K., 2002. Transient simulation of a vapour-compression packaged liquid chiller. International Journal of Refrigeration 25, 597–610.

BS EN 14511:2013, 2013. Air Conditioners, Liquid Chilling Packages and Heat Pumps with Electrically Driven Compressors for Space Heating and Cooling. British Standards Institute, London.

BS EN 14825:2013, 2013. Air Conditioners, Liquid Chilling Packages and Heat Pumps with Electrically Driven Compressors for Space Heating and Cooling – Testing and Rating at Part Load Conditions and Calculation of Seasonal Performance. British Standards Institute, London.

Chen, Y., Halm, N.P., Groll, E.A., Braun, J.E., 2002a. Mathematical modelling of scroll compressors – part I: compression process modelling. International Journal of Refrigeration 25, 731–750.

Chen, Y., Halm, N.P., Braun, J.E., Groll, E.A., 2002b. Mathematical modelling of scroll compressors – part II: overall scroll compressor modelling. International Journal of Refrigeration 25, 751–764.

Ciat, 2007. HydroCiat LW R134a – Model 2150-BX. Ciat Ltd, Culoz, France.
Cuevas, C., Lebrun, J., 2009. Testing and modelling of a variable speed scroll compressor. Applied Thermal Engineering 29, 469–478.
Duprez, M.-E., Dumont, E., Frère, 2007. Modelling of reciprocating and scroll compressors. International Journal of Refrigeration 30, 873–886.
Eames, I.W., Milazzo, A., Maidment, G.G., 2014. Modelling thermostatic expansion valves. International Journal of Refrigeration 38, 189–197.
Gupta, R., Irving, R., 2013. Development and application of a domestic heat pump model for estimating CO_2 emissions reductions from domestic space heating, hot water and potential cooling demand in the future. Energy and Buildings 60, 60–74.
Han, D.H., Lee, K.J., Kim, Y.H., 2003a. Experiments on the characteristics of evaporation of R401A in brazed plate heat exchangers with different configurations. Applied Thermal Engineering 23, 1209–1225.
Han, D.H., Lee, K.J., Kim, Y.H., 2003b. The characteristics of condensation in brazed pate heat exchangers with different chevron angles. Journal of the Korean Physical Society 43, 66–73.
Hepbasli, A., Erbay, Z., Icier, F., Colak, N., Hancioglu, E., 2009. A review of gas engine driven heat pumps (GEHPs) for residential and industrial applications. Renewable and Sustainable Energy Reviews 13, 85–99.
Hermes, C.J.L., Melo, C., 2008. A first-principles model for the start-up and cycling transients of household refrigerators. International Journal of Refrigeration 31, 1341–1357.
Jin, H., Spitler, J.D., 2002. A parameter estimation based model of water-to-water heat pumps for use in energy calculation programs. ASHRAE Transactions 108 (1), 3–17.
Kays, W.M., London, A.L., 1984. Compact Heat Exchangers, third ed. McGraw-Hill, New York.
Koury, R.N.N., Machado, L., Ismail, K.A.R., 2001. Numerical simulation of a variable speed refrigeration system. International Journal of Refrigeration 24, 192–200.
Li, W., 2013. Simplified steady-state modelling for variable speed compressor. Applied Thermal Engineering 50, 318–326.
Liu, J., Chen, J., Ye, Q., Chen, Z., 2007. A new model for depicting mass flow rate characteristic of electronic expansion valves. Experimental Thermal and Fluid Science 32, 214–219.
MacArthur, J.W., Grald, E.W., 1987. Prediction of cyclic heat pump performance with a fully distributed model and a comparison with experimental data. ASHRAE Transactions 93, 1159–1178.
Murphy, W.E., Goldschmidt, V.W., 1986. Cyclic characteristics of a typical residential air conditioner – modelling of start-up transients. ASHRAE Transactions 92 (1A), 427–444.
NIST, 2015. REFPROP – Reference Fluid Thermodynamic Ad Transport Properties, Version 7. National Institute of Standards and Technology, Gaithersburg.
Payne, W.V., O'Neal, D.L., 1999. Multiphase flow of refrigerant R410A through short tube orifices. ASHRAE Transactions 105 (2).
Park, Y.C., Kim, Y.C., Min, M.-K., 2001. Performance analysis on a multi-type inverter air conditioner. Energy Conversion and Management 42, 1607–1621.
Radermacher, R., Hwang, Y., 2005. Vapour Compression Heat Pumps with Refrigerant Mixtures (Chapter 8). Taylor and Francis, London.
Rice, C.K., 1987. The effect of void fraction correlation and heat flux assumption on refrigerant charge inventory predictions. ASHRAE Transactions 93, 341–361.
Sami, S.M., Dahmani, A., 1996. Numerical prediction of dynamic performance of vapour-compression heat pump using new HFC alternatives to HFC-R22. Applied Thermal Engineering 16, 691–705.

Sami, S.M., Duong, T.N., Mercadier, Y., Galanis, N., 1987. Prediction of the transient response of heat pumps. ASHRAE Transactions 93 (1), 471−490.

Tangwe, S., Simon, M., Meyer, E., 2014. Mathematical modelling and simulation application to visualise the performance of retrofit heat pump water heater under first hour heating rating. Renewable Energy 72, 203−211.

Underwood, C.P., 1998. HVAC Control Systems − Modelling, Analysis and Design. E & FN Spon, London, pp. 232−235.

Underwood, C.P., 2014. On the design and response of domestic ground-source heat pumps in the UK. Energies 7 (1), 4532−4553.

Underwood, C.P., 2015. Fuzzy multivariable control of domestic heat pumps. Applied Thermal Engineering 90, 959−969. http://dx.doi.org/10.1016/j.applthermaleng.2015.07.068.

Vargas, J.V.C., Parise, J.A.R., 1995. Simulation in transient regime of a heat pump with closed-loop and on-off control. International Journal of Refrigeration 18 (4), 235−243.

Venkataraman, P., 2009. Applied Optimisation with Matlab Programming, second ed. John Wiley & Sons, New Jersey, pp. 266−271.

Winandy, E., Saavedra, C., Lebrun, J., 2002. Experimental analysis and simplified modelling of a hermetic scroll refrigeration compressor. Applied Thermal Engineering 22, 107−120.

Wang, F.Q., Maidment, G.G., Missenden, J.F., Tozer, R.M., 2007. A novel distributed method for dynamic refrigeration system simulation. International Journal of Refrigeration 30, 887−903.

White, S.D., Yarrall, M.G., Cleland, D.J., Hedley, R.A., 2002. Modelling the performance of a transcritical CO_2 heat pump for high temperature heating. International Journal of Refrigeration 25, 479−486.

Zhang, C., Ma, S., Chen, J., Chen, Z., 2006. Experimental analysis of R22 and R407C flow through electronic expansion valve. Energy Conversion and Management 47, 529−544.

Zhao, Y., Haibo, Z., Zheng, F., 2007. Modelling and dynamic control simulation of unitary gas engine heat pump. Energy Conversion and Management 48, 3146−3153.

Geothermally activated building structures

15

L. Helsen
University of Leuven (KU Leuven), Heverlee, Belgium; EnergyVille, Genk, Belgium

Abstract

Geothermally activated building structures (GEOTABS) refers to the combination of geothermal heat pumps and thermally activated building systems: a sort of 'perfect marriage' provided that the global system design and control is part of a holistic approach. Within this global system approach the dynamic interaction between all components (ground, heating, cooling and ventilation system, building, users, control) is crucial, leading to a broad range of time scales to cope with, which in turn increases the importance of dynamic simulations, control system design as well as monitoring and commissioning. The building design needs to match the typical characteristics of GEOTABS systems and efficient control tries to satisfy multiple objectives: thermal comfort at minimal cost on the short term combined with sustainable use of the ground on the long term. Model predictive control (MPC) is capable of meeting the challenges induced. However, large-scale market implementation of MPC still needs advances in the automation of the MPC tool chain.

15.1 What are geothermally activated building structures (GEOTABS)?

15.1.1 GEOTABS concept

Heat pumps and energy storage systems are gaining prominence in the road to nearly zero-energy buildings (NZEB) and increasing share of renewable energy sources (RESs). Within the category of heat pumps, ground-source heat pumps (GSHP) are best of the class thanks to their high energetic performance, the potential for direct cooling and seasonal energy storage. Combined with thermally activated building systems (TABS), such as concrete core activation (CCA), they give rise to the geothermally activated building structures (GEOTABS) concept, an impressive implementation of energy efficiency and smart use of thermal energy storage (TES), provided that design and control of the global GEOTABS system is part of an integrated holistic approach. GEOTABS systems have an energy saving potential of 20−70% compared to conventional heating and cooling systems − typically gas boilers and air-conditioning systems (Ground-Reach, 2008; Helsen et al., 2008; Lund et al., 2004; Zogg, 2001). However, these savings potentials are rarely reached in practice, due to suboptimal

Advances in Ground-Source Heat Pump Systems. http://dx.doi.org/10.1016/B978-0-08-100311-4.00015-7
Copyright © 2016 Elsevier Ltd. All rights reserved.

system integration and inefficient control (Sourbron et al., 2009; Thermac, 2008; Tian and Love, 2009; Todtli et al., 2009).

GSHPs as well as TABS are state-of-the-art technologies. Their energy saving potential is a result of both the ground coupling (at the evaporator side) and the combination with low temperature heat emission systems and/or high temperature cold emission systems (at the condenser side), such as floor heating and TABS. This allows keeping the temperature difference between source and emission low, which in turn leads to high heat pump performance. Moreover, thanks to seasonal storage the cold stored in the ground during winter can be used for direct cooling during summer by use of a simple heat exchanger (HEx). Fig. 15.1 shows the three operating regimes: heating (by using a GSHP), direct cooling (by using a HEx) and active cooling (by using a chiller which can be the heat pump in reversed mode).

To guarantee long-term sustainable operation, attention should be paid to the long-term thermal balance of the ground. For small projects natural regeneration by conduction in the ground could be sufficient, but for larger projects the combination of (and preferably balance between) heat demand (ground heat extraction) and cold demand (ground heat injection) is crucial for efficient and sustainable operation (see Fig. 15.2). Depending on the ground properties (geological formations characterized by thermal and hydraulic properties such as thermal conductivity, capacity, temperature of undisturbed ground, moisture content, permeability, groundwater flow, etc.) heat dissipation will occur to a greater or lesser extent, influencing the degree of seasonal storage (using the ground as a battery) and the regeneration of the ground. When imbalance between heat injection and extraction exists, supplementary production systems (such as gas boilers and cooling towers) or injection of thermal energy from solar collectors or exhaust process heat can assist in keeping the annual thermal ground balance.

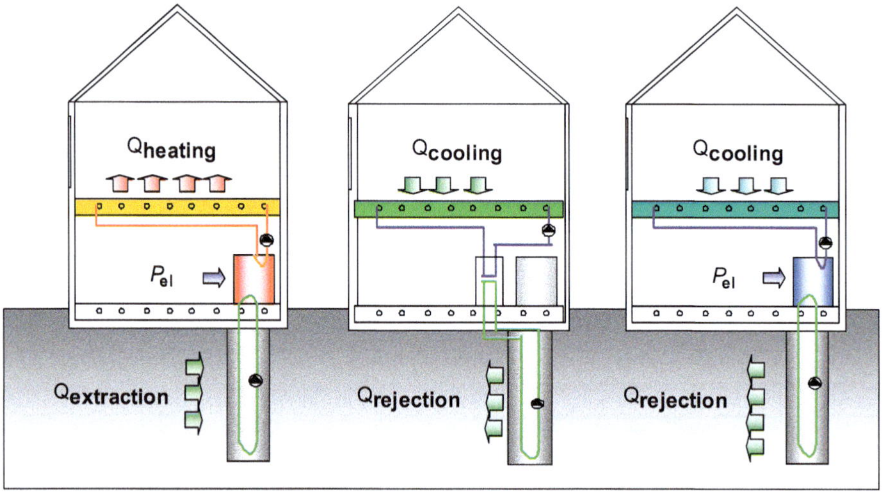

Figure 15.1 Geothermally activated building structure concepts for heating, direct cooling and active cooling (Verhelst, 2012).

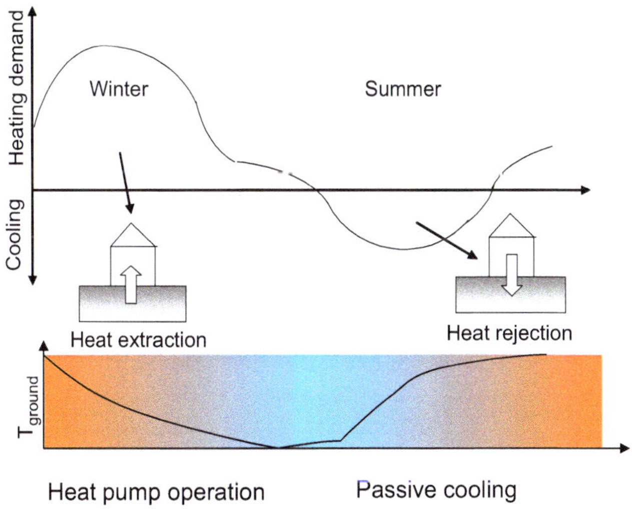

Figure 15.2 Thermal balance of the ground by heating and cooling with a ground-source heat pump (Verhelst, 2012).

15.1.2 Thermally activated building systems

Different types of TABS exist, among them CCA, which is the type often referred to in the context of GEOTABS. Concrete floors/ceilings of a building are activated by embedding pipes in the concrete and pumping warm/cold water through the tubes to achieve thermal comfort in the building (see Fig. 15.3). These large thermally activated surfaces allow high temperature cooling (>15°C) and low temperature heating (<35°C). Their ability to heat and cool at moderate temperatures, close to room temperature, makes them the perfect match for heat pump technology and results in a substantial degree of self-regulation (the concrete automatically starts cooling as soon as its temperature is lower than the room temperature and vice versa). On the other hand, small temperature differences also mean low heating and cooling heat transfer rates. Thanks to the large thermal capacity of the concrete, which represents useful thermal storage capacity, flattening the thermal load peaks is facilitated, resulting in lower installed thermal production capacities and thus lower investment costs. However, the large thermal inertia leads to large time constants (an order of magnitude of 5−10 h) and does not allow quick system reactions like changing room temperatures

Figure 15.3 Concrete core activation (CCA) systems being installed.
Adapted from Boydens Engineering.

as a response to sudden disturbances such as increasing internal heat gain. As a consequence, the room temperature in TABS buildings may drift during the day. According to ISO 7730 the differences in drifting temperatures in the building should not exceed $\pm 2°C$ during 1 h. Recent research shows that this limit can be increased to $\pm 4°C$ per hour within the comfort range (Toftum et al., 2008), but usually the temperatures in TABS buildings fluctuate in a range of $\pm 0.5-1°C$ per hour (Olesen, 2012).

CCA and floor heating/cooling differ. Unlike floor heating/cooling, where the embedded pipes are separated from the building structure by a layer of insulation, within CCA the pipes are embedded in the building structure. Thereby, the thermal mass of the building (which in modern buildings is mainly located in the floor) is actively heated or cooled. It is crucial that for radiant cooling systems in general, surface temperatures or even pipe temperatures below the indoor air dew-point are avoided (a particular concern in humid climates). To ensure good thermal conditions in the building and energy efficient operation, special control strategies have to be implemented (Sourbron et al., 2009; Todtli et al., 2009).

15.2 Geothermally activated building structures as a global system concept

The optimal GEOTABS concept requires careful incorporation in the overall building design, adaptation in a building's operation and integration with other systems. Successful application is constrained by specific boundary conditions: technical and constructional aspects, systems operation as well as user comfort and limited individual control (REHVA, 2013). Table 15.1 lists the internal and external constraints that typically need to be considered in a feasibility study.

Table 15.1 Internal and external boundary conditions to be accounted for in a feasibility study

Internal constraints	External constraints
• Functional specification • Level of indoor environmental quality (eg, desired comfort level, individual control possibilities) • Level of sustainability and energy performance • Aesthetics (eg, percentage of window area) • Flexibility and modularity for future use (eg, room division and function alteration) • Financial budget	• Climate and geography • Geology and hydrogeology (presence and characteristics of aquifer, soil and rock character ization, and presence and movement of moisture) • Ground availability for ground heat exchanger and building ground floor area • Environmental and legal issues (permits and limitations, fees, energy performance, share of renewable energy sources, ground temperature, borehole depth, brine type) • Technical aspects (eg, drilling possibilities, possibilities of geothermally activated building structure technologies) • Economical aspects

Based on REHVA, 2013. Guidebook No 20, Advanced System Design and Operation of GEOTABS Buildings, ISBN 978-2-930521-12-1.

Table 15.2 Strength, weakness, opportunity and threat (SWOT) analysis for geothermally activated building structures

Strength	Weakness
1. Exploiting environmental energy 2. Seasonal storage, increasing the energy per formance for applications with both heating and cooling load 3. Water supply temperatures (<35°C for heating, >15°C for cooling) 4. Thermally activated building system (TABS) thermal storage capacity (load shifting, peak shaving) 5. Increased thermal comfort by radiation heat exchange 6. Reduced building height compared to build ings with air-conditioning 7. No visible installations in the room, no chimney	1. Limited average thermal power output of TABS (40–50 W/m^2) ⇒ allow temperature drifts within the specified comfort range or provide an additional fast reacting system 2. Limited individual room control 3. Not useful to change set-point during a day 4. Acoustic comfort – limited sus pended ceiling 5. Drilling in the ceilings (and floors) should be carried out with care
Opportunities	**Threats**
1. Suitable for low-energy, passive or zero-energy buildings 2. Increased share of efficiently generated heat and cold 3. Increased share of a renewable energy source 4. Less energy and more energy efficient form of cooling than air-conditioning 5. Thermal energy storage – increase in flexi bility of energy system – decouples heat/cold production and heat/cold emission	1. Careful system integration and con trol design is required a. limited thermal power b. ground thermal balance c. integration of all components 2. Additional fast reacting system must not overrule or counteract the TABS operation

Based on REHVA, 2013. Guidebook No 20, Advanced System Design and Operation of GEOTABS Buildings, ISBN 978-2-930521-12-1; Sourbron M., September 2012. Dynamic Thermal Behaviour of Buildings with Concrete Core Activation, (Ph.D. text). KU Leuven, Leuven (Belgium), ISBN 978-94-6018-572-4; Verhelst C., April 2012. Model Predictive Control of Ground Coupled Heat Pump Systems for Office Buildings, (Ph.D. text), KU Leuven, Leuven (Belgium), ISBN 978-94-6018-497-0.

The GEOTABS concept creates benefits but also faces limitations. The strength, weakness, opportunity and threat (SWOT) analysis, presented in Table 15.2, can be a useful instrument to evaluate the feasibility for the implementation of GEOTABS in specific buildings at defined locations (REHVA, 2013; Sourbron, 2012; Verhelst, 2012).

15.3 Building design as a crucial part in the geothermally activated building structure concept

The suitability of GEOTABS for a specific building varies with the building function, location and design. Due to the structural character of TABS, implementation is

Table 15.3 **Main building characteristics impacting the suitability of geothermally activated building structures**

Façade	High performance envelope • high insulation quality (low U-values) • effective solar shading and/or reduced percentage of glazing (less daylight) • proper building orientation
Appliances	Energy efficient devices
Lighting	Energy efficient lighting systems Demand control of artificial lighting Thermally activated building system (TABS) heat transfer area not covered
Finishing materials	TABS heat transfer area not covered (ceiling and floor not covered with insulating materials)
Acoustics	Integration of acoustic material TABS heat transfer area not covered, attach to walls and furniture, use vertical baffles

Based on REHVA, 2013. Guidebook No 20, Advanced System Design and Operation of GEOTABS Buildings, ISBN 978-2-930521-12-1.

mainly in new buildings and deep retrofit of existing buildings. Only buildings with solid construction, characterized by high thermal storage capacity are suitable for effective implementation of TABS. Moreover, GEOTABS are mainly used in non-domestic buildings where cooling is relatively dominant (high energetic performance of direct cooling is required) and where the usage pattern is rather uniform and constant. The building design should aim at minimizing heating and cooling energy demand by reducing heat transfer between the building interior and the external environment (transmission and ventilation losses and external gains). Also the role of internal gains (amount and variability) in the performance of TABS is substantial (REHVA, 2013). Table 15.3 lists the main building characteristics that impact the suitability of GEOTABS.

15.4 System integration

A global system approach is of utmost importance in all building phases, from design to operation, taking all interactions between ground, building, heating/cooling, ventilation and air-conditioning (HVAC) systems, control strategies and users into account in all switching and operation modes. The high thermal inertia of the TABS means that the building, ground and the system rarely reaches steady-state operation. The transient behaviour thus needs to be considered in sizing and energy performance calculation, which makes dynamic simulations a very important tool. Building simulation software tools, such as TRNSYS (2010), EnergyPlus (2012), ESP-r (2015) or Modelica (2015), are suited for this purpose. Within these simulation tools the complex physical reality is modelled by a computable set of equations, thereby making some simplifications (balancing accuracy and calculation time), the most important being the single-node thermal building zones (neglecting temperature stratification) and one-dimensional heat transfer in building components. Output of these dynamic simulations usually includes: thermal comfort parameters, energy use, peak power, daily average power, reaction time of components to variations in boundary conditions and ground temperature for a chosen borehole array configuration.

In the early design phase, benchmarks are used to (roughly) estimate the heat extraction and injection rate to the ground, as presented in Table 15.4. However, the actual ground properties depend on different conditions and can vary greatly from values given in the literature. Therefore, for large projects it is recommended to perform a thermal response test (TRT) to determine ground characteristics, needed for detailed sizing (see previous chapters), which is usually based on multiple-year simulations for thermal response of different configurations, taking interference between boreholes and thermal ground balance into account. Decoupled simulations (eg, in Earth Energy Design (EED) (Hellström et al., 2010); Programm zur Dimensionierung von Erdwärmesonder (EWS) (Huber, 2015); COMSOL, 2015; FEFLOW, 2015) use demand profiles or injection/extraction values (average, peak or time-dependent) as input for ground simulations, taking the risk that sizing problems might not be detected. Coupled simulations (eg, TRNSYS, 2010; ESP-r, 2015; EnergyPlus, 2012; Modelica, 2015) simultaneously simulate ground, building, HVAC system, control and users and thus need detailed information about the building and building services. These simulations, which require smaller time steps, are crucial for unbalanced injection and extraction and for undersized borehole arrays (to evaluate system performance). However, due to the high number of unknown parameters it is difficult to obtain a solid model in the design phase.

The limited thermal power of the GSHP (taking investment cost into consideration) and the need for appropriate control design of GEOTABS (due to large thermal inertia)

Table 15.4 Benchmarks for heat extraction and injection power to the ground (REHVA, 2013)

Ground system	Values
Well	5–6 kW/(m^3/h)
Borehole heat exchanger	30–100 W/m
Energy piles	10–80 W/m

may drive system designers to provide an additional production and/or emission (fast reacting) system, resulting in hybrid systems. Supplementary production systems play an important role in the annual thermal ground balance. The interaction between all components within the system should be handled with care. For hybrid heat pump systems different operation modes can be chosen (REHVA, 2013):

1. Bivalent-alternative: GSHP has first priority, when higher power needed, then other source only
2. Bivalent-parallel: GSHP has first priority, when higher power needed, then both sources
3. Bivalent-partially parallel: GSHP has first priority, when higher power needed, then both sources till limit, then the GSHP switches off
4. Monoenergetic: GSHP is very few hours per year assisted by resistance heating

Minimum and maximum antifreeze/water flow rate constraints of the heat pump should be investigated and if possible optimized. A variable antifreeze/water flow rate can be considered as a function of active compressor capacity step and maintaining minimum temperature difference over the borehole array (REHVA, 2013).

Moreover the TABS should deliver the main part of the thermal power, while the supplementary emission system is designed to compensate for thermal peaks, unexpected loads and individual room control. At moderate ambient air temperatures, when day-to-day changes between heating and cooling might still be frequent, excessive heating or cooling can be avoided by only using faster systems and to activate the TABS when the building is constantly in heating or cooling mode.

A well thought out hydraulic concept sets the basis on which advanced control strategies can be built to guarantee energy efficient and cost-effective plant operation (REHVA, 2013). Ideally, optimal design and control are developed in an integrated way, since design impacts control and vice versa. However, appropriate frameworks for integrated optimal design and control do not yet exist. Each change in design may impact the hydraulic concept, making this integration very challenging. Fig. 15.4 shows the basic hydraulic scheme for GEOTABS, where the storage tank is added to avoid heat pump cycling (frequent switching).

In hot climates an additional cooling source, usually a cooling tower (in parallel or in series to the borehole array), is added to the heat pump. Fig. 15.5 shows the corresponding hydraulic scheme for cooling tower mode operation.

Similarly, in moderate or cold climates an additional heating system, such as district heating or gas boilers, might be integrated. Fig. 15.6 shows the corresponding hydraulic scheme with connection to a thermal network. One may question the use of a heat pump when a thermal network is available. It is needed, though, to cool the ground in winter such that direct cooling (at very high energetic performance) can be applied in summer. Efficient control strategies are crucial to supervise and select the right operation mode. Otherwise supplementary systems might run excessively, which is usually not detected in winter (since there are no comfort complaints), but will lead to insufficient cooling power and possibly serious derogation of thermal comfort in summer for which the other building services cannot compensate.

The development of efficient control strategies requires knowledge of the dynamic system behaviour (Sourbron, 2012). GEOTABS systems are characterized by a broad

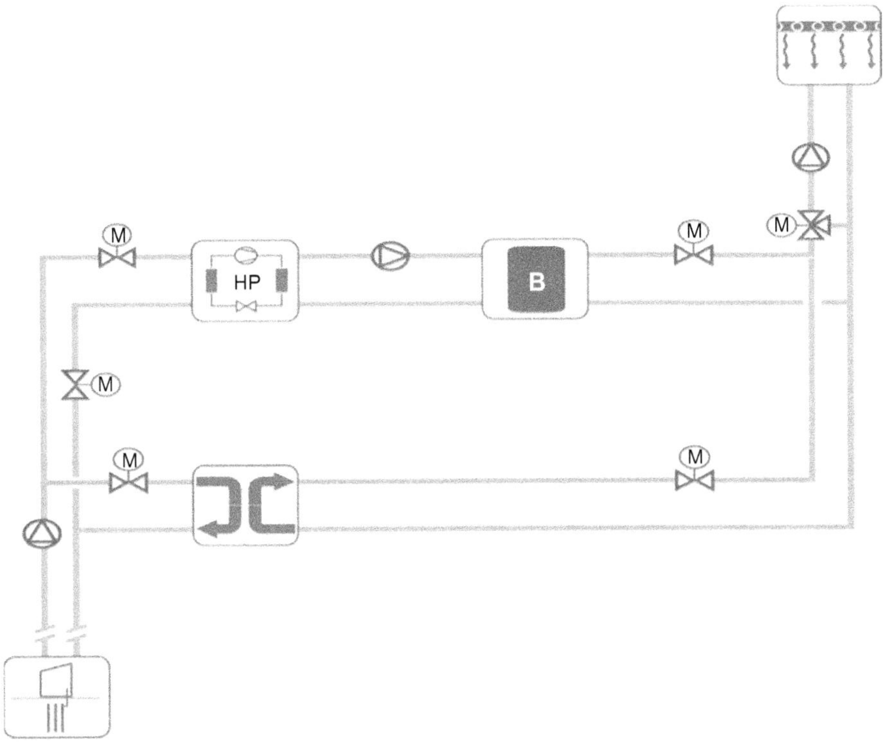

Figure 15.4 The basic geothermally activated building structure hydraulic scheme (REHVA, 2013).

range of time constants: small values for disturbances such as internal gains (by occupancy, electrical appliances, lighting) and solar gains, up to 10 h for TABS, and even multiple years for the thermo-hydraulic processes in the ground. The complex interaction of processes and broad range of time constants ask for a rethinking of control strategies (Armstrong et al., 2006; Tian and Love, 2009). Several studies have shown that the use of conventional control strategies results in suboptimal system operation due to the mismatch between the time constants of the disturbances and the TABS reaction time (Sourbron et al., 2009; Thermac, 2008; Tian and Love, 2009). TABS should guarantee thermal comfort in the building, thereby incorporating the knowledge that they have a limited steady-state thermal power and a large time constant. Therefore, typical on–off control based on feedback of zone temperature will inevitably lag behind. TABS should be controlled according to the outside temperature (heating/cooling curve (HC/CC)) but with a much smaller gradient of the curve compared to faster radiant systems. Consequently the water temperature must be controlled within a very limited temperature range. Instead the time and operation period are more important. No switching between heating and cooling is allowed within 24 h since that would lead to a waste of energy not being used to condition the room. Supply temperature and mass flow control is realized by three-way valves and speed controlled pumps.

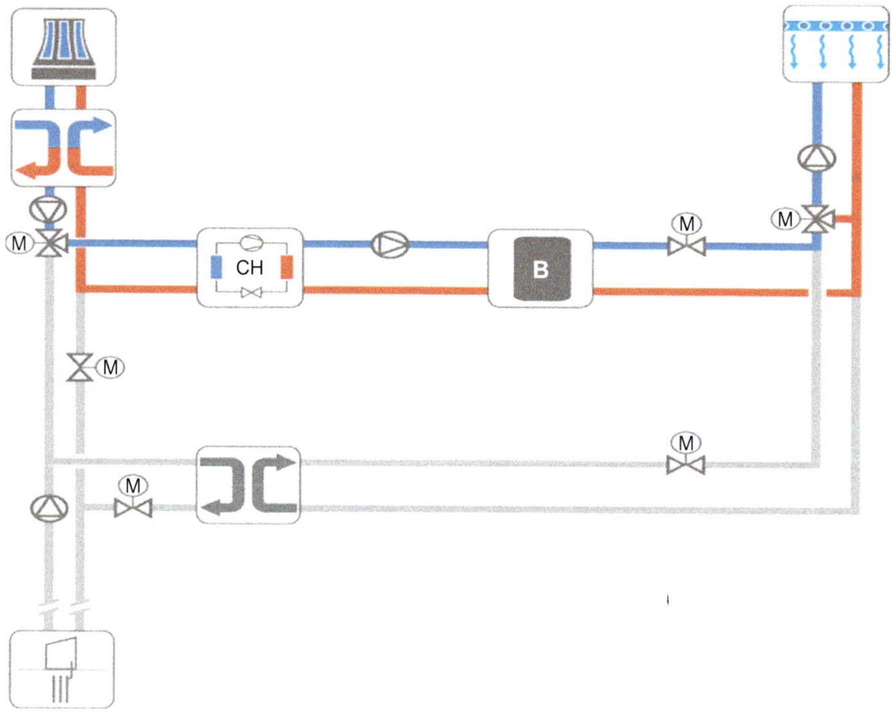

Figure 15.5 Geothermally activated building structure hydraulic scheme with additional cooling tower (REHVA, 2013).

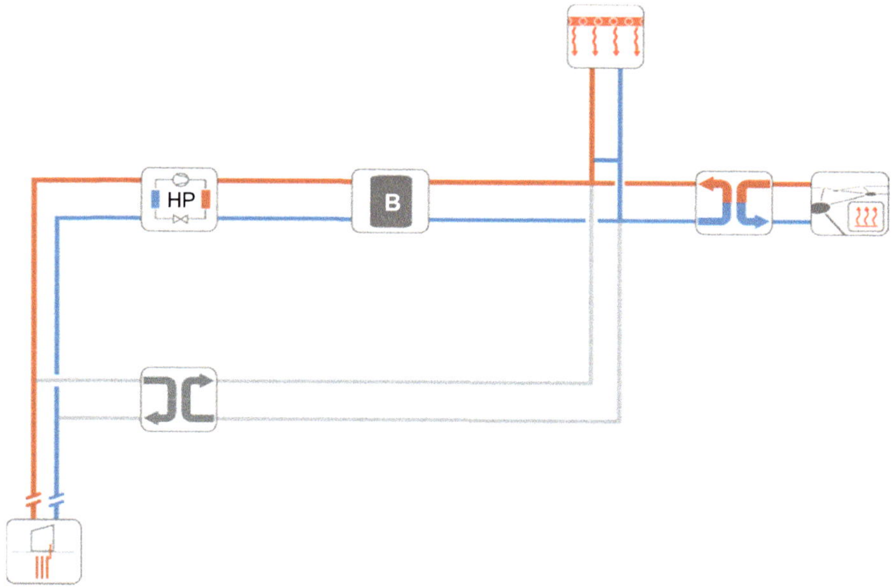

Figure 15.6 GEOTABS hydraulic scheme with integrated district heating (REHVA, 2013).

A number of different control strategies have been investigated (REHVA, 2013):

1. night time operation (based on night tariff for electricity) – results in lower cost, however also higher thermal discomfort;
2. weather dependent supply/average water temperature – supply/average water temperature as a function of (a 24–48 h average) outdoor temperature (HC/CC) leads to good thermal comfort and energy use (Olesen et al., 2002);
3. intermittent pump operation combined with HC/CC – leads to further reduction of energy use while maintaining a similar level of thermal comfort (Burkard, 2004);
4. constant concrete core control (C^4) – 'nearly' optimal control can be translated into keeping the concrete core temperature at $T_{comfort,min} + 1°C$ during the whole year (to be tuned depending on zone heat gains) (Sourbron and Helsen, 2014) or keeping the concrete temperature on a constant level of 22°C for low heating or cooling loads (Olesen and Liedelt, 2000).

At plant level, the main question is how to size the heat pump system (and the supplementary system) to guarantee that the heating and cooling demand can always be met, without oversizing the system. Oversizing the heat pump system should by all means be avoided, as the cost related to borehole array drilling constitutes the main part of the investment cost (energy piles can also be used but not horizontal ground heat exchangers). To improve the economic feasibility of heat pump systems, design guidelines suggest sizing the borehole heat exchanger (BHE) array to cover only the smallest of both loads, which results in smaller and thus cheaper borehole arrays. Current control strategies, however, assume that the BHE array is large and can be continuously operated. If the BHE temperatures exceed the lower or upper temperature bounds, the supplementary system is used. While this is a sound operation in case of sparse borehole arrays with limited thermal interaction between boreholes, for compact borehole arrays (characterized by larger thermal interaction between boreholes) this may not be the case. Such a strategy could result in borehole array thermal depletion or thermal build-up. The operation should guarantee a long-term sustainable operation with the annual building heating and cooling demand being delivered at the lowest cost (De Ridder et al., 2011; Vanhoudt et al., 2010). This requires an optimal use of the limited resources, ie, of the amount of heat and cold stored in the borehole array. The decision on when to switch between the GSHP system and the supplementary system should thus depend on the available amount of heat and cold in the borehole array, the future heating and cooling load, the efficiency of the heat and cold production devices and the electricity price profiles (Verhelst, 2012).

Although GEOTABS is a 'simple' concept, the integrated system requires a high quality level in design, construction and operation. Many GEOTABS buildings have been constructed over the last decades; some examples can be found in REHVA (2013) and on the GEOTABS European research project Website (www.geotabs.eu/Database) and show a diversity of implementations. The variety in location, building type and geometry, energy supply systems and energy demand for cooling and heating gives a sample of operational experiences in real cases with different user profiles. As a general conclusion we would like to emphasize the need for dynamic simulations for optimal design, intelligent control strategies for optimal operation and monitoring and commissioning to avoid malfunctions. Research is progressing and continuously moving further towards optimal design and control of GEOTABS buildings.

15.5 Advances in (optimal) control of geothermally activated building structures

Some work has shown that there is a need for further development of control strategies for GEOTABS buildings in general, and Hybrid GEOTABS (Hy-GEOTABS) systems in particular, to enable guaranteed thermal comfort at the lowest cost. To this end, the TES capacity of the TABS at building side and the seasonal energy storage capacity of the borehole array at source side should be optimally exploited. Conventional control strategies are not suited for this purpose. They are based on static building models. The time delay of the system response is accounted for by heuristic rules, requiring a lot of trial-and-error to tune the control parameters. Model predictive control (MPC), however, optimizes the control variables online. For each control time step the control variables are selected, which minimize a given cost function and take into account the system dynamics (through a controller model) and constraints. In the case of GEOTABS buildings, the cost function could be a weighted sum of energy cost and thermal discomfort (cost). The controller model could include both the dynamics of the building and the dynamics of the borehole array (Verhelst, 2012). A schematic presentation of the MPC framework is shown in Fig. 15.7. The information flow is as follows:

1. the current state of the system is predicted based on the measurement feedback of the system (state predictor);
2. the future system behaviour over a certain time horizon is predicted by using a system model, based on the current state, predictions of future disturbances and an operation profile (future inputs);
3. this operation profile is iteratively optimized (optimizer) such that the predicted system behaviour does not violate any of the constraints and the cost function is minimized;
4. the first value (optimal input for the next time step) of the optimal operation profile (future inputs) is applied.

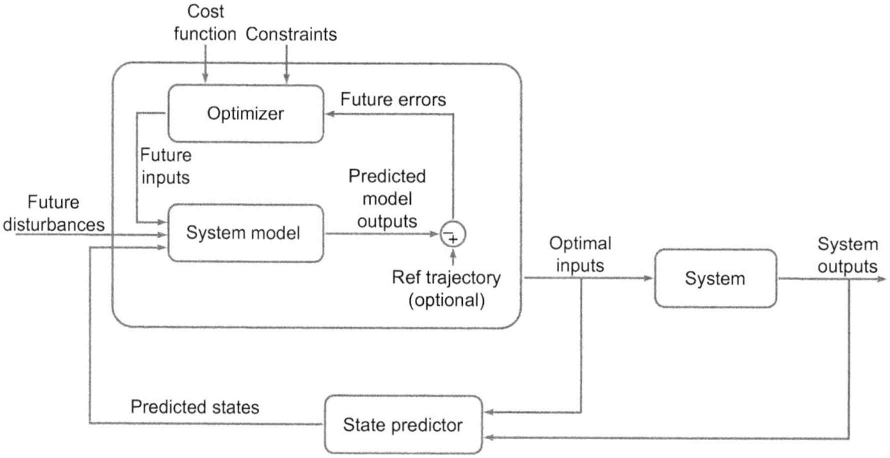

Figure 15.7 Schematic presentation of the model predictive control framework (REHVA, 2013).

When using a receding horizon the entire optimization procedure (1−4) is repeated the following control time step.

MPC is already well-established in the chemical industry, where it has originally been developed. Meanwhile, MPC has been proven successful in other areas such as aerospace, automotive and power systems (Qin and Badgwell, 2003). The basis of MPC is the solution of an optimal control problem (OCP) at each time step, using updated system information obtained by monitoring, as shown in Fig. 15.7. Whereas MPC popped up in the 1970s, optimal control theory was already developed and implemented in the 1950s for use in the space industry. In these early years, the optimal solution was found analytically, requiring substantial simplification of the system description and boundary conditions. With the introduction of computers, numerical optimization solvers have been developed. Most of them are gradient-based methods: the minimum is found iteratively by moving into a descending direction. The most well-known methods are the steepest-gradient and the Newton-based methods, which are very successful for solving convex optimization problems (Nocedal and Wright, 2000). For problems where the gradients are difficult to derive, direct optimization methods are used as well (Kolda et al., 2003). Some examples of direct optimization problems are particle swarm optimization (Kennedy an Eberhart, 1995), Nelder−Mead simplex (Nelder and Mead, 1965) and genetic algorithms (Goldberg, 1989) and these are available in commercial software.

MPC allows explicitly incorporating control objectives (minimal energy use or cost, optimal thermal comfort), constraints on the control inputs and/or state variables and system dynamics. An OCP is repeatedly solved in an online framework, as such combining the benefits of feedforward and feedback control. At each control time step, the control profile for the next control time steps is optimized using knowledge of the current state (feedback), the building dynamics and future disturbance predictions (feedforward). Only the first control time step is applied, and after this time step the optimization process is repeated, using updated system information and disturbance predictions. The combination of feedforward and feedback results in a good control performance even in the presence of model mismatch and prediction errors (Maciejowski, 2002). This is an important asset for practical implementation, as it allows the use of simple controller models and simple disturbance prediction methods.

The development of MPC for GEOTABS systems faces some challenges. First, a dynamic system model must be developed, both for the building and for the borehole array. A lot of simulation environments exist to model the building and/or the borehole array in detail, based on first principles (eg, Crawley et al., 2000; TRNSYS 17, 2010). Those are, however, far too complex to be incorporated in the OCP formulation. One needs to develop simple dynamic models for control. The challenge in system identification of a building lies in the large amount of unmeasured disturbances acting on the system (solar gains, internal gains, ventilation losses) and the small number of sensors and limited room for applying good excitation signals when the building is occupied. The challenge in system identification of a borehole array lies in the very broad dynamic range, with time constants ranging from hours to multiple years. The broad range in time constants in the system constitutes also a challenge for the optimal control part. It is computationally impossible to guarantee thermal comfort with an hourly time scale while optimizing the use of seasonal energy storage with a time scale of several years. How to formulate the OCP in order to take both the short-term and the long-term control objectives into account, constitutes a second challenge. Third, due to the dependency of the heat pump coefficient of performance (COP) on the temperatures of the heat source and the heat sink, the optimization problem becomes nonlinear. This temperature dependency of the COP is often neglected to make the optimization problem convex. The performance loss resulting from this simplification is usually not assessed.

MPC applied to hy-GEOTABS systems has the potential to (Verhelst, 2012):

1. solve the controllability problems of TABS related to its thermal inertia by incorporating a dynamic building model;
2. minimize the energy costs by considering the efficiency of the different heat and cold production systems and the time dependency of the electricity price;
3. guarantee long-term sustainable operation of the borehole array by incorporating the long-term thermal balance condition in the optimization;
4. decrease the investment cost by exploiting the thermal capacity of TABS for load shifting and peak shaving to reduce the required installation size.

15.6 Model predictive control of hybrid GEOTABS systems: a simulation study

The entire concept of Hy-GEOTABS systems is based on sizing the borehole array to be as compact as possible, thereby reducing the investment cost. To maximize the use of the borehole array the operation will hit the upper and/or lower bounds set on the borehole array and/or antifreeze/water fluid temperatures. The fact that optimal control allows operating the system near its boundaries is identified as a crucial asset compared to conventional (rule-based) controllers. Fig. 15.8 shows the main parts of a Hy-GEOTABS system.

Intelligent control should aim for three objectives:

1. thermal comfort at building level, with slowly reacting TABS
2. minimal energy use/cost by switching between heat/cold production systems (GSHP and gas boiler for heating, passive cooler and active chiller for cooling)
3. efficient use of small borehole array, taking the long-term thermal ground balance into account

Besides these objectives, MPC also incorporates system dynamics, disturbance predictions and technical constraints, which results in the general mathematical formulation of the OCP as presented in Fig. 15.9. The main challenge is the development of controller models for borehole array, production units and building. For detailed information about controller model development the reader is referred to (Verhelst, 2012; Sourbron, 2012; De Coninck, 2015; Váňa et al., 2014).

Fig. 15.10 illustrates the simulation-based operation of a GEOTABS office building controlled with a conventional rule based controller (RBC), top, and with MPC,

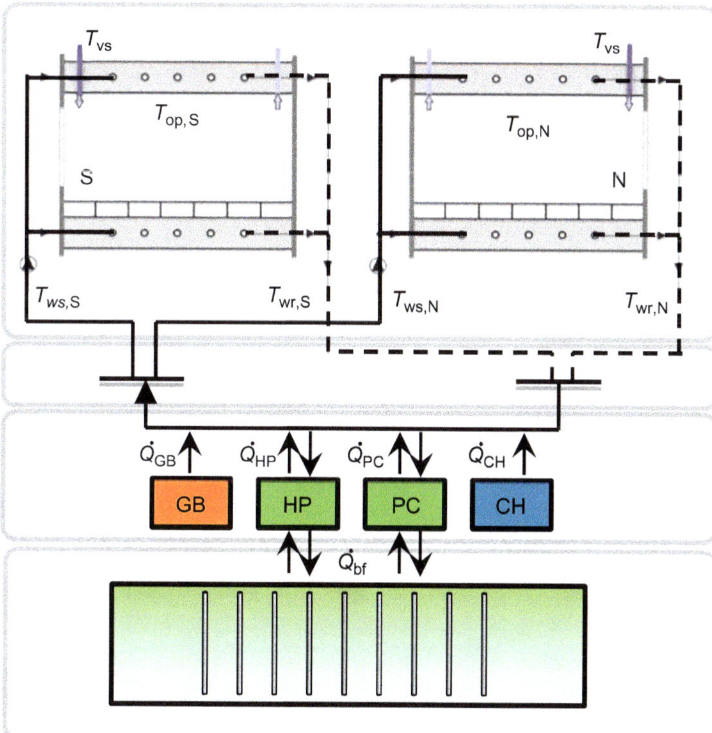

Figure 15.8 Schematic presentation of a Hy-GEOTABS system with a two-zone (south and north orientation) office building, heat and cold distribution, heat and cold production (*GB*, gas boiler; *HP*, heat pump; *PC*, passive cooler; *CH*, active chiller) and a borehole array (Verhelst, 2012).

$$\min_{u(t)} J = \int_0^{t_{end}} L\,(u(t), x(t), t)\,dt + M\,(u(t_{end}), x(t_{end}), t_{end})$$
$$\dot{x} = f\,(x, u, t)$$
$$c_{path}(x(t), u(t)) \leq 0$$
$$c_{boundary}\,(x(t_0), x(t_{end})) = 0$$

Figure 15.9 Mathematical MPC formulation with objective function, system dynamics, disturbance predictions and technical constraints, with u the inputs, x the states and t the time (Verhelst, 2012).

bottom, for a typical week in autumn (Verhelst, 2012). The RBC is a well-tuned HC/CC control strategy. Fig. 15.10 shows that both RBC and MPC satisfy the thermal comfort requirements since the operative temperatures remain within the thermal comfort bound during occupancy hours. It is clear, however, that the underlying operation of the heating and cooling devices significantly differs. In autumn the RBC strategy gives rise to alternating heating and cooling of the TABS, which is

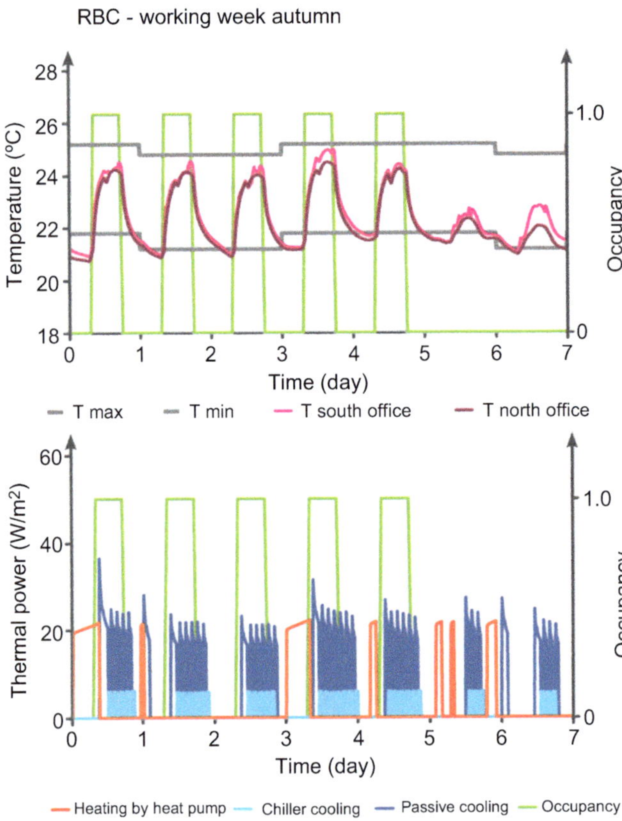

Figure 15.10 Comparison between rule based controller (RBC) (top) and model predictive control (MPC) (bottom) in a geothermally activated building structure office building for an autumn week; within each subfigure the operative temperature in the south and north office at the top and the thermal power of the production units at the bottom.

detrimental for the energy performance. In addition, there is cycling of passive cooling and a substantial use of the supplementary chiller. MPC, on the contrary, delivers the right amount of cold at the right time and a rate that can most of the time be covered by passive cooling only.

The benefits of MPC over RBC can be summarized as: (1) good anticipation by better estimation of building heating and cooling loads of next hours and days; (2) optimal use of TABS thermal capacity to minimize the use of supplementary heating and cooling devices by peak shaving and to benefit from low-price electricity by load shifting; and (3) improved control performance characterized by similar or better thermal comfort and lower energy cost. Energy cost savings (MPC versus RBC) depend on the installation size and range from 26% for large borehole arrays to 40% for small borehole arrays, as presented in Fig. 15.11 by the simulation-based performance clouds for both RBC and MPC strategies.

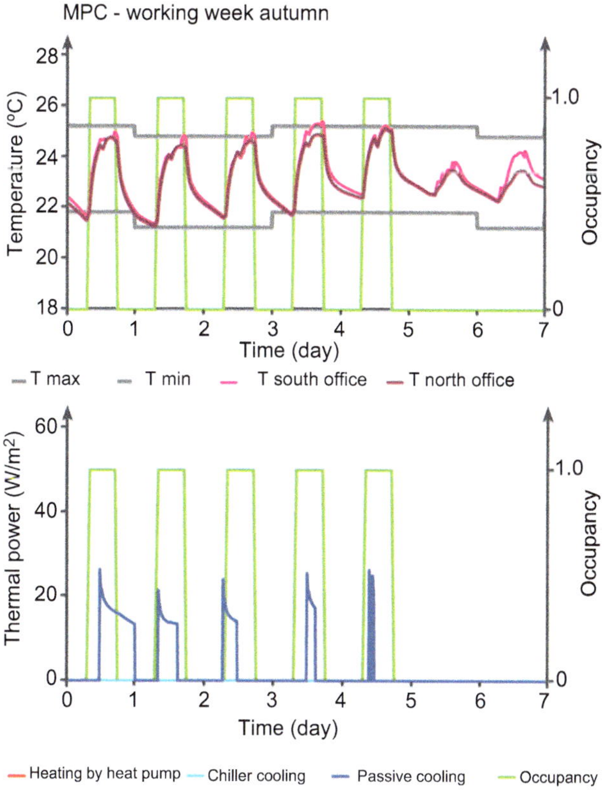

Figure 15.10 Continued.

The performance cloud in Fig. 15.11 shows that the more compact the installation (DC3), the higher the RBC performance sensitivity on control parameters and the larger the deviation from the OCP. The smaller borehole arrays (DC3) were not always able to reach the thermal comfort limit (expressed in degree-hours) of 100 Kh discomfort in the case of RBC. The RBC strategy must be fine-tuned for each individual case, not only as a function of the building characteristics but also as a function of the installation size. MPC, on the contrary, automatically adjusts the control action as a function of the installed capacities, and allows the system operating near its boundaries, without trespassing them. Small borehole arrays are interesting from an economic point of view (lower investment cost) and exactly for these systems MPC shows a large potential to increase control performance.

15.7 Model predictive control of geothermally activated building structure offices: a case study

Within the European FP7-EraSME GEOTABS project MPC was developed and implemented in a real office building, Hollandsch Huys in Hasselt (Belgium)

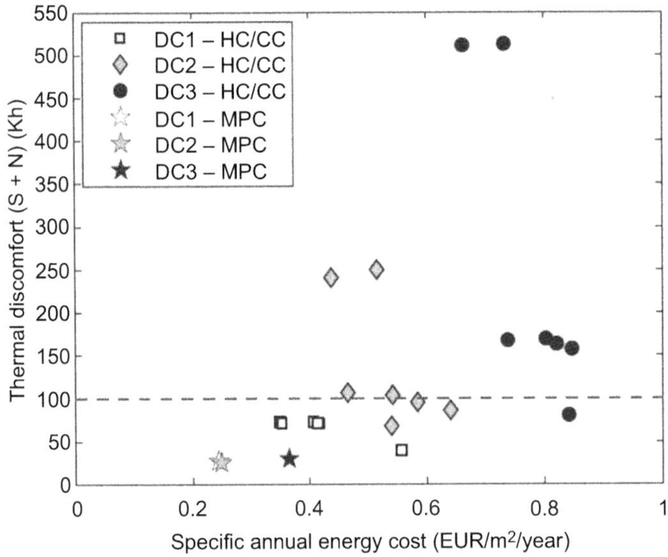

Figure 15.11 Performance cloud of the rule based controller (RBC) strategies (using heating/cooling curve) for the different hybrid geothermally activated building structure design cases (DC1 (large borehole array – 24 borehole heat exchangers (BHEs)), DC2 (medium borehole array – 15 BHEs) and DC3 (small borehole array – 7 BHEs)) obtained by varying the RBC settings. The performance of the model predictive control strategy for each design case is added as a reference. The thermal comfort limit of 100 Kh is indicated by the dashed line (Verhelst, 2012).

(Váňa et al., 2014). More information about the building, the energy system and the hydraulic scheme can be found on the GEOTABS Website (http://www.geotabs.eu/Database/Buildingflyer_HollandschHuys_120921.pdf). Design and application of advanced control in buildings requires an interface between numerical tools and the building management system (BMS). Fig. 15.12 presents the communication interfaces used in this case study. The building, located in Hasselt (Belgium), is equipped with a Priva BMS. Numerical tools within the Matlab environment are used, and the RcWare client is considered for process data acquisition (resulting in an RcWare SoftPLC connected to the Priva BMS in Hasselt and an RcWare Database and RcWare weather forecast server connected to the MPC server in Prague). Weather forecast data are obtained from NOAA Agency (United States) that provides updates each 4 h. These were not corrected by local measurements.

Details about the grey-box state-space controller building model (including model structure design and system identification) and the control concept (including high-level control, low-level control and optimization problem formulation) can be found in Váňa et al. (2014). The grey-box state-space approach provides a good multistep ahead predictor and matches the physical structure of the real system well. A two-level control scheme is proposed, in which the MPC at the high-level computes optimal heat fluxes supplied to the TABS and then the optimal heat fluxes are interpreted by a

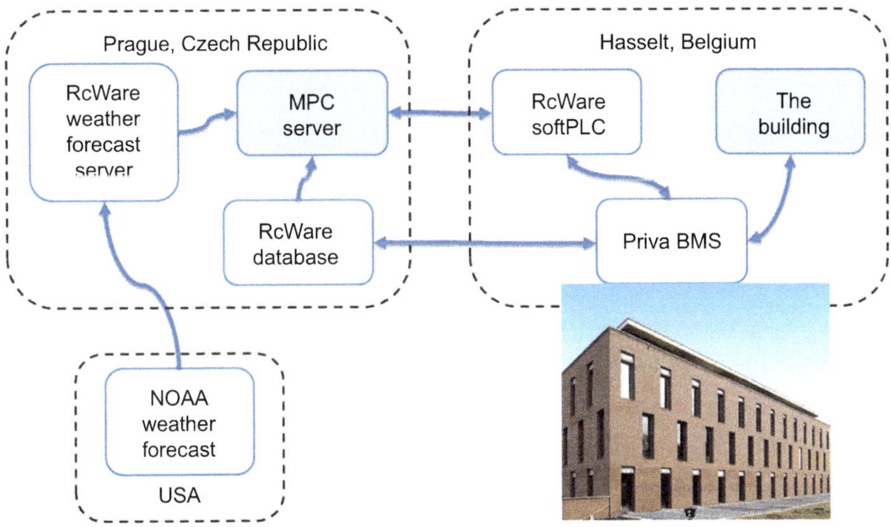

Figure 15.12 Communication interfaces for MPC implementation in Hollandsch Huys (Airdeck, Belgium) (Siroky, 2013).

straightforward function that computes the schedule for mass flow rates and supply water temperature, turning the MPC problem into a convex optimization.

In a first experiment, two similar winter periods are compared: one period using RBC and one period using MPC. Table 15.5 lists the main characteristics of both periods and the system operation by using RBC versus MPC, showing that MPC saved more than 30% electricity use.

Table 15.5 Two similar winter periods used for the comparison between rule based controller and model predictive control in Hollandsch Huys (Siroky, 2013)

	RBC	MPC
Number of days (–)	23	23
Minimal outdoor temperature (°C)	−3.9	−3.7
Maximal outdoor temperature (°C)	12.5	11.4
Mean outdoor temperature (°C)	4.7	4.4
Heat extracted from the ground (kWh)	14,304	9645
Electricity used by the heat pump (kWh)	4890	3330
Number of on–off switches (–)	336	183
Heat pump operation time (h)	204	134
Thermal discomfort (EN15251 class B) (Kh)	12.8	0.56

Table 15.6 Comparison of controller performance for three similar winter periods in Hollandsch Huys (Váňa et al., 2014)

	Mean outside temperature (°C)	Thermal discomfort (Kh)	MPC energy saving (%)
MPC	2.14	2.88	
RBC – period 1	2.02	17.50	22.77
RBC – period 2	3.02	10.40	11.37

A second experiment is presented in Table 15.6. Again, a winter reference time period is selected during which the building was operated using MPC. This period is compared with two other similar periods when the building was operated using RBC. Length of all compared time periods was 25 days and difference in average ambient temperature was not larger than 1°C.

From Tables 15.5 and 15.6, it is clear that MPC outperforms the RBC approach. MPC reduces energy use by, on average, 17%, and additionally reduces thermal discomfort. Less on–off switches are also observed in heat pump operation – potentially improving efficiency and service life. When the objective is to minimize energy cost, more heat supply is seen during the cheaper night periods (day/night electricity tariff). Analysis of one-week data, as presented in Fig. 15.13 (Váňa et al., 2014), provides insight in the system operation. MPC anticipates on future predictions and starts heating sooner, knowing that the building reacts slowly. This reduces comfort criteria violations. Due to reduction of overheating and weekend setback, MPC also uses less energy.

It should be stressed that experimental comparison is very difficult, since boundary conditions are never the same, even not in well-chosen similar time periods. Referring to the results in Table 15.5, MPC is applied at the start of the winter while RBC operates in the middle to the end of the winter, which means that MPC could use a higher ground temperature (leading to a higher heat pump COP). Furthermore, a good comparison needs longer time periods. Nevertheless this practical implementation shows clear potential (even under these severe conditions characterized by lack of measurement data and absence of time for a long-term identification experiment) and is already a step forward compared to theoretical simulation studies, which still have their added value.

The case study described above applies MPC to the hydronic TABS only (the ventilation air-based system was controlled separately). Moreover the borehole array behaviour was not explicitly taken into account in the MPC. Both aspects are considered in the following subsections.

15.8 Optimal exploitation of ground thermal energy storage on the long term

In most European countries office buildings require both heating (in winter) and cooling (in summer) to maintain the requested thermal comfort throughout the entire year. The system can act like a borehole thermal energy storage (BTES) system, thereby

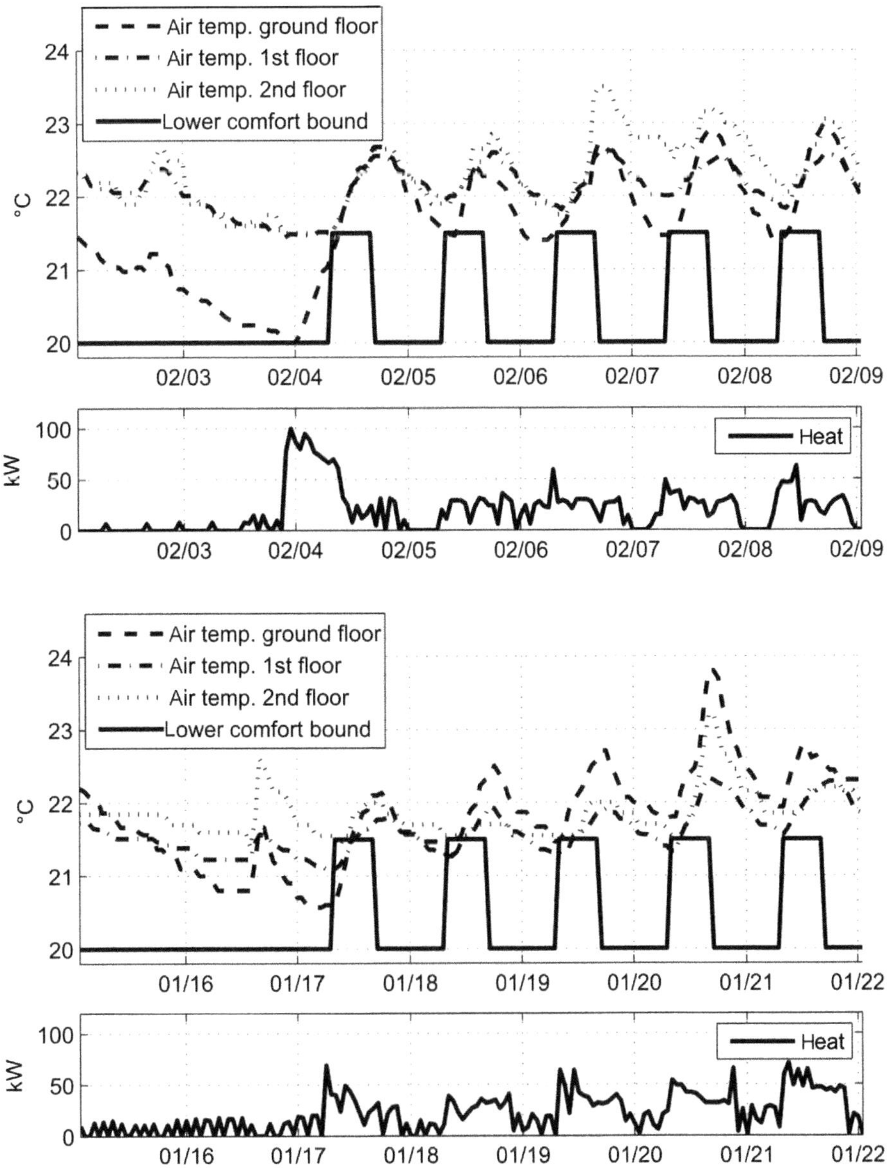

Figure 15.13 Comparison of typical weekly winter profiles of air temperature in Hollandsch Huys and heat flux supplied to the TABS, top for MPC and bottom for RBC (Váňa et al., 2014).

extracting and injecting heat from and to the ground by means of vertical BHEs. This way, the ground can be exploited as seasonal thermal energy storage (STES). However, the building heating and cooling loads are usually not in balance. The question arises how to operate the system such that a long-term sustainable ground energy storage operation is guaranteed. Most research on the optimal control of these systems focuses either on the short-term objectives (ie, thermal comfort and short-term energy cost, making abstraction of the ground storage dynamics; eg, Bianchi, 2006), or on the long-term objectives (ie, ground thermal balance and annual energy cost, making abstraction of the building dynamics; eg, De Ridder et al., 2011).

Antonov et al. (2012) presented an innovative optimization strategy which combines the short and the long-term objectives (short- and long-term optimization (SALTO)). The thermal dynamics of both the building and the ground storage volume are integrated to answer three questions:

1. How are the optimal heating and cooling duties (supply to TABS office zones) distributed over the heating/cooling devices in order to achieve yearly cyclic ground thermal behaviour while satisfying thermal comfort requirements?
2. What are the differences between two installation concepts for passive cooling (ground coupled, primary cooling) and active cooling (air coupled, back-up cooling) devices when they run simultaneously in serial connection?
3. How does the downsizing of the active cooling installation influence the system operation and the primary energy use?

Fig. 15.14 shows a schematic presentation of the system investigated with GSHP and gas boiler for heating, geothermal passive cooling (direct cooling) and active cooling connected to cooling tower for cooling.

Developing and solving a SALTO allows analyses of the distribution of heating and cooling duties between the corresponding devices, guaranteeing a long-term thermal balance of the ground storage volume, and thermal comfort inside the building at a minimal energy cost. The long-term optimal solution – ie, the control profile which satisfies the imposed thermal comfort level at the lowest annual energy cost and which can be repeated unaltered year after year (for the given weather and occupancy profile) – is found by imposing cyclic boundary conditions on the ground storage temperatures. The results first indicate that for a cooling dominated building the optimal operation is characterized by an annual net heat injection to the ground and mean ground storage temperatures which are higher than the undisturbed ground temperature. Second, passive

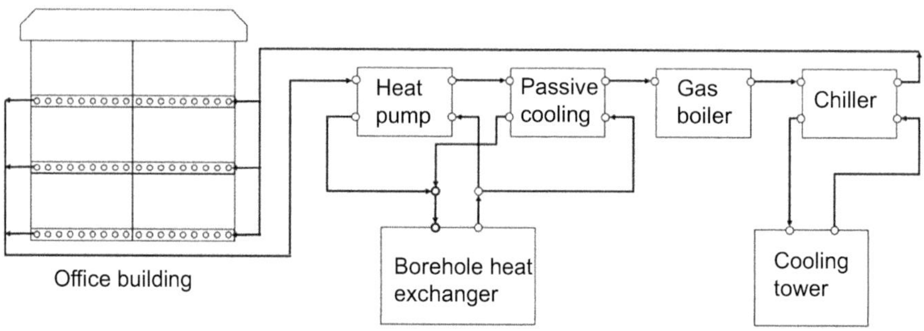

Figure 15.14 Schematic presentation of the investigated hybrid geothermal heat pump system consisting of office building, geothermal heating and cooling devices (heat pump, passive cooling) and supplementary heating and cooling devices (gas boiler, chiller) (Antonov et al., 2014).

cooling is constrained by the upper limit on the antifreeze/water fluid temperature. Third, the corresponding optimal ground storage temperature distribution is reached during the first year of operation (ie, starting from ground storage at undisturbed ground temperature). These conclusions hold for all borehole array sizes, as well as for all sizes of the supplementary cooling device (active cooler), considered in the study. A trade-off could be calculated between investment cost for BHEs, investment cost for supplementary cooling installation and operation cost of the hybrid system. Engaging passive cooling as first installation and active cooling as second installation in the serial connection is the cheaper of the two options when both installations are needed to run simultaneously.

Antonov et al. (2013, 2014) furthermore addressed the question of to what extent the long-term time scale influences the optimal system operation for the hybrid ground-coupled heat pump system (presented in Fig. 15.14 for a cooling-dominated office building). Should we prevent borehole array thermal build-up in the case of a cooling-dominated building? Should we restrict the use of passive cooling in order to balance the annual heat injection and extraction loads? Is it advantageous to extract more heat by the heat pump during the heating season in order to have a colder borehole array at the start of the cooling season? Should we inject more heat than strictly necessary during the summer in order to have a higher borehole array temperature at the start of the heating season? Should we take into account the seasonal variation and the intraday variation of the chiller coefficient of performance? To this end, the optimal operation – found by solving the OCP for a prediction horizon of 1 year – was compared to the operation obtained with MPC with a receding horizon of one week. The results show that the operation with the one-week horizon is identical to the one found with a 1-year horizon, which indicates that for the considered system there is no optimization potential at the seasonal time scale. The study shows that STES is not an optimal control strategy for the specified system. The use of the supplementary chiller (active cooler), which could be started at any time and can be controlled with a short-term strategy, is a cheaper alternative. The seasonal cold storage in the ground is less efficient because the main part of the stored cold is lost in the far field before it could be used. An artificial decrease of the chiller energy efficiency ratio (EER) below the value of 0.4 leads to an increase of the total annual extraction loads (see Fig. 15.15), which is an indication for additional cold storage. All other parameters being equal, a chiller EER value of 0.4 can thus be defined as the critical EER value below which additional seasonal underground TES can become important (values in the range of 2.5–4 being more typical of real equipment). Therefore, we can conclude that for realistic values of the performances of the ground-coupled devices (heat pump and passive cooling) and supplementary devices (gas boiler and chiller), it is energetically better to combine the use of passive and active cooling for cooling-dominated buildings than to impose passive cooling only by precooling the ground in winter through additional GSHP operation.

It was concluded that for long-term optimal operation of the system the MPC should have a prediction horizon, which is as long as the largest time constant at building side. A long-term constraint in the controller of the investigated system is not needed to achieve ground thermal balance or to guarantee long-term sustainable performance. For the considered hybrid system, the ground is naturally recharged from

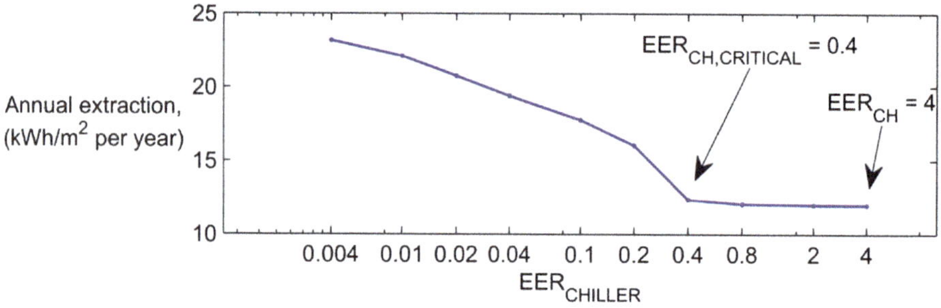

Figure 15.15 Additional seasonal underground thermal energy storage would be economical if the energy efficiency ratio (EER) of the air-coupled active chiller was at least 10 times lower than in reality (Antonov et al., 2014).

the far field. The efficiency of the underground BTES is only about 54%, and therefore not economical without external sources of 'free' heat or cold, such as, for instance, solar assistance or waste heat.

15.9 Joining the forces of slow thermally activated building systems and fast air handling units

TABS have a large potential for energy use reduction in office buildings, because they fit nicely with low-exergy heat and cold production such as GSHP systems. However, due to the large thermal capacities, TABS control should be carefully designed in order to avoid low control efficiencies. Moreover, due to the limited thermal power and the low controllability, TABS cannot always react adequately to disturbances (peaks in office building heat gains). On the contrary, the air handling unit (AHU), which is installed anyway to guarantee indoor air quality, is a fast-reacting system but requires high-exergy heat and cold production and is thus less energy efficient. The question is how these two systems should cooperate when they both provide thermal input to the building zone. Sourbron and Helsen (2013) investigated MPC to control both TABS and AHU to guarantee thermal comfort at the lowest possible energy cost. Coupled TRNSYS–Matlab simulations were used to assess the MPC behaviour. For details the reader is referred to (Sourbron and Helsen, 2013). Figs. 15.16 and 15.17 compare the system behaviour in a 2-day period (Sunday–Monday) for a conventional HC/CC control strategy with optimal settings (called constant concrete core control or C^4) and MPC for heating and cooling regimes, respectively.

The MPC allows operating the system near its boundaries and thus fully exploits the temperature band provided by the thermal comfort range, thereby taking into account the dynamics of TABS and building. The MPC will not always deploy the TABS, but chooses the best to operate based on the overall system situation. The AHU is used to compensate heat gain peaks in a cooling regime (eg, due to solar

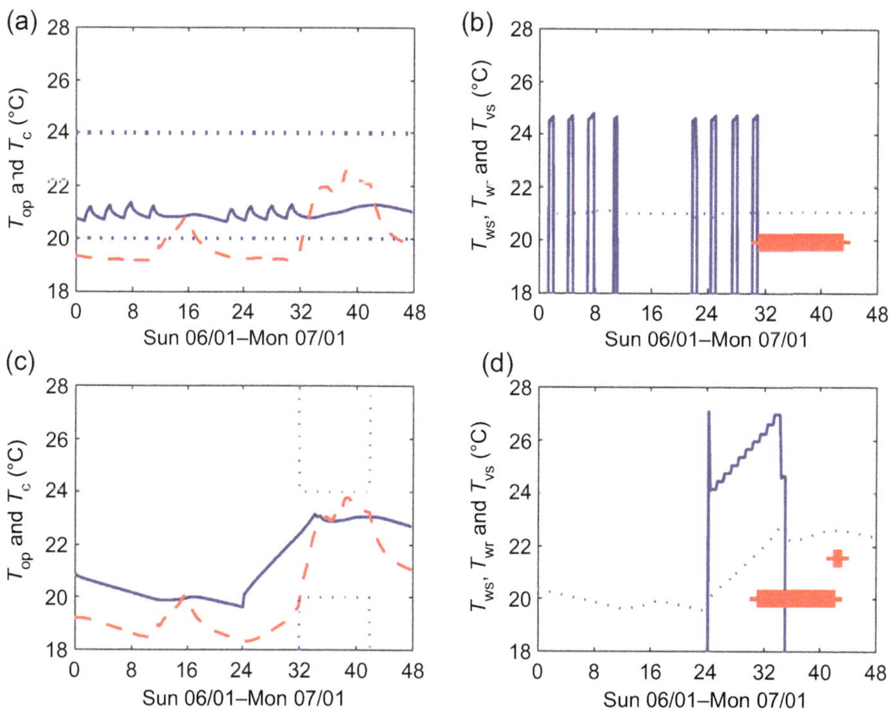

Figure 15.16 System heating behaviour for the C[4] controller (a)–(b) and for the MPC (c)–(d): (a)–(c) operative temperature in *dashed line* and concrete core temperature in *solid line* (the comfort band is shown by the *dotted lines*); (b)–(d) water supply temperature in *solid line*, water return temperature in *dotted line* and *+signs* for ventilation air temperature (Sourbron and Helsen, 2013).

radiation) while the TABS is used during night to precool the building. September may be characterized by relatively low outdoor temperatures which allow cheap free-cooling by the AHU. In heating mode the AHU is used to achieve thermal comfort during the first office hours. The concrete core temperature is not kept constant by the MPC, which was, however, the result of the conventional controller C[4] (Sourbron, 2012; Sourbron and Helsen, 2014). MPC, on the contrary, only heats or cools the TABS as much as required (but not more) to obtain thermal comfort (usually near the boundary). Both energy use and thermal discomfort can potentially be reduced by MPC compared to conventional HC/CC controllers, both in heating and cooling mode. However, implementation in a real building introduces important errors, due to building model mismatch, cost function mismatch, measurement errors and weather prediction uncertainties. Understanding these errors is crucial for a successful deployment of MPC and using MPC to its full potential. Nevertheless, MPC is shown to be the required link to unite TABS and AHU into one low-energy office building comfort system.

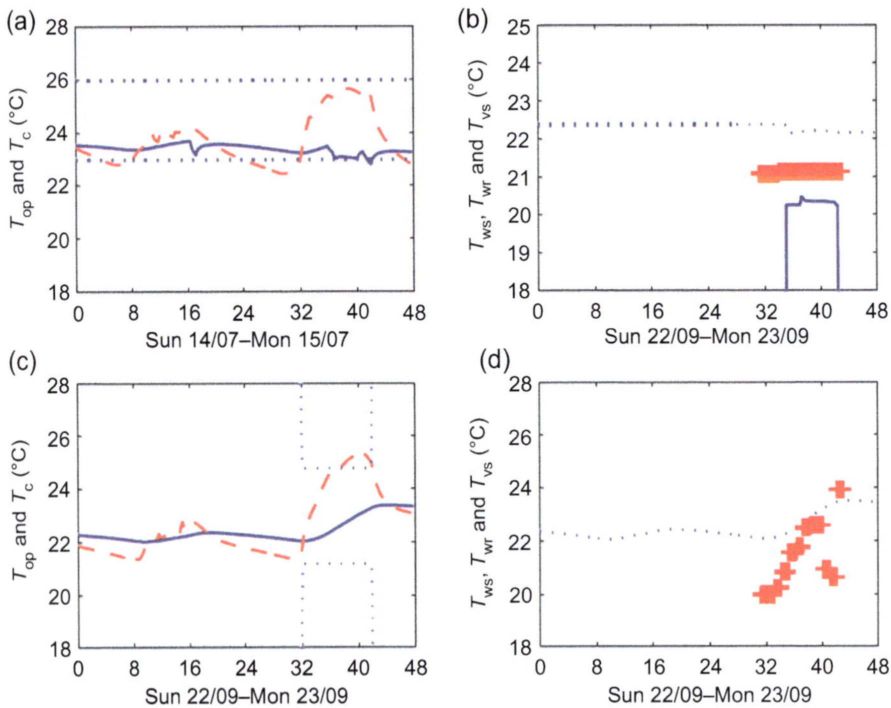

Figure 15.17 System cooling behaviour for the C^4 controller (a)–(b) and for the MPC (c)–(d): (a)–(c) operative temperature in *dashed line* and concrete core temperature in *solid line* (the comfort band is shown by the *dotted lines*); (b)–(d) water supply temperature in *solid line*, water return temperature in dotted line and +*signs* for ventilation air temperature (Sourbron and Helsen, 2013).

15.10 Conclusions and outlook

With the development of smart grids and intelligent control, GEOTABS can play a very important role. GSHPs upgrade low temperature heat (an RES) to usable heat in a very efficient and sustainable way by using electricity. The thermal energy stored in the TABS furthermore represents storage capacity that can be taken advantage of by the smart grid. As a consequence, intelligent control can help by stabilizing the electricity grid and matching supply peaks (by intermittent RESs) and demand. Consequently, thermal and electrical energy fluxes can be decoupled and the potential for RES integration increases significantly.

MPC has been proven to offer significant energy savings together with equal or even better thermal comfort compared to conventional control (based on static heating and cooling curves) for GEOTABS buildings. The first MPC implementations have been tested in real time; however, the tool chain within this MPC framework still presents some challenges, such as: (1) (automatic) controller model development (tuned to the specific case with large amount of unmeasured disturbances, small number of sensors, limited freedom for applying good excitation signals when the building is occupied and broad range of time constants); (2) solution of the OCP with a formulation as simple as possible, which needs to be embedded in cheap control hardware; (3) uncertainties due to model mismatch, prediction errors (weather data, user profiles) and measurement errors; and (4) state estimation for unmeasurable states. These challenges present a clear research focus of several teams worldwide and need to be tackled before large scale commercial implementation can be realized. The tests performed so far strengthen the confidence in the potential of MPC, in particular for GEOTABS buildings, thanks to the predictive nature.

Often an optimal controller is developed for a given design or an optimal design is made for a given controller. However, the design influences the optimal control and vice versa. Therefore, recent research efforts are focussing on integrated optimal design and control whereby both are optimized simultaneously. Hybrid system installations in office buildings are characterized by high complexity, which makes this global optimization problem difficult to solve without simplification. Reduced order models have a crucial role here.

Knowledge is power; the role of monitoring and commissioning cannot be overestimated. System integration offers a huge potential for energy savings; however, this integration requires that high standards of quality are monitored and checked.

Acknowledgements

The contribution of the KU Leuven SySi members, in particular Maarten Sourbron, Clara Verhelst, Jan Hoogmartens, Stefan Antonov and Damien Picard, to the research results is gratefully acknowledged, as well as the fruitful collaboration with Veronique Houben and Ghislain Gielen of Airdeck, Wim Boydens of Boydens Engineering and the researchers of TU Prague: Lukas Ferkl, Jan Siroky, Jiri Cigler, Samuel Privara, Zdenak Vana and Eva Zacekova. The author is also grateful to the European Union and the Agency for Innovation by Science and Technology in Flanders (IWT) for funding the GEOTABS project, 'Towards optimal design and control of geothermal heat pumps combined with thermally activated building systems in offices', within the FP7 EraSME programme; to the IWT and WTCB for the financial support of the IWT-VIS Traject SMART GEOTHERM focusing on integration of thermal energy storage and thermal inertia in geothermal concepts for smart heating and cooling of (medium) large buildings; to the IWT for funding the PhD Fellowship 'Model Predictive Control of Ground Coupled Heat Pump Systems for Office Buildings' of Clara Verhelst; to the KU Leuven for funding the PhD Fellowship 'Dynamic Thermal Behaviour of Buildings with Concrete Core Activation' of Maarten Sourbron; and to the Research Foundation Flanders (FWO) for funding the research project entitled 'Black-Box Model Based Predictive Control of Ground Coupled Heat Pump Systems'.

References

Antonov, S., Verhelst, C., Helsen, L., May 16–18, 2012. Control of ground coupled heat pump systems in offices to optimally exploit ground thermal storage on the long term. In: Innostock 2012. Lleida, Spain art.nr. INNO-U-47.

Antonov, S., Verhelst, C., Helsen, L., June 16–19, 2013. Optimal operation of ground coupled heat pump systems: should We take the seasonal time scale into account? In: Clima 2013, 11-th REHVA World Congress & 8th International Conference on IAQVEC. Prague, Czech Republic.

Antonov, S., Verhelst, C., Helsen, L., 2014. Should the optimization horizon in optimal control of ground coupled heat pump systems cover the inter-seasonal time scale? ASHRAE Transactions 120 (2), 346–356.

Armstrong, P.R., Leeb, S.B., Norford, L.K., 2006. Control with building mass - part I: thermal response model. ASHRAE Transactions 112 (Part 1). CH-06-5-1, 1-13.

Bianchi, M.A., 2006. Adaptive Modellbasierte Prädictive Regelung einer Kleinwärmepumpenanlage (Ph.D. thesis). ETH, Zürich (in German).

Burkard, T., 2004. Die Nutzenübergabe von Systemen zur Bauteilaktivierung. HLH (in German).

COMSOL Multiphysics® - Multiphysics Modeling, Finite Element Analysis, and Engineering Simulation Software. http://www.comsol.com/. USA: COMSOL, Inc. (consulted on October 2015).

Crawley, D., Lawrie, L., Pedersen, C., Winkelmann, F., 2000. Energy plus: energy simulation program. ASHRAE Journal 42 (4), 49–56.

De Coninck, R., June 2015. Grey-Box Based Optimal Control for Thermal Systems in Buildings – Unlocking Energy Efficiency and Flexibility. KU Leuven, Leuven (Belgium) (Ph.D. text).

De Ridder, F., Diehl, M., Mulder, G., Desmedt, J., Van Bael, J., 2011. An optimal control algorithm for borehole thermal energy storage systems. Energy and Buildings 43 (10), 2918–2925.

EnergyPlus, 2012. Energy Simulation Sofware, Version 7.2.0. U.S. Department of Energy – Energy Efficiency & Renewable Energy, USA. www.eere.energy.gov/buildings/energyplus (consulted on October 2015).

ESP-r, Building Energy Software Tools, IBPSA-USA, http://www.buildingenergysoftwaretools.com/software/esp-r (consulted on October 2015).

FEFLOW, Berlin, Germany: DHI-Wasy GmbH Berlin. http://www.mikepoweredbydhi.com/products/feflow (consulted on October 2015).

Goldberg, D., 1989. Genetic Algorithms in Search, Optimization, and Machine Learning. Addison-Wesley.

GROUND-REACH, 2008. Reaching the Kyoto Targets by Ground Source Heat Pumps. Project Report D7 Technical, environmental and economic feasibility of ground coupled heat pump technologies under defined conditions. Arsenal Research. Available from: http://www.groundreach.eu/script/tool/forg/doc820/D7%20Feasibility%20of%20GCHP%20technology.pdf (consulted on 15.03.10.).

Hellström, G., Sanner, B., Eskilson, P., Claesson, J., Blomberg, T., July 2010. EED - Earth Energy Designer EED 3 (Version 3). Blocon Sweden, Blocon USA, Lund, Sweden; Reading, MA (USA). http://www.buildingphysics.com/ (consulted on October 2015).

Helsen, L., Verhelst, C., Sourbron, M., November 2008. Grondgekoppelde warmtepompen als bron voor betonkernactivering, Kluwer MilieuTechnologie. Tijdschrift Veiligheid en Milieu 15 (10-extra) (in Dutch).

Huber, A., Programm EWS - Erdwärmesonden. Berechnung von Erdwärmesonden. Version 5.0. Zürich, Switzerland : Huber Energietechnik AG. http://www.hetag.ch/ews.html (consulted on October 2015).

ISO 7730: Ergonomics of the Thermal Environment – Analytical Determination and Interpretation of Thermal Comfort Using Calculation of the PMV and PPD Indices and Local Thermal Comfort Criteria.

Kennedy, J., Eberhart, R., 1995. Particle swarm optimization. In: Proceedings of the IEEE International Conference on Neural Networks, vol. 4. IEEE, pp. 1942–1948.

Kolda, T.G., Lewis, R.M., Torczon, R.M., 2003. Optimization by direct search: new perspectives and some classical and modern methods. Society for Industrial and Applied Mathematics 45 (3), 97.

Lund, J., Sanner, B., Rybach, L., Curtis, R., Hellström, G., September 2004. Geothermal (ground-source) heat pumps - a world overview. GHC Bulletin, 1−10.

Maciejowski, J., 2002. Predictive Control: with Constraints. Pearson Education.

Modelica, Linköping, Sweden: Modelica Association et al., www.modelica.org (consulted on October 2015).

Nocedal, J., Wright, S.J., 2000. Numerical Optimization, second ed. Springer Verlag.

Nelder, J., Mead, R., 1965. A simplex method for function minimization. The Computer Journal 7 (4), 308.

Olesen, B.W., Liedelt, D.F., 2000. Cooling and heating of buildings by activating their thermal mass with embedded hydronic pipe system. In: CIBSE/ASHRAE Joint Conference "20/20 Vision", Dublin, Ireland.

Olesen, B.W., Sommer, K., Düchting, B., 2002. Control of slab heating and cooling systems studied by dynamic computer simulations. ASHRAE Transactions 108 (2), 698−707.

Olesen, B.W., 2012. Thermo Active Building Systems Using building mass to heat and cool. ASHRAE Transactions 34 (2), 44−52.

Qin, S.J., Badgwell, T.A., 2003. A survey of industrial model predictive control technology, Control Engineering Practice 11 (7), 733−764.

Guidebook No 20 REHVA, 2013. Advanced System Design and Operation of GEOTABS Buildings, ISBN 978-2-930521-12-1.

Siroky, J., 2013. Practical Aspects of MPC Implementation. Presentation Available from: jan.siroky@energocentrum.cz.

Sourbron, M., De Herdt, R., Van Reet, T., Van Passel, W., Baelmans, M., Helsen, L., 2009. How efficiently produced heat and cold is squandered by inappropriate control strategies: a case study. Energy and Buildings 41, 1091−1098.

Sourbron, M., September 2012. Dynamic Thermal Behaviour of Buildings with Concrete Core Activation (Ph.D. text). KU Leuven, Leuven (Belgium), ISBN 978-94-6018-572-4.

Sourbron, M., Helsen, L., June 16−19, 2013. Slow thermally activated building system and fast air handling unit join forces through the use of model based predictive control. In: Kabele, K. (Ed.), Proceedings of the CLIMA 2013 Congress. CLIMA2013 REHVA World Congress. Prague, Czech Republic. Czech Society of Environmental Engineering (STP), Prague, pp. 1−10 art.nr. 160.

Sourbron, M., Helsen, L., 2014. Sensitivity analysis of feedback control for concrete core activation and impact on installed thermal production power. Journal of Building Performance Simulation 7 (5), 309−325.

THERMAC, 2008. Handboek voor het verwarmen en natuurlijk koelen van Thermisch Actieve Gebouwen. resultaat van het IWT-TETRA project THERMAC 50108, W&K De Nayer (in Dutch).

Tian, Z., Love, J.A., 2009. Energy performance optimization of radiant slab cooling using building simulation and field measurements. Energy and Buildings 41 (3), 320−330.

Todtli, J., Gwerder, M., Lehmann, B., Renggli, F., Dorer, V., 2009. TABS Control − Steuerung und Regelung van thermoaktiven Bauteilsystemen. Faktor Verlag, Zurich (in German).

Toftum, J., Olesen, B.W., Kolarik, J., Mattarolo, L., Belkowska, D., 2008. Occupant responses and energy use in buildings with moderately drifting temperatures. ASHRAE Research Project 1269 Final Report, Atlanta, GA, USA, p. 145.

TRNSYS 17, 2010. TRaNsient SYstem Simulation Program. Version 17. University of Wisconsin-Madison (USA). Solar Energy Laboratory. http://www.trnsys.com/ (consulted on October 2015).

Vanhoudt, D., De Ridder, F., Desmedt, J., Van Bael, J., 2010. Controller for optimal cost operation of a borehole thermal energy storage system. In: 10th REHVA World Congress for Building Technologies - CLIMA, Antalya, Turkey.

Verhelst, C., April 2012. Model Predictive Control of Ground Coupled Heat Pump Systems for Office Buildings (Ph.D. text). KU Leuven, Leuven (Belgium), ISBN 978-94-6018-497-0.

Váňa, Z., Cigler, J., Široký, J., Žáčeková, E., Ferkl, L., 2014. Model-based energy efficient control applied to an office building. Journal of Process Control 24 (6), 790–797.

Zogg, M., 2001. Gekoppelte Kälte- und Wärmeerzeugung mit Erdwärmesonden. Bundesamts für Energie, Zwitserland (in German).

Index

'*Note:* Page numbers followed by "f" indicate figures, "t" indicate tables.'

H

CPI Antony Rowe
Chippenham, UK
2016-06-15 18:15